Longwall-Shortwall Mining, State of the Art

R.V. Ramani, Editor

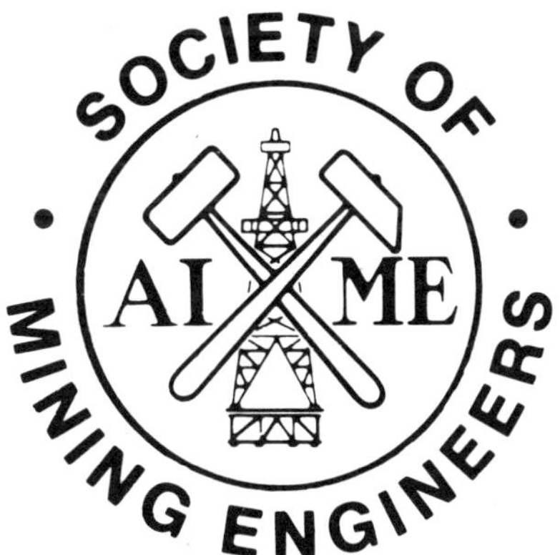

Society of Mining Engineers

of

The American Institute of Mining, Metallurgical, and Petroleum Engineers, Inc.

New York, New York • 1981

**Copyright © 1981 by
The American Institute of Mining, Metallurgical,
and Petroleum Engineers, Inc.**

Printed in the United States of America by
Edwards Brothers Inc., Ann Arbor, Michigan

**Library of Congress Catalog Card Number 81-67436
ISBN 0-89520-288-3**

ORGANIZING COMMITTEE

Raja V. Ramani, Chairman
Professor of Mining Engineering
The Pennsylvania State University
University Park, PA 16802

John P. Baugues
President
Mountain Inc.
Knoxville, TN 37902

James A. Bloom
Chief Engineer
Y&O Coal Company
Martin's Ferry, OH 43935

Lanny L. Richter
Manager, Mining Engineering
Old Ben Coal Corporation
Benton, IL 62812

Howard E. Rutherford
Engineering Manager
Monterey Coal Company
Carlinsville, IL 62626

SESSION CHAIRMEN

I. LONGWALL EQUIPMENT

Claude A. Goode
US Department of Energy

Lanny L. Richter
Old Ben Coal Corporation

II. LONGWALL ENVIRONMENT

Joseph Cervik
US Bureau of Mines

James A. Bloom
Y&O Coal Company

III. UNIT OPERATIONS

John P. Baugues
Mountain, Inc.

IV. CASE STUDIES

William H. Mullins
Freeman United Coal Mining Company

V. RESEARCH AND DEMONSTRATIONS

Joseph J. Yancik
National Coal Association

Howard E. Rutherford
Monterey Coal Company

Foreword

For over a decade energy problems and the need for greater utilization of domestic energy sources have been a focus of American thought, media coverage, and political activity. Mining management with its support by technological experts, particularly in the coal industry, has done much more than just talk. It has steadily worked on the problems associated with improving the ways of producing an important energy product. One phase of this effort has been the research, development, and implementation of longwall-shortwall methods for mining coal. Nor has there been only concentration on production and equipment improvements. The welfare of those engaged in producing coal by these methods and the impact of such activity on the environment around operations have been important aspects of the total effort. All aspects of this total effort are well documented in this volume.

The Society of Mining Engineers of AIME has also played an important role. A primary function of SME is the advancement of the engineering and science of mining through the conduct of professional meetings. By organizing, presenting, and publishing the results of specialty forums, such as this one on longwall-shortwall mining, SME-AIME is assisting in the dissemination of information to those who can apply it for better service to the public of energy and products.

It is my pleasure and privilege to prepare the Foreword for this important work. On behalf of the Society, its management, and members, I commend Raja V. Ramani and his able committee consisting of John P. Baugues, James A. Bloom, Lanny L. Richter, and Howard E. Rutherford for organizing the sessions at the 1981 Annual Meeting and for this volume, a most valuable contribution to the technical literature in the minerals engineering field.

Alfred Weiss, President
Society of Mining Engineers
of AIME

April 28, 1981
Stamford, CT

Preface

The evolution of longwall systems in the United States has been slow and steady. There are indications that the modifications and adjustments made to adapt European technology to American conditions are paying off. In recent years, longwall mining has accounted for only 5% of underground production; however, the growth in longwalls has been dramatic. The number of faces in operation has increased from under 60 in 1975 to more than 110 in 1981. The trend is definitely toward increased use of this method. By 1985, there may be as many as 200 faces in operation producing well over 15% of the underground tonnage.

Several problems, however, must be overcome if longwalls are to make a major contribution to the predicted rapid increase in coal production. Some of these problems are unique to US longwalls having to do with multientry systems, health and safety regulations, and mining conditions. Reexamination of the regulations and more research and development are needed to overcome the limitations to production and productivity. Longwalls are expensive. The total cost of a face can range from $1.25 to $1.75 million per 100 linear feet of face. For cost effectiveness, high productivity and production rates must be achieved.

To provide a forum to discuss these problems, progress, practices, and research and development efforts and needs in longwall mining, the Coal Division technical program committee early in 1980 decided to have a number of sessions on longwall mining at the SME-AIME annual meeting in Chicago in 1981.   At the annual meeting, February 23-26, 1981, in five technical sessions, 24 papers were presented by professionals including operators, manufacturers, and government personnel. The session papers covered equipment developments, longwall environment, case studies, longwall unit operations, and research developments and demonstrations. This volume contains the proceedings of those sessions. In addition, other selected papers on longwall and shortwall applications have been included.

An effort of this kind depends for its success on the commitment of a number of individuals. Assistance and support from Trevor J. Jones, Jeffrey Mining Machinery Div., Dresser Industries; Louis Kuchinic, Jr., Penn Virginia Resources Corp.; Robert E. Murray, North American Coal Co.; E. Minor Pace, Inland Steel Co.; Eugene R. Palowitch, US Department of Energy; James Paone, US Bureau of Mines; Robert Stefanko, The Pennsylvania State University; John W. Straton, Gates Engineering Co.; Edwin B. Wilson, Bethlehem Steel Corp.; Joseph J. Yancik, National Coal Association; and David A. Zegeer, Mining Consultant, during the programming of the technical sessions are gratefully acknowledged. Thanks also are due to the members of the longwall organizing committee and session chairmen for their efforts in arranging and conducting an excellent technical program. The contributions of the authors who gave so generously of their time and experience are sincerely appreciation by the Coal Division and the Society. Permission from McGraw Hill, Inc. and Maclean Hunter Publishing Co., publishers, respectively, of *Coal Age* and *Coal Mining and Processing,* to reproduce selected articles is gratefully acknowledged.

Lastly, publishing this proceedings volume has required a large investment of time and effort. Sincere appreciation is extended to Marianne Snedeker of the Society of Mining Engineers for her many contributions to this end.

Raja V. Ramani
The Pennsylvania State University

University Park
April 20, 1981

Table of Contents

I. Equipment

Claude A. Goode
US Department of Energy
Lanny L. Richter
Old Ben Coal Corporation

Chapter 1

GROWTH OF LONGWALL TECHNOLOGIES IN THE UNITED STATES

William E. Souder

Professor of Industrial Engineering
Director of the Technology Management Studies Group
University of Pittsburgh
Pittsburgh, Pennsylvania

and

Eugene R. Palowitch

Acting Manager
Pittsburgh Mining Technology Center
U.S. Department of Energy
Pittsburgh, Pennsylvania

I.   INTRODUCTION

The longwall method of mining coal underground is now a highly developed and accepted mining technology. However, it was only through a long history of successes and failures that this technology now has the potential for reversing the declining trend in underground mining productivity in the United States.

From its early beginnings in the United States during the 1800's, several species of longwalls have evolved. The evolutionary record consists of a series of overlapping birth-death eras, with each era giving way to another unique longwall technology. Typically, an era begins when some new longwall technology appears in response to a new technical opportunity or economic need. This new technology is adopted initially by only a few innovators. Later, other imitators get on the bandwagon and the technology diffuses throughout the user population. But evolving technical demands and socio-economic circumstances eventually cause many of the adopters to discontinue its use.

This paper chronicles seven eras in the history of the evolution and utilization of longwall technologies in the United States. Some of the important factors that have influenced this history are discussed. Some distinctions are drawn between the operators who were first to adopt longwalls (innovators), and the imitators who followed. Statistics are presented which show that longwall mining is being adopted at an exponentially increasing rate, substantiating the forecast for a significant number of new longwall installations in the future. Potential barriers to this increased utilization of longwalls are noted.

II.   THE LONGWALL TECHNOLOGY ERAS

Figure 1 shows the census track of longwalls in service in the United States from 1875 to 1977. Seven distinct technology eras are identifiable within this track. The labels, "Early Style," "Old Style," "Coal Plow," etc, are coined here to describe the nature of the prevailing technologies. The broken vertical lines denote the approximate year of appearance of each new technology. However, as it takes several years for a new technology to replace an older one, the broken line does not necessarily represent the disappearance of an era. Technology birth-death cycles typically overlap.

A.   Early Style (Wood Props) Longwalls Era

By the late 1800's, a "longface" method which originated in Wales during the 1700's was introduced in the southern Illinois (LaSalle) region by emigrating miners. Figure 2 depicts one variant of this Early Style longwall technology. The face was undercut by hand to a depth of 2 to 2-1/2 feet. Knock-out wood props or sprags were used to support the undercut, and the roof was supported by packwalls in the

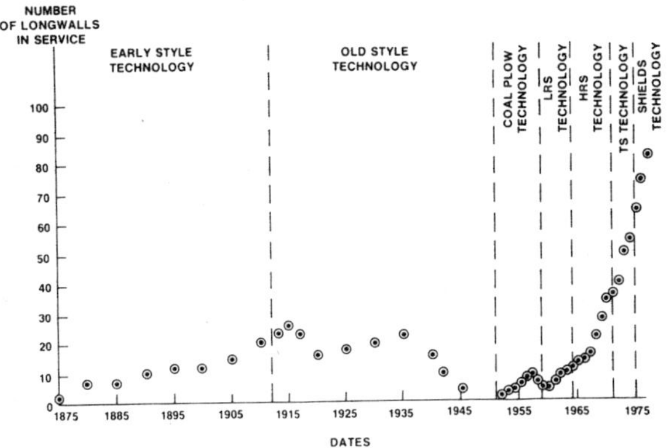

FIGURE 1.   SCATTER PLOT OF NUMBER OF LONGWALLS IN SERVICE

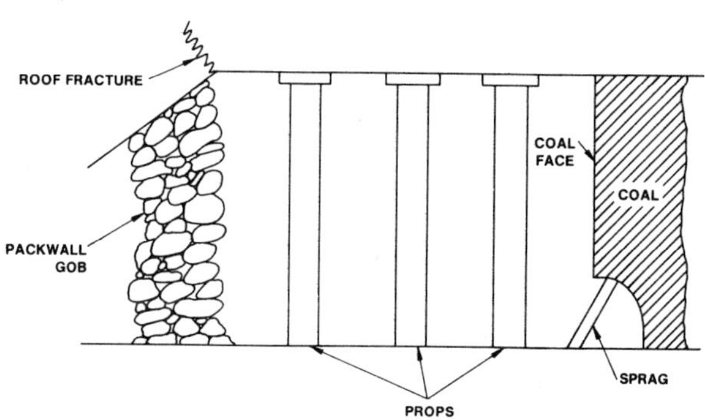

FIGURE 2.   EARLY STYLE LONGWALL WITH WOOD PROPS
(PILLARS) AND PACKWALL

gob. Under ideal conditions, the face would break under the weight of the roof when the sprags were removed. Apparently, these ideal conditions were rare. Roof falls were often reported to be "violent and unexpected," and the methods were reported to involve "many unpredictabilities."[5] In spite of these shortcomings, by 1915 the Early Style longwalls had been adopted as far west as Washington state and as far south as Alabama. The method was attractive because it provided the increased supply of coal needed for the Nation's rapid growth during the 1875-1914 period and it was well-adapted to the geologic conditions of those regions.

In contrast, the 1915-1930 period was one of overcapacity and economic hardship for the Nation and the coal industry. By the 1920's, mines in thin seams could no longer compete with mines in the thicker-seam southern coal fields with their cheaper labor and mechanized room-and-pillar technologies. However, Early Style longwalls did not disappear overnight, and a few of these smaller mines continued to operate through the 1950's.

Figure 3 presents the adoption and diffusion patterns for the Early Style longwall technology.* "Adoption" refers to the acceptance and the commitment by a single firm to use an innovation. For this study, adoption is considered to have occurred if the firm installed and operated the equipment for at least one year. "Diffusion" refers to the percentage of adoption among a population of users.[1] In Figure 3, the curve labeled "F(x)" gives the percentage of the ultimate total population that had adopted Early Style longwalls by the end of each year of the era. The shape of the F(x) curve describes the rate with which the technology spread among the users. For instance, the curve shows that during the 30-year period from 1870 to 1900, Early Style longwalls achieved only 30 percent adoption. This is considered a relatively long <u>incubation</u> period. It took only about 10 years (from 1900 to 1910) for the technology to go from 30- to 80-percent adoption. This is considered a relatively long <u>establishment period</u>. And about 20 more years elapsed (from 1910 to 1930) before the technology achieved 100 percent adoption. Again this is considered a relatively long <u>maturity</u> period.

The F(x) statistic does not indicate whether the population continued to use the innovation or abandon it after their first year's commitment. For this information we need to examine the f(x) statistic, or the net diffusion. The differences between the values of the F(x) and f(x) statistics are labeled the "% Discontinuances." This statistic shows the percentage of adopters who subsequently discontinued their previous commitment to the innovation. As the % Discontinuance data in Figure 3 show, by 1940 most of the users had discontinued their use of Early Style longwalls.

B. Old Style (Steel Jacks) Longwalls Era

In 1912, the 200-ton capacity friction-yielding Langham jack was invented. It was subsequently followed by such other designs as the Lorraine and the H-beam. These steel jacks introduced the modified caving theory which quickly caught the interest

---

* The use of percentage statistics rather than the raw numbers normalizes for varying population sizes across the eras, thus facilitating comparative analysis.

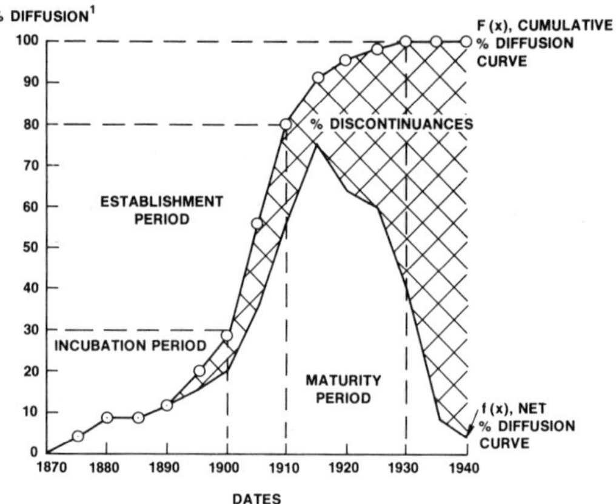

¹ % DIFFUSION = NUMBER OF ADOPTERS AS A PERCENTAGE OF THE TOTAL POPULATION THAT ULTIMATELY ADOPTED

F (x) = CUMULATIVE % DIFFUSION = CUMULATIVE NUMBER OF ADOPTERS AS A PERCENTAGE OF THE TOTAL POPULATION THAT ULTIMATELY ADOPTED

% DISCONTINUANCES = PERCENTAGE OF ADOPTERS WHO DISCONTINUED THEIR EARLIER ADOPTION

f (x) = NET % DIFFUSION = F (x) MINUS % DISCONTINUANCES

FIGURE 3. DIFFUSION CURVES FOR THE EARLY STYLE (WOOD PROPS) LONGWALLS ERA

of many mining engineers. Though the jacks were highly unreliable they had the potential for substantially increasing worker productivity.

By 1935, the installation of Old Style longwalls had spread throughout the eastern and mid-western coal regions. The idea of using steel supports with a modified caving concept was so appealing that some prominent mining authorities began to question whether room-and-pillar methods could continue to florish. But, the demise of these Old Style longwall technologies occurred rapidly during the 1940's. The spread of unionization, the introduction of mobile loaders, the development of surface mining technologies, and the decline in coal demand during the 1930's made thin-seam mining increasingly less profitable, and effectively sealed the fate of the Old Style longwalls. Figure 4 presents the diffusion curves for the Old Style technology. A comparison of Figure 4 with Figure 3 shows that the diffusion patterns of these two technologies are similar. However, the incubation, establishment, and maturity periods are all shorter for the Old Style Technology.

C. Coal Plow/Friction Jack Longwalls Era

The increased demand for coal during the war years of 1938-1945 provided little stimulus for the utilization of longwalls, as the room-and-pillar thick-seam mines and surface mines provided adequate coal. But, as the industry moved into the post-war period, the metallurgical coal producers who were seeking more efficient systems for mining high quality thin seams recognized the potential of the newly developed German coal plow.

In 1951, the U.S. Bureau of Mines was instrumental in arranging a number of friction jack/coal plow longwall mining trials involving mines owned by Eastern Fuel and Gas Associates and equipment supplied by Mining Progress, Inc. Though much was learned on these early installations, the trials were

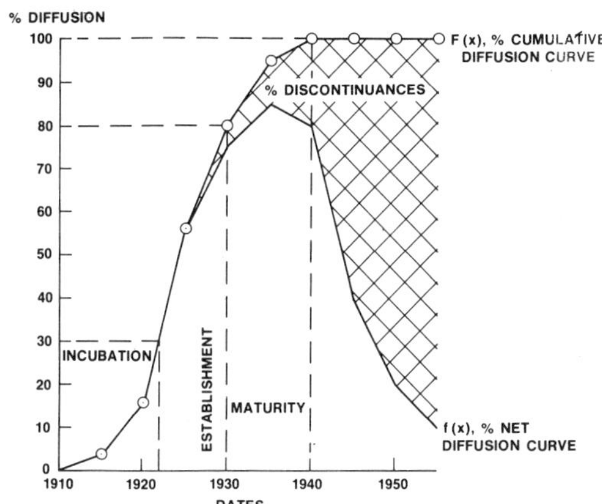

FIGURE 4.   DIFFUSION CURVES FOR THE OLD STYLE (STEEL JACKS) LONGWALLS ERA

plagued by roof and water problems, crib and jack failures, and a lack of spare parts; these problems tended to diminish with subsequent panels.

During the 1953 to 1960 period, several imitators in Pennsylvania, West Virginia, and Arkansas experimented with longwalls using the plow and friction type roof supports.[5] But by the 1960's most of these operators had discontinued their longwall operations, either because the coal was too hard for the plow to cut, the roof failed to cave properly, or the capacity of the roof supports were inadequate. In most cases, the average Coal Plow/Friction Jack longwall productivity levels were below those being achieved by conventional methods, due mainly to the long periods of downtime and the high manpower requirements to advance and service the jacks.

Figure 5 presents the diffusion patterns for the Coal Plow/Friction Jack longwalls. Relative to the Early Style and Old Style longwalls (Figures 3 and 4), the Coal Plow/Friction Jack longwall technology diffused very rapidly. Only 7 years were required to achieve 100 percent diffusion. But as the f(x) curve in Figure 5 shows, nearly all of these installations were short-lived. By 1961, or about 10 years after the technology first became available, there remained few users. Thus, relative to the Early Style and Old Style technologies, the Coal Plow/Friction Jack longwall had a very short birth-death cycle.

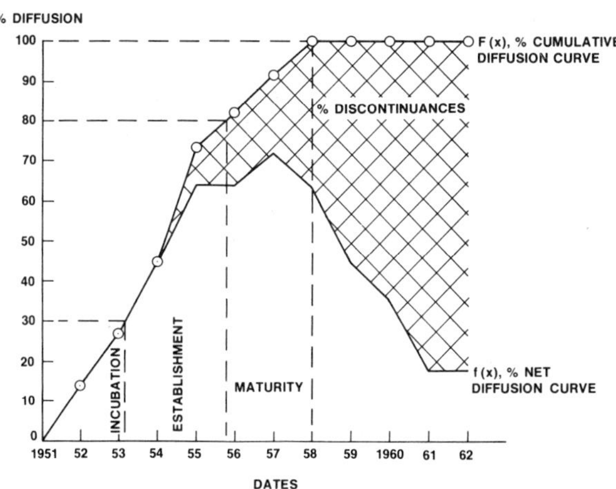

FIGURE 5. DIFFUSION CURVES FOR THE COAL PLOW/FRICTION JACK LONGWALLS ERA

D. Low Resistance ($\leq$ 100 tons) Supports (LRS) Longwalls Era

The experience gained during the coal plow era reinforced the viewpoint by many that the longwall method was a rather specialized, last-resort method, useful only for those thin-seam mines where room-and-pillar methods failed. This attitude was reinforced by the fact that further improvements in the continuous miner and the growth of surface mining technologies had proven capable of supplying most of the Nation's needs at competitive costs. However, latent interests in longwall technologies were revived with the exhibition of British and European-made hydraulic-yielding, self-advancing roof supports at the American Mining Congress Coal Show in Cleveland in 1959.

Early in 1960, Eastern Gas and Fuel Associates installed a plow cutter-loader in their Keystone Mine with the new hydraulically operated self-advancing frame-type roof supports. This installation used a variable-height tandem plow, on a 340-foot wide face; only 4 men were needed to operate the roof supports. Excellent roof control and productivity were achieved. Writers and mining authorities have referred to this first experience with powered roof supports as the "event that opened the floodgates." In 1961, Kaiser Steel Company installed a longwall face in their Sunnyside Mine using self-advancing supports and a new type of shearer-loader. This was the first United States trial of a modern-type longwall at depths approaching those in Europe, and under strata and seam conditions considered too adverse for any other mining method.

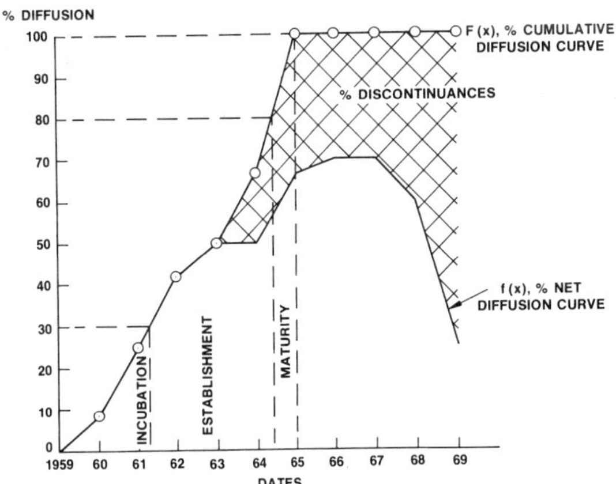

FIGURE 6.   DIFFUSION CURVES FOR THE LRS (LOW RESISTANCE SUPPORTS) LONGWALLS ERA

From 1960 to 1965, imitators in West Virginia, Illinois, Pennsylvania, Colorado, and elsewhere installed Low Resistance Support (LRS) longwalls. Though some successes were achieved, many failures occurred because of massive overlying strata and soft bottoms. As Figure 6 shows, a substantial number of these adopters began to discontinue their LRS longwall operations after 1967.

E. High Resistance (> 100 tons) Supports (HRS) Longwalls Era

Some prominent mining engineers felt that many of the failures of the early coal plow longwalls could have been prevented with higher-capacity roof supports. It took the LRS experience to prove to the European manufacturers that the unavailability of higher-resistance supports was a major impediment to expanding the market for longwalls in the United States. When higher-resistance roof supports were made available to the industry, Eastern Gas and Fuel Associates and Island Creek Coal Company immediately adopted them for their mines under deep and massive overburden. Subsequently, several imitators jumped on the bandwagon. Many successes were scored during this era and several world production records were established.[6]

Figure 7 presents the diffusion patterns for the High Resistance Supports (HRS) Longwalls Era. These patterns differ from those in Figures 3 through 6 in the following three major respects: First, the HRS Era was dominated by a few innovators,** each of whom adopted more than one longwall during the 1964-1977 period. These innovators had gained valuable experience during the Coal Plow and LRS longwalls eras, and had already made the decision to adopt longwalls if and when adequate roof supports became available; second, the HRS Era was characterized by relatively fewer technical failures and discontinuances than any of the preceding eras; and third, the HRS Longwalls Era included a large number of users who adopted more than one longwall. Thus, the HRS Era was characterized by a preponderance of LRS users who were committed to the concept of longwall mining, but who had not found roof supports sufficiently adequate for their physical conditions. As high resistance supports became available, the old LRS users were quick to upgrade and expand their operations with the more rugged HRS technology.

F. Total Systems (TS) Longwalls Era

It was not until the 1970's that the longwall technology came to be accepted as a fully engineered mining system. The cutter-loader subsystem, the articulated face haulage subsystem, and the hydraulic self-advancing roof control subsystem came to be perceived as a complete mining system that could be specified, engineered, and matched to the mine conditions. More rugged and dependable logistical and support systems became available, and hydraulic and electric components were developed specifically for United States longwall conditions. By 1973, improved guidance systems, more efficient cutting heads and variable height winning systems were readily avail-

____
** An innovator is a firm that adopts the new technology during the first year of its availability.

able. By 1975, the state of the art in longwalls had advanced to the extent that, given the wide variety of equipment available, longwalls could be engineered to fit the specific needs of individual mines. This Total Systems (TS) Era was the time when longwalling "got it all together."[7]

Figure 8 presents the diffusion pattern for the TS Era. Several interesting features uniquely characterize this era. First, as shown in Figure 8, relatively few discontinuances occurred. In fact, the percent discontinuances are lower during this era than during any of the previous eras. Second, about 53 percent of the adopters who discontinued their earlier LRS and HRS longwalls readopted the TS longwalls. Clearly, great strides are being made when once-unsuccessful users will try again! It is especially important to note that the TS Era is unique in another respect: for the first time in the history of longwalls, the small operators played a major adoption role. During the TS Era, many operators with no prior longwall experiences adopted their first longwall. A total of 86 percent of these inexperienced operators would normally be considered as "small" by most standards. This entry of the small, imitative operator suggests that the last bastions of resistance to the further diffusion of longwalls have now been overcome. Those firms which have historically waited on the sidelines have now played a major role in adopting the TS longwalls.

G. The Shield Longwalls Era

The advent of shield-type supports has been chronicled elsewhere.[2] Because their introduction is so recent, it would be premature to attempt to develop any adoption analyses for this technology at this time.

H. Technological Progress

Table 1 presents some selected statistics for the seven longwall technologies studied here.[3] Overall, a consistent trend toward increased roof support capabilities is evidenced as one moves from the Early Style (wood props) to the present Shields Era. This trend parallels the trends toward increased adoption and diffusion of longwalls among the coal mining industry. Table 1 also shows that the maximum depth of cover over longwalls has increased from 750 to over 3,000 feet.

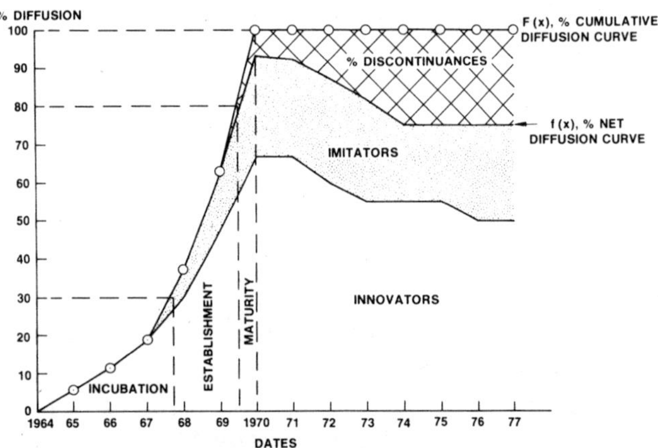

FIGURE 7. DIFFUSION CURVES FOR THE HRS (HIGH RESISTANCE SUPPORTS) LONGWALLS ERA

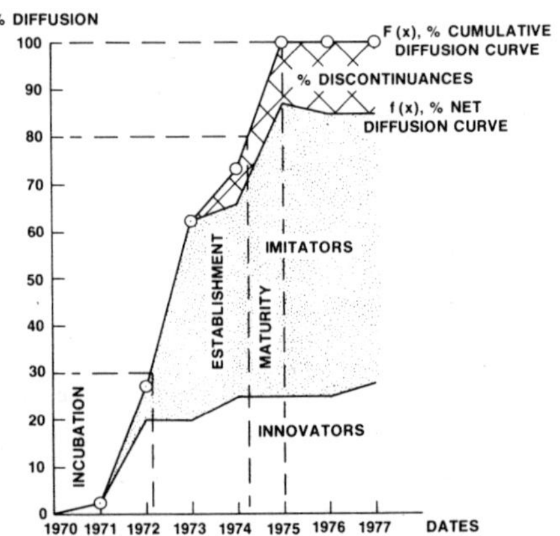

FIGURE 8. DIFFUSION CURVES FOR THE TS (TOTAL SYSTEMS) LONGWALLS ERA

TABLE 1
TYPICAL STATISTICS FOR THE LONGWALL TECHNOLOGIES

	LONGWALLS						
	EARLY STYLE	OLD STYLE	COAL PLOW	LRS	HRS	TS	SHIELD
MAXIMUM DEPTH OF COVER, FEET	750	500	500	2000	2500	3000	3000
MAXIMUM OUTPUT, TONS PER DAY	750	885	700	980	6000	8000	12000
CREW SIZE, NUMBER OF MEN	42	25	15	12	12	12	10

III. ANALYSIS OF THE ADOPTION AND DIFFUSION PATTERNS

A. Time to Spread Among the Users

Table 2 presents some comparative statistics which support the conclusion that today's longwall technologies diffuse much more rapidly among the user population. Today, the durations of all three life cycle stages---incubation, establishment, and maturity---are substantially less than during the Early Style and Old Style Eras.

A closer analysis of these data show that during the Early Style Era both the innovators and the imitators were slow to take up the technology. It took more than 30 years for the first 30 percent of the users (incubation period) to adopt Early Style longwalls. It took another 30 years for the remainder of the users to adopt it. Even at the end of 40 years (end of the establishment and incubation periods), 20 percent of the potential users still had not adopted.

The Old Style longwall was clearly an improvement over the Early Style longwall and the innovators responded by taking up the new technology more rapidly. Only 12 years were required for the first 30 percent of the users to adopt it. The imitators also picked up the technology more rapidly. Only 18 years were required for the remaining 70 percent of the users to adopt Old Style longwalls, compared with the 30 years required for Early Style longwalls.

The Coal Plow faces represented the beginning of modern longwall technology. As Table 2 shows, both the innovators and the imitators were quick to pick up this new technology. Only 7.1 years were required for the entire user population to take up the Coal Plow longwall. However, this longwall technology did not fulfill all the users expectations, and it died out rapidly. Subsequent modern longwalls---the LRS, HRS, and TS technologies---were much improved. These modern longwalls were taken up relatively rapidly, as the Total Life Cycle column in Table 2 shows. It is interesting to note how rapidly these modern technologies spread among the remainder of the population once they were established. As the Maturity column of Table 2 shows, less than a year was required for the "laggards" (the last 20 percent of the population) to take up these technologies.

Thus, it can be anticipated that it will take only about 5 to 7 years for a new longwall technology to spread throughout the entire population of users. This is in dramatic contrast to the 60 years that was required for the first Early Style longwalls.

B. Discontinuances

In general, the user's degree of success with longwall technologies has substantially increased in recent years. This is indicated by the statistics in Table 3. The second column of Table 3 shows that discontinuances occur later in the life cycles of the modern longwalls (the LRS, HRS, and TS longwalls). For instance, in the case of the Early Style longwalls, discontinuances began relatively early--- within the first third of the era. By contrast in the TS longwalls, discontinuances did not start until the era was 60 percent complete.

The last column of Table 3 shows the percentage of the users who eventually discontinued their adoptions within each era. It is clear from these data that the most recent technologies, i.e., the HRS and TS longwalls, are the only sustaining ones. All the former technologies are essentially abandoned.

C. The Adoption Process

Table 4 presents data on the time lapses within the adoption process, for both innovators and imitators. The awareness-to-trial stage covers the time lapse from the adopter's first awareness of the innovation to his first trial. The trial-to-adoption stage covers the lapse from his first trial to his decision to go to full scale implementation.[1]

TABLE 2
COMPARATIVE DIFFUSION PATTERNS (YEARS)

	PERIODS			TOTAL LIFE
	INCUBATION	ESTABLISHMENT	MATURITY	
EARLY STYLE	30	10	20	60
OLD STYLE	12	8	10	30
COAL PLOW	2	3	2	7
LRS	2	3	1	6
HRS	3	2	1	6
TS	2	2	1	5

TABLE 3
COMPARISON OF DIFFUSION STATISTICS

	PERCENT ERA TIME WHEN DISCONTINUANCES BEGAN	PERCENT DISCONTINUANCES AT END OF ERA
EARLY STYLE	30	98
OLD STYLE	25	90
COAL PLOW	28	82
LRS	30	91
HRS	40	25
TS	60	13

The following observations may be made on the basis of the data in Table 4: (1) Relative to other imitators with other innovations, longwall imitators are slow to trial and slow to adopt; (2) Relative to other innovators with other innovations, longwall innovators are quick to trial and adopt; (3) The time lapses for imitators have not declined substantially over the years; and (4) The time lapses for longwall innovators have declined over the years, especially in the awareness-to-trial stage. Thus, today's longwall innovators adopt new technologies rather quickly. There is only a 3-year lag between their first awareness and adoption. By contrast, longwall imitators take about 7 years to go from awareness to adoption, or nearly 50 percent longer than the average for other imitators with other innovations. Clearly, the challenge here is to find ways to reduce the time it takes for the imitators to become aware of and try new technologies.

D. Innovators and Imitators

Innovative and imitative firms play reciprocal roles in the adoption/diffusion process. The innovators effectively debug and demonstrate the new technology, thereby reducing the risk of failure for the larger population of imitators. But, the subsequent diffusion of the innovation among this larger population erodes the innovator's markets. Thus, he is stimulated to look for additional innovations, and the cycle repeats.

Table 5 lists five factors which were found to distinguish longwall innovators from longwall imitators during the 1870-1978 period. In the innovative firms, the decision makers attended more conventions and trade associations, visited more competitors' mines, traveled abroad more frequently, read more journals, and generally had more diverse contacts outside the firm. They were on a first-name basis with their counterparts in several equipment manufacturing firms, and they generally had a more favorable attitude toward research and development activities. In several cases, the top-level executives actively sought outside-the-industry technologies which they might transfer to their operations. In the most innovative firms, this attitude pervaded all levels of the company, and even the lower levels of the firm were open to trying out new ideas.

Mixed results were found for financial factors such as profits, sales, asset position and return on investment. In some cases, it appeared that the firm's financial health was a result of its early successful adoption of a longwall, rather the cause or main reason why it was an early adopter. However, it was also clear that only those firms that can afford longwalls are able to innovate with them. The five factors in Table 5 must thus be considered as a total cluster. If all of the factors are present, the firm is more likely to be an innovator than an imitator. But all of them must be present; if only the first three are present, the firm may not be able to afford the luxury of being an innovator.

IV. THE FUTURE

A. Trend Projections

Standard trend-forecasting techniques were used to predict the number of longwalls that might be in operation over the coming decade. The results indicate that the number of longwalls will rise from the current (1978) level of about 95 units, to about 124

units in 1980, to about 247 units in 1985, etc., as shown in Figure 9. The fourth-power equation used to extrapolate the actual pre-1978 data in Figure 9 was found to provide an excellent fit to the 1950 to 1978 data base, as shown by the goodness of fit statistics at the bottom of the figure. It must be noted that these projections are extrapolative (trend forecasts). The curve fitting methodology does not take into account any changing economic factors (e.g., the energy crisis) or political influences (e.g., federal energy policies, mining laws, etc.), except insofar as these influences may have already tempered the 1950-1978 data base.

B. Some Obstacles

In a recent survey of coal mine operators and longwall equipment makers,[4] eight major categories of barriers were identified which could upset the optomistic forecasts presented in Figure 9. Table 6 presents these categories.

A lack of real appreciation and comprehension of longwall technologies was found among some of the smaller operators. They believed that longwalls were only appropriate for: (1) those mining companies

TABLE 4
AVERAGE YEARS OF ELAPSED TIME

STAGES	LONGWALLS		OTHER INNOVATIONS
	1945-65	1968-78	
AWARENESS-TO-TRAIL			
INNOVATORS	5.5	1	1.5
IMITATORS	5	5	3.5
TRIAL-TO-ADOPTION			
INNOVATORS	2.5	2	3
IMITATORS	2.5	2	1
AWARENESS-TO-ADOPTION			
INNOVATORS	8	3	4.5
IMITATORS	7.5	7	4.5

TABLE 5
DISTINGUISHING CHARACTERISTICS:
INNOVATOR vs IMITATOR FIRMS

1. OUTSIDE INFORMATION SOURCES
2. NUMBER OF "FLOOR-SHOP" RESISTANCES
3. FIRM'S WILLINGNESS TO ACCEPT OUTSIDE HELP
4. FIRM'S ASSET POSITION
5. FIRM'S PROFIT AND SALES GROWTH RATE

TABLE 6
BARRIERS TO INCREASED USAGE OF LONGWALLS

1. LACK OF APPRECIATION AND COMPREHENSION
2. FELT RISK AND UNCERTAINTY
3. GENERAL LACK OF KNOW-HOW
4. "BELIEVABILITY"
5. LACK OF COMPREHENSIVE INFORMATION
6. CULTURE OF THE COAL MINING COMMUNITY
7. THE "ALL-OR-NOTHING" CHARACTERISTIC OF LONGWALLS
8. GENERAL SUPPLY AND DEMAND UNCERTAINTIES

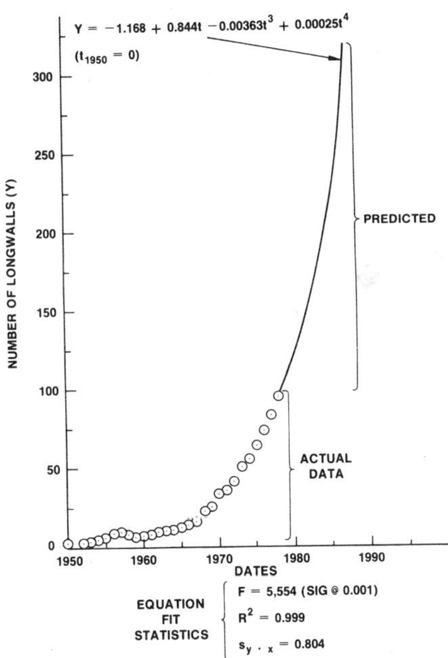

FIGURE 9. PREDICTED GROWTH IN NUMBER OF LONGWALLS

holding vast reserves of coal, and (2) for mining coalbeds under conditions where conventional mining methods had previously failed. Throughout the mining community, there was some lack of appreciation for the fact that the longwall system requires extensive planning and careful attention to details.

The size of the investment, the long payback period, the high breakeven point, the required commitment, the potential regret if coal demand falls—all these factors combine to make longwalls appear to be a risky venture in the eye of some mine operators. The technical uncertainty that longwalls will succeed under their particular conditions also dissuades many potential users.

Several "believability" barriers were found with regard to the outstanding results being publicized for a few specific longwall installations. Most of the subjects did not believe these results to be representative of typical long-term outcomes. The lack of long-term comparative data on different longwalls and the absence of predictive information about ground mass movements were frequently cited barriers.

The coal mining community was found to be a rather heterogeneous collection of economic entrepreneurs, each of whom faces somewhat unique market requirements and technical problems. A brotherly culture prevails; technical information is openly shared and exchange-mine visits are common. However, the uniqueness of physical conditions at various operators' mines often impedes real technology transfer. Moreover, there is apparently no one operator who is an influential authority within the mining community, from whom the other operators can solicit advice and information. The openness of the community precludes any long-term technology advantage or monopoly profit for an innovator, which may account for the fact that there is little research and development activity within the community. The acceptance of technologies from outside the mining industry are generally viewed with apprehension by the community.

Physically, the longwall is a total system which is not easily trialed in piecemeal fashion. This "all-or-nothing" configuration represents too large a commitment for many mine operators. A longwall also requires high add-on investment costs, in terms of higher capacity coal handling facilities.

Many operators felt that the demand for coal in the future will depend upon complex ecological and political forces which are now unfolding. These uncertainties cloud the feasibility of long-term commitments to longwall technologies at this time.

V. CONCLUSIONS

Throughout history, longwall technologies have promised coal mine operators lower labor and operating costs, higher coalbed recovery, greater outputs and improved safety. But in the early years, the adoption/diffusion of longwall technologies was inhibited by: the apathy of operators to change their handicraft practices; the primitive state-of-the-art; inadequate equipment; and a low demand for the coal output. In later years, as equipment improved, the demand for coal rose, and coal operators gradually became more willing to experiment with innovations, roof control remained the major technical barrier.

Looking back over the history, Early Style (wood props) and Old Style (friction steel props) longwalls had only limited potential from the start. The longwall method did not really emerge as a viable technology until the advent of self-advancing hydraulic roof supports, mechanized cutter-loaders, and articulated face conveyors. These developments reduced the large labor complement that was formerly required on the face, and provided the potential for large volume outputs. But, this potential was not realized until reliable and adequate equipment could be developed and made available for the conditions extant in the United States. About the time when this point was reached, coal supply began to exceed market demands. It took an energy crisis beginning in the early 1970's to provide the demand needed to take up this excess capacity.

Today, the modern longwall technology stands as a highly developed, tested, and proven innovation. It is ready, willing, and able to contribute significantly to the projected doubling of coal production over the next decade.

REFERENCES

1. Rogers, Everett, _Diffusion of Innovations_, Free Press: Glencoe, Illinois, 1962.
2. Von Derlinden, Walter, "Summary of Performance of Shield Type Supports in the U.S.," paper presented at the SME-AIME Fall Meeting and Exhibit, Denver, Colorado, September 1-3, 1976 (preprint No. 76-F-321).
3. Palowitch, E. R. and J. W. Corwine, "The Bureau's Program to Automate Longwall Mining," _Coal Age_, July 1975, pp 152-164.
4. "1976 USBM Census of Operating American Longwall Installations," _Coal Age_, January 1977, pp 99-107.
5. Schmellenkamp, Manfred, "Progress Report on Longwall Mining," _Mining Engineering_, No. 15, June 1963, pp. 55-58.
6. Watson, Brian, "Longwall Produces Record Tonnage," _Coal Age_, Vol. 26, No. 3, March 1971, pp. 64-75.
7. Kuti, Joseph, "Longwall Mining in America," paper presented at the 1978 SME-AIME Fall Meeting, Nassau, Bahamas, September 14-15, 1978 (preprint No. 78-AU-336).

Chapter 2

LONGWALL MINING - SHEARERS AND PLOUGHS
AND SYSTEM CONSIDERATIONS

Robert Stefanko, Ph.D.

Professor of Mining Engineering and Associate Dean
College of Earth and Mineral Sciences
The Pennsylvania State University
University Park, Pennsylvania 16802, USA

INTRODUCTION

Longwall mining which has a long history abroad, was used only on a limited scale in the United States until less than 20 years ago. Modern longwall mining in this country can be said to have begun in 1960 with the first installation of self-advancing props in a coal mine in southern West Virginia.

In Europe, because of adverse natural conditions, longwall mining is frequently the only practical method, but, in the United States, the system has had to compete economically with highly productive room-and-pillar systems. However, because it is inherently continuous, it has greater productive potential than room-and-pillar systems, and, therefore, although only about 5% of the total U.S. underground production today is the result of the longwall system, its growth potential is great. As mining, by necessity, progresses deeper and must be done under poorer natural conditions, less opportunity will exist for the room-and-pillar system, and the longwall system will have to be used.

Besides its continuity, there are other advantages that may be anticipated with longwall:

(1) Improved safety resulting from complete overhead steel face support and minimum moving of equipment.

(2) Higher extraction rates and thus better conservation.

(3) Greater flexibility in mining under poorer roof, at greater depths, and in multiple seams.

(4) Better subsidence control.

(5) A great many collateral benefits from the elimination of the need for bolting on the face, rock dusting, and many ventilation controls.

Of course, longwall mining has disadvantages also:

(1) More pronounced and thus more critical delays.

(2) Higher capital costs for equipment.

(3) High moving costs.

(4) Inflexibility where there are want areas or gas wells, in seams of variable thickness, and where bottom and top are too soft.

(5) Possible impracticability under massive top at shallow depth because of the size and cost of support required.

As the depth of mining increases, the excavation must become narrower and the pillars wider to minimize destructive stresses. Finally, the point is reached at which the excavation is so narrow that efficient mining cannot be done, or the difficulty of supporting an entry is so great and expensive, irrespective of how wide it is, that a new system becomes necessary. That system is longwall.

LONGWALL SYSTEMS

Longwall mining consists of driving one or more gates or entries approximately 300 to 600 feet apart, providing an interconnection, and then mining the rib of the interconnection on a longwall, hence the name, longwall mining. The longwall can be mined either on retreat as shown in Figure 1 or on advance as in Figure 2.

The Retreat Longwall

In U.S.-style retreat longwall, the face is retreated away from barrier or bleeder entries, Figure 1. There usually are unmined chain pillars on both sides of the gob and the solid coal face itself.

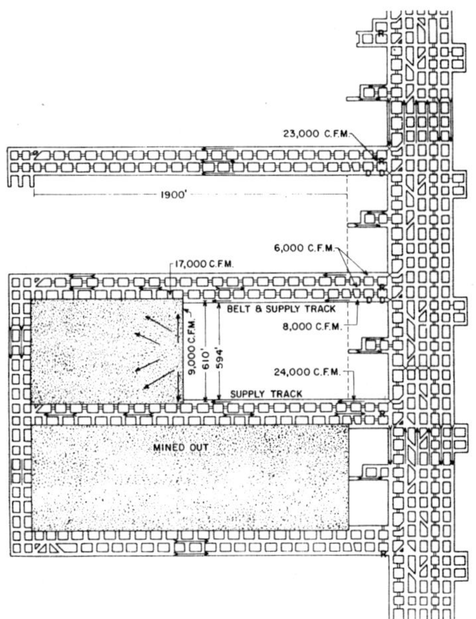

Figure 1. Typical Longwall Retreat System.

In effect, then, as the face is retreated and until caving begins, the mined-out void is covered by the equivalent of a plate. As the weight of the over-burden loads the plate, the plate bends on all four sides sides, transferring the weight of the unsupported strata to pillars or coal on the ribs, and these become abutments.

As the size of the opening increases, the load on the abutments keeps increasing until something gives-- the roof, if all goes as planned. The trick is to get a good first fall or cave and then to maintain caving at the waste edge of the props while preventing the overriding of the supports and the breaking of the roof at the face.

The Advance System

The advance system is used most widely abroad whereas the retreat system is used almost exclusively in the United States. Abroad, at the great depths mined (up to 4,000 feet), it is impossible to drive and maintain entries of the required width. There-fore two entries (gates) are maintained through the gob as the face advances, the mothergate on the solid side containing the belt and providing the intake air and a tailgate completing the loop for ventilation.

The advantage of the advancing system is that coal production is realized almost at once, and thus development costs are not deferred. However, overall costs will be increased because of the large cost of maintaining the gates during the life of a panel.

Once the first panel has been driven to its limit, a second panel will be advanced with the mothergate of the first panel becoming the tailgate of the second one, whenever possible. This obviously saves time and costs because only one entry has to be created for the second and each subsequent panel. However, if gob pressures are so great that they preclude this, a solid barrier equal in width to 1/10th the depth of the seam is left, and the second panel is driven identically to the first one shown in Figure 2.

(1) The belt is on intake air.

(2) Multiple mining machines are on a single split of air.

(3) There is no intake escapeway provision.

However, an experimental advancing longwall system is being tried in Colorado, the only such installation in the United States known at this time. In the days of cheap energy, such a system would have been rejected outright as economically unfeasible but, under present conditions, its development is being watched very closely by the industry.

In the United States, gates or entries have not been maintained through gob areas. In the commonly employed practice, Figure 1, entry driving is divorced completely from face work, and the entries are not maintained through gob areas, simplifying support and ventilation, and reducing development costs.

Alternate Advance and Retreat Systems

To minimize equipment moving costs, alternate advance and retreat systems have been tried in the United States, Figure 3. However, it should be noted that this advance method of mining differs from that practiced abroad. Haulage does not have to be main-tained through the gob, and, therefore, support costs and ventilation losses are not as critical. Neverthe-less, alternate systems have been largely abandoned in the United States because of the difficulty in maintaining the entries. The chain pillars near the gob areas have a tendency to squeeze, cracking stoppings and disrupting air flow. Cost savings

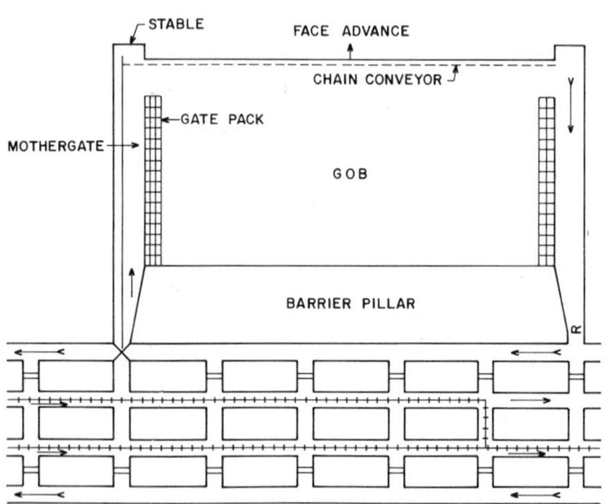

Figure 2.   Example of a Typical Longwall Advance System.

It should be recognized that this advance system would be illegal in the United States today for a variety of reasons:

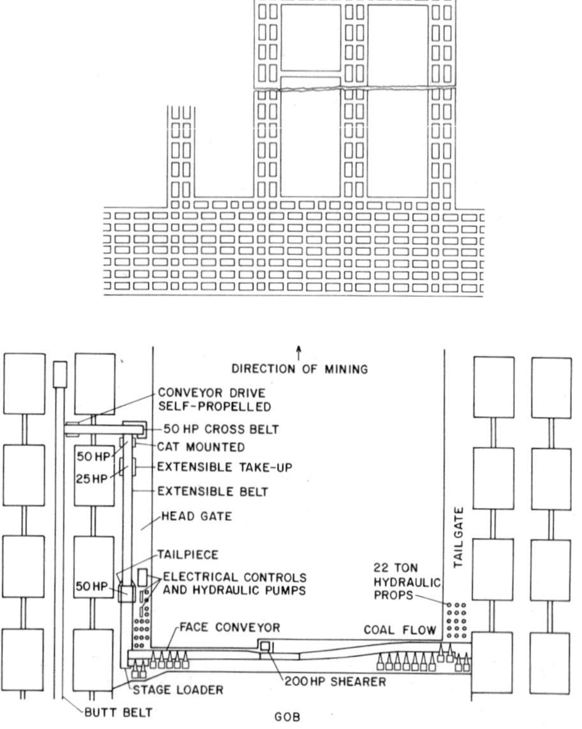

Figure 3.   An Alternate Advance and Retreat System.

achieved by minimizing moving distances are counteracted by the need for rearranging the equipment
(conveyor and supports) because any panel will be
basically a mirror image of either the preceding or
subsequent one. In fact, the confusion this moving
of equipment causes may well wipe out all anticipated
benefits. So during initial installations of longwall by a company, the alternate advance and retreat
systems should not be attempted until mechanics have
become completely familiar with the equipment
components and the system.

LONGWALL SYSTEM COMPONENTS

The longwall system consists of a combination of
three basic equipment components:

(1)　A support system.

(2)　A coal-mining machine.

(3)　A haulage system, Figure 4.

Figure 5 shows the arrangement of a long prop-free
front face with continuous transportation by a chain
conveyor which also acts as a track, or a guide, for
a shearing machine. The shearing machine pulls itself
back and forth on a stationary chain, cutting slices,
or webs, from the face. This protected area is
created by cantilevered roof beams that extend from
powered supports. By providing a longer and more
substantial cantilever frame, this one-web-back
system permits the machine operator to travel between
the conveyor and the front props, a costly convenience
under some conditions.

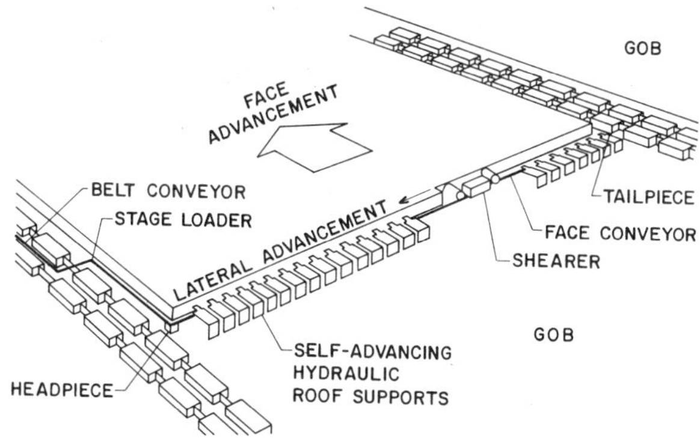

Figure 4.　Basic Equipment Components on a Longwall
　　　　　　Face.

Because roof pressures usually make it possible to
maintain only a limited cantilevered top, the actual
working space is generally limited to 10 to 12 feet,
Figure 6, and the machine operator and other workers
must travel in the confined area between the props.
Designing highly productive equipment to fit such
limited space, maintaining it properly, and providing
for adequate movement of men are no simple tasks.
Obviously, the space for mining machine, conveyor,
and supports must be kept as narrow as possible. But
since the web mined with each pass varies from as
little as 2 inches with some planers to as much as
36 inches with some shearers, the equipment must be
capable of being moved quickly as the face retreats.

The coal-winning machine can be located between the
conveyor and the face resting on the bottom or it can

Figure 5.　Prop-free Face One-web-back Shear
　　　　　　Operation.

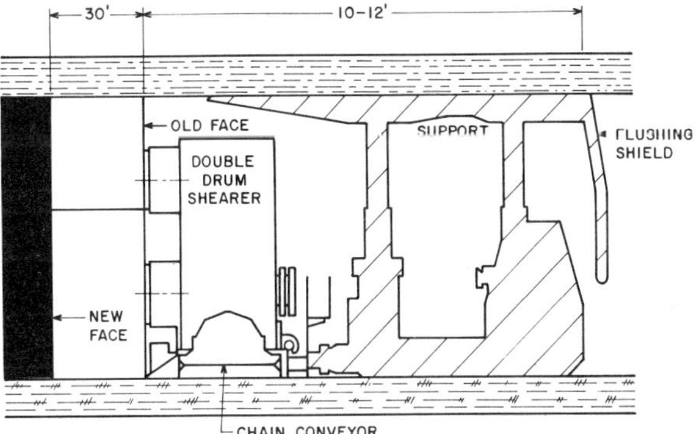

Figure 6.　Conventional Double Drum Shearer Face
　　　　　　Profile.

be mounted so that it rides on top of the conveyor.
Since the planer is narror and takes a thin slice, it
rests on the bottom. Shearers until recently were
always conveyor-mounted, decreasing conveyor life,
but in-the-web shearers have now been introduced.
Chain conveyors used with planers have about five
times the life of those used with shearers. With
either machine, coal is now mined in both directions
or bidirectionally.

COAL-WINNING MACHINES FOR LONGWALL MINING

General

Two types of machines are utilized on longwall
faces: 1) shearers, and 2) plows. While there are
many criteria to consider in deciding whether to
select a plow or a shearer, the thickness of the seam
is probably one of the most important. With seams
that are less than 42 inches thick, clearances are too
critical for the use of a modern high-potential
shearer, and plows are generally used. On the other
hand for seams that are more than about 72 inches
thick, the plow becomes too unstable and cannot be
maintained in an upright, effective cutting position.
Here the shearer which is designed to mine up to 120
inches is used. Support instability due to tangential
forces above 6 to 7 feet is a real problem in utilizing chocks and has resulted in the failure of several
installations. Shields appear to have overcome this
problem in higher coal, and today some very thick-
seam, shearer-shield combinations (up to 19 feet) are

in operation or are in the planning stage. Thus it can be seen that the plow is a thin-seam miner, and the shearer is used in thicker seams. Choosing the mining equipment for seams of intermediate thicknesses requires careful consideration of many other factors that will be discussed later.

There are, however, other criteria besides seam thickness that affect the choice between the shearer and plow. There are summarized in Table 1.

Table 1. Comparison of Conditions Applicable for Shearers and Plows

Plow	Conditions	Shearer
Poor	Poor Seam Parting	Immaterial
Poor	Hard Coal & Impurities	Flexible
Low	Bit Cost	High
Low	Dust Production	High
Poor	Dust Control	Good
Low	Methane Liberation	High
Equal	Entry Widths Needed (Controversial)	Equal
Better	Roof Control (Controversial)	Poorer
11 or 12	Manpower Needed	9 or 10
Lower	Initial Investment (Controversial)	Higher
Low	Maintenance	High
Lower	Consistency of Output	Greater
Better	Supervision and Organization	Less
No	Operator Accompaniment	Yes

Changes have been very rapid in longwall mining machine development and these sections are not intended to be a comprehensive history of this equipment. Rather, an attempt will be made to show basic developments and list reasons for changes with primary emphasis on the latest techniques.

Shearers

Shearers are relatively narrow machines that ride on the face conveyor and take a web or cut that is usually from 24 to 36 inches deep. They require a prop-free face that is achieved by a cantilevered bar that extends from a cluster of self-advancing hydraulic props, Figure 5.

When the first European longwall systems were introduced in this country, shearers were unidirectional in operation and took relatively narrow webs (approximately 20 inches deep). They would cut coal while traveling from the head drive to the tail drive and then plow the coal from between the conveyor panline and the face on their way back to the head.

To be competitive in the United States, this system had to be modified. The drum was redesigned into a spiral arrangement to assist in moving the coal to the conveyor, Figure 7. To complete the spiral housing and create a screw conveyor, a cowl was attached to the machine at the back end of the drum to prevent most of the coal from falling into the

newly mined zone between the conveyor and the face instead of onto the conveyor. To make bidirectional mining feasible, it has to be possible to reposition the cowl at each end of the face to trail the drum. This requires headroom in the face entries. The use of two hanged deflector doors offers an alternative to a cowl. The deflector door in the direction of the drum travel is rendered inactive by turning it parallel to the conveyor, Figure 8, while the trailing door is turned perpendicular to the conveyor to act as a cowl. In the reverse direction, the doors are repositioned in the opposite manner. However, the greater degree of confinement of the dust to the drum area where it can be more effectively sprayed makes the cowl more effective for dust control. To help ramp plates were put on the front end of the conveyor to bulldoze the coal into it as it was pushed (snaked) into position, Figure 6. A further refinement was the placing of bits on the face end of the drum so it could be sumped in at the face ends. This is done to avoid cutting out into the entry and permit attacking the buttock on the return run. This allows much narrower entries for improved roof control.

Use of the single fixed drum is limited to seams that are uniform in thickness and relatively free of undulations. Otherwise, either rock must be cut or coal must be left in place; both alternatives are expensive. Therefore, ranging drums of either the single or double variety have become widely used. The single ranging drum whose diameter is equal to the thickness of the seam, usually mines a web in a single pass. However, the drum must not be indiscriminately ranged because of the effect that offsets in the floor and roof would have on the support and conveyor movement. The machine can be equipped with a drum whose diameter is approximately two-thirds the average thickness of the coal and then utilized to mine the entire seam in two passes, the first pass taking the upper portion of the seam while the second pass mines the lower portion. While this arrangement provides flexibility, an obvious evolution was to use a double ranging drum to mine the entire seam in a single pass. This maintains the desired flexibility without any loss of production.

The double drum shearer such as one shown in Figure 7 has two main advantages in retreat longwall mining:

(1) It can cut out the ends of the longwall faces, permitting the use of narrow entires

(2) By ranging both drums, the cutting height can be adjusted to accommodate variable seam thickness while maintaining the high productivity made possible by its capability of mining the entire seam height in a single pass.

Theoretically, during each pass, the leading drum should be raised and the trailing drum lowered to provide maximum cleanup. Also, with this procedure the operators are in a good position to maintain horizon when rolls are encountered. Therefore, this is the more commonly recommended technique. However, the leading drum is sometimes used in the lowered position and the trailing drum raised, the argument being that less dust is produced with this arrangement, and that by using specially activated ramp plates, cleanup is not necessary. However, these ramp plates, used widely, are virtually extinct today.

Figure 7. Double-drum Shearer with Cowls

Figure 8. Use of Deflector Doors on Ranging Single-drum Shearers

Since the lead drum will mine almost two-thirds of the coal, clearance must be provided under the machine for the coal to pass by on the conveyor. It is usually impossible to design adequate clearances in seam heights of less than 55 inches. This machine has been used in seams down to 45 inches, but this is possible only by throttling its speed and cutting down production.

Each drum of the double-drum shearer is equipped with a water-cooled motor with a rating of from 150 to 500 HP, depending upon its duty and voltage. The trend to 950-volt motors to minimize cable size for higher horsepower is well established. Cooling water for the machine is used for dust-suppression in conjunction with water from another independent line. Spray curtains envelope each drum, the water injected through nozzles on the periphery and face of the drums and from sprays directed at the drum.

The machine is generally captivated on the armored conveyor by means of a suitable gob-side tube guidance; the two other faceside shoes ride on the lip of the conveyor. Early conveyor difficulties resulted when four captivating shoes rode on the conveyor. This caused machine binding and conveyor-pan distortion. Some of the methods used to solve these problems included making two diagonal shoes captivating while the other two were sliding shoes, providing slide wear strips on top of the conveyor, and offsetting the weight of the machine so the weight was directly distributed through the web of the conveyor.

The machine pulls itself along an approximately 1 inch diameter-wire-size chain through an internal capstan-drum arrangement. This stationary haulage chain is hydraulically tensioned and attached to the head and tail drive frames of the conveyor. Control devices sense the current drawn by the cutting motors and adjust the output of the capstan, thereby regulating the speed of the machine. Abnormal values created, for example, by tough cutting conditions will automatically regulate the hydraulic winch and slow down the machine within certain preset speed ranges.

The automatic cable handler consists of a specially designed flexible chain linkage with an open side for

inserting cable and hose. This device, mounted in a
conveyor trough on the gob side of the conveyor,
automatically handles and protects hose and cable
while the shearer moves back and forth across the face.

Plow

The coal plow (planer) is designed to be pulled
along the face in a back-and-forth shuttle fashion to
cut a layer or web of coal 2 to 6 inches thick from
the face with a planing action. The broken coal is
deflected or plowed onto an adjacent face conveyor.
The plow travels in the space between the armored
face conveyor and the coal face. It is held against
the face by the lateral pressure exerted on the
conveyor frame by pneumatic or hydraulic cylinders
which are spaced from 10 to 40 feet apart depending
upon the hardness of the coal seam. The plow is
pulled along the face by an endless chain operated
from drums in the gates. One end of the chain is
fastened to the planer and is then extended to and
passes over the head-piece drive sprocket; it returns
across the longwall face, over the tail piece
sprocket, and is attached to the other end of the
planer, making the planer the closing link of a
continuous chain. A stabilizing arm connected to the
base of the planer and extending under the face conveyor to stabilize the plow in a vertical position
was a common feature of the earlier plows.

Many plows today operate at speeds of more than
200 fpm and are usually equipped with static bits
although activated bits have been tried. Slow speed
plows (well below 100 fpm) that are 1/3 to 1/2 the
height of the seam thickness and undercut the seam
while relying on roof pressures to break down the
upper portion of the seam are most commonly used
abroad, and were initially introduced here. However,
full-seam tandem plows cutting in both directions at
the higher speeds soon became most popular in this
country.

The high-speed plow has had an interesting history
in a mine in western Pennsylvania. Use of a plow is
generally limited to softer seam conditions, but by
operating it at a very high speed (in excess of 300
fpm) it would seem that the greater kinetic energy
developed should result in efficient mining of
harder coal and partings. Unfortunately, for every
action there is an equal and opposite reaction, and
the maintenance problems encountered with the high-
speed plow in the Pennsylvania mine have been severe.

Plow guidance methods have changed through the
years. The guidance method involves the position of
the pull chain, its attachment to the plow, and how
the plow is connected to the conveyor. The oldest
form of plow guidance consisted of a guide tube which
was attached to the face side of the conveyor around
which the plow base was fitted, Figure 9A. At first,
the return side of the drive chain was encased within
the guidance tube, but a problem with the packing of
solid fines in the guide tube by the tail chain under
wet conditions resulted in the tube's being replaced
by a rectangular plank with the tail chain free
running, Figure 9B.

Face-side guidance systems are now obsolete,
having been replaced by gob-side guidance systems of
which the most common type is the hook plow, Figure
9C.

The stabilizing arm of the plow, extending under

the conveyor, quickly chewed up soft bottom, and the
friction that resulted created high chain tensions,
especially when hard coal was being cut. Therefore,
tough coal cutting, low coal, and soft bottom made
plowing conditions very unprofitable and an improvement was needed.

The Gleithobel system, shown in Figure 9D, which
was introduced recently, incorporates changes to
overcome problems encountered with the plow. The
base is removed and a ramp plate replaces the guide on
which the plow operates. Plow stability is maintained by a stabilizer arm which extends over the
conveyor and runs on a gob-side tube guide, while the
chain is captivated inside the ramp plates. As the
conveyor shifting rams are pressurized, the bits are
pushed into the coal and the base of the plow itself
is not rubbing on the bottom. The elimination of the
friction of the stabilizing arm under the conveyor as
well as of the base of the plow on the bottom provides a higher percentage of the chain drawbar pull
as an effective cutting force for tougher materials.

SYSTEM CONSIDERATIONS

Figure 10 is a plan of a typical planer face
showing the spacing of and relationship between components of the system. The conveyor is kept pressed
against the face and, as the planer is pulled across
the face between the face and the conveyor, its
cutting bits are forced into the coal.

The roof supports can be moved as much as the ram
extension system permits (up to 30 inches) but they
can be reset closer to the conveyor if roof conditions
dictate. Two props on a common base are released,
moved forward, and reset in approximately 30 seconds,
after which the other two props of a unit are moved
similarly. Each propman on the face tends a
designated number of prop units. The head- and tailgate operators move the entry roof supports and keep
their respective conveyor drive units in alignment
with the conveyor as the face is retreated.

On a planer face, the roof supports are in alignment at all times, distributing the load equally.
Because of the thicker web taken and the lag in
snaking with shearers, the unsupported triangle common
to shearer operation, Figure 11, can be avoided when
plows are used. Thus, in terms of flexibility in
moving the supports and better distribution of load,
planers can probably work under more difficult roof
conditions more safely than shearers. An added
safety feature of planers results from the face that
an operator does not need to travel the face as he
must with shearers.

One of the major problems with longwall is the
development of sufficient entries and their maintenance. With high capacity jacks on short centers,
roof control at the face is rarely a problem. The
greatest difficulty exists in supporting the tail
entries, and Figure 12 indicates why. The tail entry
at the end of the active face is subjected to great
ground pressures because of the superpositioning of
three major abutment stresses at this point. There
is the side abutment pressure resulting from the
previously mined panel upon which is superposed the
face abutment pressure from the active panel. This
frequently results in such deterioration in the tail
entries that considerable supports including cribs
and hydraulic jacks must be installed. This inter-
feres with turn-around-time and moving of the conveyor if it must reach into the entry. Therefore,

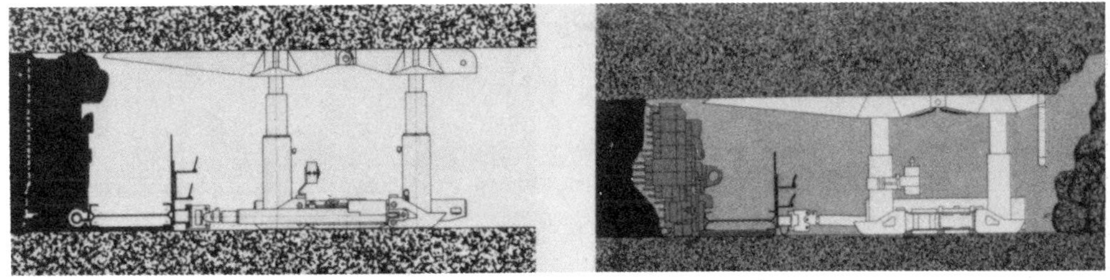

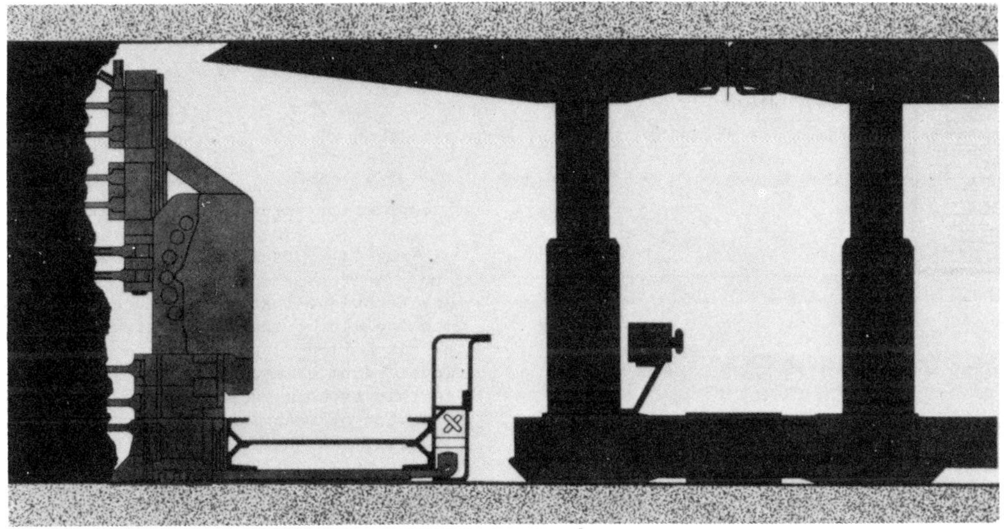

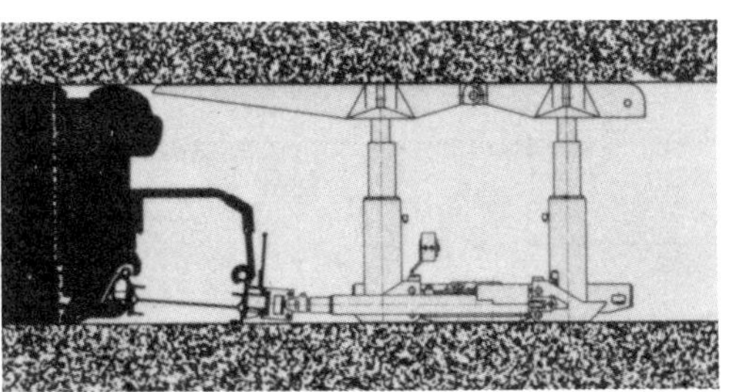

Figure 9.   Evolution of Face Guidance Systems: Tube Guidance (A),
Guide Plank (B), Hook Plow (C), and (D) Gleithobel.

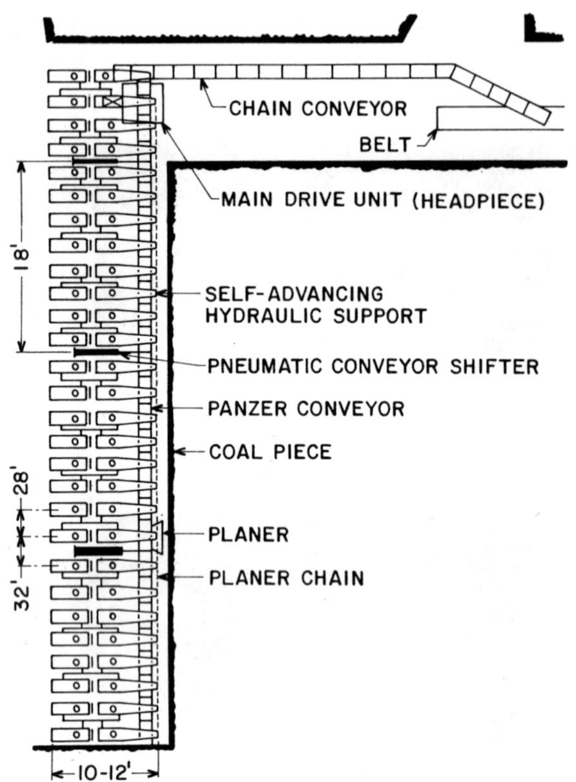

Figure 10.　Typical Planer Face.

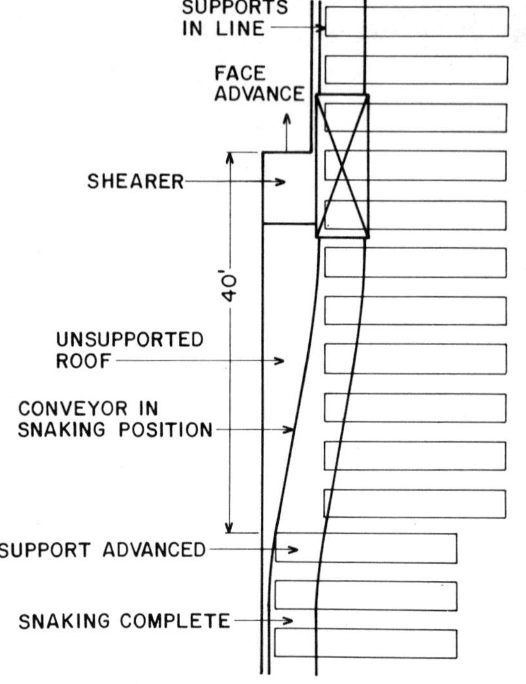

Figure 11.　Shearer Plan View Showing Unsupported Roof Triangle.

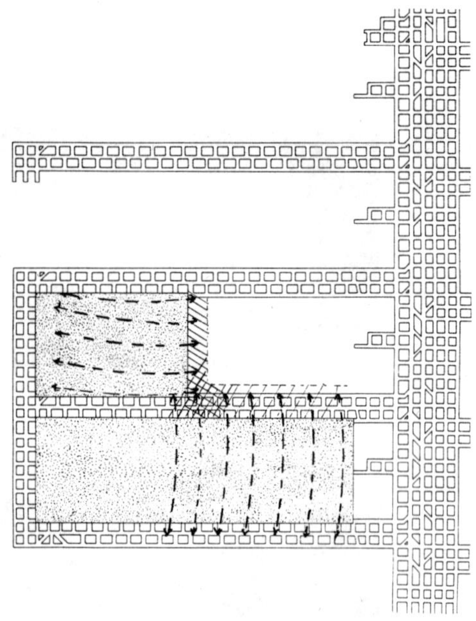

Figure 12.　Stress Concentrations on the Tail Entry Resulting from the Superposed Effect of Face and Side Abutment Pressures.

any technique that keeps the conveyor drive and mining machine out of the tail entry during turn-around will permit maximum support placement there. Using a double-drum shearer probably is the easiest solution to this problem of turnaround space. While there are a number of techniques employed by this type of shearer, the "half-face" technique, Figure 13, is one of the more popular maneuvers. With the aid of Figure 13, it will be described in detail. At the beginning of the cutting cycle (1) the shearer is at the left end of the face with its lead drum, No. 1, raised for the principal cutting action and its trail drum, No. 2, at floor level to cut the lower seam portion and load most of the material. The right half of the conveyor pan line is pushed close to the face as shown. The shearer deadheads at top speed to near point a, where, following the contour of the conveyor, it begins sumping into the coal, completing the full-cut sump by b. The shearer continues to cut at a lower speed until it reaches the right corner of the wall. As soon as drum No. 1 (top) cuts out the corner into the entry, it is lowered while at the same time the trail drum, No. 2, is raised. While this is going on, beyond point b, the left side of the conveyor is snaked over and the props advanced until position 4 exists. Now just the reverse action is employed to cut out the left side of the face to produce the situation in 6 which provides the same situation as shown in 1 but for the next cut.

Two single-drum shearers, each with cutting heads attached on opposite sides of the frame, may also be employed on the same pan line, one cutting one half of the the face and the other the remaining half. To comply with regulations, only one machine can be in operation at a time. The author is aware of three such installations where results are considered superior

There are various combinations of equipment

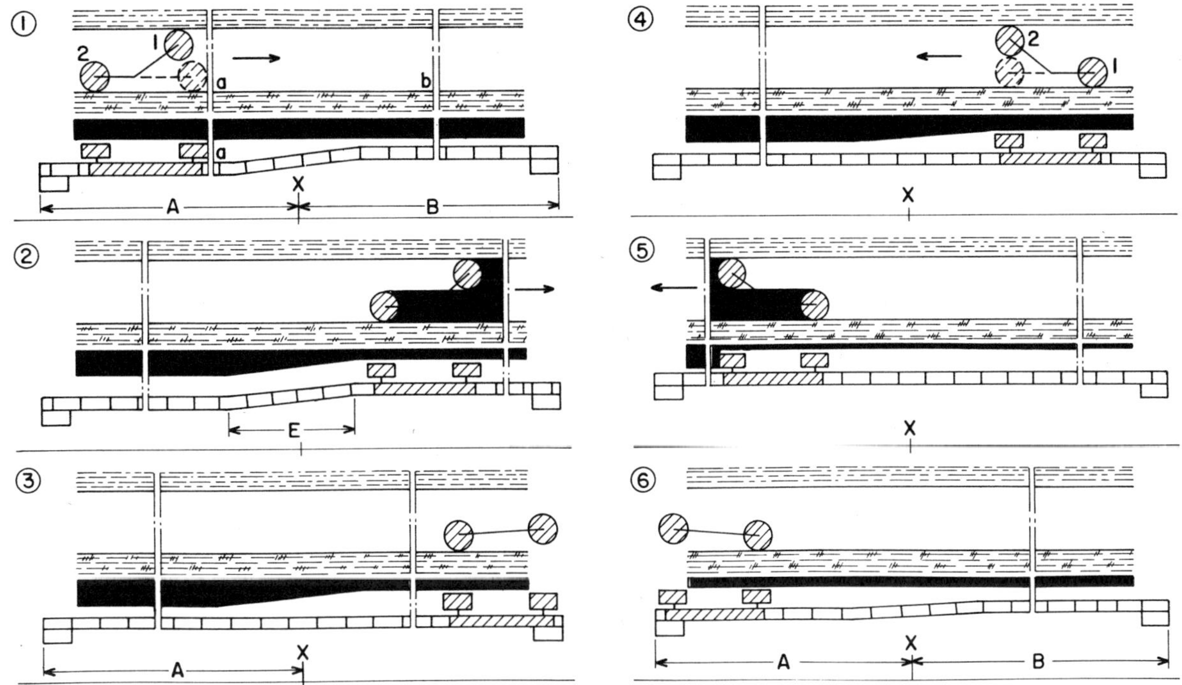

Figure 13. The Half-face Technique Used with Double-drum Shearers.

components and mining systems employed in longwall mining, and only some system design criteria have been discussed here. Results have varied widely from mine to mine and even between sections in a specific mine.

SUMMARY

At this point, a summary comparison of the shearer and planer may be helpful. Obviously, the shearer is a more complex machine than a planer and, as would be expected, has a higher initial cost and maintenance. Beyond this, some of the most important differences between the two types of equipment are the conditions of work (Table 1).

Generally speaking, planers work better in soft coals that have a minimum of hard impurities and separate easily from the top and bottom. They are, therefore, less flexible in application to a wide range of geologic conditions. The shearer is more positive in its cutting action and will cut impurities and hard coal, although not without increased costs for replacement bits. Thus shearers maintain a more consistent output. It has already been established that better roof control can be maintained with plows.

One of the more important considerations in a final selection between shearer and plow is the amount of dust produced and the methane liberated. Because of the method of attack, shearers produce a finer product and therefore create more dust. Also, since the shearer exposes a web 5 to 10 times as thick as a plow, the methane liberation will be much higher. The finer coal with its greater exposed surface area releases more methane. There are several recent

illustrations where the plow was chosen over the shearer because it was anticipated that the ventilation system could not handle the gas and dust problems. While another expert, who has had experience with both machines, admits the above statements are true, he claims that the control of dust is much easier with shearers even though the amount is greater. His reasoning is that the cowl around the shearer drum confines the dust where it can be more effectively sprayed and prevented from becoming airborne. He says that even though there is probably somewhat less dust produced by a planer, most of it becomes airborne.

Another factor to be considered in choosing a shearer and a planer is the quantity and quality of manpower each requires. A planer crew will normally consist of 1 or 2 more men than a comparable shearer crew. But even more important, greater skill is required to operate the plow. The planer must be kept on the bottom and all coal must be removed from the roof, otherwise the supports will not work properly. This calls for better supervison and organization.

The plow operates more effectively in low seams than does the shearer; it is difficult to keep it stable in an upright position during the cutting of thick coal. It is better at greater depths because it minimizes the rate of gas and dust production. Also, at greater depths, greater abutment pressures prefracture the coal and make planing much easier, but this is also true for shearers. Overall, considering the deeper seams that will now have to be mined in the United States, planer application may rise in this country.

Chapter 3

DEVELOPMENTS TO IMPROVE LONGWALL CONVEYOR PERFORMANCE

Walter von der Linden

Vice President
KB/National Mine Company
Pittsburgh, Pennsylvania

INTRODUCTION

There are three essential components for a modern longwall mining system: the coal cutting/loading machine, the self-advancing supports and the longwall conveying system. The latter consists of a face conveyor, a stageloader and a belt tail section arrangement (Figure 1). The longwall face conveyor is an integral component of the face equipment and has a threefold purpose:

1) continuous coal haulage.

2) track for cutter/loader.

3) reference/anchor rail for roof support advance.

Further requirements for the longwall conveyor are horizontal flexibility for conveyor advance without interruption of haulage and cutter/loader tramming and vertical flexibility to cope with seam undulations.

THE LONGWALL CONVEYOR SYSTEM

Figure 1 shows a system layout with its principal components. The face conveyor consists of an electro-mechanical head and tail drive station which drives the chain assembly consisting of chain and flight bars. The flights convey the coal by scraping the material over line pans each being 1.5 m (59 in.) long. The line pans are connected maintaining horizontal and vertical flexibility thus providing adaptability and robustness which are the two most outstanding features of the AFC (Armoured Flexible Conveyor). The conveyor furnishings consisting of ramp plates, spill plates and coal cutter track are bolted to the line pans (Figure 2). At the headgate the coal is transferred from the face conveyor via a stageloader onto the panel belt conveyor. A crusher mounted on the stageloader is frequently used to size big lumps of coal for belt haulage (Figure 1).

LONGWALL CONVEYOR PROBLEMS

The performance requirements and design criteria for a longwall conveyor system have produced equipment for which some compromises were necessary and which for that reason could not be optimized for its original conveying purpose. There are a number of drawbacks associated with a conveyor system which have the potential for a multitude of operational problems. A listing of these problems is as follows:

1) The scraping of the coal over the line pans by chain pulled flight bars leads to wear of the line pans, flights and chain. Service

life for today's heavy-duty versions is approximately 1.5 million tons (1.65 million short tons) for the chain and 2.5 million tons (2.75 million short tons) for the line pans. These data vary with operating conditions and the abrasiveness of the material which is mainly a function of type and content of rock. Failures due to premature wear and overload may occur.

2) The friction of moving components including coal against the line pans leads to a high specific energy consumption. Further, the friction factors vary in a wide range. High drive powers and torques are required and the various components must be designed accordingly. However, blockages and overload failures may still be experienced.

3) High loads and forces are imposed on the line pans by the roof supports especially by shield supports and by the coal cutting equipment. The flexibility of the line pans causes high stresses in the line pan connections. Overload failures used to be frequent in these areas.

4) Problems at the face ends include clearance cutting, carryback, dust generated at the transfer points and operational difficulties created by the massive drive frame assemblies.

GENERAL DEVELOPMENT TRENDS
OVER THE PAST 10-15 YEARS

Face conveyors originally developed in Germany used the outboard chain concept where the flights and chain were guided by the line pan side profile. At that time, the prop free coal face was not known and the conveyor had to be dismantled prior to relocation in the new field or web. Ease in handling and assembly were important criteria.

In the 1960's, 18 mm twin outboard chains with 178 mm and 190 mm high line pans were still universal. Mechanization of coal cutting and loading resulted in a recirculation of fines and related dust suppression measures led to briquetting of coal in the side profiles and to trapped chains.[1] In the 1970's, the introduction of shield supports in Europe exposed weaknesses of the line pans themselves. The pan connections and furnishing bolt holders could not withstand the high thrusts resulting from shield support advance. These events led to the introduction of heavy-duty line pans with higher side profiles and single or twin center chain arrangements. Larger chains with significantly increased breaking loads were used and this facilitated a significant increase in drive power.

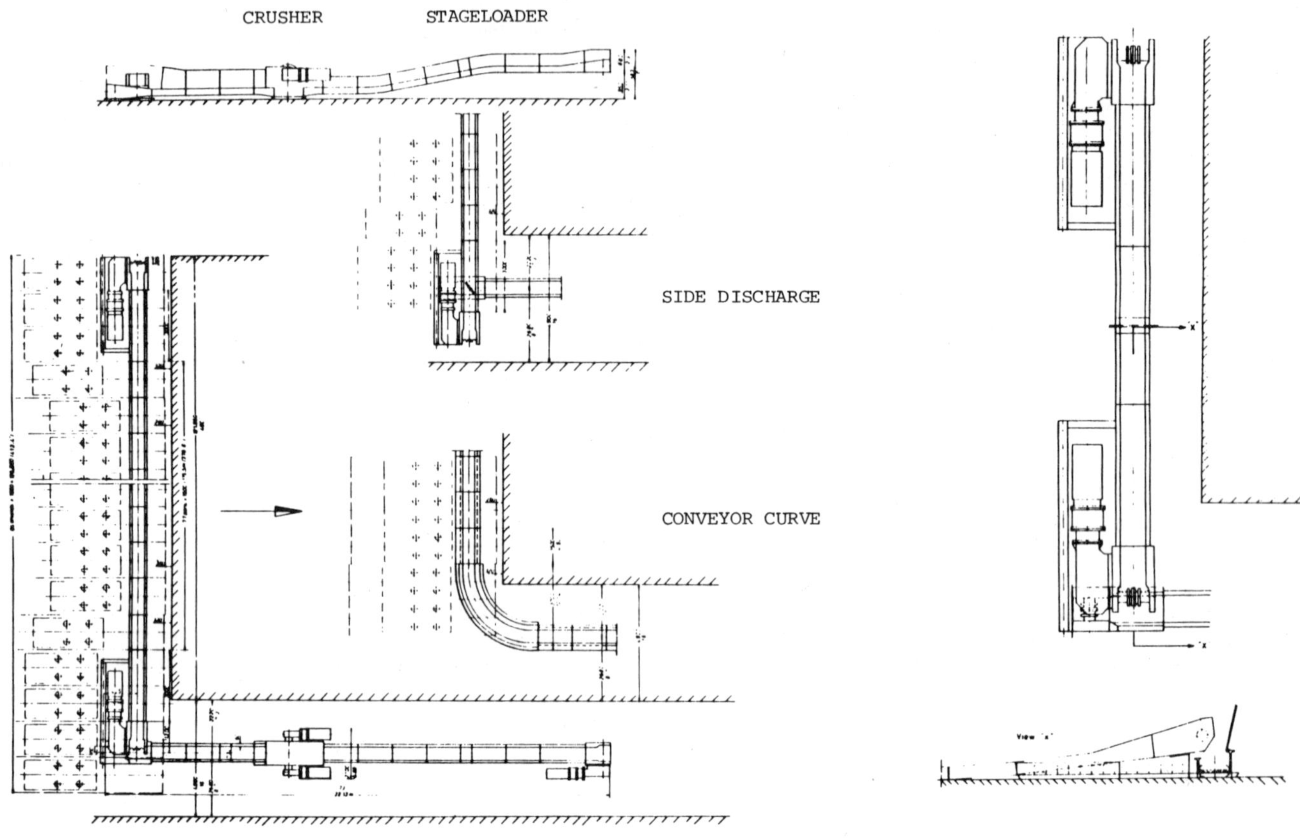

FIGURE 1   ALTERNATE LONGWALL CONVEYOR CONCEPTS

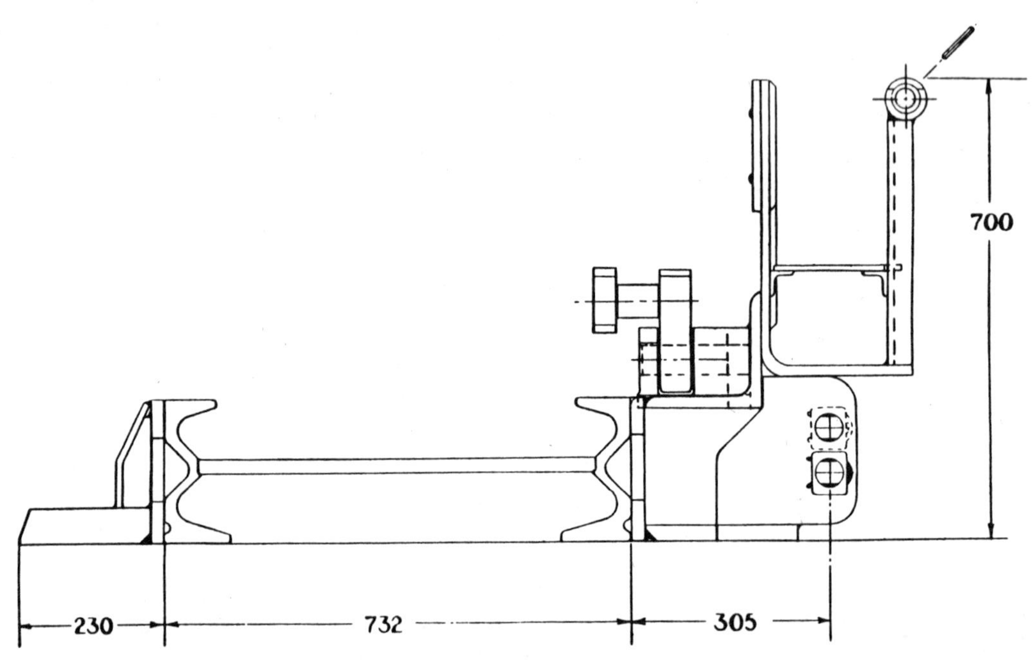

FIGURE 2   FACE CONVEYOR WITH FACE AND GOB SIDE FURNISHINGS

DEVELOPMENTS TO IMPROVE CONVEYOR PERFORMANCE

In the following paragraphs, developments to improve conveyor performance are described for the various conveyor components: linepans, chains and flights, drive assemblies and face end equipment.

Line Pans

Line pans are the basic structural element of the AFC and are fabricated in sections from rolled steel members (side profile) welded to a thick deck plate forming a trough in which the flight bars move the coal (Figures 3 and 4). Line pan improvements have been made in the following areas:

1) Side profile.

2) Deck plate.

3) Line pan connectors.

4) Furnishing bolt holders.

5) Closed bottom line pans.

Side Profiles are distinguished by height, weight per unit length, material composition and by their shape which determines their strength and operational properties. All manufacturers claim that their side profile is a proprietory item designed with the aim to maximize strength and durability and to minimize friction of flights and moving coal in the line pans. The contour and design details of side profiles show, in fact, significant variations between various manufacturers.

Since the 1960's and early 1970's, a general trend in regard to side profile strength has been to increase the height from the then 178 mm and 190 mm profile to the now customary 220/230 mm and 250/265 mm profile. This permitted to increase material thickness, i.e., wear volume and structural strength.

The variety of side profiles can probably be categorized in three types as shown in Figure 5. All profiles have in common that their cross section resembles the shape of a capital "E" consisting of top and bottom flange with connections to the center section to which the deck plate is welded. The distinct differences between the three profiles shown here are the inside contours. The design criteria are: (1) application as outboard chain conveyor versus inboard chain conveyor; and (2) minimizing of friction.

The PF III profile was primarily designed for use with outboard chains. The area enclosed by the top and bottom flange is maximized for use of largest possible chain size. Typically, the contour is formed by two 45° angles at the center and a 9° angle at the top and bottom flange. The profile 250V was primarily designed for application as center chain conveyor. Here the profile contour is shaped to enclose the flight tip only and typically this profile is formed by two 30° angles and two 20° angles. In an outboard chain application the chains would not be covered by the profile flanges and the chain spacing would be slightly narrower. The UFv 34 profile offers a combination of the previous two approaches. This profile is suitable for center chain and outboard chain application. The angles of this so-called universal profile are 41° and 18°.

The friction between flight and profile is determined by the contours of flight and side profile itself and by the likelihood for coal and debris material to briquet inside the profile contour which would increase the resistance to movement. In this regard, all manufacturers equally claim superiority of their profile design. It has been argued that the 250V profile is less prone to briquetting of coal due to its larger angles at the top and bottom flange. On the other hand, the PF III profile claims to offer less friction between flight and profile in the presence of lateral flight forces. In this regard, the flight contour for the 250V profile may be wedged into the profile in the presence of lateral flight forces which may increase the resistance to flight movement. The newer type UFv 34 profile tries to optimize this aspect by combining a larger flange angle with a flight contour which is designed to avoid wedge friction.

The material composition of the side profile used to be carbon steel with manganese steel end sections. Today several manufacturers use manganese steel throughout the entire side profile length. A typical material analysis includes contents of C = .42 - .48% Si = .15 - .35% and Mn = .85 - 1.05%.

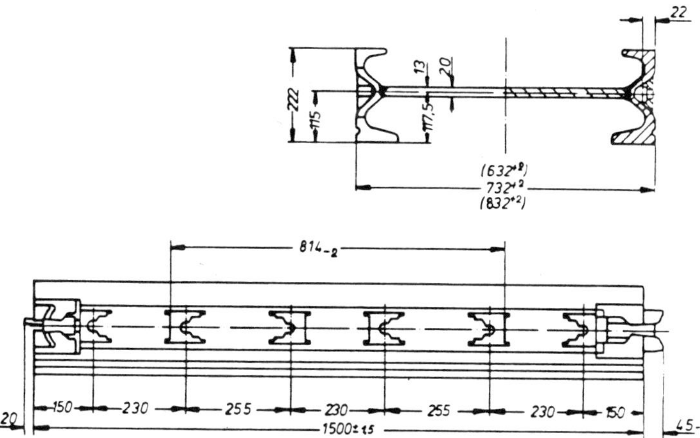

FIGURE 3   LINE PAN WITH 222 mm PROFILE FOR 6 FURNISHING BOLTS (UFv 26 KLOECKNER-BECORIT)

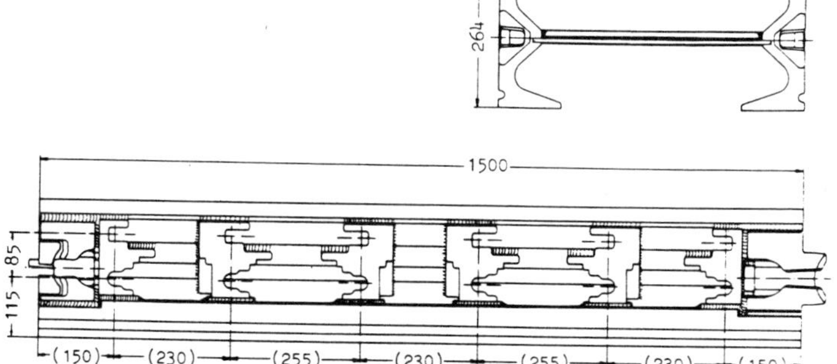

FIGURE 4   LINE PAN WITH 264 mm PROFILE FOR 12 FURNISHING BOLTS (UFv 34 KLOECKNER-BECORIT)

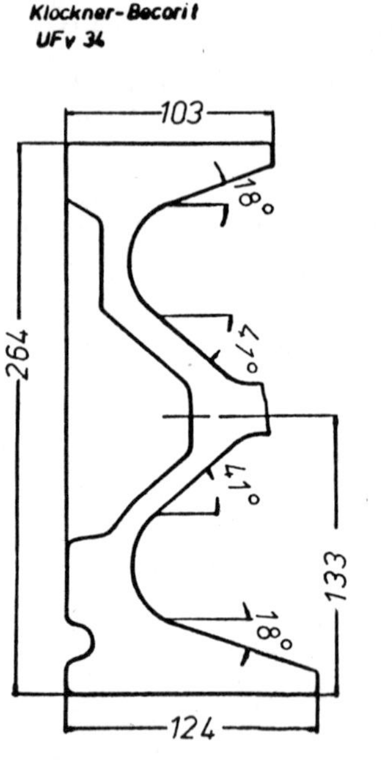

WEIGHT 78.1 kg/m

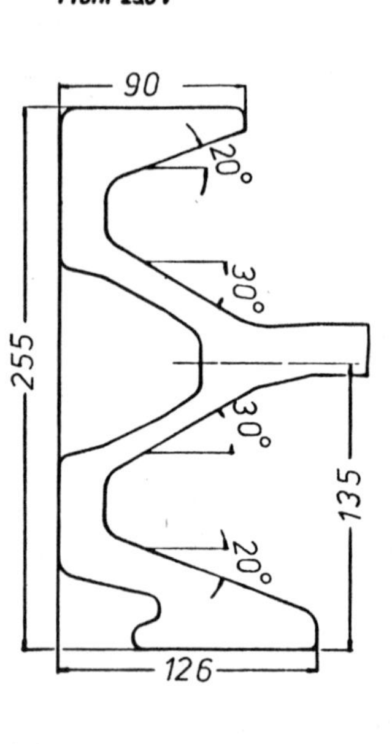

83 kg/m

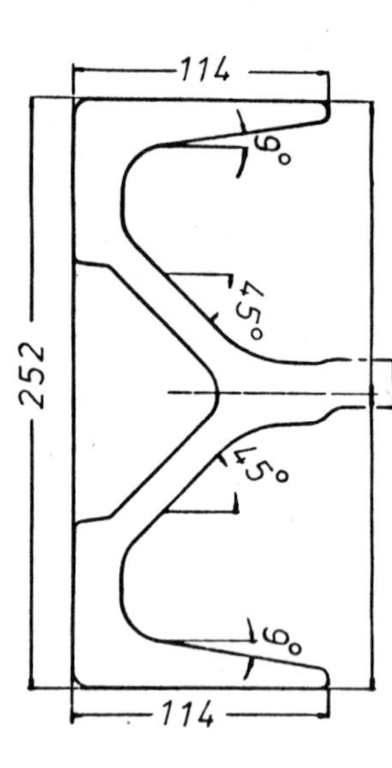

73 kg/m

FIGURE 5   SIDE PROFILES

Line Pan Deck Plates used to be 10 to 16 mm thick. Today, deck plates of 20, 25 and 30 mm are common. The end sections of the deck plates are normally composed of the same manganese steel as used for the side profile. The contour is such that they provide an overlap between two adjacent deck plates.

The Line Pan Connectors couple two adjacent line pans and must be designed to avoid lateral and vertical shift but maintain a flexibility of approximately 2° horizontal and 6° vertical between adjacent line pan sections. Line pan connectors used to consist of 20 - 30 mm high strength bolts with a breaking strength of up to 50 tons (55 st). Today, a type of line pan connector is common which consists of a "dogbone" shaped forged alloy steel with a breaking strength of up to 150 tons (165 st). Resistance against lateral and vertical shift is provided by matching male and female end sections (Figure 6). The quality and strength of a dogbone connection depends on the avoidance of bending forces when the dogbone connection is under tensile stress.

A different pan connector design approach was taken by Halbach & Braun. These line pans have a flush end contour with a large bore through which the so-called 3-D coupling is inserted (Figure 7). It is important to note that this 3-D coupling must accept the tension and shearer forces while the previously described "dogbone" design needs to accept tensile forces only. The quoted breaking strength of the 3-D coupling is therefore more theoretical and in practice, it is only used to avoid lateral and horizontal shift between line pans. The tension connection is provided by additional connecting elements in the conveyor side trimmings.

Furnishing Bolt Holders. The face conveyor furnishings, the ramp plates on the face side and the spill plate assembly on the gob side, are bolted to the line pan profile by hammer head bolts which are inserted in bolt holders welded to the profile section. Normally, six 24 mm high strength grade 9 bolts are being used. For higher seams, line pans with two rows of 6 bolt holders each are being offered (Figure 4). These can better accept the additional forces resulting from the moments imposed on this bolt connection. For further optimization of the bolt connection, the UFv 34 profile has raised top and bottom edges to maximize the effective clamping length.

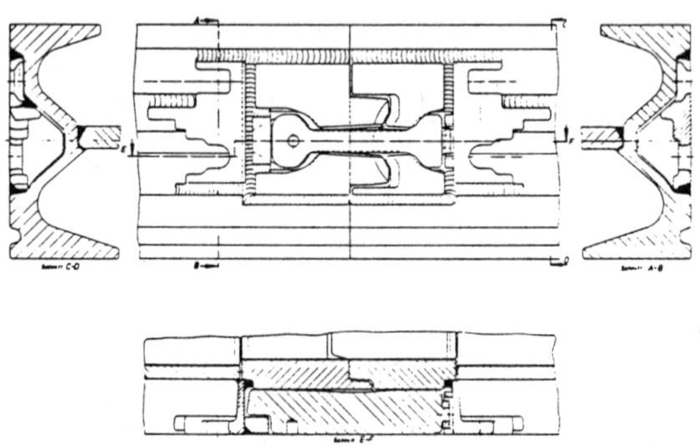

FIGURE 6   LINE PAN WITH DOGBONE CONNECTOR

Closed Bottom Line Pans are being used frequently in the United Kingdom, and in a few cases, also in Germany and in the United States. The objective is to reduce the resistance to movement of the chain and flights in the bottom race by protecting this area by a bottom plate against interference from soft floor and spillage. A critical problem is access to the bottom race in case of chain breakage. Usually, this is done by access doors fitted in

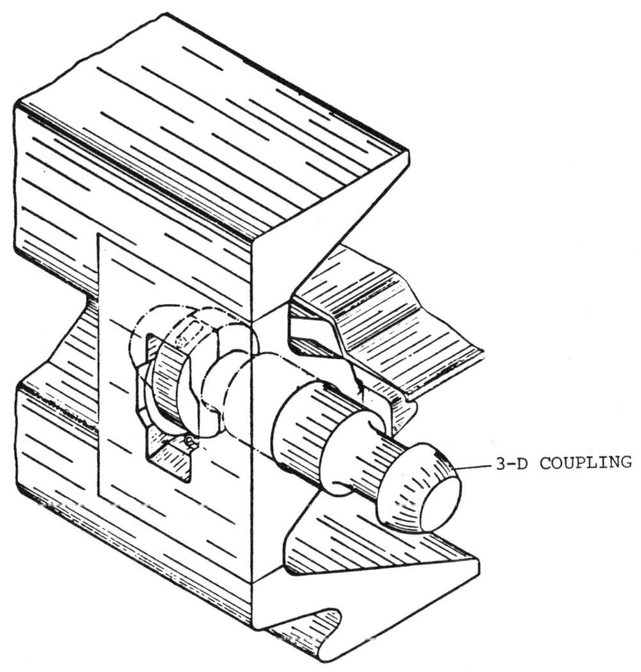

FIGURE 7　LINE PAN WITH 3-D COUPLING
(HALBACH & BRAUN)

machined recesses in the deck plate of approximately
every fifth line pan. More recently, closed bottom
line pans with removable bottom plates are being
used which eliminates the complicated access door
fitting in the deck plate (Figure 8). In this case,
access to the bottom race throughout the conveyor
is available after unbolting of the bottom plate and
tilting of the line pans.

Chain and Flights

Simultaneously with the enlargement of the line
pan profiles, larger chain dimensions were introduced. The 18 mm chains used in the 1960's were
adequate for shorter face length and power requirements of up to 200 hp. Today, single center 30 mm
or twin center 26 or 30 mm chain arrangements are
in common use. The longwall conveyor chains generally comply to specification DIN 22252 Class 2.
The 26 mm chain has a breaking strength of 85 tons
(93 st) and the 30 mm chain of 115 tons (126 st).
Twin chain arrangements have a sufficient safety

factor for high capacity faces with up to 600 to
900 hp. For example, the safety factor against
breaking load is 2.4 for twin 26 mm chain at 600 hp
and 2.15 for a twin 30 mm chain at 900 hp. These
safety factors are based on a chain speed of 1.07 m/s
(210 fpm) and a maximum motor torque of 230%.

The life of the chain which determines the reliability and economy of a longwall operation to a
great extent can be prolonged by protecting the
chain against corrosion. The chains are either
coated with a zinc plastic mix or galvanized. The
uniform surface contour of the chain links which is
maintained with the aid of the corrosion protection
reduces friction between chain links and prevents
formation of stress risers.

The flights are designed to sustain high loads
and bending moments, particularly apparent with
center chain flights. Figure 9 shows flight chain
assemblies of single center, twin center and twin
outboard chains. Adequate clearances from chain to
deck plate and to the seam floor are important for
reduced chain and deck plate wear.

Twin outboard chains used to consist of short
chain strands with shackles for flight connection.
These connections are the weak links in a chain.
A new development for outboard chains uses long
strand chains up to 100 m (328 ft) long where the
flights are hooked into the chain links (Figure 10).

Most longwall chain conveyors installed in recent
years in Germany and in the United States have single
or twin center chains. Twin outboard chains are
preferred in pitching seams where resistance to
down-drifting coal lumps is critical. Twin outboard
chains should also have an advantage in stageloaders
due to better utilization of the available line pan
cross section.

In the United Kingdom, twin outboard chains are
preferred which require lower pretensioning forces
and appear to be better suited for closed bottom
line pans.[2]

The movement of chain and flights causes a noise
level which may become an area of concern in the
future. A recent underground noise level measurement by a German research institute of an empty
running conveyor, UFv 34 with twin outboard 26 mm
long strand chain, showed an noise level of 75 dB(A)
which was approximately 15 dB(A) lower compared to
other conveyors operating under comparable operating
conditions. [3]

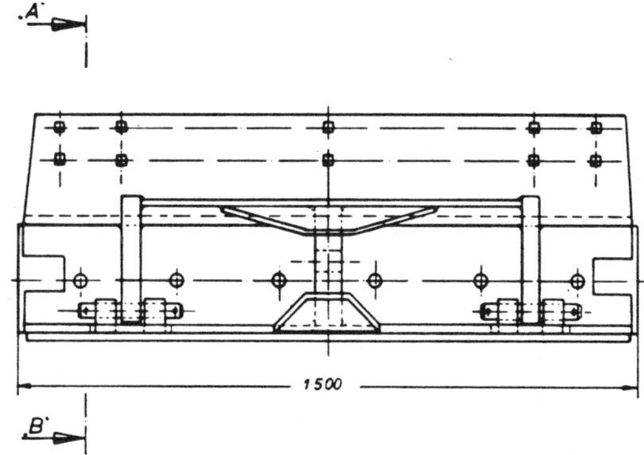

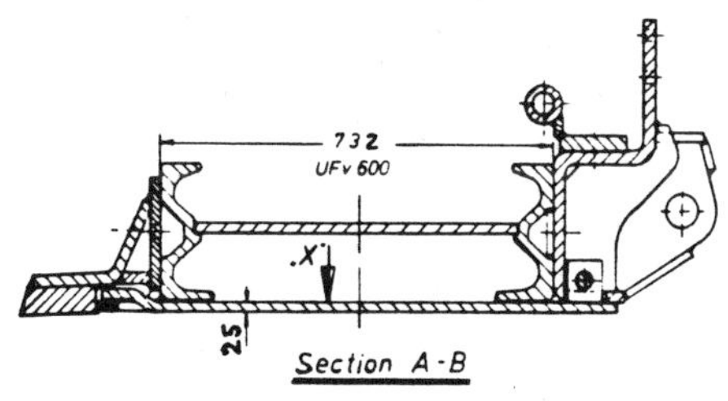

FIGURE 8　CLOSED BOTTOM LINE PAN (REMOVABLE BOTTOM PLATE)

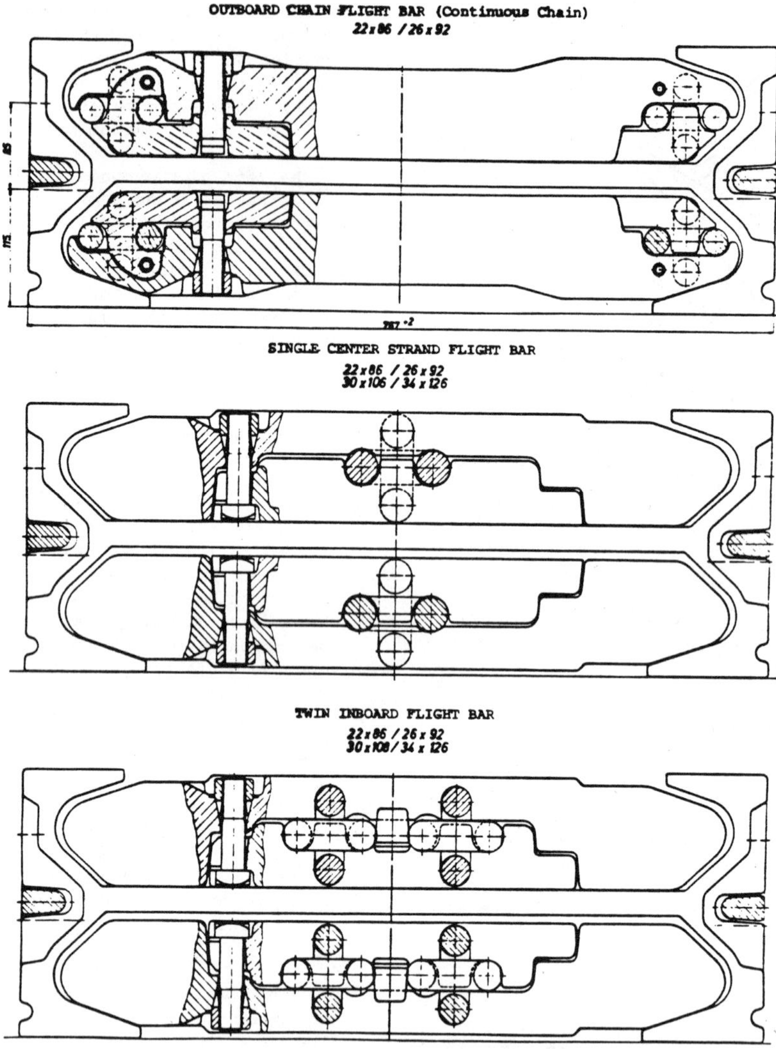

FIGURE 9   LINE PAN UFv 34 WITH TWIN OUTBOARD,
INBOARD AND TWIN INBOARD FLIGHT BAR
(KLOECKNER-BECORIT)

Drive Assemblies

In the United States, the standard drive assembly is made up of an air-cooled 1800 rpm induction motor, hydraulic coupling and bevel spur gear. Typical torque requirements for the motors are shown in Figure 11. The actual torque curves depend on starting voltage drops and may be significantly lower. The electrical installation must be designed to minimize this aspect.

In Germany, the use of dual-speed double-wound, water-cooled motors has become the rule. The conveyor is started with one-third of nominal speed which eliminates the need for hydraulic couplings. Starting currents can more readily be limited and acceptable starting torques be maintained.

Important for a satisfactory chain life is the matching between chain and chain sprockets. The so far known chain sprockets were concave-shaped to match the round contour of the chain. Recent research has shown that this type of chain sprocket requires a close chain tolerance in order to maintain optimum matching conditions. A different type of chain sprocket was developed which uses a V-shaped tooth profile (Figure 12). This eliminates the need for close tolerance chains by maintaining optimum matching conditions also for under and over-sized chains.[4] Wear at the outside of the chain links is reduced (Figure 13). A further advantage of this sprocket is reduced friction between chain and sprocket and reduced stresses in the chain link when entering and leaving the chain sprocket. In laboratory tests, it was found that the maximum chain stresses were reduced by up to 35% when using a V-type sprocket.[5]

The employment of heavier drive assemblies and chains made necessary the proper pretensioning of the chains both from a safety and chain wear aspect. Research has shown that the chain life can be severely shortened if chains are over-tensioned. Peak loads caused by vibrations may exceed the elastic load limits.[6] Two methods are available

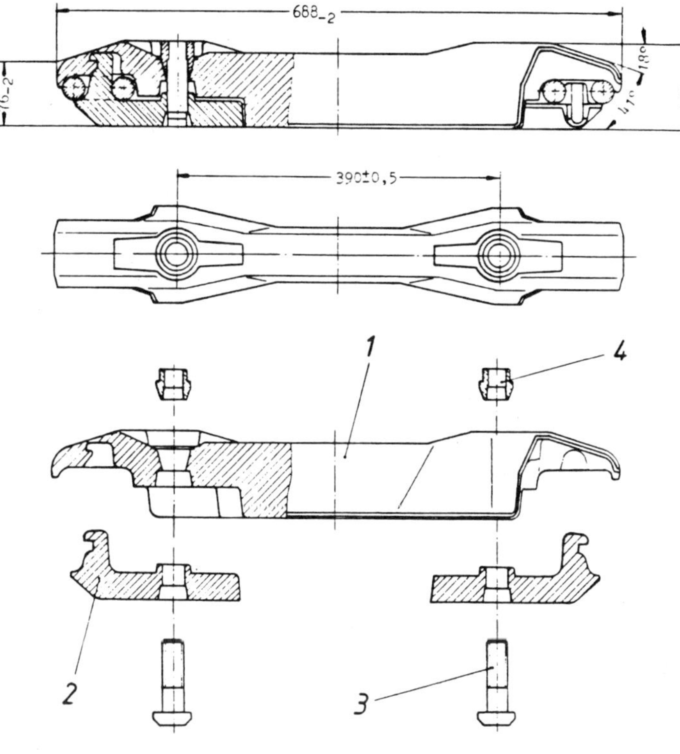

FIGURE 10   FLIGHT BAR FOR LONG STRAND
TWIN OUTBOARD CHAIN

for safe and easy chain splicing and tensioning.
Mining Progress supplies a hydraulic motor gear
assembly operated by the roof support fluid and
mounted on the gearcase input side. The chain can
be tensioned with slow speed and controlled force.
An alternate method consists of a disk brake assembly also to be mounted at the gearcase input side.
The chain is tensioned by the electric drive motor
and then a controlled tensioning force for the chain
is maintained by the disk brake.

Problems and Improvements
at the Face/Headgate Junction

A weak point in the United States longwall
installations is the face/headgate junction area
with a number of drawbacks which frequently prevent
smooth operation and adversely affect productivity
and production. There is a massive face conveyor
drive frame with gearbox and electric motor requiring critical space in making the roof support task
more difficult. The transfer of coal is an additional dust source and lumps of coal and rock may
cause blockages in this area. A large amount of
fine coal is recirculated which requires additional
horsepower capacity due to the friction by the fine
coal in the bottom chain race. This problem is
enhanced in the United States installations where
face and entry have the same floor elevation.[1]

A Face Conveyor Side Discharge reduces the above
stated problems. However, this concept requires
a longwall layout where the entry floor has a lower
elevation than the face floor. The deflection of
the material leads to problems when face and entry
have the same floor elevation as is the case in
practically all United States longwalls. For these

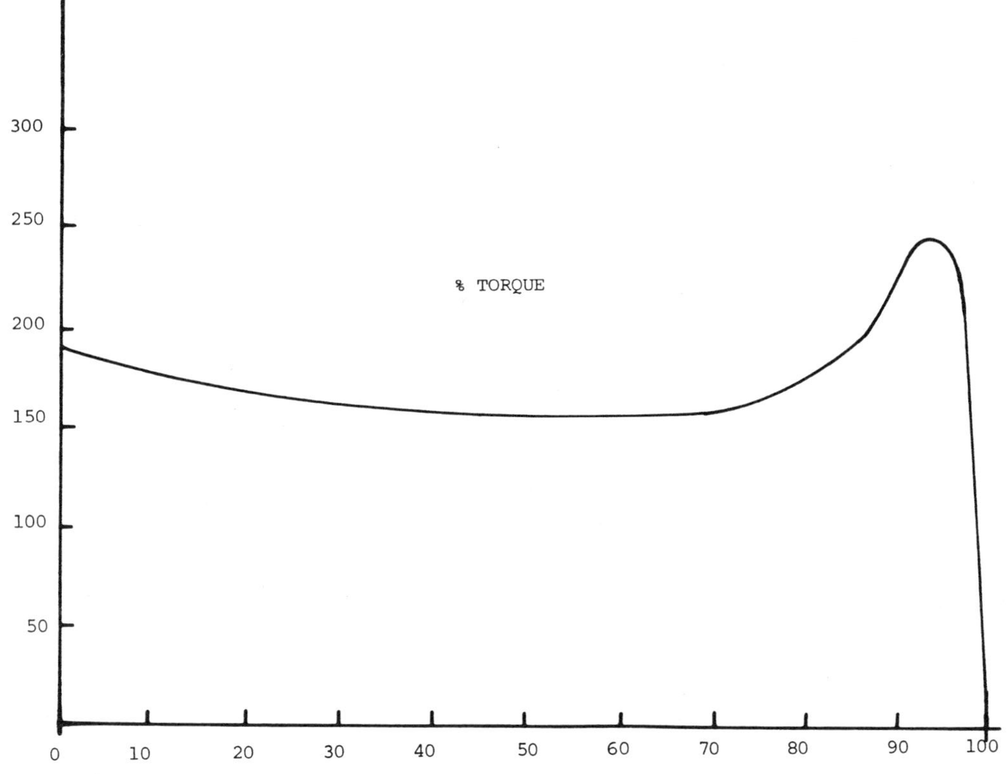

FIGURE 11   TYPICAL MOTOR TORQUE FOR LONGWALL CONVEYOR

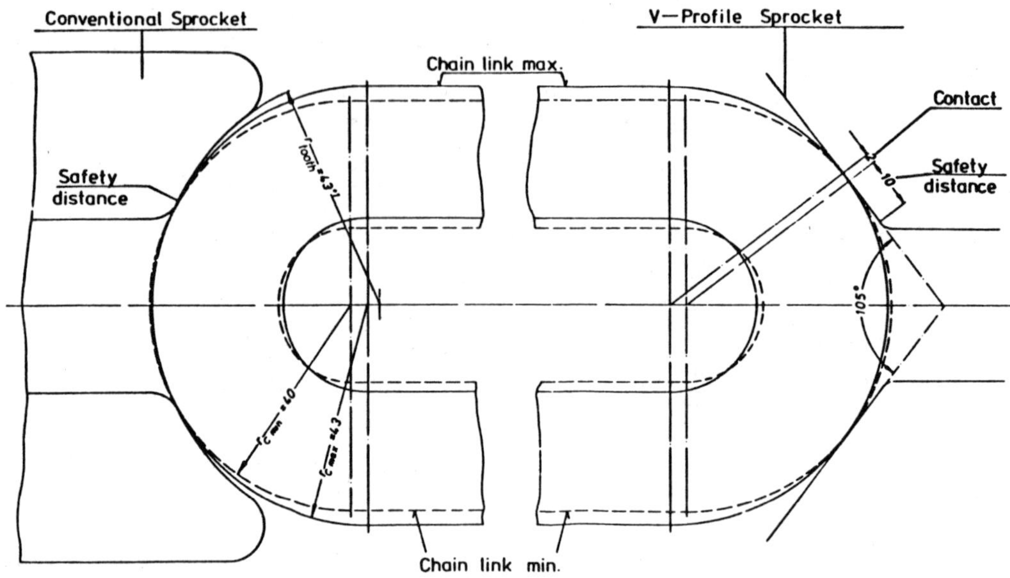

FIGURE 12   CONVENTIONAL VERSUS V-PROFILE CHAIN SPROCKET

conditions, a side discharge cross frame was developed by Kloeckner-Becorit which has been operating successfully for several years in a South African longwall installation (Figure 14). The chain strands of face conveyor and stageloader are interwoven and there is no face conveyor deck plate where the material is transferred to the stageloader. This permits a free fall of the material onto the stageloader without the danger of entrapment between deck plate and deflector plate. The cross frame arrangement permits horizontal deflections of ± 7° between face conveyor and stageloader.

The Roller Curve Conveyor introduced some years ago was a promising step in order to improve all aspects of the face/headgate junction. In this concept, the top and return chain strands are deflected around the corner by 90° by means of two counter-rotating sheaves (Figure 15). This method combines face conveyor and stageloader into one unit with the head drive being located at the discharge to the belt conveyor. Spillage, dust and recirculation of fines are eliminated. However, the approximately 5 ft diameter sheaves still require valuable space and there are some other disadvantages as well. A single chain arrangement can only be used for this roller curve limiting the drive power to approximately 500 hp for a 30 x 108 mm chain. The 90° deflection requires a 4-link flight spacing with 2.5 times the normal number of flights. Practical experience has shown that the chain/flight curve kinematic causes additional wear to the extent that the chain service life in some cases was reduced to approximately 50% of what was normally experienced.

A new roller curve concept is now being introduced by Kloeckner-Becorit. This system was extensively tested under simulated underground conditions and underground tests are now under way. The engineering approach is unique for underground chain conveyors. The chain concept consists of a balanced flight and off center twin chain assembly (Figure 16). The systems pitch is equal to the flight spacing. Hydraulic tensioning cylinders are an integral part. Flight moment differentials do not lead to differ-

ential tensions in the twin chain assembly. The 7-link chain strands are hooked to the flight bars. Conventional chain connectors are obsolete.

One end of the flight is shaped to match the rollers located on the inside of the curve and substituting for the side profile. The flights support themselves on the rollers and at any time each flight traveling around the curve is supported by at least two rollers (Figure 17). Moments and forces of the flight chain assembly traveling the multi-roller curve are closely defined. The torque transmission from the conveyor drive shaft to the chain uses a new concept as well. In conventional conveyors the torque transmission takes place between the teeth of the sprocket and the flat chain links. The new Kloeckner-Becorit concept uses a 2-tooth sprocket which transmits the torque onto the flight bars (Figure 18). The chain itself is only being used for tension forces and is kept free from outside forces resulting from the torque transmission.

Transfer to Panel Belt Conveyor. The equipment for material transfer to the panel belt conveyor should be designed to facilitate uninterrupted transfer also when face conveyor, stageloader or belt tailpiece are being advanced. In most United States longwalls, there is a telescoping arrangement between stageloader and belt tailpiece, i.e., the stageloader overrides the tailpiece by approximately 3 - 4.5 m (10 - 15 ft). The transfer arrangement between face conveyor and stageloader normally requires that face conveyor and stageloader are being advanced simultaneously. For belt alignment, it is important that the tailpiece can be steered while it is being advanced. Skid-mounted tailpieces with hydraulic advancing devices and self-propelled crawler-mounted tailpieces are used.

SUMMARY AND CONCLUSIONS

The longwall conveyor is an integrated component of longwall face equipment with the purpose to convey coal, serve as track for the coal cutter loader and as

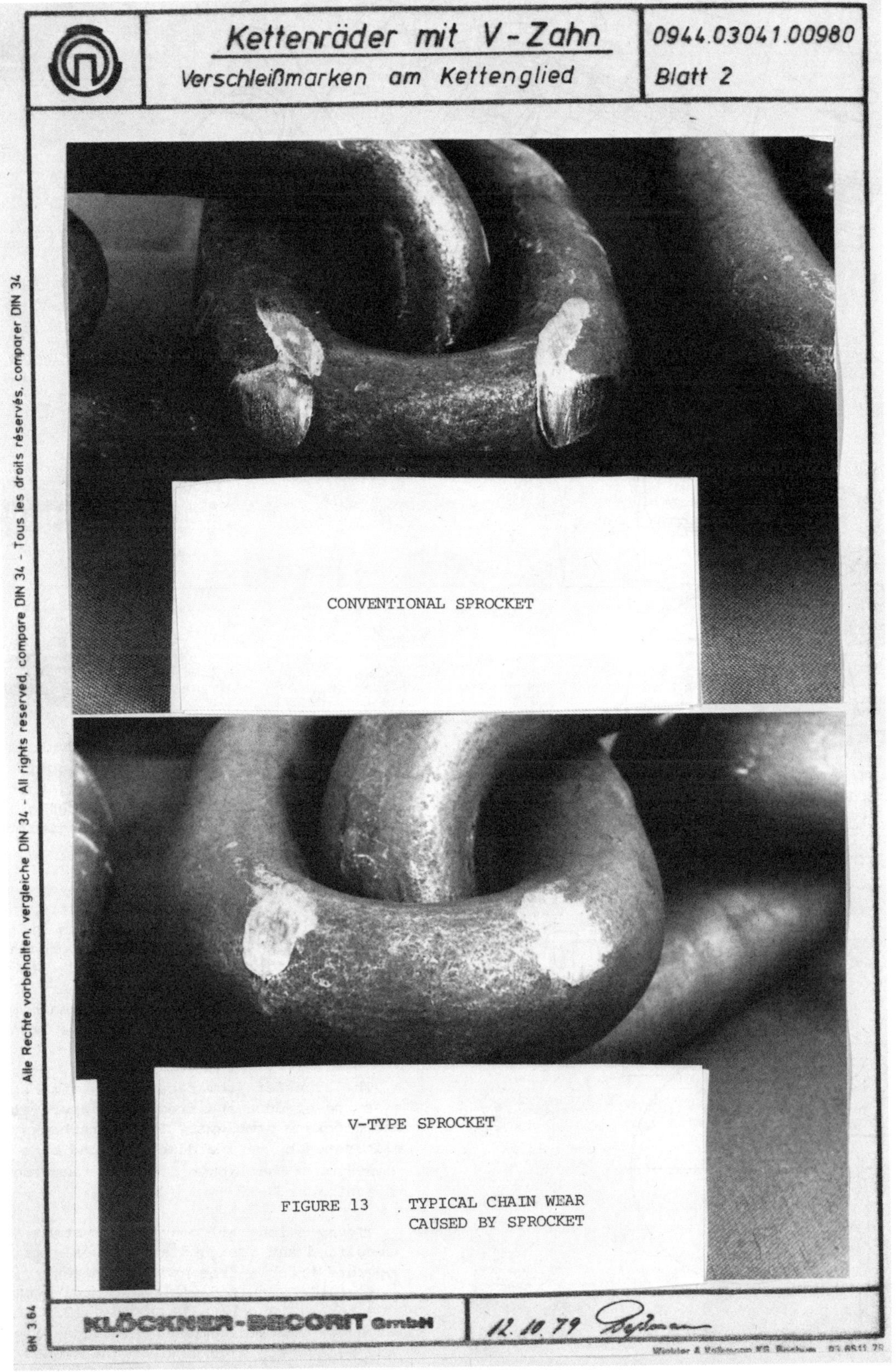

Kettenräder mit V-Zahn
Verschleißmarken am Kettenglied
0944.03041.00980
Blatt 2
CONVENTIONAL SPROCKET
V-TYPE SPROCKET
FIGURE 13   TYPICAL CHAIN WEAR
CAUSED BY SPROCKET
KLÖCKNER-BECORIT GmbH
12. 10. 79

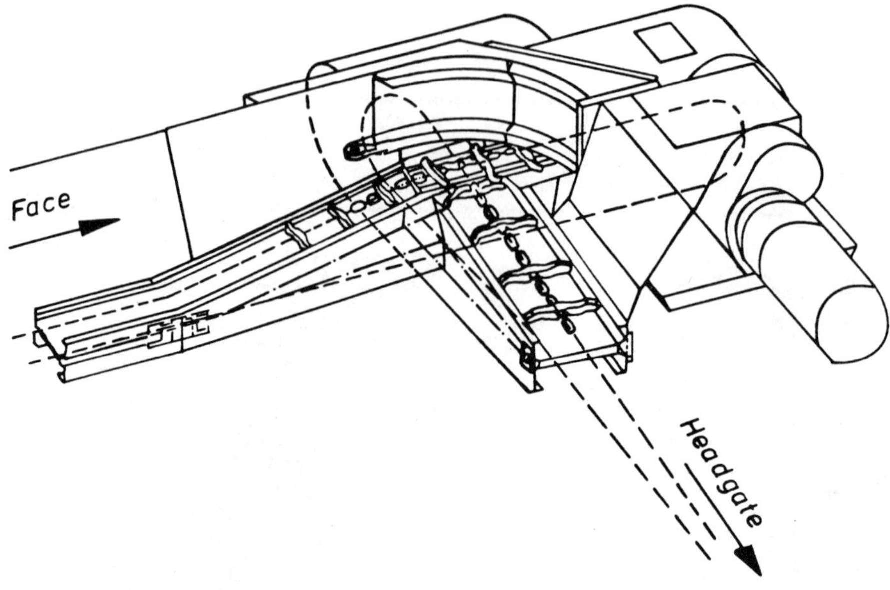

FIGURE 14   SIDE DISCHARGE
             CROSS FRAME

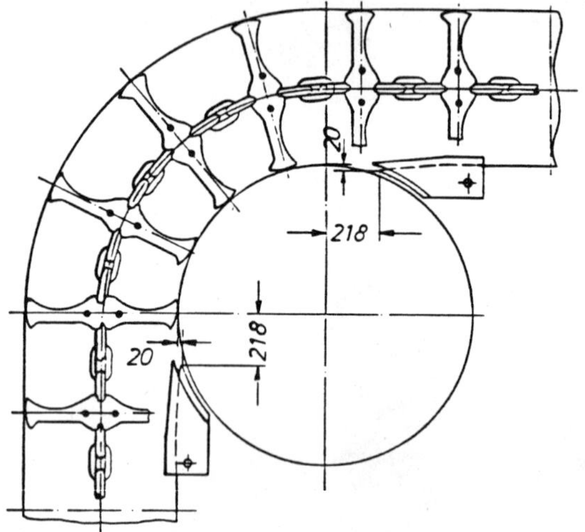

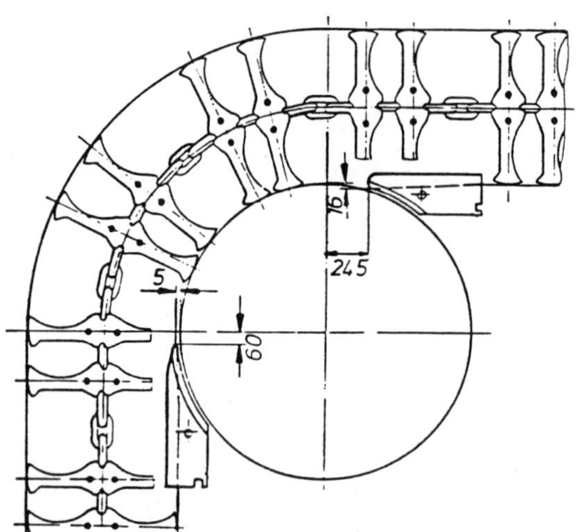

FIGURE 15   ROLLER CURVE
             (HALBACH & BRAUN)

reference rail for roof support advance. Some conveyor design criteria inhibit the optimization of
the conveying task and as a result, there are a
number of operational drawbacks such as wear, noise,
high energy consumption and susceptibility to malfunctioning.

The general development trend over the last 10 -
15 years has been to enlarge and strengthen all components in order to improve service life and operational reliability. Systematic approaches have been
made to optimize side profiles and flight contours
in order to reduce friction and wear. Further
developments were directed to drive frame and chain
sprocket design in order to improve transmission to
the conveyor chain.

The transfer from face conveyor to stageloader is
a source of dust and frequently there are carryback
and blockage problems. These drawbacks can be
eliminated by a side discharge and by a roller curve
conveyor. The latter combines face conveyor and
stageloader.

Today's longwall conveying systems, adequately
specified and designed and properly maintained,
operate trouble-free and have a conveying capacity
compatible with today's coal loading machines of
700 tons (770 st) to 1200 tons (1320 st) per hour.
However, limited component service life due to wear
and relatively high power requirements and costs
are remaining drawbacks and require further
development.

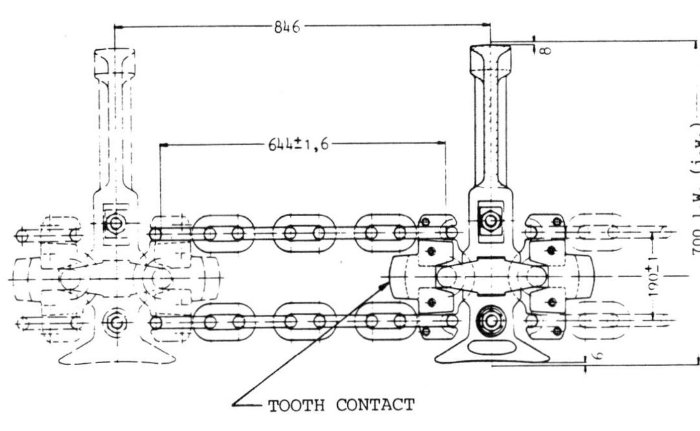

FIGURE 16   CHAIN/FLIGHT ASSEMBLY FOR
ROLLER CURVE CONVEYOR
(KLOECKNER-BECORIT)

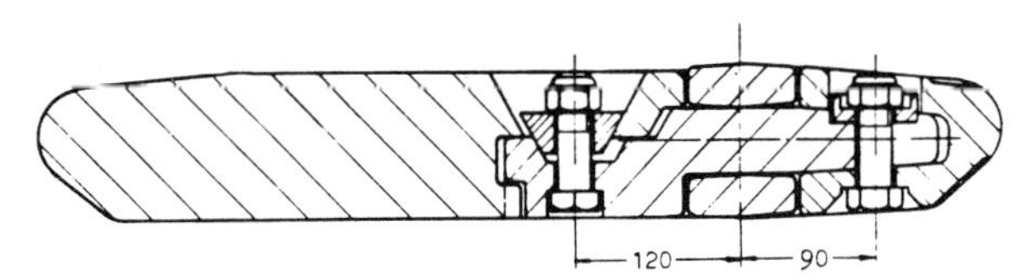

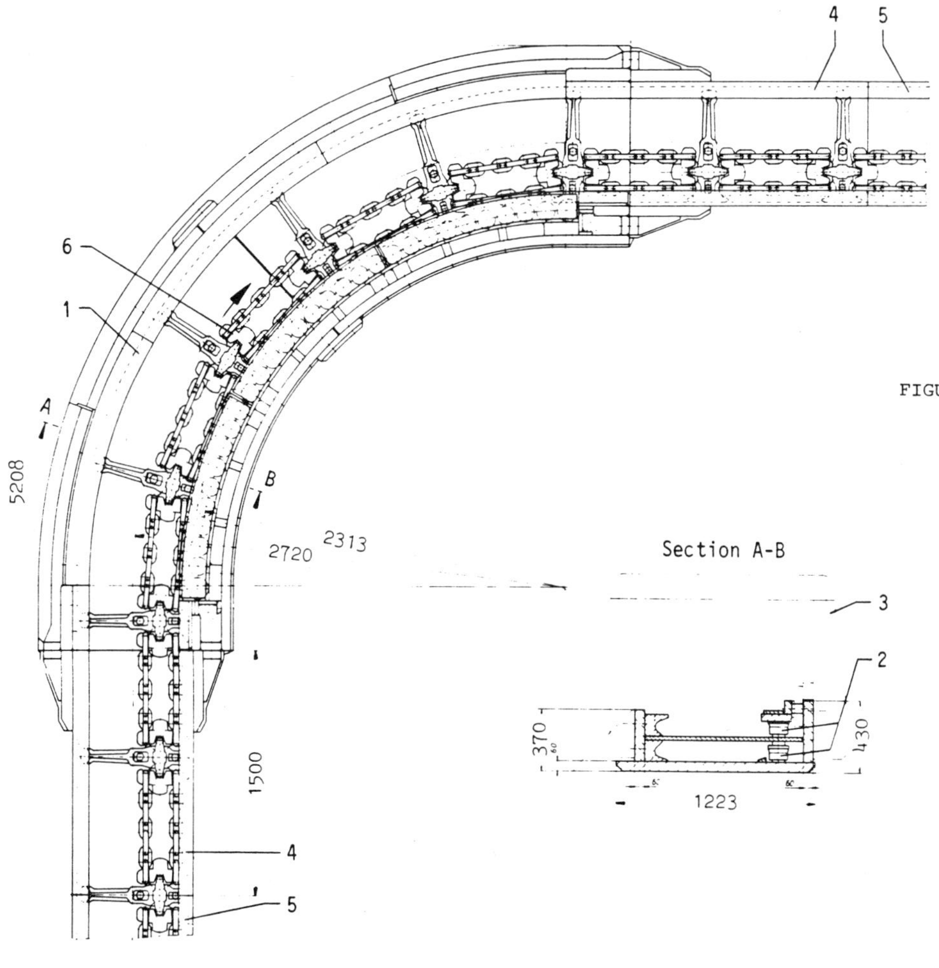

FIGURE 17   ROLLER CURVE CONVEYOR
(KLOECKNER-BECORIT)

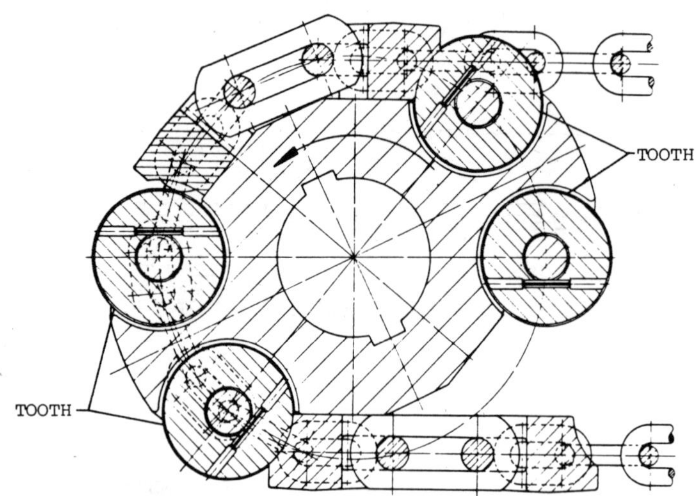

FIGURE 18   2-TOOTH SPROCKET FOR
              ROLLER CURVE CHAIN

REFERENCES

1) Kuti, J., 1979, "Longwall Mining in America",
    Mining Engineering, Nov., pp. 1593-1602.

2) Anon., 1980, "Report on Noise Level Measure-
    ments of Conveyor UFv 34" (transl.), May,
    Bergbau AG Westfalen

3) Rawlinson, R., 1980, "Changes in Coal Getting
    Equipment", Colliery Guardian Annual Review,
    August, pp. 339-342.

4) Goette, E., Dr., 1979, "New Sprocket for Chain
    Conveyor" (transl.), Glückauf, pp. 27-29.

5) Anon., 1980, "Einfluß der Zahnflankenform von
    Kettenrädern auf die Bean-spruchung der
    Kettenglieder", Aug., Bergbau-Forschung.

6) Anon., "Betriebsempfehlungen Nr. 15: Vermeidung
    von Kettenrissen an Strebförderern durch
    maschinentechnische Maßnahmen", Verlag
    Glückauf.

Chapter 4

ROOF SUPPORT DEVELOPMENTS IN LONGWALL

Joseph Kuti

Mining Progress, Inc.
605 Boulevard Tower
Charleston, W. Va. 25301

ROOF SUPPORT DEVELOPMENTS IN LONGWALL

(1) European longwall mining technology developed towards its modern form in relatively deep mines. With increasing depth the stability of openings driven for ventilation, travel or transport becomes paramount in mine design considerations. As a result the number of roadways and entries, as well as their width must be a minimum. (1)

Another important aspect of European mining is the high density of mineable seams, as opposed to the wide lateral extension of a single seam in the U. S. A. Multi-seam mining requires strict planning and near 100% extraction, since with increasing depth, the effect of remnant pillars can be devastating.

As a result the face layout of a modern European longwall panel is based on single panel entries of moderate width and great height. The mode of extraction is generally a mixture of the advance and retreat system which is called a Z-system.

It is obvious, therefore, that the supports developed for the entries and for the intersection areas are more complex than those in the U. S. A.

Mine design layout and face-entry intersections affect the development of face conveyor drives, transfer points, freecut at the panel corners and the design of face-end supports. The main factors affecting the actual face supports such as required support resistance and functional parameters (operating range, travelway, etc.) should, however, be similar in Europe and in the U. S. A. Any apparent differences are rooted in the accidents and vagaries of "technology transfer". (2) (3)

The initial European development of self-advancing supports took place in mines of medium depth in seams overlain by bedded and friable strata and in the absence of massive sandstone. Initial experience indicated that in order to control the prefractured immediate beds, a support density in excess of 3 tons/sq. ft. could, in fact, be harmful. It was - and in these conditions still is - considered more critical to operate with a minimum canopy tip to face distance and to advance the supports as soon as possible in order to control the immediate weak shale or drawslate beds.

The value-engineering approach at that time suggested, therefore, the use of small bore props safeguarded against an excessive rise in pressure by yield capsules set at values between 500 and 550 bars. At the same time, however, the pump stations were operated at pressures between 60 and 170 bars, so that the freshly-set supports attained an extremely low setting load in comparison with the ultimate yield load. In this way, for instance, a chock rated at 360 tons of resistance was only offering a resistance between 40 and 120 tons to initial bed separation.

It was noted, of course, that in certain instances, such as close to the outcrop and under massive beds in the Northeast of England and in multi-seam mines in Germany where the face operated under worked-out areas of overlying seams, the self-advancing supports proved to be totally inadequate. In both instances cracks parallel to the face appeared in the roof and the immediate beds slipped downwards between these cracks, leading to a characteristic step formation.

It is useful to note that in both cases we are dealing with a condition of relatively low roof pressure, yet in both cases the self-advancing supports failed because they did not develop an adequate resistance.

Even though experience related to these cases existed at the time of introducing self-advancing supports to the United States, such experience appeared to be the exception rather than the rule. (4)

(2) <u>Review of longwall support development prior to the "shield era".</u>

The development of two-leg shields

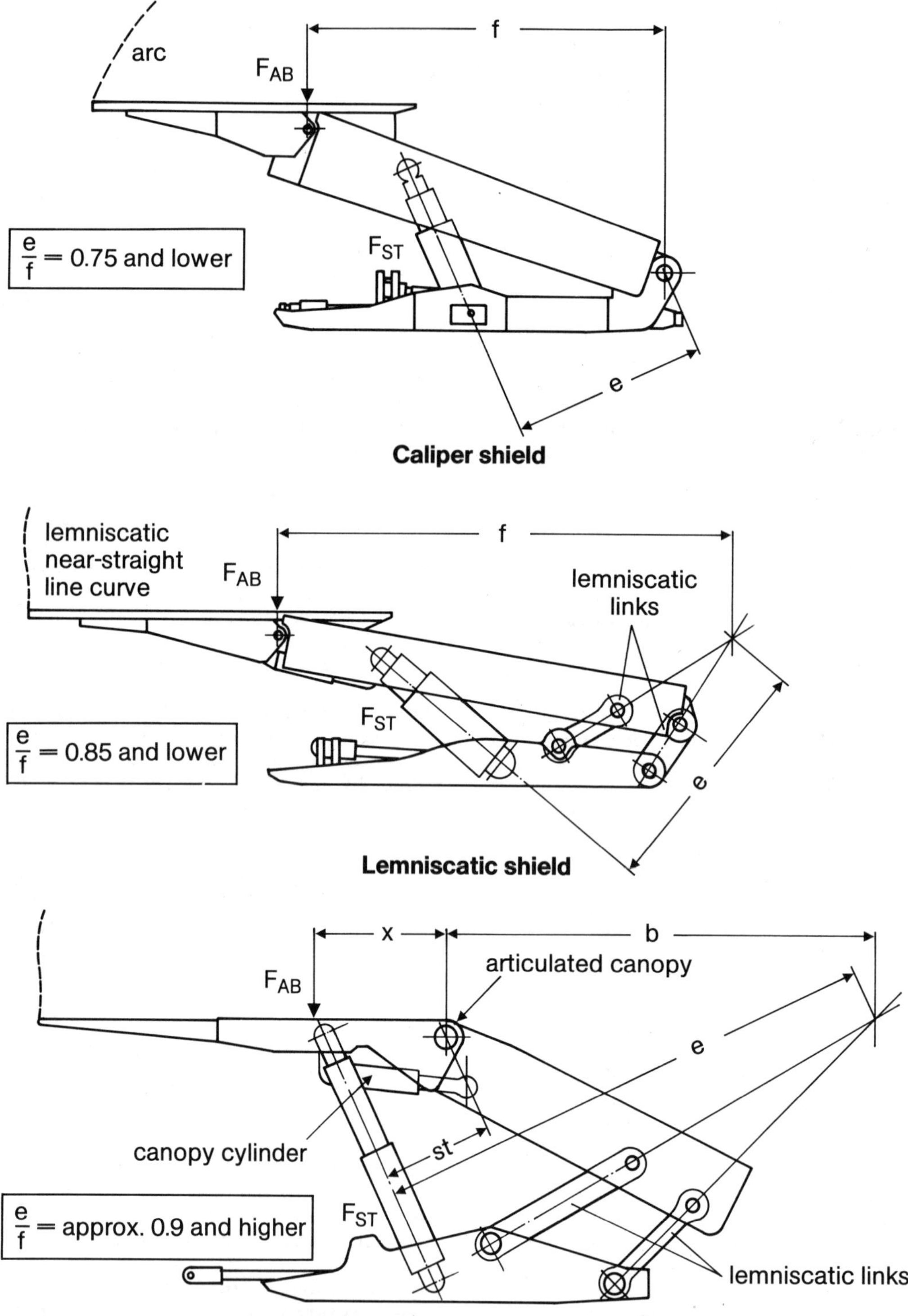

Fig. 1. The development of two-legged shields.

As a result, the early installations of self-advancing supports were either glowing success stories or utter failures at this first stage of technology transfer from Europe. It requires little analysis to sort out the wheat from the chaff: the successful installations were generally at depths of 300 ft. and more and the main roof beds consisted of friable shales. The failures were in seams close to the outcrop and under massive sandstone beds. Often the problem was aggravated by the fact that the immediate beds between sandstone and coal seam were brittle and fractured mudstones, causing stability problems with the light frame - or chock-supports.

At the same time, experience has shown that where the thickness of the friable beds exceeded 5 to 7 times the mining height, reasonable roof control could be maintained even in the presence of a massive sandstone bed. This experience appears to be validated by the findings of Salamon et al. (5) in the South African Coal Mines.

Two schools of thought developed as a result of this early mix of success and failure. According to one, longwalling had to be restricted to seams with overlaying friable, shale beds. Such a restriction was, however, not accepted by others and their goal was the development of more stable supports of high density. The first heavy-duty supports with a support resistance of 560 to 700 tons were introduced in 1966 and were crowned with immediate and continued success. Many of these early installations continue to function today.

Mining Progress, Inc. was the innovator with heavy-duty Westfalia frame-supports and chock supports, followed by Dowty introducing their first heavy-duty thin-seam chock supports in the U. S. A. Other well-known manufacturers of supports were either adhering to the concept of six-leg chocks, as was Gullick or were still strangers to the U. S. coal industry as Hemscheidt at that time.

The application of heavy-duty frame and chock supports cover a period of ten (10) years between 1966 and 1976 and was directly responsible for the first significant expansion in the longwall mining system in the U. S. A. (6) Yet, in Europe even the mention of heavy-duty supports was received with skepticism, if not with cynicism, at that time. The motto of "everything has to be bigger in America" was voiced by many otherwise capable mining engineers.

Yet the stage was being set in Europe for a more far reaching and more radical change in support design, than our modest departure from standard frames and chocks. Kloeckner-Ferromatic has been supplying the Hungarian mines for several years with hydraulic props and valves used in a Hungarian-version of the Russian shield supports. These supports were operating successfully in mining conditions where the roof and floor were weaker than the coal being mined.

The keynotes of these shields were the very narrow supported roof span and the large area base skids extending all the way under the conveyor. The support density was not much greater than that of contemporary frame or chock supports but the stability of the shields, particularly against lateral thrusts was superior.

These Hungarian shields, known as caliper shields were introduced into the captive German mines of Kloeckner-Ferromatic in the early 1970's. (7) The first installation met with success in mining conditions which were formally regarded as unsuitable for self-advancing supports. As a result, the number of shield installations in Germany rose to 100 in six (6) years between 1970 and 1976.

(3) Shield types in current use and in the development stage.

On April 14, 1975 Consolidation Coal Co. began to operate a longwall face in their Shoemaker mine using caliper shields manufactured by Kloeckner-Ferromatic (Fig. 1). Shortly thereafter, Kaiser Steel Corporation, in cooperation with the Bureau of Mines, started up a caliper shield face supplied by Hemscheidt Corp. at the York Canyon Mine. (8) & (9)

Early experience with these and other shield faces was positive. As a result, the interest of operators switched from frame and chock supports to shields almost overnight. This was quite remarkable since adequate roof control has been attained in many cases with the U.S.-style chocks and frame supports. Even recently, during the month of June 1980, a face supported by Westfalia chocks installed nearly ten (10) years ago at the Federal No. 2 Mine of Eastern Associated Coal Corporation, several shifts produced in excess of 4,000 tons of material and the average production was 1,934 tons of clean coal per shift for the month of June. (10)

It is important, therefore, to list the advantages of the early caliper shield supports in comparison with the frame and chock supports to understand the wholesale shift in interest. These were:

The development of four-leg shields

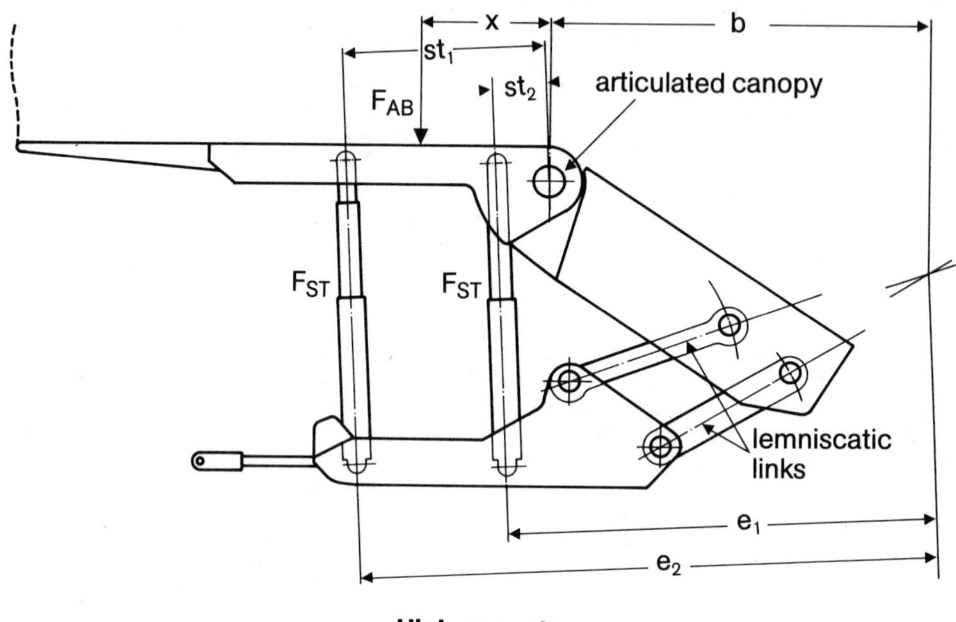

Fig. 2. The development of four-legged shields.

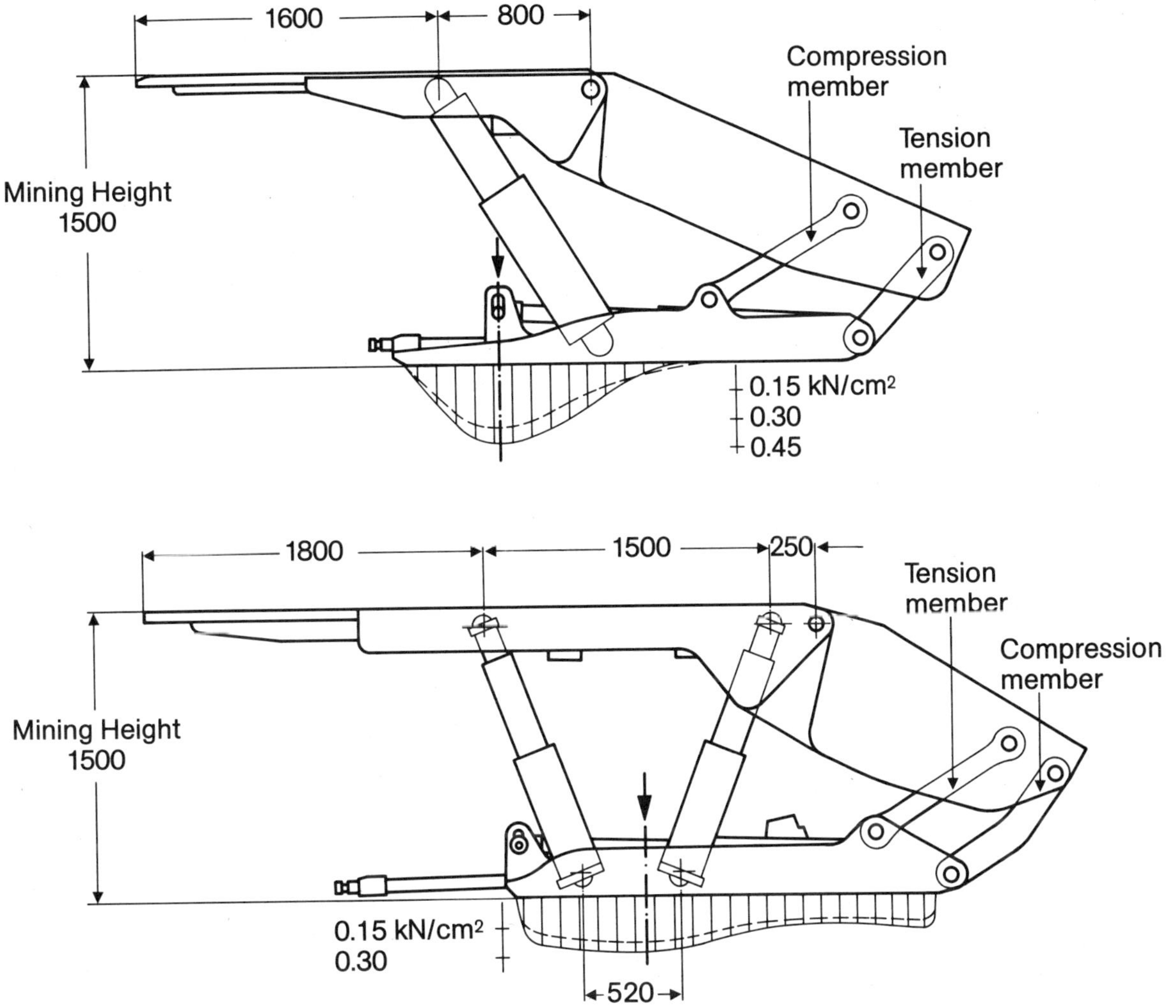

Fig. 3. Floor pressure diagrams.

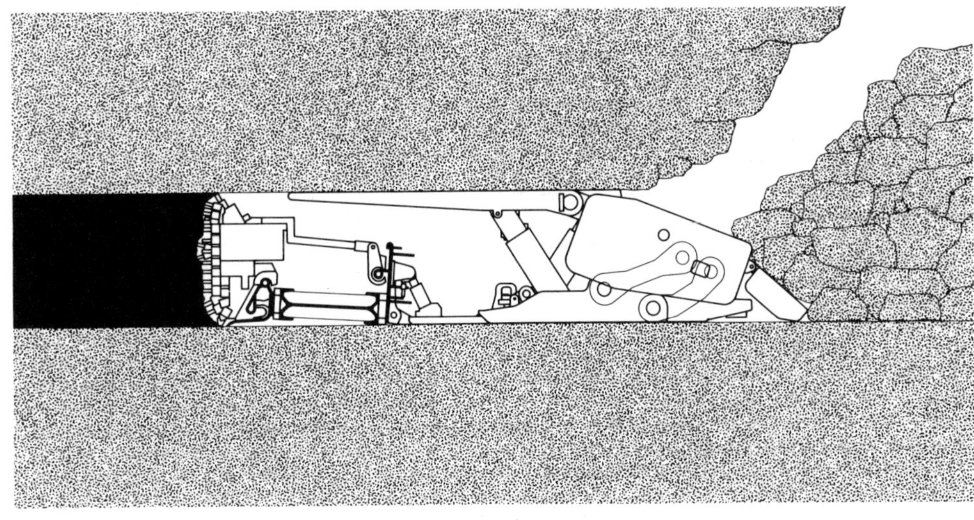

Fig. 4. X-type shield.

(1)   Large operating range of the shield coupled with a low transport height.

(2)   Complete protection from the caved debris.

(3)   Great stability and resistance against lateral loads.

(4)   Reduction in damage to cables, hoses and in minor cuts/bruises to jacksetters.

(5)   The capability to mine under an incompetent roof without the need to leave roof coal.

(6)   A reduction in the supported span with subsequent improvement in drawslate control.

Nevertheless, several basic weaknesses of the original single pivot or caliper shields detracted from their overall success during the rapid expansion in their use.

These were:

(a)   The low support density due to the fact that only a fraction (ratio: e/l on Fig. 1) of the prop force is support-effective.

(b)   The angle of the caving shields – particularly in low seams - was inferior to the angle of repose so that a high column of debris rested permanently on the caving shield, further reducing the effective support density.

(c)   Due to the arc described by the tip of the canopy, as the shield was being raised, the tip to face distance increased.

(d)   Rock could lodge between the rear portion of the canopy and the caving shield, limiting the articulation or causing a permanent "droop" on the canopy.

(e)   All of these weaknesses were either eliminated or their impact reduced in the successive stages of development. (1.2) on Fig. (1) show the lemniscatic shield which offered an improved utilization of the prop forces and a near straight line canopy tip movement in that the single pivot between caving shield and these skids was replaced by two pairs of lemniscatic links.

(1.3) on Fig. (1) represents the modern form of the two-leg shield where the props are inserted into the roof canopy instead of the caving shield for optimum utilization of the prop forces. The roof canopy and the caving shield are joined by an articulation at a distance "X" from the resultant F_A. At one stroke the angle of the caving shield becomes steeper and the rock is excluded from the area of the articulation.

With this type of shield it is necessary, however, to introduce one or two hydraulic cylinders between the canopy and caving shield to control the attitude of the canopy in case a cavity develops above the articulation. These cylinders must be protected against a possible over-extension as well as against momentary pressure peaking.

The base skids are equipped with extended tips compared with earlier types to minimize the effect of the peak floor loading. (Fig. 3)

Overshadowed in the great revolution caused by the two-leg shields, four-leg shields were first introduced in the West European coalfields by the innovators - Westfalia Luenen - in 1974. They already exhibited the desirable features of lemniscatic links and articulated canopy. (Fig. 2)

The only disadvantage experienced with these shields was the reduced range between closed and extended height, a ratio of 1 to 2.2 compared with 1 to 3 with two-leg shields, In order to improve the available range, particularly in low seams, the props were later inserted in a V-form (2.2 on Fig. 2) between base skids and canopy. It is clear, of course, that this improvement in the working range automatically increased the width of the supported span, and this is a serious disadvantage if the immediate roof is incompetent.

A further variant of the V-type leg arrangement overcomes this problem by an overlap of the legs in X-form at the base. This ("X"-shield, Fig. 4) is particularly suitable for seams of low or medium height where the main roof is massive.

If the required prop size exceeds 100 Mp, the overall width of an X-shield would require shield spacing at centers greater than 1500 mm. In those cases, therefore, it is possible to seek yet another alternative of the V-shield. In this one the rear props are inserted into the caving shield similar to the old style caliper shields and the front legs are attached to the roof canopy. The net result is a decrease in the supported span compared with the conventional V-shields. The innovators for this shield type were Dowty.

It is not possible to list every single variant of the four-leg or even the V-type shields, since every new project introduces some new aspect to design detail. There is one variant, however, which is gaining favor, partially to take issue with the champions of two-leg shields. In this variant, the front props are of higher resistance than the rear ones. This variant with its high tip load-

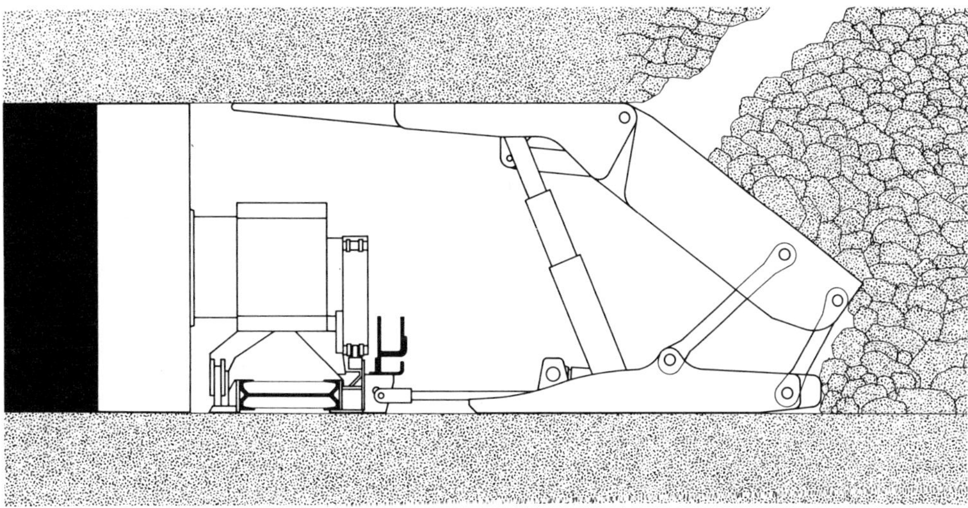

Fig. 5. Typical two-legged shield with articulated canopy in the Pittsburgh No. 8 seam.

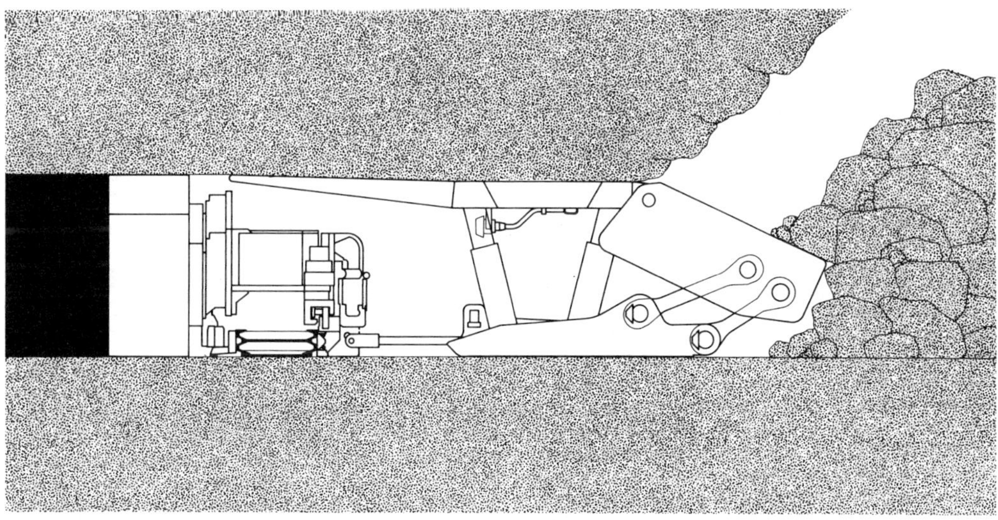

Fig. 6. Typical four-legged shield at medium seam height.

ing characteristic is of interest in thick seams in which some of the top coal may roll out of the face prior to cutting, thus exposing additional roof.

(4) Review of Support Parameters.

With the rich harvest of shield designs it is not surprising that the approach to selection is often associated with some confusion and uncertainty about the essential features of a shield for a particular application.

The selection process is also bedevilled by an often rigid polarization of concept between proponents of two-leg and four-leg shields.

In general, there are very few instances where all the requirements laid down by a sober and experienced operator may not be met with either type of shield. The choice will, then, generally lie within the economic or ecological advantages of one design over the other.

It is wrong to ask " do I need more than two props", it is much more relevant to attempt an ideal solution to the problems of roof control, travel, ventilation, safety and functional suitability with at least two alternate types of shields and, then, select the one which meets all the requirements in the most cost-effective manner.

The listing and analysis of the desirable shield parameters for the various mining

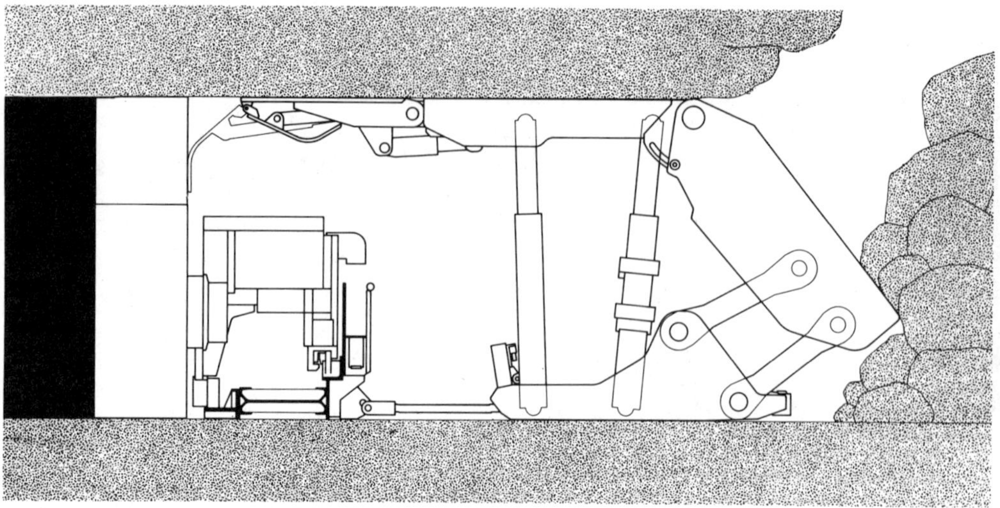

Fig. 7. Typical four-legged shield in a thick seam.

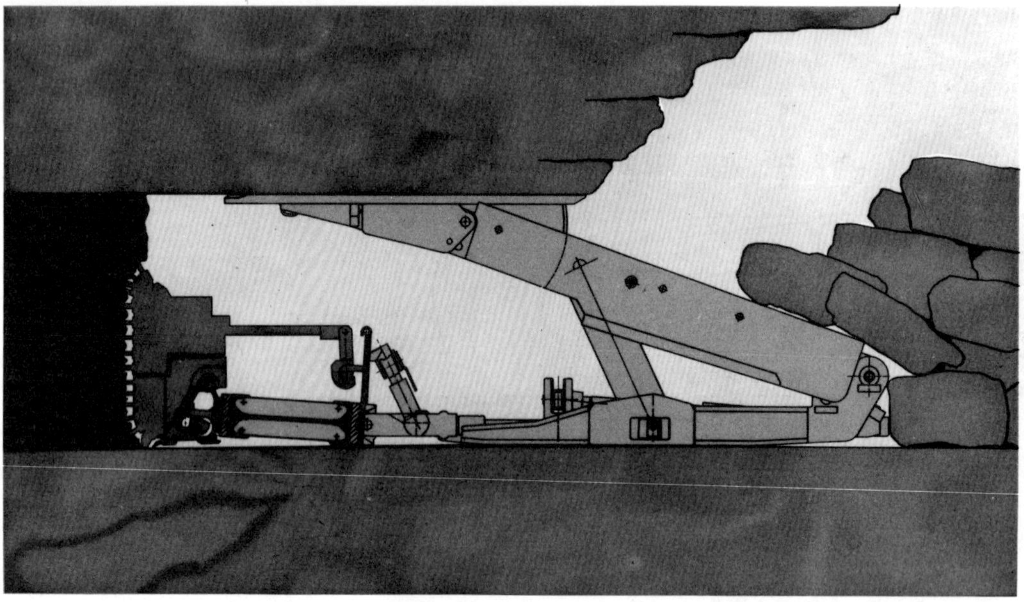

Fig. 8. Fully integrated mining and support system. Gleithobel, outrigger steering, caliper shield.

applications would require more space and time than is available at the present time. A base outline of the required steps may only be presented:

(1) The first step in all cases is a <u>thorough observation of the underground mining conditions,</u> associated with simple but suitable measurements.

(2) The second step is to gather up and <u>evaluate all geological data</u> such as borehole logs, information on anomalies, relevance of surface geography, particularly related to rivers, creeks, lakes, etc.

(3) The third step is to <u>determine the desirable mine layout plan</u> for a longwall system and with that to lay down the sizes of the panel, of the panel entries and of the chain pillars.

(4) Next comes the <u>selection of the mining-conveying system</u> for the longwall extraction. In very broad terms, the dividing line between plows and shearers is height-dependent. In seams less than 1.6 M. in height, plow systems may be preferred for higher production, easier compliance with respirable dust and independence from operator fatigue. In seams thicker than 1.6 M. the problems of vertical clear-

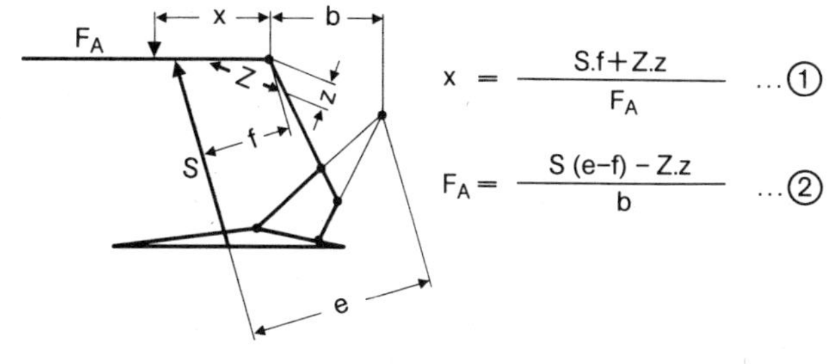

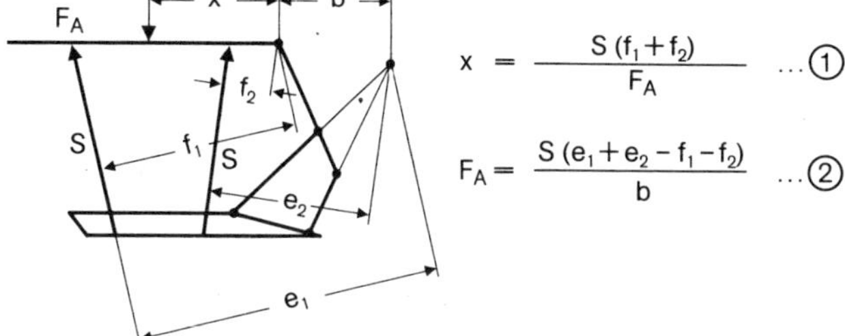

Fig. 9. Calculation of support resistance comparison of various types of shields.

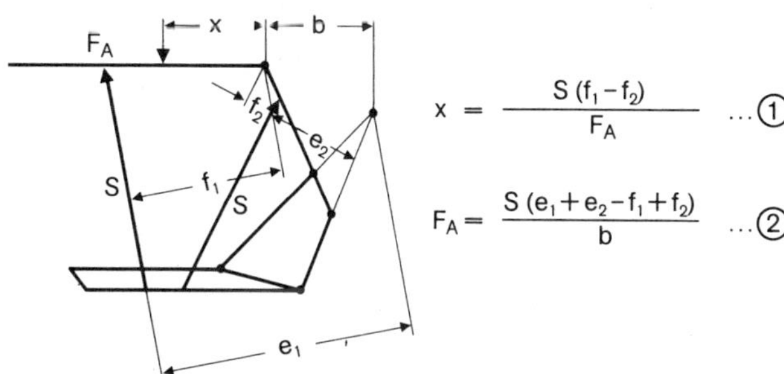

ances and travel facilities diminish to the point that the organizational and functional advantages of a shearer system are generally preferred unless specified problems, such as a high rate of methane emission and an excessive production of fines force the operator's attention to a plow system.

The proper analysis of the mining system is crucial for the long-term success of the longwall system, but it is not always possible to foresee all the factors at the time of selection. In marginal cases, therefore, the support system should be so specified that it may be used with a plow or a shearer-system with a minimum of modification.

(5) All the above steps may have been taken jointly with one or more manufacturers of shield supports. In some cases the same manufacturers are, then, required to project a shield support system suited for the conditions.

In other instances, adequate know-how exists to evaluate and interpret the results of the first four steps to write down a list of specifications, indicating the priorities to be observed. It is essential in such cases to remain open to suggestions and argument. An operator's priority may be an excessively low transport height, for instance, which may conflict with the optimum design of the shield for the mining height. Travelway requirements may prejudice essential support density in another case.

A third alternative is, therefore, for the operator to gather up and evaluate the relevant information and, then, submit it to the shield manufacturer in some form similar to the attached "Pre-specification" list.

Based on this list, and with proper computer-programs established, it takes a relatively short time to produce alternate shield designs with all the relevant technical data in print-out form for detailed consideration.

In spite of the access to computer programs, there is still a general interest in comparing the available, effective support resistance of the various types of shields. The attached table is an attempt to provide the relevant information.

(6) <u>Hydraulic controls and the road towards automation.</u>

The existing hydraulic control systems of shield supports may be summarized as follows:

(a)   With cam-operated poppet-valves

(b)   With single lever rotary valves

(c)   With multi-lever valve banks

6.2.   <u>Pilot-operated controls</u> (direct or adjacent) using coaxial, multi-duct hoses

     (a)   With single lever rotary selectors

     (b)   With piano-key or push-button selectors.

The "adjacent control" mode may be arranged to be unidirectional or for a limited number of functions bi-directional. (1)

6.3.   <u>Batch control or group automation</u>

     (a)   A hydraulic impulse triggering the sequential ("auto"-) operation of a single shield or a group of shields.

     (b)   An electro-hydraulic system of sequential operation, where the hydraulic impulse is replaced by an electrical impulse but the pressure-dependent sequencing of the individual functions is retained.

     (c)   An electro-hydraulic system with electrical impulsing and sequencing, using micro-processors. These control systems are generally designed to be triggered and observed by a shield-operator, who is positioned on the fresh-air side of the mining machine but has visual control over the support advancement. (12)

6.4.   <u>Remote control and machine dependent automatic sequencing</u> of full-face or part-face support advancement.

These systems are the latter-day variants of the early "Rolf" and "Rebecca" concepts and are still in the conceptual-development stage provoking thought and discussion.

On modern high production shearer faces, particularly in seams of great height, a so-called modified half-face method is being practiced. With this method the shields are advanced and conveyor is rammed over during the rapid clean-up pass of the shearer, moving from the head to the tail. Accoringly, an unusually high number of users are being opened to the pumps within a very short period of time. The results are a collapse of the hydraulic system, at worst, slow operations near the tail and very low setting pressure , at best.

An attempt to overcome this problem by adding a third pump and still chanelling the increased delivery through the existing valve system of the shields is counter-productive and may lead to erosion and back-pressure problems in the system. One simple and effective solution out of this dilemma is the separation of the conveyor advancement, after introducing a third pump. The delivery of this pump is piped along the face with a T-connecting branch supplying the annulus of each shield ram. A single valve for a group of ten shields would, then, fill the annulus of the rams, causing a "bank" of ten pans to advance smoothly and without any adverse effect on the shield operations.

Once introduced, this "bankpush" system becomes very popular with the operators, since it is simple and virtually maintenance free.

It is important to restate that modern hydraulic valves and hydraulic supply-control systems should be designed to pass large volumes at relatively high line pressure. Pump pressures - and, therefore, setting pressure - of 300-350 bar are common today.

The setting pressure and the ultimate yield pressure should be closely linked. The props of the shield should be selected in such a way that the rated or yield load of the props does not exceed the setting load by a margin greater than 35%.

Auto-set valves form part of batch control and other automation systems. These valves are designed to keep the piston chamber of the props connected to the pump until the predetermined setting pressure has been reached. At first sight, the attainment of uniform and high setting loads appears to be a worthwhile objective; nevertheless, a wholesale application of auto-valves cannot be recommended.

The provision of clean pressure fluid at near-constant operating pressures and in large volumes is best attained by setting up a central pump station outby in a well-lit, well-ventilated pump chamber, catering for run-off, emulsion-mixing and housing standby components. The pressure fluid is supplied from such pump stations by means of high pressure steel pipes to the successive longwall panels.

(7) Recent Shield Developments

The development of standard and special purpose shield-types continues at an amazing speed. Most of the development occurs in Europe but the innovative techniques will eventually affect us all, as the same mining methods or similar geological conditions must be applied and faced.

A few of these developments are listed as examples, without any claim for completeness.

7.1. Shield centers in pitching and in thick seams

Practically all shields are installed presently at 1.5 meter centers so that the thrust rods of the shields are attached to the center of the conveyor pans which are also 1.5 meters long. On a pitching seam floor, particularly at greater seam height, the stability of a shield may be enhanced by designing it for wider centers. After all, conveyor pans and their furnishings may be made in units of 1.75 meters or 2.0 meter lengths, rather than the customary 1.5 meters.

Several faces are being operated successfully with shields at 1.75 meter centers and with reverse-mounted rams arranged externally to the shield bases. These faces are being operated in seams of 0.9 - 1.1 meter thickness, with Westfalia Gleithobel systems. The external attachment of the rams insures a great flexibility in the steering of the plow and the open tunnel between the shield bases provides the required clearance for spillage and fines.

7.2. Heat-insulated shields

The mean depth of coal mines in the Ruhr District of West Germany reached 860 meters in 1978, and the average descent in depth is about 12 meters per year since 1969, resulting in an increase in the mean rock temperature of 0.4 K. In 1978 already 54.5% of all the faces were regarded as "hot" with dry bulb temperatures exceeding 28 degrees C (82 degrees F).

It is not surprising, therefore, that the first steps towards climate control were undertaken some time ago, in particular, by the provision of piped cooling water at the head and along the longwall face.

Since shield supports are provided with wide, undivided canopies and caving shields, it is logical to consider the utilization of these structures for the installation of heat-insulating materials. At the same time, provisions are made for the attachment of air-conditioning units to the individual shields.

The experiments with such shields at a West German Mine (13) indicate the great effectiveness of insulation.

7.3. The Hemscheidt Troika-Shield

There are a number of viable shield designs for pitching seams. An original concept is undoubtedly the one developed by the Hermann Hemscheidt Company. (14)

With this concept, the face need not be equipped with a conveyor, as is customary with ram-plow systems at gradients in excess of 45 degrees, for instance. These two-leg shields form a group, the base skids of which are anchored within a beam-structure. This structure doubles as a guard against falling coal from the face and may be equipped with steps to form a safe walkway.

One face in the U. S. A. will be equipped with such Troika-shields in the near future.

7.4. Special shields for sublevel caving, multilift extraction and pneumatic stowing

All of these shields have a common feature in that the caving shield is replaced by an open structure but without sacrificing the lemniscatic principle of interconnecting canopy and base.

In sublevel caving a special, hydraulically retractable shield guards a gobside armored chain conveyor (13) and (14) when the roof beds, consisting of coal, begin to cave, the shield is retracted, exposing the conveyor for transporting away the caved roof coal. The shield is run down to curtain the conveyor once the coal beds have fallen and the overlying rock beds begin to cave.

A pipeline, carrying the crushed rock must be suspended under the gobside projection of the shield canopy on faces where pneumatic stowing is being practiced.

(8) <u>Conclusion</u>

It is clear that a paper of this type cannot do justice to the wide subject of roof support developments. Questions such as entry supports, supports combined with a full-face mining system, such as the Russian AK 3 system, hydraulic controls developed for mountain-bump conditions, various types of forepole extensions, the advantages and disadvantages of close-up and I.F.S. systems, etc., certainly deserve full but separate treatment.

The task to define and describe support developments, as they affect us in the U.S.A. was relatively simple because it could be confined to the introduction of shield supports.

The task to detail the operational implications of support systems and to enter the realm of forecasts of future techniques can now be assigned to another author at an appropriate meeting in the future.

"Pre-Specifications" List

(Re: Section 5)

(1) Thickness of overburden.

(2) Composition of the first 100' of roof strata, or to the next seam above.

(3) Projected major bed separation planes above the seam.

    No. 1   ft. above seam.
    No. 2   ft. above seam.
    No. 3   ft. above seam.

(4) Assumed mode of roof failure behind shields.

    4.1. Overhang   ft.

    4.2. Close fracture.

    4.3. Swell factor.

(5) Composition of the first 10' of floor strata.

(6) Experience with roof and floor.

    6.1. In entry drivages.

    6.2. In pillaring.

(7) Seam height and rock inclusions.

    7.1. Minimum height: typical section: frequency.

    7.2. Maximum height: typical section: frequency.

(8) Separation between roof and seam and between floor and seam.

(9) Grindability (Hardgrove Index) volatile matter, cleat system and other factors affecting the mineability of the seam.

(10) Calculations for height of caving, and required support resistance (static and dynamic approaches) for stable roof conditions.

(11) Determination of max. volume and velocity of ventilating air current to calculate minimum sectional area required to pass the ventilation through the shield-supported face.

(12) Survey of transport route and inspection of height restrictions (belt conveyors, trolley wire, overcasts, etc.).

(13) Pitch, rolls, faulting and other geological factors.

(14) Presence and significance of ground or roof water.

(15) Temperature, humidity, corrosive or abrasive elements.

(16) Proposed mine design.

REFERENCE

(1) Irresberger, H.
Gebirgsbeherrschung in Streb und Strecke bei grosser Teufe
Glueckauf 116 (1980), Nr. 5 Pages 195-200

(2) Kuti, J.
Longwall versus Shortwall Systems
Mining Congress Journal, August 1975

(3) Beckmann, K.
Ausbau in Streb und Strecke
Bergbau 1/19/79, Pages 13-18

(4) Kuti, J.
Longwall Mining in America
Mining Engineering, 11/79, Pages 1593-1602

(5) Salamon, M.D.G., Oravecz, K. I. and Hardman, D. R.
Rock Mechanics Problems associated with Longwall Trials in South Africa
Fifth International Strata Conference 1972

(6) Kuti, J.
     The Status of the Longwall Mining
     System in the United States
     Coal Age, August 1972

(7) Kohlgruber, W.
     Shield Supports in Longwalls of West
     German Coal Mines
     American Mining Congress, Pittsburgh,
     May 1974

(8) Elkins, H. and Sikes, W.
     Longwall Mining at York Canyon Mine
     American Mining Congress, Detroit
     May 1976

(9) Von der Linden, W.
     Summary of Performance of Shield Type
     Supports in the U.S.A.
     SME Transactions, June 1977, Pages 156-164

(10) Federal No. 2 Longwall - The Record
     Breakers
     Eastern Coal News - Issue No. 4 - 7/80

(11) Hahn, L.
     Neue Entwicklungen der Steuerungstechnik
     fuer Schreitausbau
     Glueckauf 116 (1980, Nr. 8, Pages 381-387)

(12) Morgan, D.
     In Support of Longwalling
     Colliery Guardian, July 1980, Pages 28-33

(13) Boldt, H., Prof. Luerig, H.J.
     Waermeisolierter Schildausbau auf dem
     Verbundbergwerk Rheinland
     Glueckauf 116 (1980) Nr. 15, Pages 759-765

(14) Irresberger, H.
     Werksausstellung und Symposium der
     Maschinenfabrik Hermann Hemscheidt
     Glueckauf 116 (1980) Nr. 3, Pages 128-131

Chapter 5

LONGWALL LIGHTING

W. H. Lewis

Bureau of Mines
U.S. Department of the Interior
Pittsburgh Research Center

J. C. Yingling

Bituminous Coal Research, Inc.
Monroeville, PA

M. H. Leon

Bureau of Mines
U.S. Department of the Interior
Pittsburgh Research Center

ABSTRACT

This paper describes the present state-of-the-art and status of longwall lighting in the U.S.A. Specific subtopics discussed are as follows: review of MSHA lighting regulations and requirements for longwalls; overview of available lighting hardware; installation procedures and costs; analysis of installations to date, including operator and management acceptance, complaints, benefits, and maintenance considerations.

INTRODUCTION

Lighting technology, like longwall technology, is relatively new on the American mining scene. Like most major changes, the introduction of lighting systems in underground coal mines has met with mixed acceptance. Longwall lighting, in contrast to room and pillar applications, has received almost unanimous management and miner acceptance. This acceptance is largely responsible for the dramatic growth since 1978, when most illuminated longwall installations were experimental in nature and federally sponsored. Since then, the number of longwalls equipped with lighting systems has increased to where currently 65% are illuminated.

LIGHTING REGULATIONS

The Federal Coal Mine Health and Safety Act of 1969 recognized the essential need for better illumination systems in the working environment of miners and

...charged the Secretary of the Interior with the authority and responsibility for developing and prescribing minimum illumination standards for working places in coal mines. These standards became effective on July 1, 1978, as mandatory regulations under Section 75.1719 through 75.1719-4, Title 30, Code of Federal Regulations. These regulations (1) require that designated surfaces within the miner's normal field-of-vision be illuminated to 0.06 footlambert while self-propelled mining equipment is being operated in the workplace; (2) specify requirements for lighting fixtures and circuits; (3) specify

methods of making light measurements; and (4) specify additional requirements intended to provide increased visibility for miners in working places.[1]

Definitions

"Working place" is defined in the Act as "the area of a coal mine inby the last open crosscut."

"Miner's normal field-of-vision" refers to those surfaces that can be readily seen by a miner from any position in the working place that his duties require him to be, while self-propelled mining equipment is being operated. Floor surfaces under the machine or surfaces behind line curtains, etc., are not included.

"Self-propelled mining equipment" is defined in Section 75.1719-1 (b) as

...equipment which possesses the capability of moving itself or its associated components from one location to another by electric, hydraulic, pneumatic, or mechanical power supplied by a source located on the machine or transmitted to the machine by cables, ropes, or chains.[2]

Lighting Requirements for Longwalls

Section 75.1719-1 (e) (4) requires that the following areas be illuminated to 0.06 footlambert while longwall mining equipment is operated in the workplace:

...the face, and the areas between the gob side of the travelway and the side of the block of coal (Figure 1) from which coal is being extracted for the length of the self-advancing roof support system, and the control station and the headpiece and tailpiece of the conveyor, and the roof and floor for a distance of 5 feet horizontally from the control station, headpiece, and tailpiece.[2]

The implementation of these longwall lighting requirements translates into the installation of "typically" one lighting fixture for each support and an additional three lighting fixtures on the control station, head, and tailpiece. A more detailed description of "typical" installations is discussed in a later section.

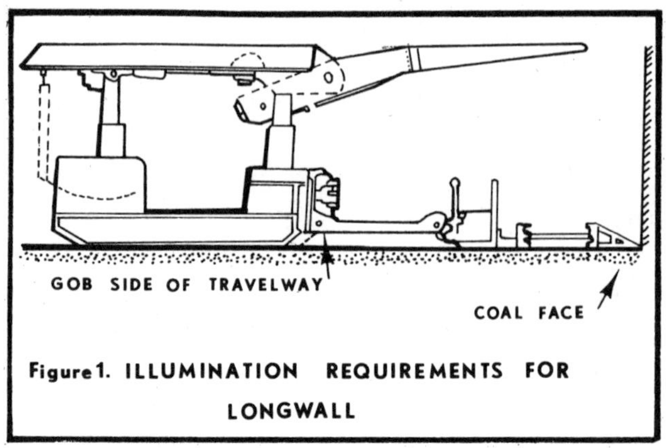

Figure 1. ILLUMINATION REQUIREMENTS FOR LONGWALL

STATUS OF LONGWALL LIGHTING IN THE U.S. (1980)

Table 1 outlines the distribution of illuminated longwalls in the U.S. by seam height. The illuminated installations represent a good cross section of conditions and technology found across the country and demonstrates that longwall illumination is not restricted to any particular mining environment. Currently, there are 88 active longwall faces in the U.S. of which 57 (65%) are illuminated. An additional 13% of the active faces have systems on order or testing programs in progress.

AVAILABLE LIGHTING HARDWARE/SYSTEMS

Currently, six illumination hardware manufacturers (Table 2) offer systems for longwall applications.

At least two other manufacturers, currently offering illumination systems for mobile face equipment, are developing new hardware or investigating the feasibility of adapting their existing hardware to longwall applications.

A variety of light sources are available, including fluorescent, high-pressure sodium, mercury vapor, and incandescent. Popularity of particular types of light sources and systems varies with the application and individual preferences.

Table 2. LONGWALL LIGHTING MANUFACTURERS IN THE U.S. (1980)

MANUFACTURER	TYPE OF SYSTEM
Crouse Hinds	High Intensity Discharge (FIXTURE CAN BE EQUIPPED WITH INCANDESCENT)
McJunkin	X/P Fluorescent
Mine Safety Appliance	X/P Fluorescent
National Mine Service	X/P Fluorescent
Ocenco	Intrinsically Safe Fluorescent
Service Machines	Intrinsically Safe Fluorescent

Table 1. DISTRIBUTION OF ILLUMINATED LONGWALLS IN THE U.S. BY SEAM HEIGHT (1980)

	SEAM HEIGHT				
	LOW < 48"	MEDIUM 48"-84"	HIGH > 84"	TOTAL NO.	%
Total Active Faces	15	64	9	88	
Illuminated Active Faces	9	43	5	57	(65%)
Faces With Systems On Order	0	6	0	6	(7%)
Faces With Testing Programs In Progress	2	3	0	5	(6%)
Totals	11	52	5	68	(77%

* Includes systems presently being installed

Table 3. DISTRIBUTION OF LIGHTING SYSTEM TYPE BY SEAM HEIGHT   (1980)
(CURRENTLY ILLUMINATED AND ON-ORDER SYSTEMS)

SYSTEM TYPE	CURRENTLY ILLUMINATED				ON - ORDER SYSTEMS			
	SEAM HEIGHT				SEAM HEIGHT			
	LOW ≤48″	MEDIUM 48″-84″	HIGH >84″	TOTAL NO. (%)	LOW ≤48″	MEDIUM 48″- 84″	HIGH >84″	TOTAL NO. (%)
Intrinsically Safe Fluorescent	8	34	1	43 (76%)	0	4	0	4 (66%)
Explosion Proof Fluorescent	1	4	2	7 (12%)	0	1	0	1 (17%)
High Intensity Discharge	0	5	2	7 (12%)	0	0	0	0 (0%)

All of the lighting systems can be categorized into 2 basic types, intrinsically safe and explosion-proof systems. As shown in Table 3, intrinsically safe fluorescent systems are are most popular and more widely used. High intensity discharge (HID) lamp systems (e.g. high-pressure sodium and mercury vapor) are more widely used in high coal applications where the high light-output fixtures are desirable. Presently, no HID systems are installed in low coal, which is appropriate, considering that HID fixtures are typically larger in profile and more prone to glare complaints.

System Description

This section summarizes the characteristics of currently available illumination systems used in longwall lighting. In the following tables where descriptive data are given, it must be noted that the information only applies for the "typical" system. Manufacturers can and will modify the systems for the particular application and local statutes that apply, as well as customer wishes. Pertinent luminaire data and specifications are displayed in Table 4 for easy comparison by prospective consumers.

Component requirements for each of the manufacturers' longwall lighting systems are presented in Table 5. The data shown are for a "typical" installation with the following requirements: (1) 126 face supports (approximately 500-600 feet depending on the support width); (2) all systems use three luminaires on the last two roof supports at both head and tail ends of the support line; and (3) all systems will use three luminaires on the headgate entry. Cable requirements are not shown because this is highly dependent on the particular face in question.

When using the tables to make comparisons of the various manufacturers' component requirements, it is not always appropriate to consider systems with fewer components as being more desirable. For example, increased use of connectors increases modularity and facilitates installation and maintenance. Increased use of luminaires may result in better light distribution and fewer problems when luminaire failure occurs.

Figures 2, 3, and 4 show a "typical" wiring layout for each of the manufacturers' systems. As indicated in the figures, all systems utilize a power center (typically a transformer with overcurrent, short circuit, and ground fault protection, along with ground monitoring) to provide 120-volt, three-phase AC power for distribution along the face. The 120-volt, 3-phase power is connected to intermediate power conditioning or distribution units which subsequently deliver power to the luminaires. A functional description of each of the manufacturers' system is given below:

Ocenco (Figure 2). The main power is transformed, rectified, and filtered to 13-16 VDC, which is fed to each of two identical control circuits or channels which provide intrinsically safe 12-VDC power to the luminaires. Each channel limits power output to 12 volts at 10 amp and insures that intrinsically safe conditions are maintained in the event of failure. The current limit restricts the number of lamps that may be driven to 10 Model 15/3M or 8 Model 30 luminaires per power conditioning unit.

Service Machine (Figure 2). The main power is rectified to approximately 160 VDC by a three-phase bridge rectifier. DC power is fed through a thermal protective relay package to each of three inverter modules which convert the DC to high-frequency, intrinsically safe power used to drive the lamps. Each module drives two lamps for a total of six per power conditioning unit.

McJunkin (Figure 3). To maintain load balance, distribution units are connected across alternate phases of the main feed-through line. Two 120-volt, single-phase feeder lines originate at the distribution box and provide power for up to 10 feed-through Model 100/30 luminaires per line, or a maximum of 20 lamps per distribution unit. An intrinsically safe interlock unit is provided for connector deadfacing/ground-fault protection. To offset line losses, an autotransformer voltage boost is used in units distant from the system power center.

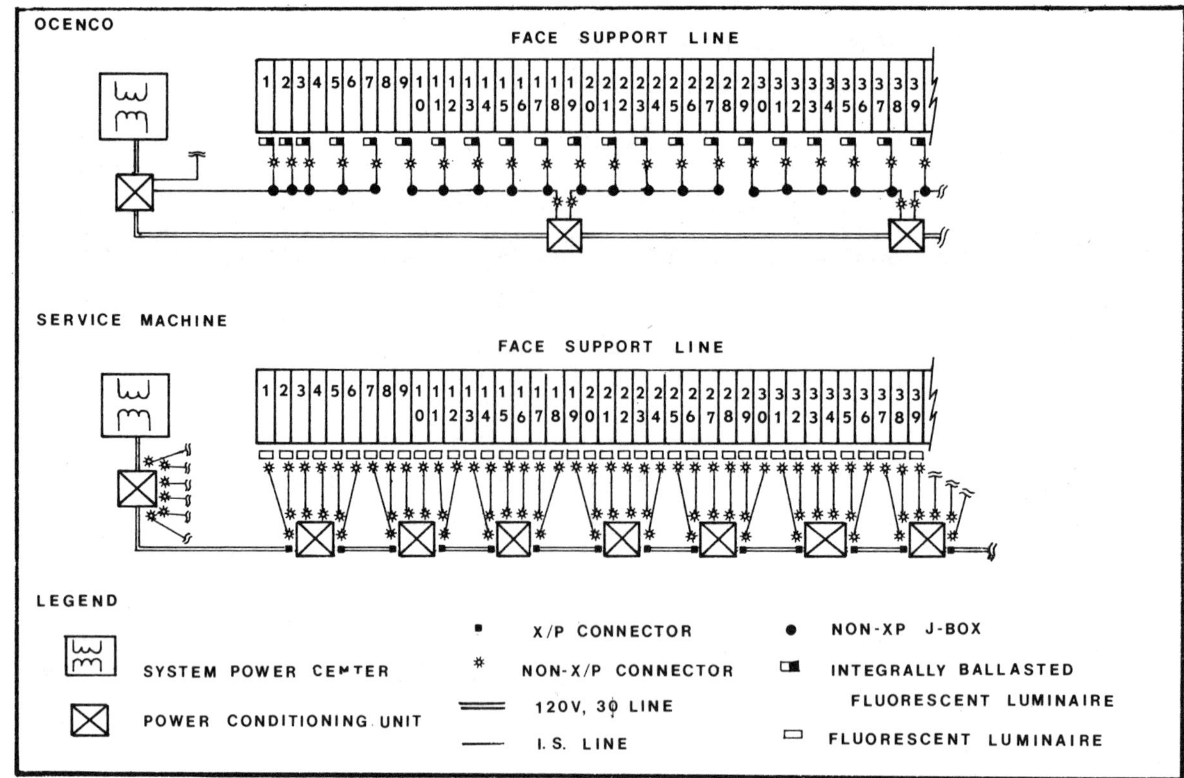

Figure 2. TYPICAL WIRING SCHEMATIC FOR INTRINSICALLY SAFE FLUORESCENT LONGWALL LIGHTING SYSTEMS

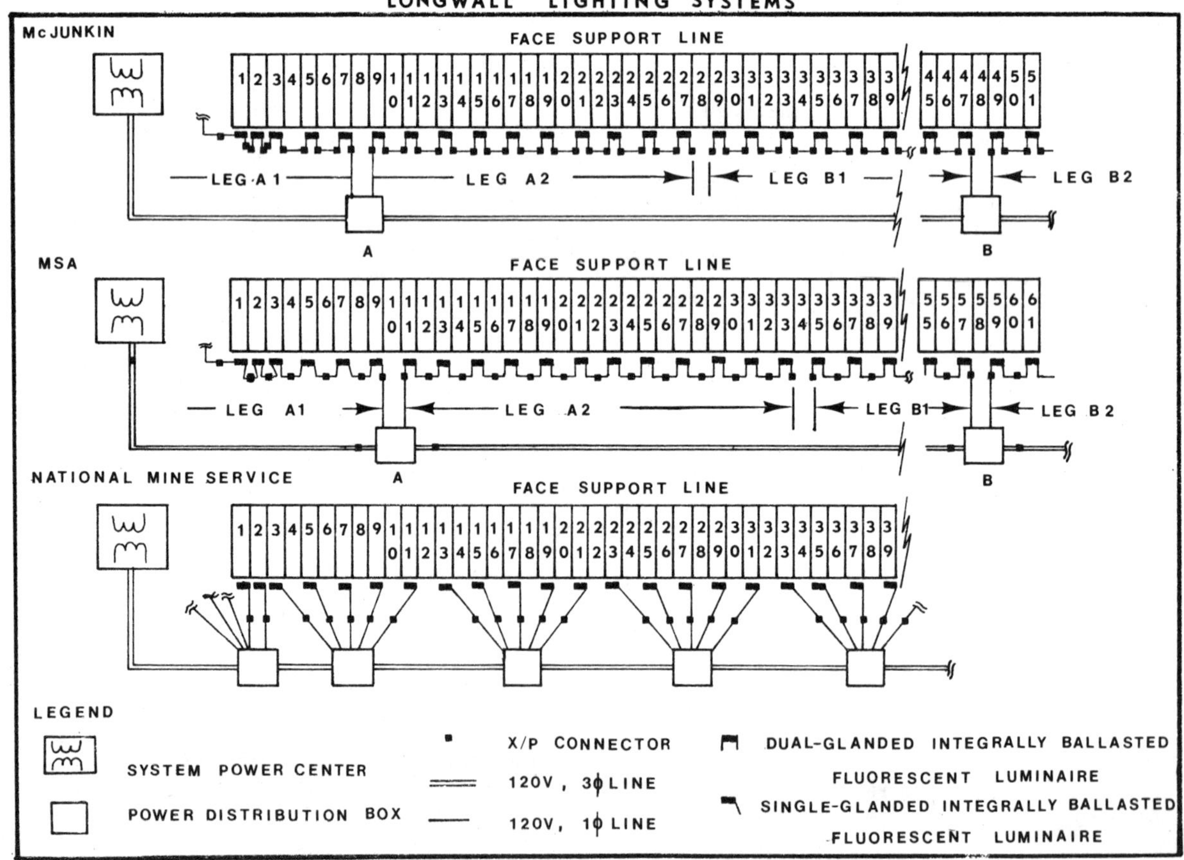

Figure 3. TYPICAL WIRING SCHEMATIC FOR X/P FLUORESCENT LONGWALL LIGHTING SYSTEM

Table 4 LUMINAIRE SPECIFICATIONS FOR LONGWALL LIGHTING SYSTEMS

SYSTEM TYPE	INTRINSICALLY SAFE (FLUORESCENT)				EXPLOSION PROOF (FLUORESCENT)							EXPLOSION PROOF HIGH INTENSITY DISCHARGE
MANUFACTURER	OCENCO			SERVICE MACHINE	MINE SAFETY APPLIANCE GENERAL ENERGY		McJUNKIN			NATIONAL MINE SERVICE		CROUSE-HINDS
MODEL IDENTIFICATION	15/3M	30	16	ISC	LX 2401	LX 1600	100/30	100/15	100/6	65 WATT	50 WATT	ME/H
DIMENSIONS (HxWxL) (INCHES)	3x3x28	3x3x40	3x3x14	$2\frac{3}{16} \times 2\frac{3}{8} \times 26$	$4\frac{1}{2} \times 4\frac{1}{2} \times 33$	$4\frac{1}{2} \times 4\frac{1}{2} \times 27$	$4\frac{1}{4} \times 5\frac{1}{4} \times 48$	$4\frac{1}{4} \times 5\frac{1}{4} \times 29$	$4\frac{1}{4} \times 5\frac{1}{4} \times 15\frac{1}{2}$	$3\frac{1}{2} \times \frac{5}{16} \times 34$	$3\frac{1}{2} \times \frac{5}{16} \times 28$	$7 \times 7\frac{1}{8} \times 16$
INTEGRALLY BALLASTED	YES	YES	YES	NO	YES	YES	YES	YES	YES	YES	YES	NO
LENS MATERIAL	POLY-CARBONATE	POLY.	POLY.	POLY.	POLY.	POLY.	POLY.	POLY.	POLY.	POLY.	POLY.	TEMPERED GLASS
CAGE INCLUDED	NO	NO	NO	NO	YES	YES	YES	YES	YES	NO	NO	YES
WEIGHT (LBS.)	8	7	$2\frac{1}{2}$	$4\frac{1}{2}$	24 (STEEL) 12 (ALUMIN.)	21 (STEEL) 10 (ALUMIN)	18	14	10	12	10	53
HOUSING MATERIAL	-	-	-	-	STEEL OR ALUMINUM	STEEL OR ALUMINUM	STEEL	STEEL	STEEL	ALUMINUM OR BRONZE	ALUMINUM OR BRONZE	FERALOY[R]
LAMP(S) TYPE	F15T12/CW	F25T12/CW	F8T5/CW	F13T5/CW	F24T12/CW HO	F18T12/CH HO	F30T12/CW	F15T8/CW	F6T5/CW	F24T12/CW VHO	F18T12/CW VHO	70 WATT HPS 100 WATT HPS 100 WATT MV
PACKING GLAND CONNECTOR ARRANGEMENT	NOTE 1	NOTE 1	NOTE 1	NOTE 2	NOTE 3	NOTE 3	NOTE 4	NOTE 4	NOTE 4	NOTE 5	NOTE 5	NOTE 6

Notes:

(1) Integral pigtail with line-installed connector attached to each luminaire. No packing glands.

(2) Connector is integral to luminaire. No packing glands.

(3) Dual-glanded fixture is typically equipped with single line-installed connector between adjacent luminaries.

(4) Dual-glanded fixture is typically equipped with line-installed connector on both incoming and existing cables.

(5) Single-glanded fixture with line-installed connector.

(6) Single-glanded fixture with slip-fit gland assembly.

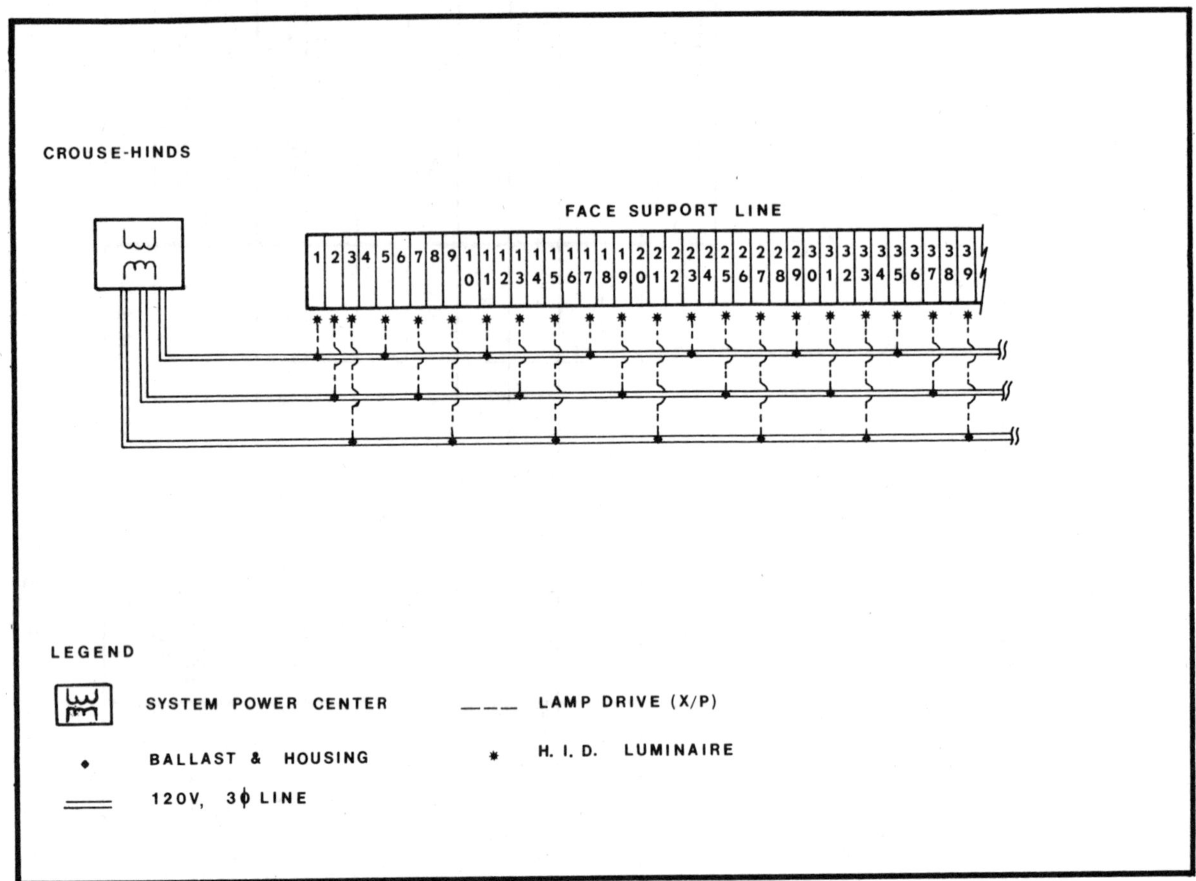

Figure 4. TYPICAL WIRING SCHEMATIC FOR HID LONGWALL LIGHTING SYSTEM.

Table 5.   COMPONENT REQUIREMENTS FOR TYPICAL LONGWALL ILLUMINATION SYSTEMS

TYPE OF SYSTEM	MANUFACTURER	ASSUMED SUPPORT LUMINAIRE RATIO**	NUMBER OF COMPONENTS								
			LUMI-NAIRES	PWR. CON-DITIONING UNITS	POWER DIST. BOXES	BALLAST/HOUSING	CONNECTORS			X/P PACK GLANDS	NON-X/P J-BOXE
							X/P	3Ø X/P	1Ø ISC		
ISC	OCENCO MODEL 15/3M	2:1	70	7	N/A	N/A	0	N/A	84	27	70
FLUO-RESCENT	OCENCO MODEL 30	2:1	70	9	NA	N/A	0	N/A	97	34	70
	SERVICE MACHINE	1:1	134	23	NA	N/A	45	N/A	268	0	0
X/P	NATIOAL MINE SERVICE	2:1	70	N/A	14	N/A	0	70	N/A	167	N/A
FLUO-RESCENT	MINE SAFETY APPLIANCE	2:1	70	N/A	3	N/A	5	70	N/A	145	N/A
	McJUNKIN	2:1	70	N/A	4	N/A	0	132	N/A	149	N/A
HID	CROUSE-HINDS	2:1	70	N/A	N/A	70	0	N/A	N/A	219	N/A

<u>Mine Safety Appliance</u> (Figure 3). Two branch feed lines originate at the distribution unit by connecting across phases of the main feed-through and provide 120-volt, single-phase power to a maximum of 12 feed-through Model LX2401 luminaires per line for a maximum of 24 per distribution unit. To maintain load balance, the branch phase connections alternate between phases along the face. A different unit with a pilot system is provided for meeting the Commonwealth of Pennsylvania deadfacing requirements.

<u>National Mine Service</u> (Figure 3). Distribution units are connected across alternate phases of the main feed-through line. Five feeder lines originate at each distribution box with each line supplying 120-volt, single-phase power to a single luminaire for a total of five luminaires per distribution unit. Three different distribution units are available, providing options concerning pilot circuit configurations. One unit makes no provision for a pilot circuit and basically consists of a terminal board in an X/P enclosure. A second unit provides a pilot circuit to each luminaire with a single relay at the distribution box. The third option provides a separate pilot circuit and relay for each luminaire, enabling decoupling of one luminaire while maintaining the operation of the remaining luminaires on the distribution unit.

<u>Crouse Hinds</u> (Figure 4). Typically, three 120-volt AC, three-phase, feed-through lines are run the entire length of the face, with each line providing power to one-third of the luminaires. Ballasts, housed in a separate X/P box, are connected across one phase of the line, and each drives a single HID lamp. An X/P box and ballast is required for each luminaire.

An important consideration in comparing the systems is the cable arrangement from the power conditioning unit or distribution box to the luminaire. Table 6 summarizes the various cable arrangements and typical practices followed by each manufacturer. The use of connectors and junction boxes is an arbitrary matter, and the table only describes what is typical.

TABLE 6. TYPICAL CABLE ARRANGEMENT FROM
POWER SUPPLY/DISTRIBUTION BOX TO THE LUMINAIRES

Description	Manufacturer(s)
A. Separate cable to each luminaire with intermediate connector(s)	National Mine Service Machine
B. Separate cable to each luminaire with slip-fit glands on ballast boxes and luminaires.	Crouse-Hinds
C. Cable is connected feed-through to string of dual-glanded luminaires. Intermediate connectors utilized.	MSA McJunkin
D. J-boxes are installed on the feeder line for each luminaire. A cable with line-installed connector runs from each J-box to each luminaire.	Ocenco

Typical System Costs

In an effort to determine approximate longwall lighting system costs, each of the six manufacturers were requested to quote on a "typical" longwall lighting application. To insure that each manufacturer was quoting on a comparable installation, specifications, drawings, and system requirements were prepared. The hypothetical "typical" system was assumed to be a 500-foot longwall section consisting of 100 Dowty 2-leg shields on 60-inch centers. Seam height was specified at 72 inches and constant. Available power was assumed to be 1000 VAC, 3-phase. Quoting requirements assumed that all installation labor and necessary mounting hardware would be supplied by the coal company. Installation and start-up supervision would be provided by the supplier. Explosion-proof fluorescent lighting systems were required to have a pilot circuit between junction boxes and luminaires.

Of the six companies contacted, four responded. Prices for the equipment and specified services ranged from $26,000 to $55,000 for an average of $40,000.

ANALYSIS OF ILLUMINATED LONGWALLS

The following discussion is a summary of the findings of a recent analysis conducted by Bituminous Coal Research, Inc., for the U.S. Bureau of Mines under Contract J0308040, "Development of Guidelines for Installation and Maintenance of Mine Illumination Systems."

During the project, 14 companies were contacted which operate a total of 24 illuminated longwalls. In-depth interviews were conducted with company representatives to assess performance of their lighting systems. When possible, underground visits were made to observe the installations firsthand and talk directly with the personnel working on the illuminated face or maintaining the lighting system. Seam heights of the longwall installations investigated ranged from 36 to 90 inches. Utilized lighting systems included intrinsically safe, permissible fluorescent and high-intensity discharge lamp systems produced by most curent manufacturers.

Personnel Acceptance of Longwall Lighting

Face crew members were interviewed to obtain their comments on the lighting systems. A total of 26 crew members were interviewed and discussions focused on four basic questions:

- What do you think of the lighting system?
- Do you like it or not?
- Are there any occasions when the lighting presents a problem for you?
- What benefits are you getting from the lights?

Based on the interview data presented in Table 7, personnel acceptance of longwall lighting is essentially unanimous; 25 of the 26 employees interviewed reacted favorably. This data includes both HID and fluorescent lighting systems, seam heights ranging from 36 to 90 inches, and employees 21 years to 60 years old. This acceptance appears to be independent of the type of lighting system used, seam height, worker age, or length of service.

Table 7. PERSONNEL ACCEPTANCE OF LIGHTING

INTERVIEWEE	NO. OF CONTACTS	REACTIONS	
		LIKED	DISLIKED
SHEARER OPERATOR	5	5	-
JACK SETTER	17	16	1
MECH./ELEC.	4	4	-
TOTALS	26	25	1

The only negative opinion was mild, with one worker expressing his opinion that "cap lamps give a better light," although he did not suggest that the lights should be removed or that they impeded his work.

Personnel Complaints

Even with the high rate of acceptance, there were some complaints as shown in Table 8 and discussed below:

Table 8. PERSONNEL COMPLAINTS

INTERVIEWEE	NO. OF CONTACTS	COMPLAINTS		
		GLARE	PHYSICAL CONTACT	WHITEOUT
SHEARER OPERATOR	5	1	1	-
JACK SETTER	17	6	6	2
MECH./ELEC.	4	-	-	-
TOTALS	26	7	7	2

Glare – Seven of the personnel interviewed (27%) had complaints about glare, but these were mild and no one wanted the lights removed to eliminate the problem. One of the glare complaints was from the shear operator who claimed the lights interfered with his observation of the drum. Conversely, other operators felt the lights aided in observing conditions and running the shear. Again, there were no requests to remove the lights to overcome this or any other problem.

Physical Contact – Another common problem, again mentioned by 27% of those interviewed, was the physical hindrance caused by the presence of the system (e.g., personnel bumping their heads or scraping their backs on the luminaires). This was especially true in low seams or medium seams using high-profile luminaires. In all cases, the personnel indicated that the problem was minimized as they became more aware of the luminaires. As with the glare problem, the longwall personnel were not willing to give up the benefits of the lighting to eliminate the problem.

Whiteout – The only other complaint was "whiteout," which is light reflecting off dust or spray, which bothered two jacksetters when they were opposite or downwind of the shear. Apparently this did not occur at all times and did not impede their work.

Benefits

Table 9 lists the various comments of the face crew members and reflects what they consider to be the benefits of longwall lighting. The comments cover production, safety, and maintenance aspects and again indicate their overall acceptance of the lighting.

TABLE 9. SUMMARY OF FACE CREW COMMENTS

* Helps keep face straight.
* Benefits safety and production.
* Makes it easier to run the shear.
* Helps in the performance of maintenance.
* Complain when lights go out.
* Best thing they ever had.

In addition to the 27 face crew members interviewed during the survey, management personnel from every company visited were asked about acceptance on the part of their face crews. In all cases, they indicated that the face crews viewed the lighting systems favorably. Management personnel were also asked what benefits they are realizing because of the lighting. Table 10 summarizes their comments.

TABLE 10. SUMMARY OF MANAGEMENT PERSONNEL COMMENTS

* Facilitates alinement of face supports and conveyor
* Improvement in face crew attitude
* Positive impact on production
* Tripping hazards minimized
* Hydraulic and electric cable pinching incidences reduced
* Maintenance problem indicators (e.g., hydraulic leaks) more easily detected
* Hazards more easily detected
* Less chance of rock falling between supports and hitting face workers
* Aids in seeing conditions at the face
* Facilitates machine operation
* Less eye strain
* More comfortable work environment

System Selection Criteria

From discussions with operators concerning their reasons for selecting a particular lighting system, and their subsequent experience with it, considerable information has been collected and is summarized below. The summary is not intended to be a complete discussion of all relevant parameters, but rather a highlight of those positive and negative aspects that appear to be of greatest concern to the operators.

Intrinsically Safe Power – These systems have fewer X/P packing glands to inspect and maintain to assure permissibility. Persons other than certified electricians can perform some common maintenance such as luminaire change-out or lamp replacement. This is

particularly important when certified electricians are
in short supply. Intrinsically safe systems are perceived to have a greater margin of safety in gassy
mines. The electronic circuitry used in the power
conditioning units is dissimilar to most other mine
circuitry, and the possible lack of familiarity could
create troubleshooting and servicing problems.

Modularity - Highly modularized systems are more
easily installed or removed, particularly for longwalls, since a partial take down is required at each
face move. Modularity also facilitates component
change-out. A change-out maintenance approach usually
facilitates troubleshooting and repair. For some
failures, modularity may decrease time for problem
isolation. This is especially important on longwall
systems for isolating cable faults, which are not
usually evident and necessitate disconnecting wiring
for continuity checks.

Size of Fixtures - Luminaires often must be located
where they are a physical hindrance to workers tra-
versing the face. Increased profile increases the
hindrance potential. Increased profile also increases
the potential for clearance related damage, particularly if mined height is reduced.

Location of power supply/distribution boxes is
often a compromise between the box becoming a hindrance problem for the workers or a servicing access
problem when repairs are necessary. Smaller fixtures
increase the availability of potential mounting
locations as well as decreasing the hindrance
potential.

Number of Fixtures - Fewer fixtures (luminaires and
boxes) result in fewer potential physical damage trou-
blespots. Fewer fixtures usually require fewer glands
to monitor and inspect and generally simplify installation.

Light Sources - The opinions that fluorescent lights
provide (1) better light distribution, (2) better
color condition, or (3) less glare potential than
HID sources were quite common at the companies
using fluorescent systems only. The survey, in general, indicated that in medium to high seams, there
was little difference in acceptance when the two types
of systems were compared, although there were no
HID systems installed in low coal seams.

Cables - Feed-through wiring facilitates routing and
reduces the amount of cable use. Cables with small-
diameter conductors have presented difficulties in
making repairs in some mines, especially when these
repairs are performed at the face. In general, increased use of cables increases the potential for
cable damage.

Number of Packing Glands - Reduced number of glands
reduces the potential for damage and time spent for
inspecting and repairing. Also, the potential for
citations concerning permissibility is decreased.

Integral Ballasts - When luminaires have this feature
it eliminates the necessity of accessing the power
supply in event of a ballast failure. This promotes
a change-out maintenance approach.

Maintenance Requirements

The majority of the operators interviewed reported
that their maintenance requirements are low or moderate with only a few experiencing major maintenance
problems. Most of the reported problems can be categorized into three major areas: (1) Cable damage,
(2) physical damage of the luminaires, and (3) electrical component failure. Cable damage and physical
damage of the luminaires are the most frequently
reported problems. Electrical problems of major significance were also encountered with several systems
shortly after market introduction, but refinements
and modifications by the manufacturers have brought
this problem to reasonable levels.

Cable Damage - It is evident that several factors or
mechanisms contribute to cable damage. In order to
develop effective preventive maintenance procedures,
these mechanisms must be taken into account. Three
general categories of cable damage mechanisms have
been identified as follows:

· Pulling damage
· Rock impact damage
· Cable pinching

PULLING DAMAGE usually occurs from the cable
catching on a rock or because inadequate cable
slack between the supports, resulting in the gland or
other terminating point being separated when the supports are advanced.

ROCK IMPACT DAMAGE usually occurs when rock falls
through the gaps between adjacent supports and impacts
the cable.

CABLE PINCHING usually occurs when the cable is
routed in the vicinity of hydraulic jacks or rams and
is pinched when the jack or rams are retracted.
Cables also get caught under the support base or when
the canopy of the support is lowered to the point that
it pinches the cable where it exits the power supply
or distribution box.

In general, companies experiencing cable pulling
damage have found the incidents to be recurring in
nature. Of the seven companies reporting pulling
damage as a problem, six have indicated that it
occurred frequently. In contrast, of the six companies reporting rock damage, only two indicated that
the problem occurred frequently. Likewise of the four
companies reporting pinching incidents, none indicated
that the occurrences were frequent. Some companies
have noticed decreases in the frequency of cable
damage as the jacksetters become more aware of the
problems and take measures to protect them.

Physical Damage of the Luminaires - Although shooting
damage in maintenance terms is not as widespread or
as serious as cable damage, several problems resulting
from face shooting were identified and are being
experienced by significant portions of the surveyed
companies. For example, if the luminaires are un-
protected, significant damage can be incurred,
although, most commonly, because of the shock wave,
lenses and housings are also damaged on occasion by
flying debris. Morever, lamps are often destroyed
because of the shock wave even if the fixture itself has been protected.

Fluorescent lamps in particular seem vulnerable to the shock wave. HID lamps seem less affected, but because of the small number of surveyed mines using HID systems, this cannot be said with certainty. In view of the fact that fluorescent lamps frequently incur damage, one would anticipate most companies would remove the fixtures from the blast zone. This seems especially true in light of the following:

(1) All fluorescent systems are highly modularized which facilitates takedown.

(2) Fabrication of a quick-release mounting configuration is easily accomplished on most systems.

(3) Many companies utilize independent repair agencies to perform repairs on damaged luminaires. Cost is significant even if the lamp replacement is the only repair necessary.

(4) Section 75.19-2 (f) of the Code of Federal Regulations requires that the lighting fixtures be removed from the line-of-site of the blast zone for a distance of 50 feet or that the luminaires be otherwise protected from flying debris.

Of the nine companies surveyed that use fluorescent systems and practice some face shooting, only two practice takedown.

<u>Electrical Components</u> - Analysis of the reported electrical problems has resulted in the following conclusions:

(1) In general, overall electrical reliability appears to be good on longwall lighting systems, considerably better than it is on the mobile-face equipment systems.

(2) The reported failure incidents differ considerably from mine to mine. No trends of consistent failure materialized.

(3) For the majority of mines, most of the reported problems were not recurring in nature. A few mines have reported turnover of some electrical components, for example, ballasts and circuit board assemblies, but these rates are low and are not considered a problem by operators.

(4) Three installations having early manufactured systems initially experienced significant electrical problems, but manufacturer modifications have brought the problems to reasonable levels.

(5) Incidents where component failure occurred in conjunction with cable damage were rather frequent.

CONCLUSIONS

The introduction of lighting systems into underground coal mines in the U.S.A. has been a controversial issue. Longwall lighting, however, has received widespread management and miner acceptance with tangible benefits being realized in both safety and production. Based on the experience that longwall operators have had with their lighting systems, the following general conclusions seem appropriate:

· A wide selection of lighting equipment and suppliers exists.
· Intrinsically safe fluorescent systems are more widely used.
· The majority of operators reported that maintenance requirements are low or moderate with only a few experiencing major maintenance problems.
· The most major maintenance problem has been damage to the system power cables.

In summary, longwall illumination technology has been proven to be in a mature state, providing many benefits in miner safety while imposing few difficulties in production and downtime. It is hoped that the information contained in this paper will be of benefit to the mining industry as an aid to specifying successful longwall illumination installations in the future and thus expedite full implementation of this beneficial innovation in U.S. mines.

REFERENCES

1) Handbook of Underground Coal Mine Illumination Requirements, U.S. Department of Labor, Mine Safety and Health Administration, Office of Coal Mine Safety and Health 1980

2) Code of Federal Regulations, Title 30, Miner Resources

Chapter 6

SOME CRITERIA FOR LONGWALL SELECTION

Anthony Sharkey

Mining Engineer
Cecil V. Peake, Inc.

Introduction

In Europe, longwall mining accounts for over 95% of coal produced from deep mines, whereas in the U.S.A., longwall accounts for less than 5% of deep mine coal production. There are many reasons for the difference, but in general:

a. U.S. mine operators have invested heavily in continuous mining or conventional equipment which continues to produce acceptable results;

b. There is a reluctance to change from a familiar (and proven) mining method to an unfamiliar one, despite promises of huge production increases;

c. Multi-seam mining in Europe has made longwall the only really viable mining method, whereas in the U.S.A. multi-seam mining (particularly in "below-drainage" mines), is relatively rare;

d. Although some spectacular results have been achieved by longwall in the U.S.A.(1), the general performance of longwall in this country has not been good enough to generate much enthusiasm for its wide application;

e. The much higher level of investment in face equipment required for longwall when compared to alternative methods (approximately $7.5 million for a medium seam longwall installation and $1 million to $1.5 million for a continuous miner section).

However, declining productivity, national energy requirements and safety legislation are creating a situation in which longwall as a viable mining method is being examined more closely and in more places than hitherto.

This paper, therefore, attempts to identify some of the principal criteria needing examination in any assessment of the feasibility of longwall application, the limitations of information for those criteria in the decision making process and the need for further research and development to assist the mine operators make the decisions.

Criteria to be Discussed

1. The need to consider advantages and disadvantages of applying longwall as the principal means of extraction on a coal property.

2. The need to equate the coal reserves available for longwall mining with the capital investment required.

3. The need to evaluate the strata conditions with respect to their suitability for longwall application. This is possibly the most important of the criteria and is also the one for which little precise data exists.

Why Longwall?

The first criterion to be satisfied is that the operator believes that the advantages of adopting longwall for his property outweigh the disadvantages.

As a mining system compared to room and pillar mining, longwall has the potential for:

o A safer environment

   - better protection from roof falls
   - easier and better ventilation.

o Higher production rates

   - a more continuous process
   - higher productivity.

o Greater resource recovery.

o Better surface subsidence control.

o As depths of workings increase, longwall becomes progressively more attractive as required pillar dimensions increase.

o Successful multiple seam mining.

o Extracting coal in difficult mining conditions

   - beds disturbed by previous mining
   - steeply pitching beds
   - deep beds.

Longwall uses fewer electric motors and gear trains per producing section than room and pillar mining. Typically, a longwall face uses seven or eight motors, while a continuous mining section uses twenty or more. Downtime due to motor failure should therefore be less of a problem with longwall.

There are fewer pieces of mobile equipment associated with longwall than with a continuous miner section and consequently, the potential for haulage accidents is greatly reduced.

Longwall is a relatively simple, repetitive process which is conceptually easier to automate.

Additionally, as depths of workings increase, longwall becomes progressively more attractive as required pillar dimensions increase.

However, longwall has some major disadvantages:

o   A high initial capital cost

    - $4.5 million to $8 million per
      face.

o   Face transfer problems

    - causing total downtime during moves
      which appear to take on average
      about twenty working days to complete.
    - causing severe haulage problems due
      to the mass of equipment which must
      be transported from a worked-out
      panel to a replacement panel (sometimes
      via the surface for overhauls).

o   A requirement for a high-skill, specialized maintenance labor force.

o   When its high production potential is approached, a severe development problem can result due to rapid panel wastage. Because of this factor, poor continuous miner performance at a mine should not be considered a major criterion for a switch to longwall.

o   Its sensitivity to haulage delays

    - there is no surge capacity in the
      longwall producing section and unless
      such capacity is built into
      the transportation system, any
      stoppage in that system almost
      immediately stops the production
      machine.

The Need for Adequate Reserves

It is assumed that the reserve base will be firmly established and the necessary quality data obtained before any consideration of mining method is undertaken. Because of the high capital cost of a longwall installation, it is recommended that proven reserves exist for at least eight years of mining at a rate consistent with acceptable costs.

Table I indicates the approximate present level of cost for a 152-meter (500 feet) face for four ranges of seam height. The variation in the seam height is determined more by

TABLE I

CAPITAL COST

Height Range meters(inches)	Supports (101 per face)	Shearer*	Face Conveyor	Stage Loader	Power Pack**	Total
1.02(40")-1.52(60")	$3,030,000	$ 700,000	$500,000	$170,000	$100,000	$4,500,000
1.27(50")-2.03(80")	$3,737,000	$1,000,000	$750,000	$270,000	$100,000	$5,857,000
1.52(60")-2.74(108")	$4,040,000	$1,500,000	$750,000	$270,000	$100,000	$6,660,000
2.29(90")-4.27(168")	$4,848,000	$1,700,000	$750,000	$270,000	$100,000	$7,668,000

*With chainless haulage

**Assumes two pumps

shearer dimensions than by the range of the
supports. Plows are still being talked
about for applications as thin as 0.81 meters (32 inches), but such thin seams are
not considered at the present time to be
longwall candidates in the U.S.A. except in
extremely isolated cases.

The costs given are an indication of the
costs likely to be incurred at the face and
for comparison purposes, the present cost of
a continuous miner section is between
$1 million and $1½ million. Panel belts are
not included in either case.

Each operator must perform his own calculations relative to the cost of the investment in the manner to which he is
accustomed or which are dictated by the
circumstances. Longer or shorter face
lengths can be used if desired and the
principal variable to be considered would
be the price of the supports. So far as
useful life of the equipment is concerned,
it can be assumed that the supports should
be written off after eight years, and the
shearer, the head and tail drives of the
face conveyor after five years with the
stage loader also lasting five years. Conveyor pans and chain should be considered as
totally renewable items after approximately
1,000,000 tons of production.

If plows are to be considered, it can be
assumed that the capital requirements will
be reduced by between 10% and 15% for the
mining system (i.e. the cutting machine, the
face conveyor and furnishings), but that the
face supports will remain the same.

Assuming an interest rate of 12%, a
depreciation rate of 12½% on the supports
and 20% on the other equipment and a required return on investment of 20%, the
annual cost of the capital invested can be
ascertained and is illustrated in Table II.

The number of shifts per day and the number of days worked per year vary with the
circumstances of each operation, but
assuming two shifts per day of production
and 220 working days per year, it is a
simple matter to arrive at required average
shift outputs if certain cost levels per ton
of capital are to be achieved and this is
illustrated by the graph at Figure 1. It
should be noted that the output figures are
the averages required and must take account
of downtime spent in moving equipment and
any idle time due to physical conditions,
markets and labor problems.

Each operator knows, or can calculate,
the costs other than face capital costs he
can expect from his own mines and to these
he must add the cost of his required capital
investment based on estimates of the likely
production performance from his installation. The figures used in this exercise
have been selected for illustrative purposes
only, but they serve to indicate that adequate proven reserves are a prerequisite to
an investment in longwall if costs per ton
are to be minimized.

$$T A B L E \quad I I$$

Height Range meters(inches)	Capital Cost $	Interest 12%	Depreciation Supports 12½%	Depreciation Shearer etc.20%	Return on Investment 20%	Annual Cost $
1.02(40)-1.52(60)	4,500,000	540,000	378,750	294,000	900,000	1,572,750
1.27(50)-2.03(80)	5,845,000	702,840	467,125	424,000	1,171,400	2,766,365
1.52(60)-2.74(108)	6,660,000	799,200	505,000	524,000	1,332,000	3,160,200
2.29(90)-4.27(168)	7,668,000	920,160	606,000	564,000	1,533,600	3,623,760

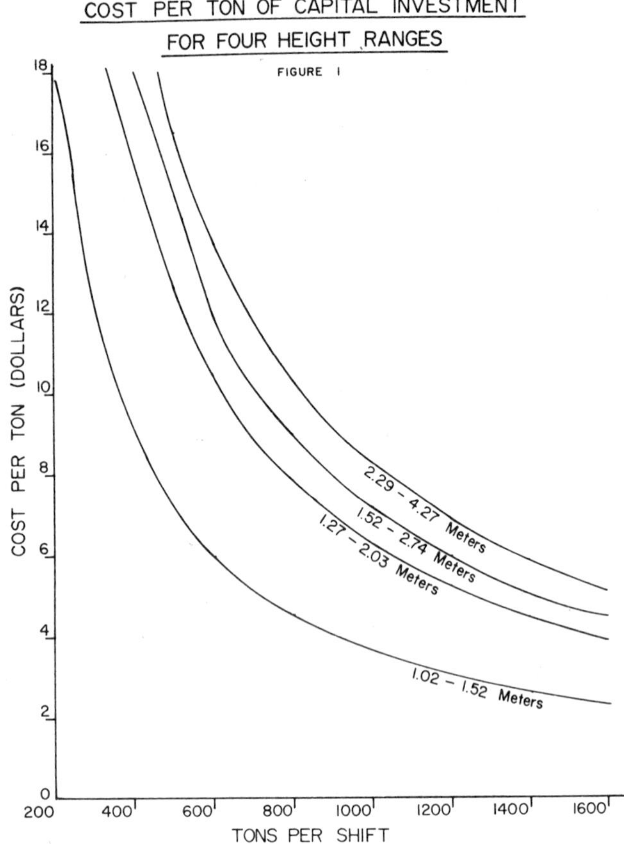

The Need for Strata Evaluation

There is no mechanical roof support in existence (nor is there ever likely to be) which can support the weight of strata between a longwall face at normal depth and the surface. The working of a longwall face is only possible because the great majority of this "cover load" is taken:

   a. by the solid coal, ahead of and to the sides of the excavation (the so-called front and side abutments), and

   b. by the consolidated waste or gob.

The mechanical roof supports on the face carry only the weight of the immediate strata within the "arch" formed by the bridging of the main beds between the front abutment and the consolidated gob.

The dimensions of the "arch" are largely governed by the height to which the roof rock breaks away from the upper beds and in so doing "bulks" or expands as it breaks. As the face moves away, falling, breaking and expansion continue until the material has bulked to an extent that no further falling can take place and the strata above this point "sits on" the bulked material and starts to compress it. As the face moves further away, further subsidence of the strata occurs and the compression of the fallen material continues, with an accelerating increase in load being taken by the

fallen material. This loading builds up to a point where the compacted material is sufficiently consolidated so that together with the solid coal ahead of the face, they take the load of the main beds above the immediate roof. The face supports carry the weight of the material within the "arch" formed by this process. Under normal circumstances the void resulting from the extraction of coal is closed by surface subsidence (60%-70% of extracted height) and by expansion of the gob material (30%-40% of extracted height).

When a strong and difficult to break bed lies at some distance above the seam being extracted, "bulking" (or expansion) of the fallen material between the seam and below the difficult bed assumes greater importance in supports design.

Authoritative sources (2,3,4,5) quantify the "bulking factor" of coal measure strata in a waste at between 1.20 and 1.5 times the volume of the same material when solid. The importance of this variation can readily be seen from Table III, where K is the "bulking factor" and H is the thickness of rock above the coal seam which must fall to fill the void created by the extraction of a coal seam, height "h".

Wilson(2) calculated "K" from observations in the U.K. where strata is known to be softer and more friable than those usually found in the U.S.A. Johnson (et al)(3) proposed 1.3 as a more realistic value for U.S. conditions. Adam (et al)(4) assumed

T A B L E   I I I

K	H	Authority
1.5	2h	(2) Wilson
1.3	3.33h	(3) Johnson and others
1.25	4h	(4) Wade
1.20	5h	(5) Adam and others

a value of 5 for "H" as a possible maximum based on calculations and French experience (the value of "K" in this case has been derived from "H"). Wade(5) chose a value of 4 for "H" as a conservative estimate of likely conditions in the U.S.A. based on the findings of other researchers(6) and "conservative" in this sense is intended to mean "erring on the safe side."

It seems quite obvious that the size of the broken pieces of rock and hence, the bulking factor will vary with the rock's resistance to fracturing.

The main intention of the paper by Wade(5) is to provide formulae for the calculation of the loads for which the supports must be designed and should be used for that purpose after an evaluation of the strata has been made.

From borehole and other data the strata column above and just below the seam to be extracted should be determined and the distance "H" from the floor of the seam to the base of any particularly strong beds ascertained. A scientifically based estimate of what is "particularly strong" cannot be made and it is suggested that a finely grained, massive sandstone, five feet or more thick, with an unconfined compressive strength exceeding 13,000 p.s.i. be considered as the point at which concern might be felt as to its ability to cave. If such a bed or a thicker or stronger bed falls within a height of 4h above the seam floor, then further investigation is needed as follows:

a. A check must be made on the strata caveability by visual observation in pillaring districts or in neighboring mines in the same seam, and

b. By determining the strength of the floor in terms of its capability to resist the loads likely to be imposed upon it by the bases of the heavy duty supports needed to support the tough beds and to induce breaking by offering a high resistance to bed separation.

Floor bearing strength tests should be carried out in accordance with Bureau of Mines recommendations(7) and comparing the indicated strength with the maximum pressure likely to be exerted by the base. If the floor is weakened in the presence of water, this should be taken into account.

Irresberger(8) describes the manner in which support load is transmitted to the floor and Figure 2 is taken from his paper. In general terms it can be said that "frame" supports have little or no application in the U.S.A., "chocks" have a cost advantage over "shields" and might still be applied in rare instances, but "shields" offer so many advantages that it is with them that we consider the problem of floor loading. Figure 2 shows that there is an irregular

distribution of load on the bases of both the two-leg and four-leg shields and it is the maximum loading with which we are to be concerned. Roof supports exist which can accept loads of fourteen tons per square foot on the roof canopy and consequently, pressures as high as 1,000 p.s.i. can be applied by the base. Clearly soft floors and very strong roof members are not compatible in longwall terms. However, it is the amount of floor penetration which must take place before the support builds up to its rated resistance (its yield load) that must be taken into account and a subjective judgement may have to be made based on the floor bearing tests and the thickness and nature of the floor rock. In many cases, if not "too thick", a soft floor can be mined with the coal to provide a good floor bearing surface at a slightly lower horizon.

Another factor to be given serious consideration when the "tough bed" situation is met is the nature of the rock immediately above the seam (at the extracted portion of the seam). If it is soft and friable, high resistance supports will crush it either when setting or when yielding. If the rock is more than two or three inches thick, considerable face problems can be caused by this crushing action.

To ascertain "H" over a large area where both seam height and the rock between the seam and the hard beds varies, it is recommended that an "Iso-Ratio" map be prepared. Figure 3 is an example of such a map with the iso-ratio lines being the values for H_{max} calculated from borehole information.

$$H_{max} = \frac{\text{height from floor of seam to base of hard rock}}{\text{extracted height of the seam}}$$

The area of the coal seam on which Figure 3 is based was in the Beckley Seam in Southern West Virginia, a soft coal with a very strong floor (in excess of 17.24 Mpa (2500 p.s.i.), as determined by plate bearing tests). The roof to the very strong Raleigh Sandstone beds was a variable mixture of well laminated shales and sandstones. The seam thickness in the area varied between 1.37 meters (54 inches) and 2.74 meters (108 inches) and the height from the floor of the seam to the base of the Raleigh Sandstones varied between 15.24 meters (50 feet) and 35.05 meters (115 feet). Examination of the iso-ratios indicates that at no place over the area did it appear that H_{max} was less than 5.8 and in general, comfortably exceeded that figure.

One further examination needs to be made of the possible effect on the longwall operations of very strong beds beyond height "H", but between the area to be worked and the surface. The possible effect referred to is that of periodic weights which are usually evident in very bad entry conditions and occasionally severe roof control problems between face supports and the coal face.

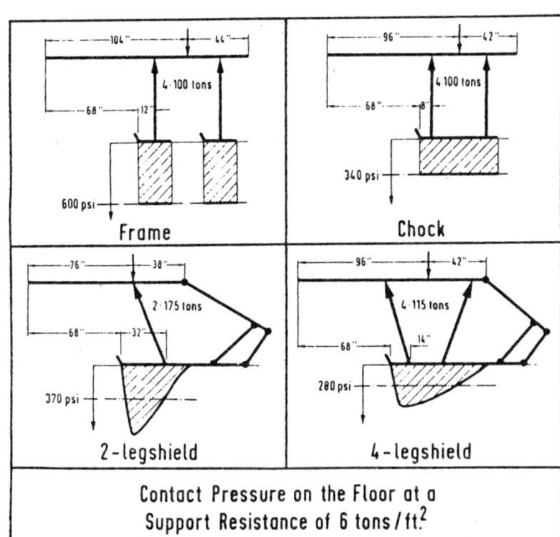

Figure 2: Example of floor pressure distribution with various support configurations. After Irresberger(8).

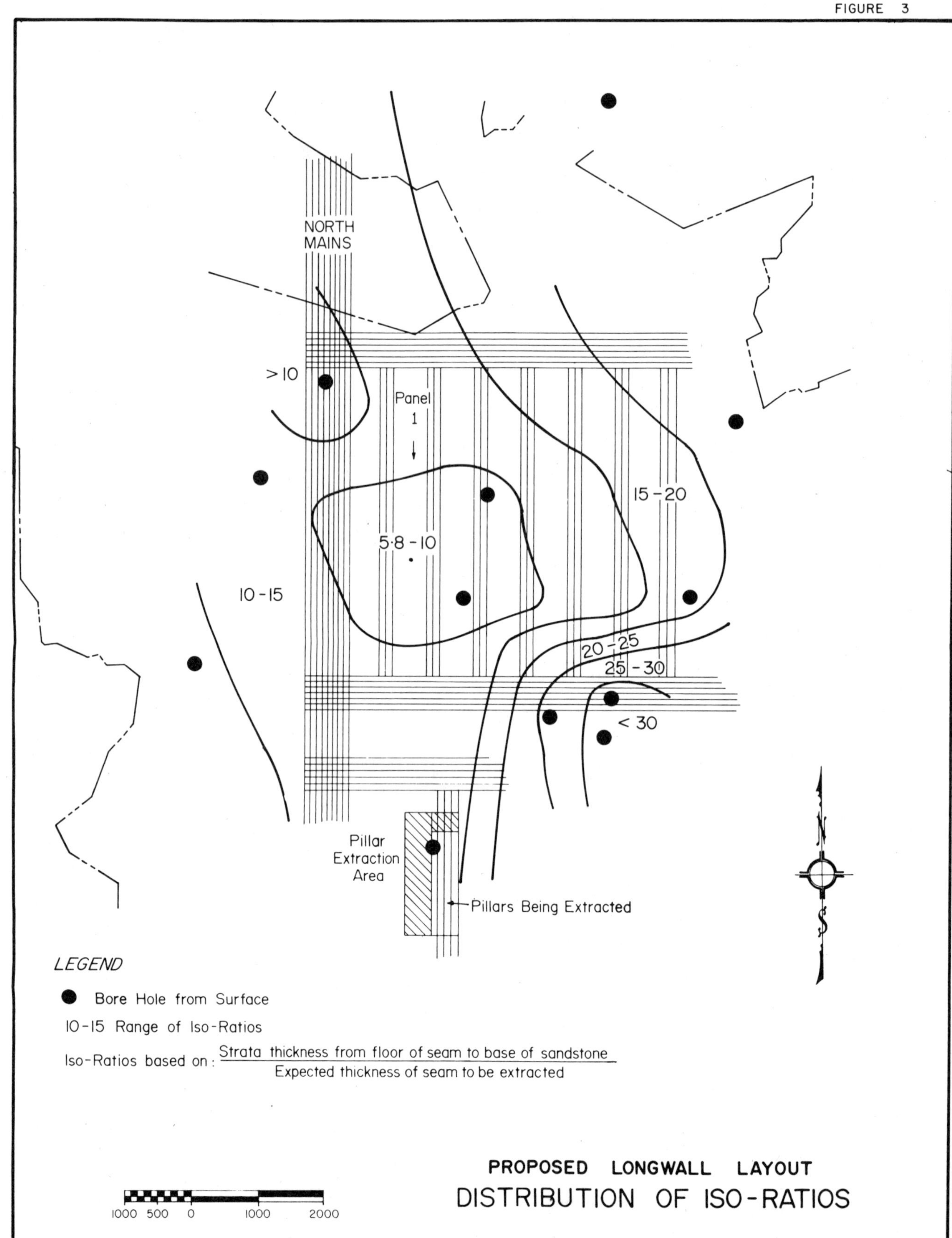

PROPOSED LONGWALL LAYOUT

DISTRIBUTION OF ISO-RATIOS

Kuti(9) makes the following statement when discussing tail entry problems:

"Quite apart from theories, it is a matter of statistical evidence that high side abutment loading develops over the tail entries starting with the second or third panel in those cases where a massive bed overlies the seam, but is separated from it by a friable rock less than seven times the mining height."

The author has personal experience of severe tailgate problems occurring on every third panel where conditions similar to those described by Kuti exist, although the precision of the "seven times mining height" has not proved to be significant.

Salamon and others(10) describe the problems of longwall working under a dolerite sill of 73 meters (240 feet) thick lying about 90 meters (300 feet) above the seam which was situated 230 meters (750 feet) below the surface. From surface leveling and borehole data it was determined that the sill [was capable of br]eaking and formulae were dev[eloped to determi]ne the face length neede[d to ensure settlin]g of the sill behind the [fa]ce and within the panel dimensions, thus avoiding a situation where a catastrophic failure of the sill could occur after a number of panels had been worked and the chain pillars between panels yielded under the accumulated load.

So far as is known, Salamon's work has not been translated into U.S. conditions, but there appears to be a strong connection between what was observed in South Africa and what has been experienced and reported on in the U.S.A. Referring to Figure 3 once more, the Raleigh Sandstone over the area occurred in two major layers, the lower being 33 to 55 meters (107 to 181 feet) thick with a compressive strength between 65.5 Mpa to 129.5 Mpa (9,500 to 18,000 p.s.i.) and the upper member 18 to 87 meters (60 to 285 feet) thick and with a compressive strength between 105.4 Mpa and 135.7 Mpa (15,300 p.s.i. and 19,700 p.s.i.) in major sections (up to 30 meters or 100 feet thick).

The mine operator was committed by development work already completed to a face length of approximately 500 feet, so that any decision to widen the panels would have resulted in considerable expense. Pillar extraction was being carried out in a section adjacent to the proposed longwall area and 100% recovery was being achieved. By the erection of surface monuments over a second and adjacent pillar area, it was established that a total surface subsidence approximating to 2/3 the extracted height of the seam was occurring after working; a certain indication that the Raleigh Sandstones were settling on the gob. It could not be deduced, however, with such a limited program, whether the sandstone was flexing elastically or breaking. Experience

to date has unfortunately not revealed whether tailgate problems are going to occur. Thin coal met late in the development phase caused the area to be abandoned as a longwall proposal.

It can be assumed that the presence of a very hard rock between the seam and the surface will cause problems of roadway stability due to periodic weighting. Solutions require a determination of the thickness and strength of the strong material, a comparison, where possible, of this data to other known situations, and the carrying out of subsidence surveys to establish strata behavioral patterns. Maximizing the face length to whatever the economics of the situation will stand (up to about 275 meters (900 feet), which appears to be the present limit on face conveyor length), may induce fracturing by increasing the unsupported span across the gob. Increasing chain pillar dimensions and leaving barriers of coal between predetermined panels sufficient to support the upper strata without collapse offers another potential solution. Future solutions would include either elimination of chain pillars in panel preparation or their total extraction during panel extraction, thus creating an area of gob of sufficient dimensions to ensure complete settlement of the whole strata column.

Conclusions

The reader will have come to one major conclusion by now, and that is that the consideration of the criteria available to him for longwall selection is almost totally an art with very little applied science. Many of the judgements required are almost entirely subjective with the only hard facts readily available to him being manufacturers quoted prices and whatever lithological data he can obtain from his own exploration.

Too little experience is available to him to enable more than an inspired guess to be made of likely productivity (and consequent profit or loss). Published data on performance seems to concentrate on the reporting of various output records with too little reporting of failures and the "ordinary" run-of-the-mill performances actually being achieved.

What can the operator, or the consultant do to determine the caveability of particular strata other than look at the compressive strength of the material? Where breakage of cantilevered layers is involved, tensile strength could be of more importance than compressive strength, but little or no published work exists to relate tensile strength to compressive strength and/or to caveability. The granular and matrix material which comprise the rock have also a significant bearing on the strength of the material and on its caving characteristics, but too little information of a practical nature exists for it to be of real value in a caveability evaluation.

Longwall appears to have a tremendous potential for good in the coal industry of the U.S.A. and conditions already exist (in the Western U.S.A.), where there seems to be no alternative to it, but given a free choice (i.e. in good conditions in Appalachia), how is the operator to be influenced by facts rather than "sales-talk?" The cost of longwall supports increases more rapidly than the designed yield load increases and in many cases operators are buying insurance against failure by the purchase of the most expensive types. In the absence of a better data base against which to measure requirements, who can blame them?

More data needs to be gathered, more research needs to be done, and it is suggested that perhaps the appropriate agencies of the U.S. Departments of Energy and Interior start by examining what is required as "Criteria for Longwall Selection", recognize the deficiencies in what is available, and thereafter, attempt to rectify those deficiencies.

In the meantime, until such research produces results, all that can be offered is that the criteria described in this paper be examined when longwall is being considered and that expert advice is solicited before a commitment is made.

References

1. "Two World Records in Longwall Production in 7 Months - Over 13,000 and Over 17,000 Tons in 24 Hours", James F. Hamlin, A.M.C. International Coal Show, Chicago, May 5-8, 1980.

2. "Support Load Requirements on Longwall Faces", A. H. Wilson, "The Mining Engineer", June 1975.

3. "Geologic Factors in Longwall Design", Johnson, Haycocks, Neall, Townsend. A. I. M. E. Annual Meeting, Atlanta, Georgia, March 1977.

4. "How to Choose a Powered Support System", Adam, Coeullet and Jacob Proceedings, 4th International Conference of Strata Control and Rock Mechanics, Columbia University, New York, May 4-8, 1964.

5. "Longwall Support Load Predictions from Geologic Information", L. V. Wade, S.M.E. Fall Meeting, Denver, Colorado, September 1976.

6. Private communication, L. V. Wade to author.

7. "In-Situ Tests of Bearing Capacity of Roof and Floor in Selected Bituminous Coal Mines", A. J. Barry and O. B. Nair, U.S. Bureau of Mines, R.K. 7406, 1970.

8. "Comparison of Longwall-Shield, Chock and Frame Supports", H. Irresberger, Coal Convention of A.M.C., Pittsburgh, Pa., May 1-4, 1977.

9. "Longwall versus Shortwall Systems", by J. Kuti, Coal Convention of A.M.C., Pittsburgh, Pa., May 4-7, 1975.

10. "Rock Mechanics Problems Associated with Longwall Trials in South Africa", M.D.G. Salamon, K. I. Oravecz and D. R. Hardman, Fifth International Strata Control Conference, London, 1972.

Chapter 7

SHORTWALL MINING - POTENTIAL AND PROBLEMS

Robert Stefanko, Ph.D.

Professor of Mining Engineering and Associate Dean
College of Earth and Mineral Sciences
The Pennsylvania State University
University Park, Pennsylvania 16802, USA

Shortwall mining represents a compromise between the room-and-pillar and longwall systems.

Figure 1 shows that the shortwall layout is very similar to that for longwall panels, the primary difference being that the shortwall is generally 150 to 200 feet wide while longwalls are 350 to 600 feet wide. The development of entries is divorced from the production of the walls with both systems. However, the depth of the web is considerably different in the shortwall system. Because a standard continuous miner is employed, a depth of 10 feet is normally mined on a single pass. This greater web thickness requires the use of a specially constructed support that can support the much longer cantilever bar that must extend over the working area. This is accomplished by not only increasing the length of the bar but also strengthening it structurally; however, the key to success lies in articulating the cantilever from the base and supporting it with a short hydraulic jack leg. This arrangement provides the additional strength required in the support. It should be apparent that a shortwall support is more costly than the type used for longwall.

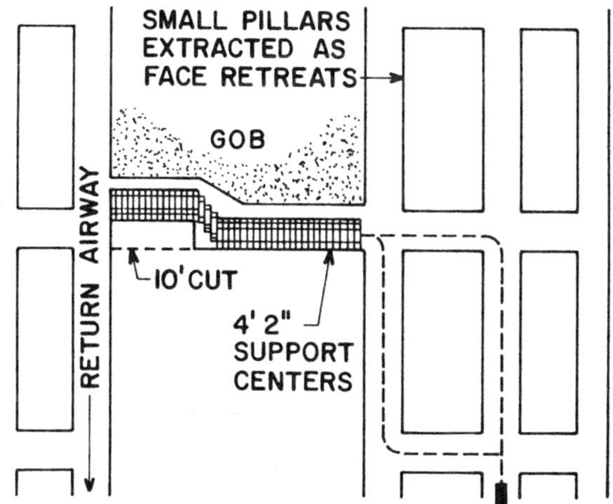

Figure 1. Shortwall Face with Shuttle Car Haulage.

Figure 2 shows the sequence of chock operations in shortwall mining. At the beginning of the miner pass, all supports are in line at the face. As soon as the miner passes a given chock, it is lowered and pulled up 5 feet to its spill plate which acts as an anchor. The spill plate is pin connected to its neighboring spill plates. The chock pressure is reset to activate the support again, and an extension in the cantilever is extended 4 feet to cover the exposed roof. This pattern of chock movement is repeated

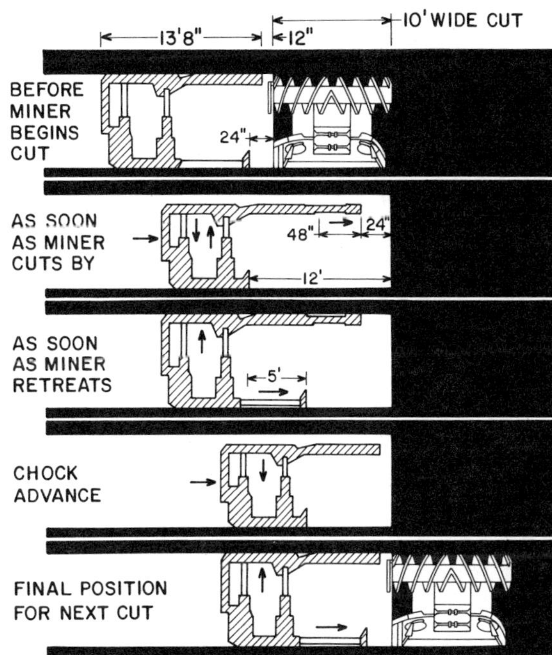

Figure 2. Sequence of Chock Operations on a Shortwall Face.

sequentially as the miner moves down the face until it finally reaches the tail entry, at which time it is backed out to the head entry. The spill plates are then hydraulically pushed forward. The chock is then moved up to the spill plate, the extension bar being automatically retracted. In the final step, the chock is reset and the spill plate advanced again. Once all chock movements have taken place sequentially the miner will be ready to begin the next pass, and the equipment will be again aligned as shown in the first step.

In Figure 3, a conveyor is shown in use between the mining machine and the panel belt. If a reliable conveyor system could be developed, this arrangement could result in a significant improvement in production. However, a satisfactory conveyor has not been demonstrated although several are presently being designed and tested.

In deciding whether or not to use the shortwall, a number of factors must be taken into consideration. One of the first is economics. If a company desires to change from a room-and-pillar system to some type of wall operation, a large capital investment is required without any assurance of the system's success. Therefore, rather than proceed directly to longwall with its larger capital outlay, a mine operator may turn first to shortwall, because only supports need to be added. The shuttle cars and

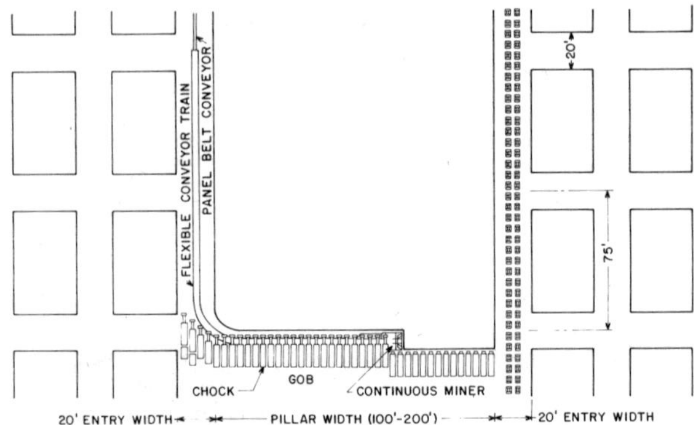

Figure 3.   Shortwall Face with Conveyor Haulage

continuous miners of the existing room-and-pillar operation will be used also in the shortwall operation.

Secondly, there may be a good technical reason for choosing shortwall over longwall. Many seams are relatively shallow and overlain by massive sandstones and limestones that cannot be easily broken. If a longwall face is established under such conditions, props of prohibitive size will be required to break the rock at the gob edge and relieve pressures on the face. There are mines where the top has never been broken under any mining conditions. Therefore, a longwall installation, dependent as it is on inducing immediate and complete falls, would, under such conditions, involve some very difficult if not impossible roof control. Rather than breaking and relieving the pressure, the roof might hang up and bend down into the gob, throwing tremendous destructive weight on the supports and making the system unworkable. However, with the shortwall method, it is possible to design the wall width so that it is less than the maximum allowable arch span, and the roof over the excavated area can be better carried on the face and side chain pillars. Therefore, from the technical standpoint, shortwall may be applicable where longwall is not.

A third condition favoring shortwall over longwall mining is related to geological or manmade situations. Since longwall faces generally are developed at a minimum width of 350 feet, a large undisturbed acreage is required for projecting longwall panels. However, if the coal seam is located in a gas-producing area, the seam will be penetrated by well casings which must be protected by a barrier of coal. Also, many seams have want areas where the coal has been replaced by rock. Under such discontinuous conditions, it becomes virtually impossible to lay out longwalls with any degree of continuity. Pressures become virtually uncontrollable if areas must be skipped since squeezes and bumps now become real possibilities. Therefore, it is much easier to establish shortwalls under such conditions. Since the maximum arch span to be supported is less, some of the problems of squeezes and bumps can be avoided.

A fourth condition favoring shortwall mining is the social incentive--improved health and safety. The shortwall system includes an umbrella of steel support over the working face similar to that in longwall mining, providing a great degree of safety from roof falls. In addition, the health hazards are greatly reduced in the shortwall method. In the longwall method, in which mining is bidirectional, some men will always be located on the downwind side of the mining machine. Because the shortwall method is unidirectional, no one ever need be on the downwind side of the machine; all face personnel work in fresh air that is dustfree.

Under the conditions just listed, high production rates and consistent output, can be best achieved with shortwall mining which provides the best compromise in technical, economic, and social factors.

Shortwall tail entry problems are similar to those encountered in longwall mining since the face and adjacent panel abutments are superposed at the tail entry. The unidirectional aspect of shortwall mining offers a real advantage over longwall, minimizing the support problem in the tail entry. As can be seen in Figures 1 and 2, the miner need not enter the tail entry; as soon as the bits penetrate the entry, a 10 foot pass has been completed and the machine is backed out. Therefore, the tail entry can be heavily cribbed since its only function is to carry return air. It also becomes apparent that any attempt to remove chain pillars must be carefully evaluated. With the tail entry full of cribs, it is not practical for the machine to proceed across the entry to attack the chain pillars between the tail entries even if the difficulty of supplying an additional haulage length is disregarded. Therefore, if chain pillars are to be pulled, those closest to the gob on the main gate side should be retreated. This can best be accomplished by using a pocket-and-wing method of attack from the center entry, Figure 1--an extraction technique that is more practical with shuttle car haulage. Where pillar removal is not the normal practice, and, where subsidence is to be prevented on the surface, these entry pillars must be designed according to barrier pillar formulas.

Figure 4 reveals that once a panel is complete and while the supports are being moved to the next retreat shortwall, the continuous miner and haulage can begin driving the next set of development entries. One operator claims that these entries can be driven so quickly that by the time the supports have been moved, maintained and installed, the next panel is ready to begin production. This minimized production losses during move time and also eliminates the need for satellite units to keep development ahead of mining.

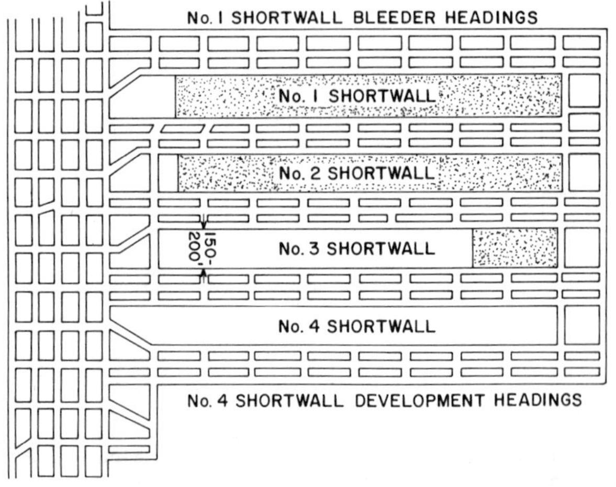

Figure 4.   One Method of Shortwall Mining.

However, at the end of the panel development, two
rooms must be driven to the right for installation of
the supports for the next panel along the outby rib of
the outby entry. This provides a bleeder entry to the
back and therefore is doubly useful. Unfortunately,
probably in the majority of mines it is not possible
to drive panel entries as rapidly as necessary to keep
up with shortwall support movement and, therefore,
panel development must precede shortwall mining as is
the general practice with longwall mining.

Most of the few shortwall units in this country
have been operating a relatively short time, and one
of these is a complete failure so far, so there is
limited statistical evidence on shortwall performance.
Unlike longwall output that averages 1000 tons per
shift, shortwalls have averaged less--600 to 700 tons
per shift. However, because the nonproductive move
time is greatly reduced in shortwall, the overall
average production for the two systems, including
move delays, are not that much different. This more
consistent tonnage results in some service savings.
A typical shortwall crew consists of 1 continuous
miner operator, 1 helper, 2 shuttle car operators, 1
prop mover, 1 mechanic, 1 general utility man, and
1 foreman. This is 1 to 2 fewer men than in the
longwall crew and is approximately the same as in
continuous mining in room-and-pillar systems.

II. Environment

Joseph Cervik
US Bureau of Mines
James A. Bloom
Y&O Coal Company

Chapter 8

LONGWALL VENTILATION PLANNING AT
JIM WALTER RESOURCES' NO. 3 MINE

John W. Stevenson
General Manager
Ventilation Department

Abstract. At Jim Walter Resources Alabama coal property, the first of twelve projected longwall mining systems started to operate in March, 1979, at the No. 3 Mine. The second longwall started to mine adjacent to the original workings in June, 1980. At the end of 1980, 2½ panels had been extracted using the same basic ventilation plan.

A single split of air ventilates each longwall panel with intake air traveling the headgate across the face to the tailgate where it is directed and controlled to a 1500 foot deep fan shaft specifically located to exhaust from the longwall area and the gobs associated with longwall extraction. The simplicity of the system makes it easy to understand, apply and control in a relatively gassy coal bed.

The No. 3 Mine is one of six underground coal mines operated in Jefferson/Tuscaloosa Counties in North Central Alabama by Jim Walter Resources, Inc. It is operating in the Blue Creek Coal Bed. Seam height is very variable but overall totals 60 inches thick. The coal bed was opened in 1972 by three vertical shafts, each 1300 feet deep -- the 22 foot diameter intake service shaft, the 20 foot diameter intake production shaft and the 20 foot diameter return fan shaft. Other shafts required since the original openings are a 16 foot diameter intake and a 16 foot diameter return shaft each raise drilled; a 20 foot diameter intake and two 20 foot diameter fan shafts. The last three shafts were sunk using conventional shaft sinking methods and techniques. The mine is ventilated with four Jeffrey Manufacturing Company 117 inch diameter fans operated in an exhausting configuration.

The original fan shaft has been upgraded and, at this location, we are now using a Jeffrey Manufacturing Company 8 HUA 117 fan operated at 900 RPM by a 2000 HP 4160 volt electric drive motor. This fan produces approximately 775,000 cfm air at a negative 10.0 inches water gauge pressure.

Alabama is the only coal mining state that has a law requiring the use of an auxiliary drive mechanism to provide 80% of the required mine air volume in the event of an electrical power failure. Consequently at this location, a Jeffrey 8HU 117 standby fan is powered by a 1000 Horsepower Caterpillar diesel engine.

At the second fan shaft site, a Jeffrey 8HUA 117 fan is operated at 720 RPM by a 1000 HP 4160 volt electric drive motor. A right angle speed reducer with a 1000 Horsepower Caterpillar diesel engine provides the auxiliary drive mechanism. This fan produces approximately 550,000 cfm air at a negative 7.5 inches water gauge pressure.

At the third fan shaft site, a Jeffrey 8HUA 117 fan is operated at 900 RPM by a 1250 Horsepower 4160 volt electric drive motor. A right angle speed reducer with a 1000 Horsepower Caterpillar diesel engine provides the auxiliary drive mechanism. This fan produces approximately 300,000 cfm air at a negative 8.5 inches water gauge pressure.

At the fourth fan shaft site which went on stream in December of 1980, a Jeffrey 8HU 117 fan is operated at 720 RPM by a 1000 Horsepower 4160 electric drive motor. Again a right angle speed reducer with a 1000 Horsepower Caterpillar diesel engine provides the auxiliary drive mechanism. This fan produces approximately 325,000 cfm air at a negative 7.0 inches water gauge pressure.

The total air volume required to ventilate 10 continuous miner development units and two retreating longwall faces is in the neighborhood of 1,950,000 cfm air. This mine operated under 1300 to 1600 feet of cover can liberate in excess of 12 million cubic feet of methane in 24 hours.

We are currently mining in our fourth longwall panel and have established two gob areas. The original longwall gob area is comprised of two and 1/4 extracted 3600 feet deep longwall panels or 140 Acres. The second gob area consists of less then 1/4 extracted 4000 feet deep longwall panel or 22 Acres. Longwall faces vary from 500 - 600 feet wide.

It is anticipated that a total of 12 longwall faces will be in operation by 1985 in Mines No. 3, 4, 5, and 7. Consequently, ventilation practices, particularly longwall ventilation practices, proven successful at No. 3 Mine will be incorporated into the design of the other mines.

There are two rather obvious features when looking at No. 3 Mine map in addition to the most obvious -- the faults crossing the mine. The first is the fan shaft site location at the back of the longwall gobs and the second is the almost continuous change in both the number of entries and the size of the pillars in the longwall development areas.

The fan shafts have been sited at the back of the gob to provide in effect flow-through or one-way ventilation. This has proven to be a very effective ventilation technique in that methane control is excellent and that methane delays on the longwall face are practically non-existant. This allows us to control the volume of air across the longwall face as well as the amount of air into the gob and the bleeder entry and return air system. Our MSHA approved Mine Ventilation System and Methane and Dust Control Plan specifies a minimum of 25,000 cfm air across the longwall face. With the cross section area under the shields, this means a velocity approximating 400 feet per minute which is excellent for control of airborn dust. At velocities less than 400 FPM experiments by others indicate that there is not sufficient air to dilute airborn dust and at velocities above 400 FPM, the air current begins to scalp dust particles from the moving conveyor. Although we attempt to control

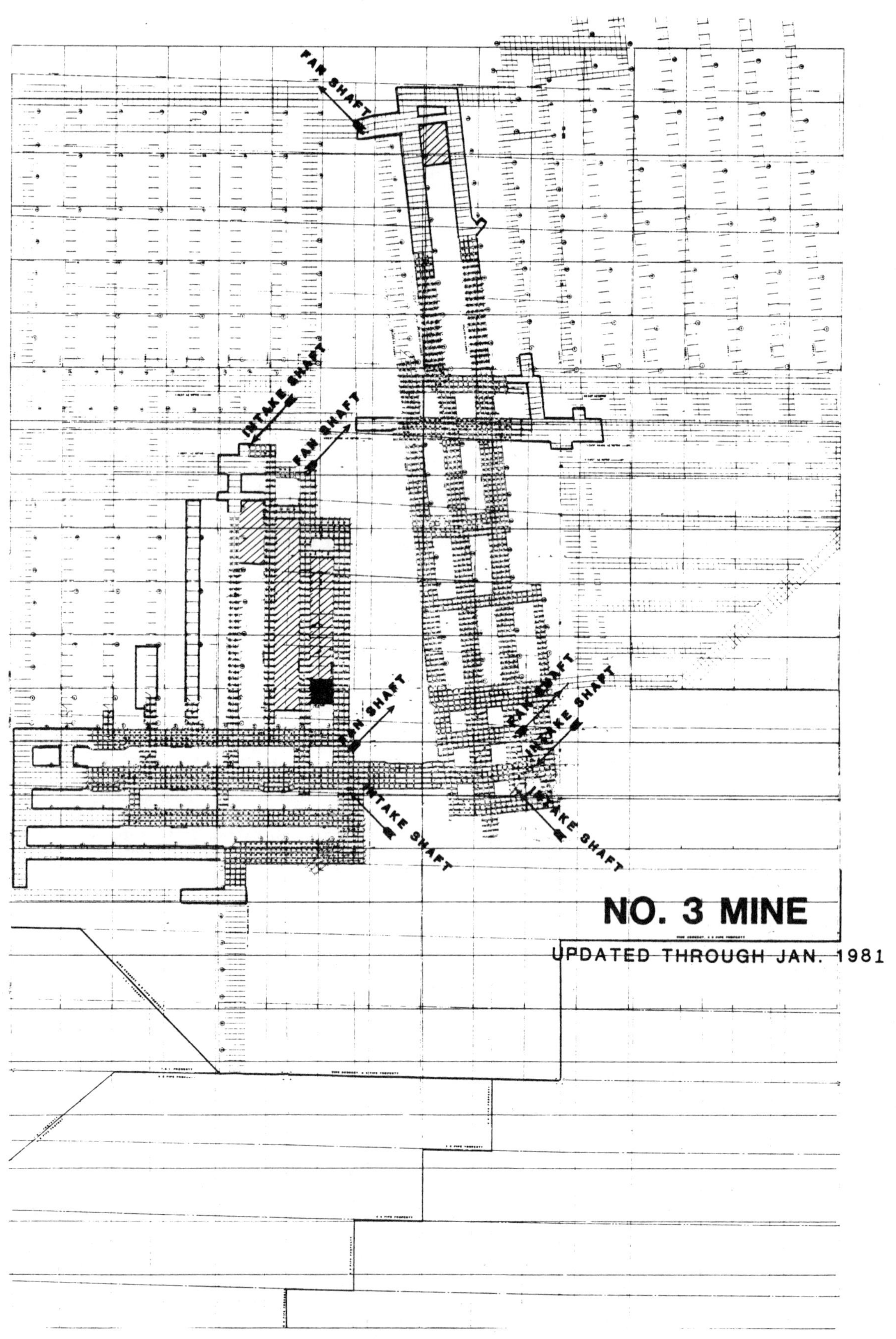
FAN SHAFT
INTAKE SHAFT
FAN SHAFT
FAN SHAFT
FAN SHAFT
INTAKE SHAFT
INTAKE SHAFT
INTAKE SHAFT
NO. 3 MINE
UPDATED THROUGH JAN. 1981

the air velocity for dust control, we have the option to increase velocity for methane control, if the need arises. A wetting agent is also used in the water spray system to assist in the control of dust. Dust control on the longwall face continues to be a problem.

The fan in the center of the property (the No. 2 Fan) is ventilating the No. 2 Longwall Face as well as maintaining a negative pressure on the larger gob area to direct airflow from the face into the gob and to the bottom of the fan shaft. This gives us good control of methane liberated in the gob, and by airflow pressure differential, gob gases migrate and channel with the ventilating air current through voids to the more negative pressure near the bottom of the fan shaft and away from the active longwall face. Additional ventilating air can also be directed through the entries into the bleeder entries and the return air courses to control gob methane liberation. At this shaft site, the fan is currently moving 300,000 cfm of air at a little more than 0.5% methane concentration.

By contrast, in the North area of the mine, the gob area is less than 1/4 longwall panel, but the methane liberation rate is higher requiring 50,000 cfm on the longwall face with a corresponding higher air velocity. In addition the newly created gob area appears to be liberating methane more profusely so that this area requires slightly more air -- 350,000 cfm -- to control the methane at the fan to about the same concentration. The suspected reason for the high methane release rate in the North area of the mine is the lack of abutment pressure on the coal beds at this time. This is partially confirmed by the fact that twice as much air is required to ventilate the longwall face in the North as in the West, to keep peak methane concentrations at the methane monitor near the tailgate below 1%. After front abutment pressure is exerted on the longwall face, crevices and cracks developed along clevage planes of the coal beds emit methane in both the Blue Creek and the Mary Lee Coal Beds at a lower but continuous rate to provide more effective dilution control of methane at lower air volumes. This appears to have occurred in the larger gob area in the West; and, we anticipate that this will soon occur in the North gob area. Based upon methane determinations at the mine fan, it also appears the methane rate of release per square foot of gob decreases with time.

The varied entry configuration in the longwall development section has been the result of efforts to create stable entry pillars under heavy cover balanced against the need to develop longwall panels at a high productivity rate together with a fast rate of advancement. The original design was to advance three entries on 100 foot entry centers by 120 foot crosscut centers.

Exhausting line brattice is used for face ventilation in twenty foot wide places with four foot maintained behind the line brattice. As the first longwall face retreated, it became apparent that abutment pressure on the supporting pillars caused them to yield and the roof to collapse and therefore it became prudent to modify entry development pillars. A two entry system with entry and crosscut centers on 200 foot spacing is being advanced as the headgate for the fourth longwall face using battery powered ram cars for haulage. Fifty horsepower auxiliary fans with rigid fiberglass 30 inch diameter exhaust tubing

is used for face ventilation. A petition for modification has been filed with MSHA to place the belt in the return air course with the results still pending.

Four entries are being developed for the fifth longwall face on varied centers. To protect head and tailgate entries, the outside blocks are on 145 foot entry and crosscut centers. To provide the fourth entry for fishtail face ventilation and to keep the solid outside rib as a return air course, the center entries are on 75 foot spacing with crosscuts likewise on 145 foot spacing.

This now appears to be the standard for longwall development entries after extraction of the second longwall panel. This plan provides a balance between strata control for head and tailgate stability during longwall extraction with ventilation for methane control during entry development. Consequently we feel this plan without Federal or State variances now optimizes continuous miner development productivity as well as the rate of advancement under heavy cover. This is at an unsatisfactory trade-off because of the reduced or lowered potential recovery of the total coal reserves of the property and that ultimately the use of two entry development systems will become increasingly popular.

All of our longwalls are ventilated with a single split of air flowing in the headgate and belt entries across the face and down the tailgate entry for several crosscuts. With stoppings partially removed, the return airflow flows through the crosscuts and reverses direction going into the gob in the centergate entry towards the fan shaft at the back of the gob. At the point where the longwall air enters the centergate it merges with either a return or intake air split. Because of the location of the fan, there is a ventilating pressure differential to induce airflow from the longwall face into and through the gob to the more negative ventilating pressure of the bottom of the fan shaft. As this air flows through the voids in the gob, it dilutes and directs methane generated in the gob away from the active longwall face towards the point of lower ventilating pressure at the bottom of the shaft.

Because of construction costs for deep shafts, large volume high pressure mine ventilating fans will become increasingly popular, together with methods to keep methane out of the ventilating airstream. Techniques and practices to successfully control and drill long horizontal holes in the coal bed in advance of mining, cross measure holes drilled above longwall gobs and piping systems and controls to safely transport methane out of the mine, dual purpose vertical drill holes to remove methane from the coalbed in advance of mining and later from above longwall gobs at commercial grade concentrations must all play a larger part in our planning of deep underground mines. Some of the above must precede coal extraction by years to be successful. However, it is becoming increasingly evident that in deep underground coal mines, control of methane by dilution with air alone is not the complete solution.

Chapter 9

METHANE CONTROL ON LONGWALLS—
EUROPEAN AND U.S. PRACTICES

Joseph Cervik

Supervisory Geophysicist
U.S. Bureau of Mines
Pittsburgh, Pennsylvania

INTRODUCTION

Common methods of controlling gob gas in U.S. mines are by means of ventilation of gob areas and gas drainage through surface boreholes. Costs of drilling surface gob holes increase as deeper coalbeds are mined. Where gob holes cannot be drilled because of topography, ventilation is the only means of control. Even then, sufficient air may not be available to dilute methane in the bleeder entries to acceptable levels. Thus, new methods of gob gas control need to be developed that are independent of surface topography, and of the mine ventilation system, and less costly than present systems.

Except for slight variations in procedures and equipment, Great Britain, West Germany, France, Belguim, Poland, and Czechoslovakia use a common methane control method that is different from the systems generally used in the United States. This Bureau of Mines paper reviews European mining methods, mining conditions, and methane control technology and discusses their applicability to U.S. mines.

ACKNOWLEDGMENTS

The material presented in this paper was obtained from discussions and observations in Poland of methane drainage practices and facilities and from a report prepared for the Bureau of Mines by the Central Mining Institute, Katowice, Poland (1).

COAL BASINS

European coal basins tend to be more tectonically disturbed than U.S. coal basins. Steeply dipping coalbeds are not exceptional and average mining depth is much greater than in the U.S. Stratigraphic columns (fig. 1) above the Pittsburgh (A), Pocahontas No. 3 (B), and above a coalbed in a coal basin in Poland (C) indicate a much closer spacing of coalbeds in both floor and roof strata in Poland. These closely spaced coalbeds, which characterize all European coal basins, present a serious methane problem during mining.

Coalbeds in the Appalachian and Illinois coal basins are flat tabular deposits. Mines are designed to operate in one coalbed. In Europe, mines are designed to operate in several coalbeds and levels (fig. 2). Crosscuts are driven through rock at several levels and because the rock strata and coalbeds dip up to 45 degrees, several coalbeds are intercepted by these crosscuts. Development entries are driven into the minable coalbeds from the crosscuts. A shaft may intercept as many as 205 coalbeds of which 10 are mined in the interval between 200 and 710 m (650 to 2330 ft) or 55 coalbeds are intercepted of which 18 are mined to a depth of 1000 m (3300 ft).

These minable coalbeds range in thickness from 1 to 3.2 m (3.3 to 10.5 ft). Because many coalbeds are exploited simultaneously, Polish collieries have developed into huge complexes. Commonly, total employment reaches 8000 with 5200 employed underground.

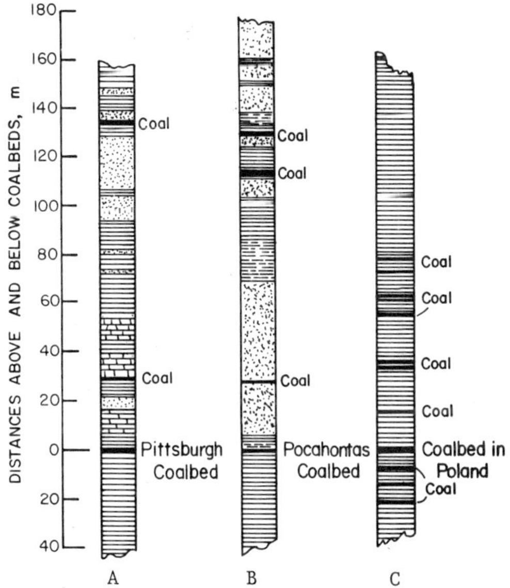

FIGURE 1. Stratigraphic columns. A, Pittsburgh coalbed; B, Pocahontas No. 3 coalbed; C, coalbed in Polish coal basin.

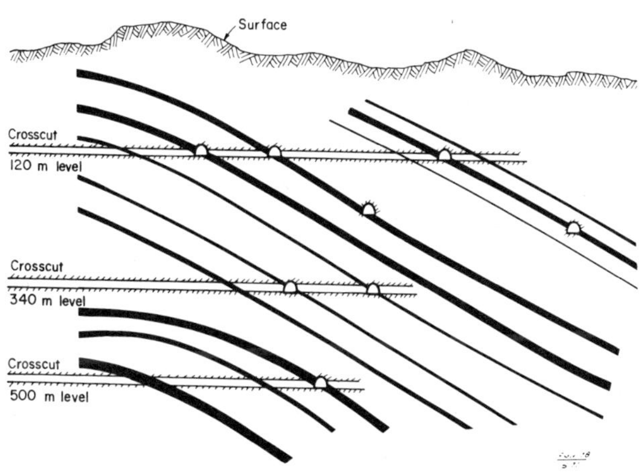

FIGURE 2. European coal basin.

Coal and methane outbursts are not a common occurrence in the United States. In the Soviet Union, Hungary, France, Poland, West Germany, and Czechoslovakia, outbursts of methane, coal, and rock are a serious problem demanding special attention. In Poland, mines are categorized not only by their methane hazard potential but also by their coal and rock outburst hazard potential. Ranking of a coalbed for its coal or rock outburst hazard is based on the tonnage of expelled material and ranking of methane hazard is based on in situ gas content. In West Germany, a methane content greater than 9 m^3/ton (318 ft^3/ton) of pure coal substance places the coalbed in a gas and rock outburst hazard category. If the same criterion is applied in the United States, the Beckley at depths between 226 to 302 m (740 to 990 ft), Pocahontas No. 3 (418 to 640 m) (1370 to 2100 ft), and Mary Lee (335 to 670 m) (1100 to 2200 ft) would fall into an outburst category.

METHANE DRAINAGE PROCEDURES

Longwall is the predominant mining system in Europe and Great Britain. Both retreating and advancing longwalls are used and the proportion varies. For example, 7 percent of Great Britain's longwalls are retreating systems (2) and 27 percent in West Germany (3).

In the United States, multiple entries are driven to develop longwall panels. In European coal basins, a single entry is used (fig. 3A). If severe methane problems are encountered in a European coal mine, a double entry is driven on one side of the panel (fig. 3B). Entries are supported with steel arches that extend into the roof strata. The distance between two arches is about 1 m (3.3 ft).

The first large-scale methane drainage and underground piping systems were introduced in West Germany in 1943 (4). After World War II, faster mining rates owing to mechanization and the mining of deeper and more gassy coalbeds forced the application of methane drainage throughout Europe. These methane drainage schemes are classified according to the development stage of the mine as follows:

(a) Advance degasification by surface boreholes.

(b) Advance degasfication of development workings.

(c) Degasification during longwall operations.

(d) Degasification of gobs.

At present, degasification during longwall operations is the most important technique applied throughout Europe.

Advance Degasification

Advance degasification by surface boreholes and advance degasification of development workings are techniques used extensively in Czechoslovakia and to a lesser degree in Poland. Both techniques are limited to depths of less than about 250 m (820 ft). At greater depths, fracture permeability of the coalbeds and surrounding sandstone reservoirs is much less, making these two techniques ineffective.

In Czechoslovakia, surface boreholes were used extensively to degasify shallow coalbeds in four coalfields. Over a 15-year period, 2.2 x 10^{9}m^3 (78 x 10^{9}ft^3) of methane was produced.

During the early stage of development of a mine, crosscuts are driven through rock to intercept minable coalbeds (fig. 2). In Poland and Czechoslovakia, the rock strata and coalbeds act as gas reservoirs, which presents a serious methane problem. In crosscuts, methane drainage holes are drilled routinely in advance of the face from niches driven into the entry walls (fig. 4). When a minable coalbed is intercepted, drilling of methane drainage holes in advance of the face is continued when driving development entries in coal. In addition to holes drilled

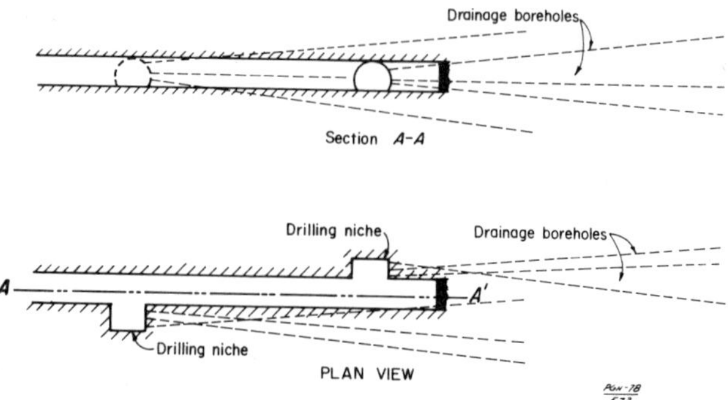

FIGURE 4. Advance degasification in crosscuts.

in advance of the face, drainage holes are drilled fanwise from the walls of the development entry into the overlying and underlying rock strata and coalbeds (fig. 5). The fan drilling pattern is repeated every 50 m (165 ft). These two degasification techniques are used in parallel and constitute the basic

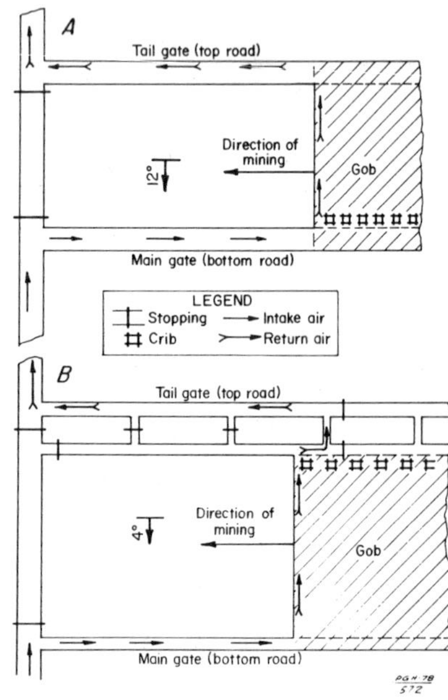

FIGURE 3. European mining plan. A, single entry
B, double entry.

methane drainage methods during the first phase of development of shallow mines.

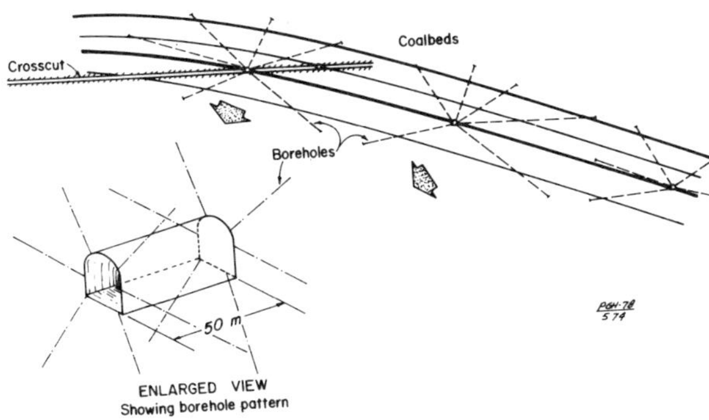

Figure 5. Advance degasification in development entries in coal.

Degasification During Longwall Operations

Below the depth of 250 m (820 ft), drainage holes drilled into the mined coalbed or into overlying and underlying strata and coalbeds are not effective because permeability is extremely low. However, coal extraction during longwall operations causes the overlying and underlying strata to become relaxed and to fracture. This relaxation and fracturing makes the coalbeds permeable, and they become the main source beds for methane, which enters the mine opening (fig. 6). The relaxed zone extends from a

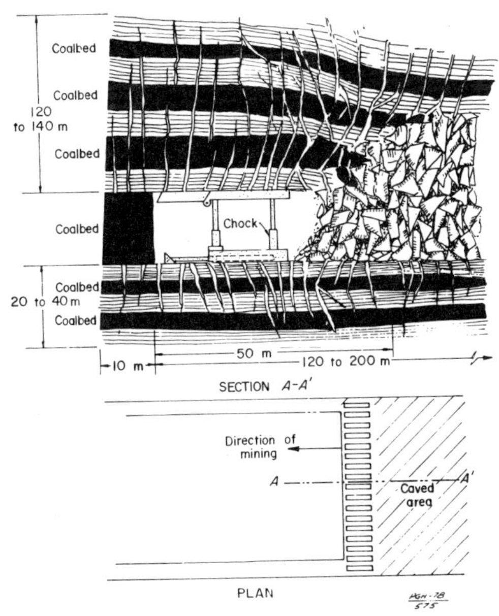

FIGURE 6. Fracture zone in roof and floor strata.

from a distance of 10 m (33 ft) inby the face to 120 to 200 m (394 to 656 ft) into the gob. Beyond this distance, the strata have settled, tightened,

and consequently gas flow subsides. The most intense methane flows occur 10 m inby the face to 50 m (164 ft) into the gob. In some mining systems, methane drainage pipelines are maintained in the gob for 50 m. Overlying coalbeds are affected by mining for distances of 120 to 140 m (394 to 459 ft), and underlying coalbeds for distances 20 to 40 m (66 to 131 ft)

Best control of methane is obtained on an advancing longwall (as the roof strata collapse) because then methane drainage is possible along the whole length of the relaxed zone (fig. 7). The retreating

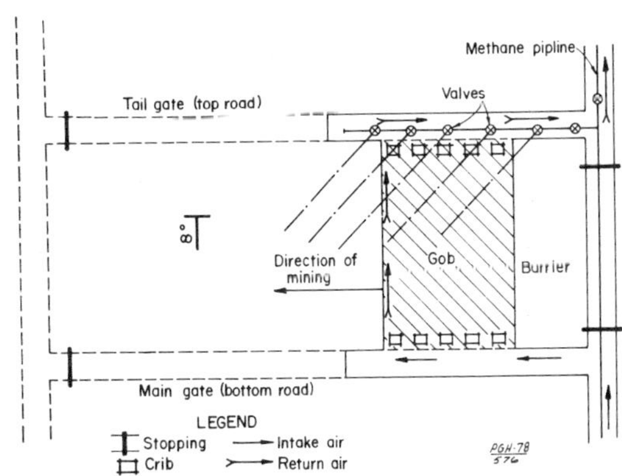

FIGURE 7. Advancing longwall with roof caving.

longwall with full caving is less effective (fig. 8). Methane cannot be drained from the first 50 m of gob because the area has collapsed. If acceptable

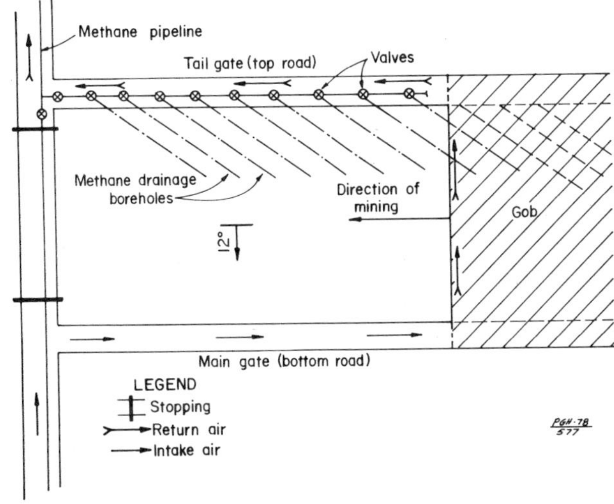

FIGURE 8. Retreating longwall with full caving.

levels of methane cannot be maintained, a double entry is driven so that methane entering the mine opening along the first 50 m of gob can be contolled (fig. 9).

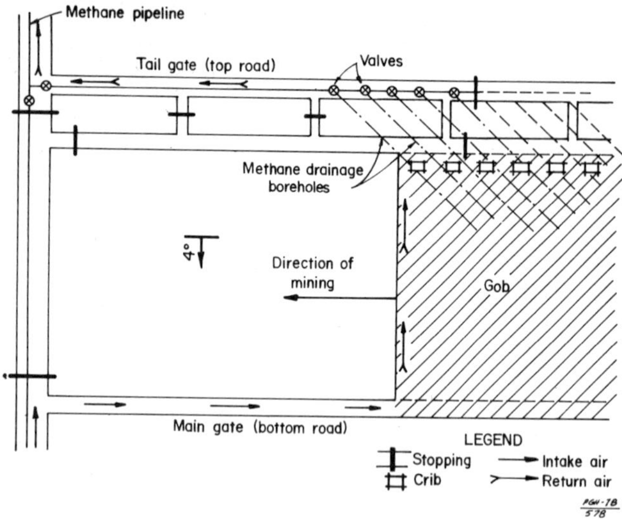

FIGURE 9.   Retreating longwall with double entry.

Generally, the methane drainage holes are drilled toward the principal methane source from the top road or return side of the longwall (fig. 10A); drilling from the bottom road or intake side is permitted in extremely gassy workings (fig. 10B).

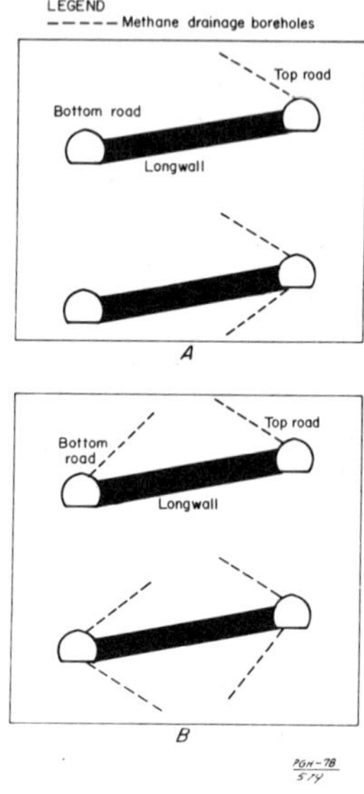

FIGURE 10.   Drilling patterns. A, gassy conditions
B, very gassy conditions.

Borehole length, which ranges from 20 to 150 m (66 to 492 ft), depends upon geological and mining conditions. Borehole diameters are variable. In West Germany, the trend is to drill fewer large-diameter holes (80 to 115 mm) (3 to 4.5 in). In Poland, hole diameters are smaller (48 to 75 mm) (1.9 to 3 in) but more holes are drilled. Small-diameter holes are easier to drill in hard rock.

Degasification of Gobs

Gobs are a source of methane that must be controlled after the longwall panel has been mined. Sealing is a common practice in European coal basins (fig. 11). These seals are 1 to 3 m (3 to 10 ft) wide and filled with a grout mixture of sand and

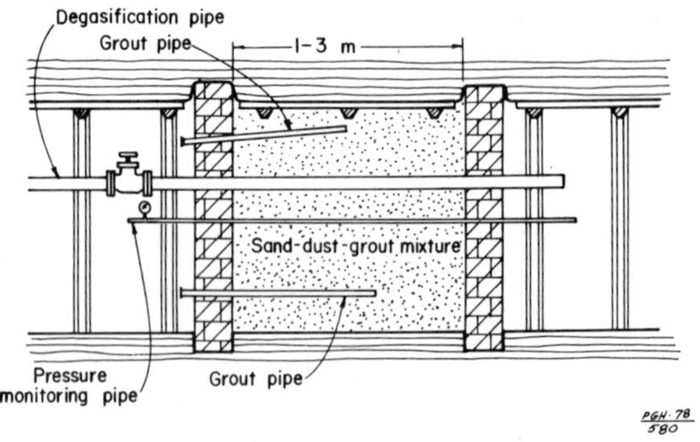

FIGURE 11.   Gob seal.

fines. Gob gas is drawn off through a pipeline that extends about 10 m (33 ft) behind the seal. Gas pressure behind the seal is monitored and controlled. Methane drainage from the gob is stopped when methane concentrations fall below 30 percent. All drainage seals are located in top roads of gobs.

In Poland, gob degasification is achieved not only from behind seals but also through drainage boreholes drilled over protective pillars at the ends of longwalls (fig. 12). These holes intercept

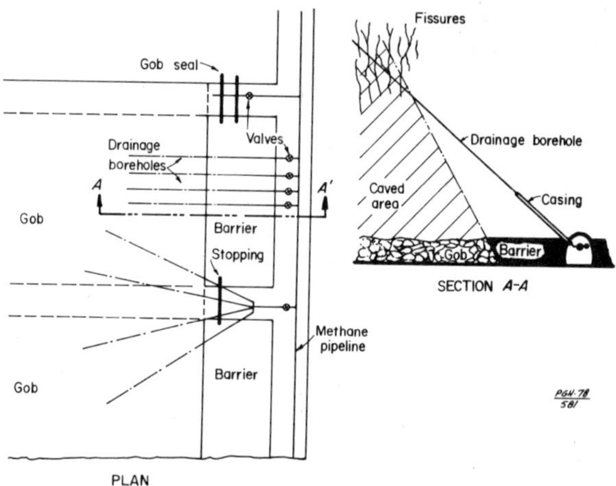

FIGURE 12.   Gob drainage from behind seal and
through drainage boreholes.

the fractured zone over the gob. The methane content of the gas mixture (62 to 93 percent) captured by this method is much greater than the mixture removed from behind seals (25 to 62 percent).

UNDERGROUND PIPELINES

Except in isolated cases where drainage from boreholes is diluted to acceptable levels underground, and controlled by ventilation, drainage holes are connected to an underground system of steel pipelines suspended from the steel arches. The pipelines are installed in return air entries, although regulations do not forbid installation in intake air entries.

Pipelines are tested section by section when installed. The volume of the test section is about 100 m^3 (3500 ft^3). A pressure of 2 N/m^2 (2 atm) must be maintained for 3 hours, and leakage must be less than 60 mm (2.4 in) water gage in 1 hour.

Pipeline diameters are determined by the gassiness of the mine. In Poland and Czechoslovakia, 400-mm-diameter (16-in) pipelines are used on the surface and in shafts. Diameters do not exceed 300 mm (12 in) underground. In West Germany, and France, 600-mm-diameter (24-in) pipelines are used underground to minimize friction. If pipeline capacity is inadequate for the gassiness of the mine, a second pipeline is installed parallel to the existing line.

PUMPING STATIONS

All underground pipelines throughout Europe are operated under a negative pressure, which is produced by rotating water ring pumps, rotary compressors with sliding blades, and piston compressors. Pumping stations are located underground or on the surface.

In Poland, both water ring pumps and rotary compressors with sliding blades are used. Capacities of water ring pumps range from 25 to 40 m^3/min (900 to 1400 ft^3/min). Exhaust pressure is about 0.3 N/m^2 (0.3 atm), which limits gas transport to short distances. These pumps are not sensitive to dust, thus filtering of incoming gas mixtures is unnecessary. Where higher exhaust pressures (1–1.2 N/m^2) (1–1.2 atm) are required to transport gas mixtures over greater distances, rotary compressors with sliding blades are used. Pump capacities are 30 to 60 m^3/min (1050 to 2100 ft^3/min).

Regardless of the type of pump used, a negative pressure of about 5000 mm (200 in) water gage is maintained on the intake to the pump to ensure a negative pressure of about 1000 mm (40 in) water gage at the most distant drainage borehole, and 100 to 200 mm (4 to 8) water gage where methane is being removed from behind the gob seals.

The pumping stations are fully automated and manned 24 hours daily. The gas by-passes the station and is automatically released to the atmosphere if the methane concentration of the incoming mixture falls below 30 percent or if a breakdown occurs. All pipelines heading into and away from the station are equipped with flame arresters. The following are recorded continuously: Methane flow rate, methane and oxygen content of the gas mixture, temperatures and pressures of the gas mixtures on the intake and exhaust sides of the pump, and calorific value of the gas mixture.

In Poland and Czechoslovakia, 90 percent or more of the gas drained is utilized, and 54 percent in West Germany (table 1). In Poland, the steel industry is the primary consumer; the remainder is used at the collieries for steam and power generation, and for driers in coal preparation plants.

TABLE 1.   <u>Degasification data for 1975</u>

Location	Methane drained, $10^6 m^3$ $(10^6 ft^3)$	Utilized pct	No. of boreholes	Total length drilled, km (miles)
Poland	207 (7 309)	90	2 802	212 (132)
Czechoslovakia	198 (6 992)	98	No data	260 (162)
West Germany	614 (21 081)	54	3 916	188 (117)

Methane drainage is a fully integrated phase of mining throughout Europe and is essential to maintain production. The length of drainage holes drilled in 1975 averaged 220 km (137 mi) for Poland, Czechoslovakia, and West Germany (table 1). In an area containing six gassy collieries in Poland, about 270 km (168 mi) of underground pipelines have been installed. This network of pipelines is maintained by a Mine Degasification Establishment, which is a separate organization or "company" from the colliery administration. The Establishment is responsible for all phases of methane drainage, including pipeline installation and maintenance, borehole drilling, pumping facilities, and sale of the gas mixture, which averages 60 percent methane. Only one advancing longwall with an integrated methane drainage program is in operation in the United States (5). However, no information has been published on methane flows and effectiveness of the holes drilled into the overlying strata.

APPLICATION TO U.S. MINING PRACTICES

Degasification of a European longwall gob requires the construction of four seals, or at most six seals when a double entry has been driven on one side of the longwall. In the United States, the number of seals required increases dramatically because of our mulitple entries around longwall panels (fig. 13).

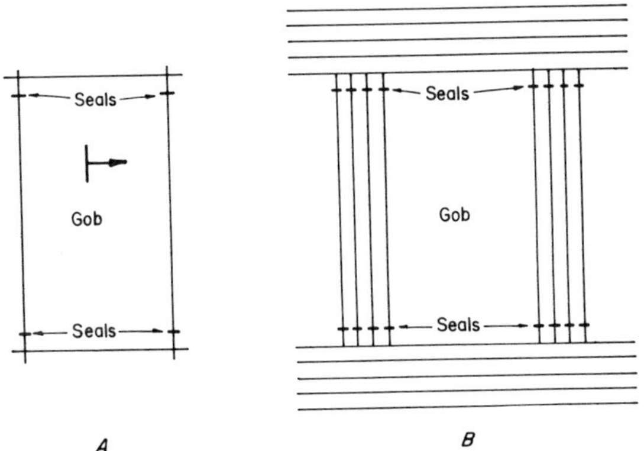

FIGURE 13. Gob sealing. A, Europe; B, USA.

Drainage holes drilled from entries diagonally over the protective pillars in our present mining system may help to maintain acceptable methane levels in bleeder entries. However, the area of influence of these boreholes may not extend more than a few hundred feet along the axis of the longwall.

The Bureau has initiated a program to investigate the application of European methane control technology to U.S. longwall mines. Studies have started to determine conditions under which large volumes of methane can be removed through underground drainage holes. The general application of European technology to U.S. mining conditions must still be determined.

REFERENCES

1. Matuszewski, J., and Sikore, W., "Report on the Technology of Degasification of Mines in Poland and Other European Countries," USBM Project No. 14-01-0001-1447, Jan. 1978.

2. Hamilton, R. J., and Frank, A. G., "Dust Control for Coal Face Machines," _Proceedings_ of the Conference on Technical Measures of Dust Prevention and Suppression in Mines, October 11-13, 1972, pp. 409-423.

3. Becker, H., "Seam Infusion in High-Speed Advance Working," _Proceedings_ of the Conference of Technique Measures of Dust Prevention and Suppression in Mines, October 11-13, 1972, pp. 273-281.

4. Bromilow, J. G., and Jones, J. H., "Drainage and Utilization of Fire Damp," _Colliery Engineering_, Vol. 32, No. 6, June 1955, pp. 222-232.

5. Reeves, J. A. Jr., "Advancing Longwall Mining at Mid-Continent," _American Mining Congress Journal_, Vol., 64, No. 7, Jul. 1978, pp. 25-29.

Chapter 10

METHANE CONTROL FOR LONGWALL GOBS

Pramod C. Thakur

Conoco Inc.

INTRODUCTION

Methane is contained under pressure within the fractures and adsorbed on the surface of the coal seams and adjacent strata. It is released into the mine atmosphere during mining of the seam creating explosive gas-air mixtures. Large volumes of air, sometimes as high as 20 tons of air for each ton of coal mined, are circulated constantly to dilute and carry away methane from coal mines. Removal of methane from solid coal, wherever high methane emissions occur, has been discussed by Spindler and Poundstone (1960) and Thakur and Davis (1977). Methane control in longwall gobs is much more complicated. As longwall mining progresses, the overburden continually subsides, and the immediate roof caves behind the roof supports. Even the underlying strata heave and open up communication channels feeding gas to the longwall gobs. The pressure of gas in the underlying and overlying strata is usually higher than the ambient air pressure in the mine. Thus the mine workings become natural pressure sinks, into which gas flows from the entire disturbed zone. Winter (1975) defines this zone as the "gas emission space". Depending on the number of gas-bearing zones in the gas emission space and their gas contents, the total methane emission from longwall gobs could vary from a few to more than one hundred cubic meters of gas per ton of mined coal. Hence the ventilation of longwall faces demands a large quantity of air. This ventilation need is further enhanced by high air losses specially on caving faces (with no stowing of the gobs). The old system of longwall ventilation, the 'U' pattern, where all the air was brought down one gate road and exited through the other gate road, loses large quantities of air through gobs giving a high methane concentration at the return end of the face. The 'Y' ventilation pattern where intake air is brought down both head and tail gate roads relieves the situation to some extent but, on most longwalls, some kind of methane control is still needed.

With the onset of longwall mining in Europe several decades ago, the initial work on methane control on longwall has naturally been done there. The basic principle of methane control has always been, and still remains, some means of "bypassing methane from the gas emission space without letting it mix with the mine air". In most cases the bypassing mechanism is either a strategically located borehole or a roadway. A successful methane control program depends on the following basic premises:

a. Determination of the geometry of the gas emission space and location of gas bearing horizons therein.
b. Estimation of the rate of methane influx into the longwall gob.
c. A scheme to bypass the gas in the most economic and efficient manner and thus prevent it from entering the mine atmosphere.

In successful methane drainage programs a high proportion, usually between 50 and 70 percent, of the total gas emissions in a working district is removed before it can enter the mine airways. Advantages of methane drainage can be summarized as:

1. Generally reduced methane levels in the mine leading to increased safety and productivity.
2. Reduced air requirements and corresponding savings in ventilation horsepower.
3. Faster rate of advance for development headings and economy in the size of headings.
4. Possible use of mine gas as an additional source of fuel.

The purpose of this paper is to briefly review the geometry of gas emission space and methane influx therein and discuss more successful European and U.S. longwall gob gas control techniques.

THE GAS EMISSION SPACE

Figure 1 shows the vertical extents of gas emission space with respect to the mine working according to various authors. Calculations of gas influx are based on the concept that there are finite limits for the gas emission zones above and below the mine workings. The smallest range of gas emission space is given by Gunther (1967) in the roof at 100 m. and by Lidin (1961) in the floor at 20 meters. The greatest range of gas emission space is given in the roof at 160 m by Lidin (1961) and in the floor at 100 m by Gunther (1967). Winter (1975) reviews the findings of the various authors thoroughly in a recent article. Based on his conclusions, there are no safe depth between the face area and the surface even if the depth of the underground working is 600 to 800 m. Thus there is apparently no finite upper limit for the gas emission space. Winter also observed heavings of 130 to 180 mm at depths of 130 to 170 meters below the mined face. This amount of heaving will create sufficient improvement in permeability for a substantial gas influx. By far the largest amount of gas encountered in longwall gobs originates from coal seams overlying and underlying the mined area. In some cases, gas produced by the comminution of coal and normal emissions from the seam being worked may also be significant.

Figure 1 also shows various estimates of the amount of gas emitted by underlying and overlying gas bearing strata. Gunther (1967) assumes that overlying seams lose all gas to the gob while the loss of gas from underlying seams varies linearly with distance from the mined seam. Lidin (1961) assumes linear relationship between the amount of gas emitted and distance for both overlying and underlying strata. Other authors show exponential and power law relationships between the percentage of total gas emitted and the distance between the gas bearing zones and the mine workings. In general, the percentage of gas lost by the gas bearing zones reduces with increasing distance from mine workings. The rate of methane flow in the gob also appears to be proportional to the rate of face advance. Hence, methane emission from longwall gob is much higher today than it was thirty years ago.

Factors which govern the rate of gas emission in the longwall gobs can be summarized as follows:

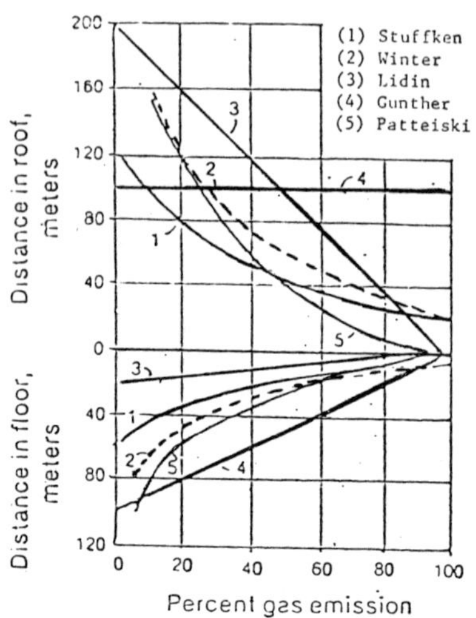

Fig. 1. Boundaries of the gas emission space.

1. The density and proximity of gas bearing zones in the gas emission space.
2. The gas content of the coal seam and other gas bearing horizons.
3. The extent to which overlying and underlying coal seams have been mined.
4. The rate of coal face advance.
5. The presence of geological disturbances and leftover coal pillars.
6. The subsidence characteristics of the area which determines the permeability of the gas emission space.

Obviously, there can be considerable variations in the gas flows into the longwall gobs because of varying combination of above factors. Experience and data obtained in one geographical area, therefore, can be transferred only qualitatively to other areas. No rigorous mathematical derivation of gas flow in gobs appears to have been done yet. This remains an important area for future study and research at present.

EUROPEAN GOB DEGASIFICATION METHODS

European coal seams are generally steeply inclined, tectonically disturbed and seated deeper than U.S. coal seams. Several coal seams are deposited in each basin and, typically, worked simultaneously. East European coal seams are not only gassy but also prone to instantaneous outbursts. Longwall mining system is the most common method of winning coal. Both advancing and retreating longwalls are employed in varying proportions. In view of the greater depth, the seam permeability is generally less than one millidarcy. Most of the methane emission takes place in the gob

areas following mining operations and strata movements leading to tremendous improvements in permeability. The front abutment is typically 10 to 15 m inby of solid coal for most longwall faces. The back abutment is 100 to 200 m from the face. The relaxed zone lying between these two abutments exudes methane. According to Cervik (1979), the most intense methane flow occurs in the zone 10 m inby of coal face to 50 m into the gob. Outside these abutments, the strata gets fairly consolidated and methane flow is curtailed substantially.

European methane control techniques can be broadly classified in the following three groups:

1. The packed cavity method and its variants
2. The cross-measure borehole method
3. The superjacent method.

Each of these techniques is discussed below separately. It has not been possible to include the regional variations but the broad principles of each method are sufficiently illustrated.

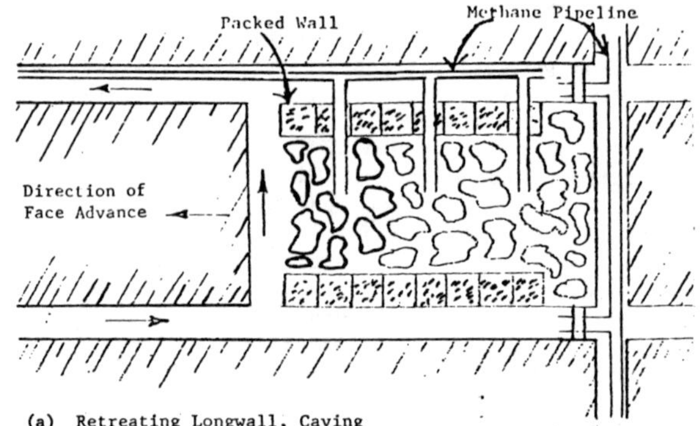

Fig. 2a. Methane drainage by packed cavity method and its variants.

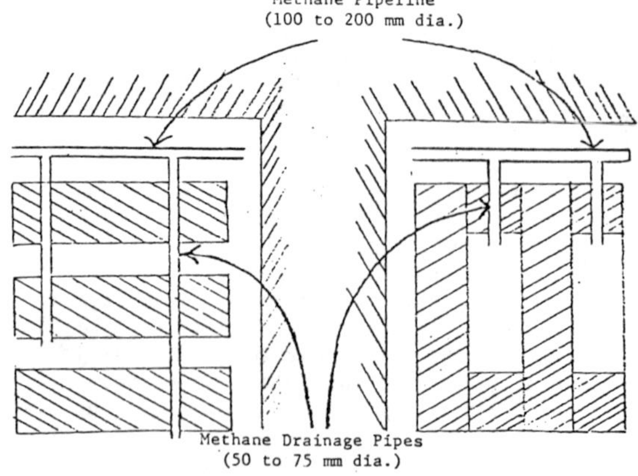

Fig. 2b. Partially filled longwall gob.

Table 1. Methane Capture Ratios for Packed Cavity Methods
(After Lidin, 1961)

Method of Mining	Method of Gob Stowing	Methane Capture Ratio (Percent)
Longwall Advancing	Caving	20 to 40
Longwall Advancing	Partial Filling	30 to 50
Room & Pillar	Complete Filling	60 to 80

1. <u>Packed Cavity Method and its Variants</u>:

Early methods of methane control consisted of simply isolating the worked out area in the mine using pack walls, partial or complete stowing, plastic sheets or massive stoppings. A network of pipeline which passed through these isolation barriers was laid in the gob and methane was drained using vacuum pumps. Figures 2 (a), (b) and (c) show typical layouts for caving, partially stowed and fully stowed longwall gobs. Lidin (1961) has reviewed several variants of this technique. Methane capture ratios quoted by him are shown in Table 1. The ratio generally seems to improve from caving (20-40%) to fully stowed longwall gobs (60-80%). Figure 2 (a) shows a caving, retreat longwall face. The gate roads are protected by a pack wall against the gob. Pipe lines are laid through the pack wall to reach nearly the center line of the gob and are manifolded to a larger diameter pipe in the gate road. Figure 2 (c) shows a fully stowed longwall gob where cavities are purposely left between alternate packs. The overlying strata in that area cracks and provides a channel for gas to flow into these packed cavities. Pipe lines are laid to connect the cavity with methane drainage mains. Methane extraction is usually done under suction. The technique is also known as "Roschen" method of methane drainage.

2. <u>Cross-Measure Borehole Method</u>:

This is by far the most popular method of methane control on European longwall faces. Figure 3 shows a typical layout for a retreating longwall face. Boreholes, 50 to 100 mm in diameter, are drilled from the top gate to a depth of 20 to 150 m. The angle of these boreholes with respect to horizon varies from 20 to 50 degrees while the axis of the borehole is inclined to the longwall axis at 15 to 30 degrees. At least one hole in the roof is drilled at each site but several boreholes in roof and floor can be drilled at varying inclinations depending on the degree of gassiness. Drill sites are typically 25 meters apart. These holes are then manifolded to a larger pipeline system and gas is withdrawn using a vacuum pump. Vacuum pressures applied vary from 100 mm to 3000 mm of water gauge. The amount of methane captured by the drainage system expressed as a percentage of total methane emission in the section varies from 50 to 90 percent. Some typical data from British mines are given by Kimmins (1971) and shown in Table II. The technique is generally more successful for advancing longwall panel than it is for retreat faces. The flow from individual boreholes is typically 0.5 m^3/min but occasionally it can go up to 3 m^3/min for deeper holes. Sealing of the surface casing is very important and is usually done with quick-setting cement. Sometimes, a liner (a pipe of smaller diameter than the borehole) is inserted in the borehole and sealed at the mouth to preserve the pro-

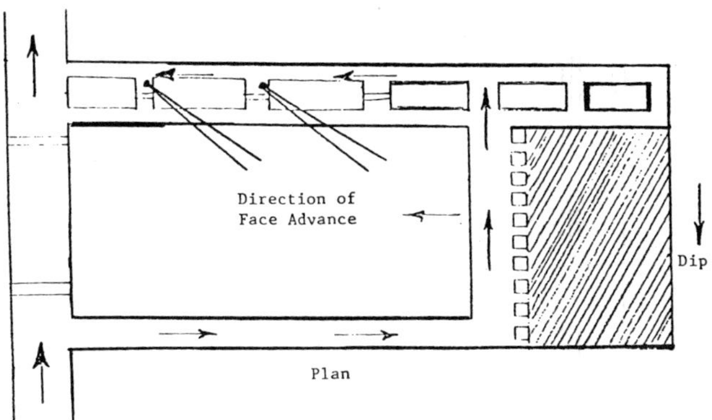

Fig. 2c. Fully stowed longwall gobs.

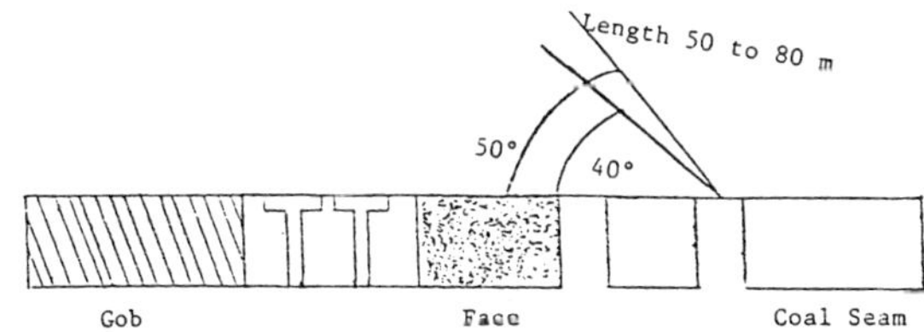

Fig. 3. Methane drainage with cross-measure boreholes.

duction from the borehole even if the borehole is sheared by rock movements.

3. <u>The Superjacent (or Hirschback) Method</u>:

This technique is mainly used for retreating longwall faces in very gassy seams. Figure 4 shows a typical layout. A roadway is driven 20 to 35 m above the longwall face, preferably, in an un-workable coal seam. The roadway is sealed and vacuum pressures up to 3 m of water gauge are applied. To improve the flow of gas, inclined boreholes in the roof and floor are drilled to intersect other gassy coal beds. If mining scheme proceeds from the top to the bottom seams in a basin, the entries in a working mine can be used to drain coal seams at lower levels. Methane flow from these entries is high, averaging 20 to 30 m^3/min for highly gassy seams. Nearly 50 percent of total emission at the longwall face have been captured (OEEC 1957).

U.S. GOB DEGASIFICATION METHODS

Longwall mining in the United States is a relatively new technique and it differs from European systems in

Table 2. Methane Capture Ratios for Cross-Measure Borehole Method
(After Kimmins, 1971)

Mine	Specific Methane Emission (m^3/ton)	Total Methane Emission, Mm^3/yr	Methane Capture Ratio (Percent) Section	Mine
Astley Green	91	42.8	60	38
Haig	85	37	59	20
Parkside	79.3	61.8	61	42.6
Point of Ayr.	161.5	53.4	68	47.7
Sutton Manor	93.5	37.1	70	40.0

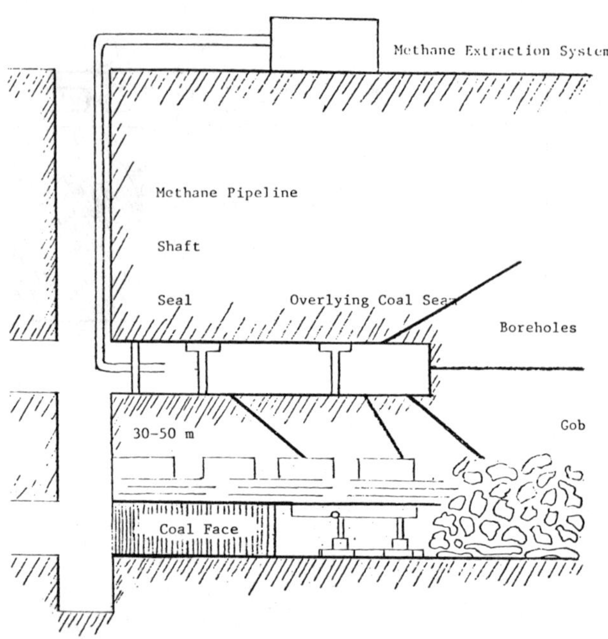

Fig. 4. Methane drainage by superjacent method.

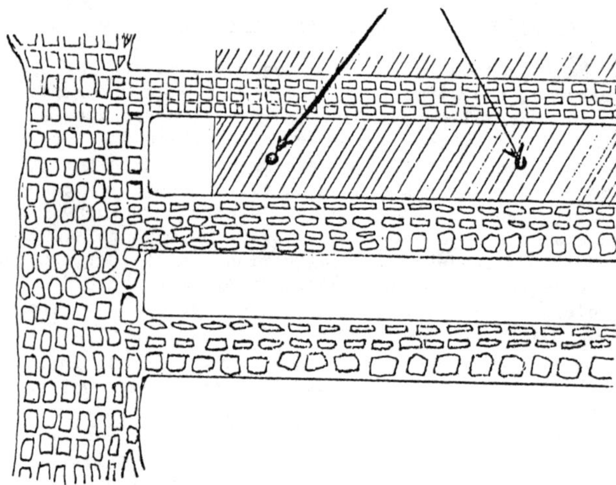

Fig. 5a. Gob degasification borehold.

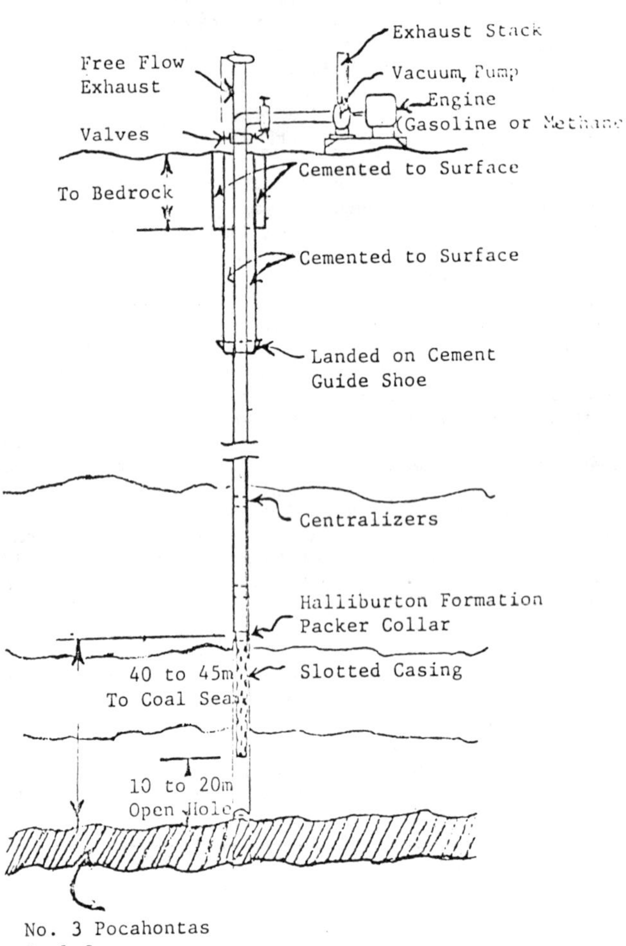

Fig. 5b. Section of a gob degasification borehole.

several ways. U.S. coal seams are generally flat, shallow and relatively more permeable. Retreat longwall mining (with few exceptions) is the only method being practiced at present. Typically, only one seam is mined in one area. Methane emission rates from gobs in various basins vary depending on the geological conditions but deep-seated longwall gobs such as those in Pocahontas No. 3 seam in Virginia and in Marylee seam in Alabama produce methane in excess of 50 m³/min. Large quantities of air are needed. Multiple (typically four) entries are driven to develop longwall panels so that necessary air quantities can be delivered to the longwall faces. In most cases, however, some sort of methane control becomes necessary.

The most popular method of methane control is to drill a vertical borehole above the longwall prior to mining as shown in Figure 5 (a). Depending on the length of the longwall panel (typically 1000 to 2000 m) one to three vertical gob degas boreholes are needed. The first hole is usually within 150 m of the line of start of longwall face. The completion procedure is illustrated in Figure 5 (b). The borehole is drilled almost to the top of the coal, cementing casing through the fresh water zones near the surface, and leaving a slotted liner over the lower open section to prevent closing of the hole by caving. These boreholes are completed prior to mining. Usually no measurable methane production is realized until the longwall face line is within a few meters from the borehole. Early experiences with this method of gob degasification have been described by Moore et al (1976) for Lower Kittanning seam, by Moore and Zabetakis (1972) for Pocahontas seam and by Davis and Krickovic (1973) and Mazza and Mlinar (1977) for the Pittsburgh seam.

Many gob degasification boreholes produce naturally when longwall face intersects them, but vacuum pumps are often added to further improve the flow and in some cases to prevent the reversal of flow. Initial methane flow rate is generally high and erratic but after nearly sixty days the flow appears to stabilize. Figure 6 shows the plot of cumulative methane pro-duction against time on a log-log scale for three coal seams. A linear relationship appears to exist in all cases. This flow law is a characteristic of all U.S. coal seams indicating that the main source of all gases is the overlying coal seams. The flow equation can be written as:

$$Q = AT^n$$

where Q is the cumulative flow in m³;

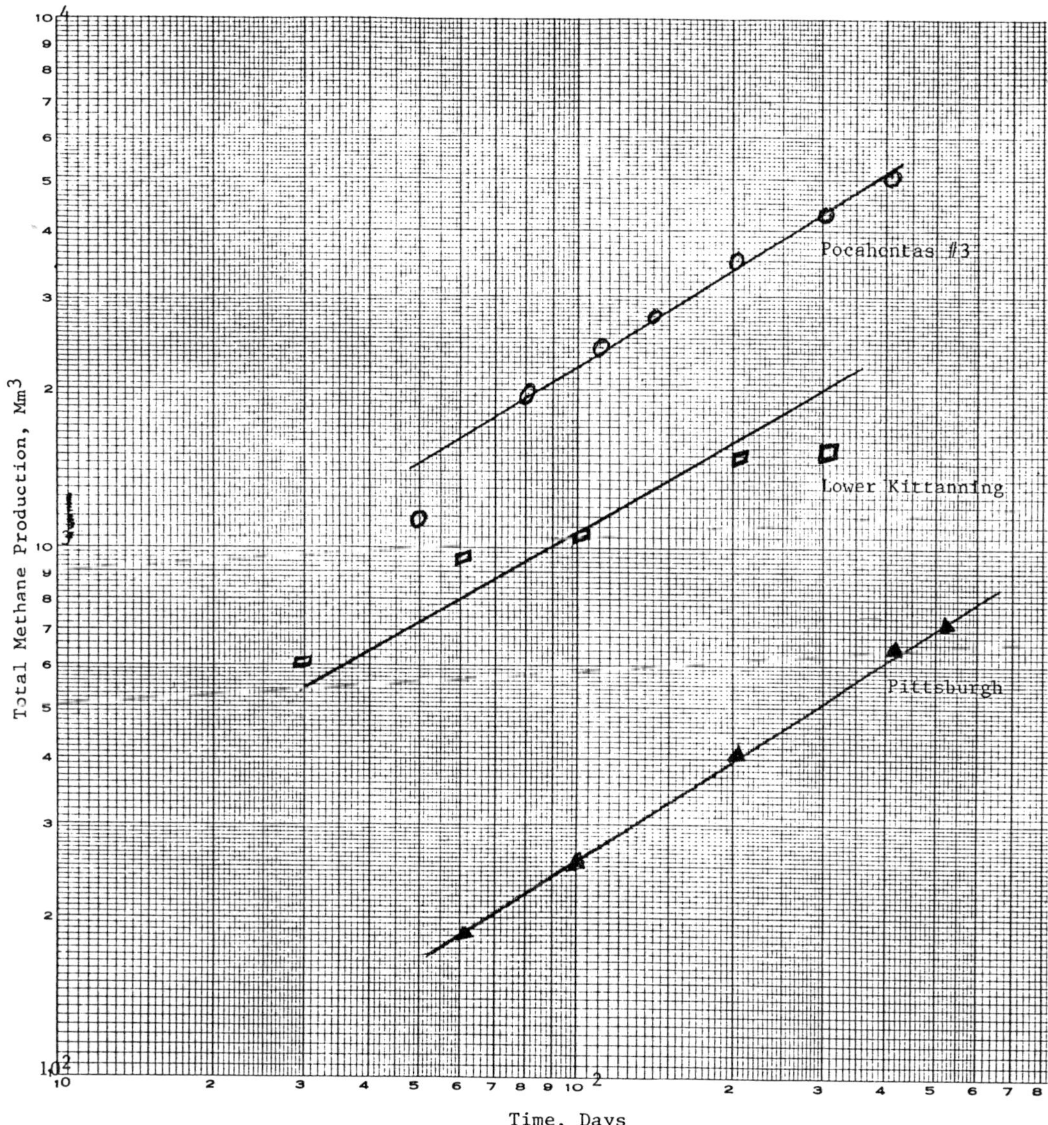

Fig. 6. Methane production from vertical gob degasification boreholes.

A is a characteristic constant for the longwall gob;

and T is the time in days.

   n is the exponent, a characteristic of the coal seams in the gas emission space.

Table III summarizes average methane production rate, and methane capture ratios for the lower Kittanning, Pocahontas No. 3 and the Pittsburgh seams using vertical degasification boreholes.

Theoretically, vertical degasification boreholes are perhaps the best way to trap methane from the gas emission space. In practice, methane capture ratios are generally less than 50 percent. This is somewhat lower than the ratio claimed for the cross-measure borehole method on European longwalls. One reason for this is that the vertical borehole does not capture any floor emissions. Besides, U.S. longwall gobs are seldom sealed and usually are ventilated with larger quantities of air. This sets up high pressure gradients across the gob and large amount of gas is lost to the ventilating air in the bleeders. Methane capture ratios of vertical degasification boreholes can be certainly improved if the gobs were isolated from mine air currents but the extra cost involved in most cases may be too high. The main advantage of this technique is that the methane control work can be kept completely separate from mining operations. The economics are good for shallow coal seams. The cost of drilling and completing a vertical gob degasification hole for a very deep longwall face is prohibitive. A limit may be reached around 600 m when the cross-measure borehole technique may prove more economical. Bad terrain and the difficulty of obtaining surface rights in many cases may also encourage the use of underground gob gas control techniques. Very limited experiments on gob gas control using cross-measure boreholes are being carried out by the U.S. Bureau of Mines in cooperation with the coal industry. If successful, this technique may find an increasing use in U.S. coal mines.

CONCLUSIONS

Longwall coal mining is becoming increasingly

Table 3. Methane Capture Ratios for Vertical Gob Degasification Boreholes

Coal Seam	Mining Method	Total Methane Emission for the Section (m^3/min)	Methane Capture Ratio (Percent)	
			Natural Flow	Vacuum Pump
Lower Kittanning	Longwall	39.3	15 to 50	15 to 20
Pittsburgh	Room & Pillar	7.0	25 to 30	30 to 80
Pocahontas #3	Longwall	55.8	25	35

popular in the United States. Methane emissions in the gob are generally high and some form of methane control is essential. European longwall methane control techniques consist of the packed cavity method, the cross-measure borehole method and the superjacent method. The most commonly used methane control technique in the U.S. coalfields is the vertical degasification borehole.

All methane control programs are essentially based on the following premises:

a. Determination of the geometry of the gas emission space and locations of gas bearing horizons therein.
b. Estimation of the rate of methane influx into the longwall gob.
c. An economic and efficient scheme to bypass the gas and thus prevent it from entering the mine atmosphere.

On the average, the cross-measure borehole technique appears to have the highest methane capture ratio. Figures above 60 percent are routinely obtained. The vertical degasification borehole method and the superjacent methods are comparable with methane capture ratios of nearly 50 percent. The packed cavity method and its variants have, in general, lower methane capture ratios.

The efficiency of vertical degasification boreholes can be substantially improved if the longwall gob can be sealed as is commonly done in Europe. A detailed economic analysis is necessary to determine if the extra cost of sealing gobs can be justified by improved methane control and reduced air quantities.

REFERENCES

Cervik, Joseph, "Methane Control on Longwalls--European and U.S. Practices", Preprint No. 79-100, presented at the 1979 AIME Annual Meeting, February 18-22, 1979.

Davis, J. G., and Krickovic, S., "Gob Degasification Research--A Case History, U.S.B.M. IC 8621 1973.

Gunther, J. and Belin, J., "Prevision du degagement du grisou en taille pour les grisements en plateure"; 12th International Conference on Mine Safety and Research, Dortmund, 1967.

Kimmins, E. J., "Firedamp Drainage in the North Western Area", Colliery Guardian Annual Review, pp. 39-45, 1971.

Lidin, G.D., et al, "Control of Methane in Coal Mines", English Translation by Israel Program for Scientific Translations, Jerusalem 1964 (original in Russian 1961).

Mazza, R. L. and Mlinar, M.P., "Reducing Methane in Coal Mine Gob Areas with Vertical Boreholes", Final Report on U.S.B.M. contract number H0322851 February, 1977.

Moore, T.D. and Zabetakis, M.G., "Effect of Surface Borehole on Longwall Gob Degasification" (Pocahontas No. 3 Coalbed), U.S.B.M. RI 7657, 1972.

Moore, T.D., et al, "Longwall Gob Degasification With Surface Ventilation Boreholes above the Lower Kittanning Coalbed," U.S.B.M. RI 8195, 1976.

Spindler, G.R. and Poundstone, W.N., "Experimental Work in the Degasification of the Pittsburgh Coal Seam by Horizontal and Vertical Drilling," SME Preprint No. 60-F-106, 1960 AIME Annual Meeting, New York.

Thakur, P.C. and Davis, J.G., "How to Plan for Methane Control in Underground Coal Mines", Mining Engineering, pp. 41-45, October, 1977.

Winter, K., "Extent of Gas Emission Zones Influenced by Extraction," 16th International Conference on Coal Mine Safety Research, Washington, D.C., pp. v 3.1 to v 3.17; 1975.

Chapter 11

LONGWALL SUBSIDENCE OVER THE PITTSBURGH NO. 8 COAL ON NORTH AMERICAN COAL CORPORATION'S EASTERN OHIO PROPERTIES

Michael S. Roscoe

Division Geologist
The North American Coal Corporation
Central Division
Powhatan Point, Ohio

INTRODUCTION

In order to more accurately predict longwall surface subsidence over the Pittsburgh No. 8 Coal in Eastern Ohio, North American Coal Corporation's Quarto Mining Company undertook or participated in several subsidence studies. Due to the changing roof geology and rapid variations in the amount of cover two separate areas were investigated at two different mines to determine the subsidence response if any to these changes. Data generated from these studies were compared with NCB derived data for accuracy and then used to predict subsidence under houses, gas lines and other critical man made structures. The following report summarizes the results of these studies as well as the subsidence case histories of gas lines, transmission towers, houses and water wells. Problems that were encountered during both phases of this process (investigation and subsequent application) are also included.

GEOGRAPHY

North American Coal Corporation's Central Division is situated along the Ohio River in the Eastern Ohio Counties of Belmont and Monroe and is comprised of The Quarto Mining Company which operates Powhatan No. 4 and Powhatan No. 7 Mines, The Nacco Mining Company which operates Powhatan No. 6 Mine, and two parent company mines - Powhatan No. 1 and Powhatan No. 3 Mines. (Figure 1) All production within the division comes from deep mines in the Pittsburgh No. 8 Coal Seam and is in excess of 6 million tons per year. Longwall production is restricted to the Quarto Mining Company which has two (2) longwall units in Powhatan No. 7 Mine and two (2) longwall units in Powhatan No. 4 Mine.

The Quarto Mining Company properties as well as the other Central Division mines are situated along the Western edge of the unglaciated Allegheny Plateau. The combination of this locally deeply disected plateau with the erosional development of the Ohio River has created a local terrain that is quite steep with a relief of 122-152 m (400' - 500') being common. Regional elevations range from 186 m (610') along the Ohio River to over 427 m (1400') in the uplands to the West with amounts of cover ranging from 61-275 m (200' - 900').

The area of longwall undermining is primarily rural and sparsely populated. The primary land use in the area is agricultural with approximately 40% of the affected ground being devoted to crops and meadow along the narrow ridge tops and the remaining 60% of the acreage is in woodland on the steep hill sides.

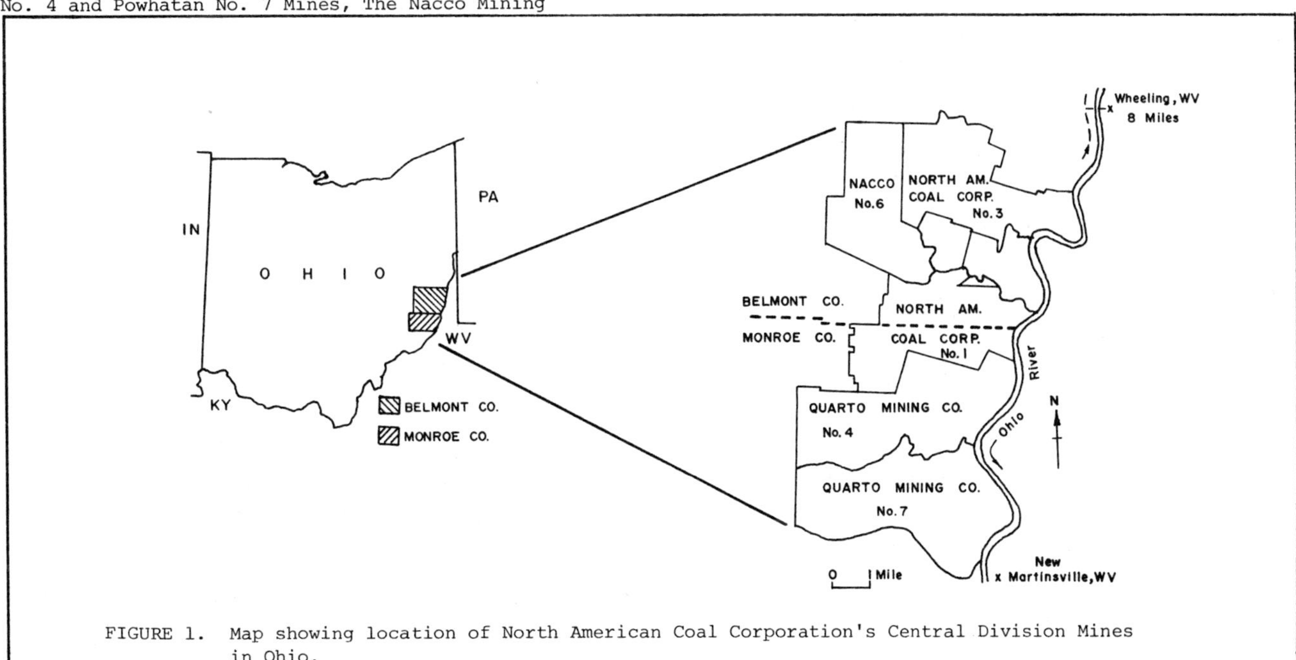

FIGURE 1. Map showing location of North American Coal Corporation's Central Division Mines in Ohio.

GEOLOGY

The Pittsburgh No. 8 Coal Seam is upper Pennsyl-
vanian in age and defines the lower boundary of the
Monongahela Group. The coal which developed along
the fringes of the Dunkard Basin extends over much of
Southwestern Pennsylvania, Northern West Virginia and
Southeastern Ohio. Quarto Mining Company's properties
lie along the southwestern edge of this basin, and
coal thicknesses in this area range from .92 - 2.44 m
(36" - 96") averaging approximately 1.52 m (60")
(including partings). The regional dip of the coal
and surrounding strata is to the southeast in amounts
of 3 - 12 m (10' - 40') per mile and is generally
consistent with very few structural irregularities.
The elevation of the coal to the southeast is approximately 107 m (350') rising to 206 m (675') along the
northwestern edge of the properties.

The Pittsburgh No. 8 Coal Seam is made up of a main
bench coal (1.22 - 1.52 m (4' - 5')) containing several thin and variable partings and a rider coal (.15-
.30 m (6" - 12")) which is separated from the main
bench coal by a shale-mudstone parting (.03 - .20 m
(1" - 18")). In general, both the rider and main
bench coals thin and become more variable to the
Southwest, reaching .64 m (2') or less immediately to
the south and southwest of Quarto's present holdings.
The rider coal grades laterally into a coaly to
carbonaceous shale in the southern and western
portions of the properties. At No. 7 Mine there is no
rider coal present while at No. 4 Mine the rider coal
is present only in the NE 1/4 of the property. The
only interruption of these regional trends is by the
occurrence of clay veins which are restricted to
certain southern portions of the properties (No. 7
Mine).

The strata immediately above the rider coal varies
from a bedded to nonbedded calcareous highly slickensided mudstone-claystone (1.8 - 3 m (6' - 10')) in the
northeast to a sandstone (Pittsburgh Sandstone 3 -
12 m (10' - 40')) to the southwest. (Figure 2) Overlying this unit is the Redstone Limestone (3 - 6 m
(10' - 18')), a bedded argillaceous limestone which
thins and rises southwest towards the area of sandstone deposition. (Figure 2) Above the Redstone
limestone, this sequence (cyclothem) is repeated with
the Redstone Coal and associated strata. Successively
higher strata (up to 244 m (800')) which comprise the
remainder of the Monongahela Group and the lower-most
portion of the Dunkard Group, do not differ much from
this generalization. The floor, a compact silty mud-
stone, (.3 - 1 m (1' - 3')) is consistent over the
entire property and is underlain by a hard calcareous
siltstone. In certain areas where the floor is
slightly weaker, this unit yields under the abutment
loading of longwall mining. The addition of water to
this unit dramatically reduces its strength thereby
causing premature gate closure in certain areas.

The variation in "roof" lithology cited above
represents a change in the depositional environment
from bay to deltaic channel deposition. The general
direction of this channel system is NW-SE with the
transport of sediment thought to be toward the north-
west. Complete erosional cutouts associated with this
channeling are not believed to exist but partial
scouring of the coal increases in the extreme southwest corner of the No. 7's properties.

Longwall production at both Powhatan No. 4 and 7
Mines is presently in areas overlain by the Redstone
Limestone. The longwall face at No. 4 Mine includes
all three members of the Pittsburgh No. 8 Coal - the
main bench, the parting and roof coal while at No. 7,
only the main bench is being mined since it is the
only member present.

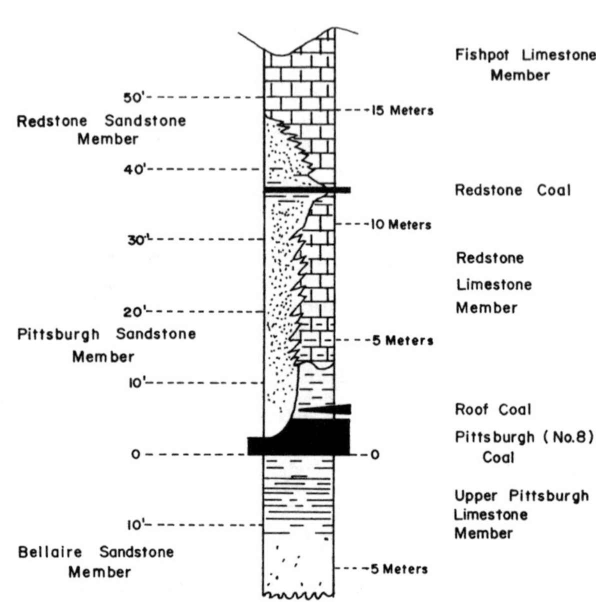

FIGURE 2. Geologic column showing Pittsburgh
No. 8 Coal and overlying strata.

ROCK MECHANICS

One of the original concerns about the initial
success of longwall mining in the Ohio Valley was
the strength of the overlying Redstone Limestone.
It was thought that its apparent strength would not
permit adequate caving to occur thereby creating
large pressures on the shields and face during initial face advancement. As it turned out, the minor
shale partings scattered throughout the limestone
unit allowed caving to occur fairly rapidly and
thoroughly. The initial massive caving associated
with panel start up generally occurs when the face
has advanced 61 m (200') into the panel. Block
caving in this strata occasionally extends up through
the No. 8A Coal Seam 4.6 - 7.6 m (15' - 25') above
the Pittsburgh No. 8 Coal.

The overlying strata involved in the caving and
subsequent subsidence consists of 52% mudstones, 13%
limestones, 10% shales, 10% siltstones, 10% sandstones, 3% clays and 2% coals. The relative
strengths of these units can be found in Table 1.

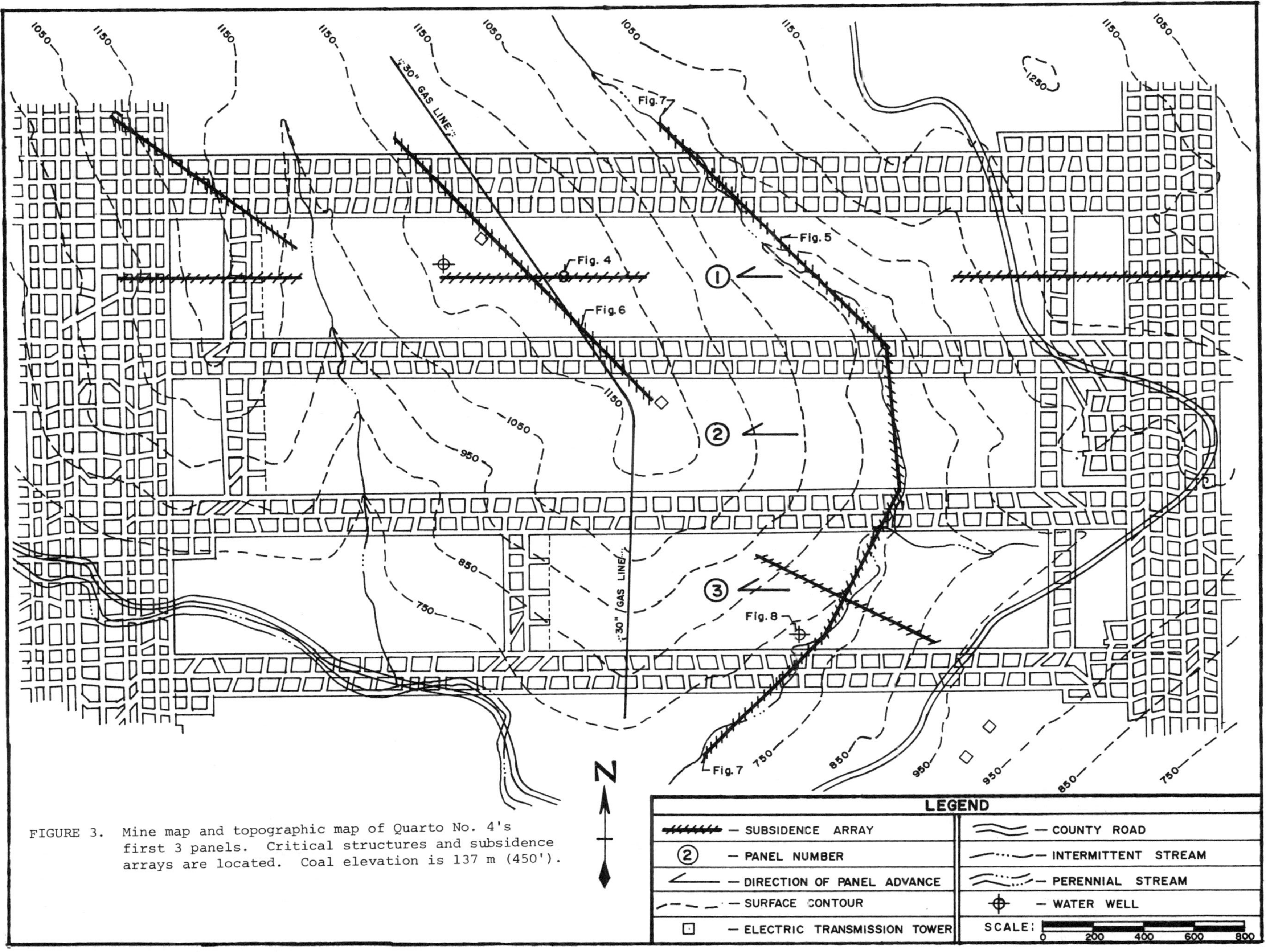

FIGURE 3. Mine map and topographic map of Quarto No. 4's first 3 panels. Critical structures and subsidence arrays are located. Coal elevation is 137 m (450').

Table 1.

Compression*

	Average		Range	
	$Pa \times 10^7$	(lbs/in^2)	$Pa \times 10^7$	(lbs/in^2)
Sandstone	3.48	(5051)	1.00- 5.81	(1334-8425)
Limestone	17.11	(24809)	7.29-21.76	(10570-31565)
Siltstone	10.45	(15153)	6.15-14.66	(8918-21257)
Mudstone	3.11	(4510)	2.46- 3.46	(3575-5017)
Coals	6.35	(9208)	4.92- 7.23	(7134-10498)

Tension*

	Average		Range	
	$Pa \times 10^7$	(lbs/in^2)	$Pa \times 10^7$	(lbs/in^2)
Sandstone	.47	(648)	.12- .82	(174-1189)
Limestone	1.82	(7234)	1.53- 2.26	(2219-3277)
Siltstone	1.65	(2393)	1.53- 1.87	(2229-2712)
Mudstone	.73	(1059)	.71- .75	(1030-1088)
Coals	.55	(798)	.45- .73	(652-1058)

*Tests run on .05 m (2") core ranging from .025 m - .075 m (1" - 3") in length.

LONGWALL DATA

The four current longwall panels at Powhatan No. 4 and No. 7 Mines have 148 m (485') faces running north and south, are 915 - 1524 m (3000' - 5000') long (E-W), and are developed with a 3 entry gate system. Two of these longwall systems are at Powhatan No. 4 Mine and 2 are at Powhatan No. 7 Mine. The original gates at both mines consisted of two rows of essentially the same sized pillars on 21 m + 24 m (70' + 80') centers with three-way intersections. One row of these pillars has since been enlarged 33.5 m x 33.5 m (110' x 110') at the expense of the other pillar 33.5 m x 15.25 m (110' x 50') as well as a reduction in entry width 5.5 m (18') to 4.9 m (16') to help control the lateral movement of side abutment pressures during and after panel mining. With this change, the panel edge to panel edge measurement within the gates remained at 53.6 m (176'), thereby reducing problems in changeover in midstream gate development. Supplemental support in the form of 2 rows of cribs (.76 m x .76 m (30" x 30")) on 2 m (6½') centers is required in the tailgate entries while advancing steel props are used in the headgate entries. No supplemental support is required in the headgate entries. After collapse, the effective panel width of this system approaches 152.5 m (500').

All four faces utilize double drum ranging arm shearers with .76 m (30") webbs and 100 - 2 or 4 legged shield type supports providing 350 - 500 tons of support capabilities. Panels advance at a rate of 6 - 12 m (20' - 40') per day depending on coal height and mining conditions with all faces advancing in a westerly direction (up dip). No zig-zagging of the panels is practiced or planned since the intent is to have these longwalls move up dip and away from possible water accumulations.

Mining height of the longwall systems varies 1.68 - 2.44 m (66" - 96") at the No. 4 Mine and 1.17 - 1.40 m (46" - 55") at the No. 7 Mine while cover over current panels ranges from 73 - 274 m (240' - 900'). The lower number of 73 m (240') represents the lowest amount of cover that can be undermined without creating continuous water problems underground. The determination of this critical number will be found in a later section of this report.

The initial longwall installation within the Quarto Mining Company occurred at the No. 4 Mine and after the initial success of this system in 1977, an additional system was added during 1978 with two more systems being added in 1979. Prior to the initial installation, the surface impact of this type of production was realized and it was decided to initiate a subsidence study over this first panel. The initial disturbance of more than 2.03 million m^2 (500 acres) per year and secondary low level subsidence affecting more than 4.06 million m^2 (1000 acres) per year by four longwall systems dictated that Quarto Mining Company be able to understand and predict subsidence at the local level.

SUBSIDENCE INVESTIGATION - QUARTO NO. 4 MINE

The initial subsidence investigation centered over the first longwall panel at Quarto's No. 4 Mine and was patterned after the study performed by Dames and Moore for the Old Ben Coal Company in Southern Illinois (1977). Due to the relatively steep terrain in the area, the pattern used by Dames and Moore could not be exactly duplicated in the field. The study consisted of over 200 points or monuments which were located to determine; 1. angle of draw in low and high cover areas, 2. percent of seam height and maximum subsidence in low and high cover areas, 3. subsidence response in the front and back barrier pillar areas, and 4. a standard profile across the panel. No monuments were specifically established to accurately monitor stress and strain and because of this, very little data of this type is presented in this report.

Approximately three quarters of the over 200 points were established by Quarto Mining Company engineering personnel by simply driving a 1 m x .01 m (3' x ½") rebar into the ground at various locations. The remaining points were established on existing features such as benchmarks, gas wells, and transmission tower legs. All of these points were monitored (leveled) on a schedule that related to face advance and proximity of the face to a certain set of points. This aray of points was subsequently extended so that the second and third panels in this 3 panel series could be monitored. (Figure 3) Monitoring is still being conducted on a reduced scale some 3 years after the initial measurements were taken.

The three panels that were monitored were 915 m (3000') long, 148 m (485') wide, 2 - 2.14 m (6' - 7') high and had amounts of cover ranging from 107 - 259 m (350' - 850'). Surface features included two drainages and an intervening ridge on which were located; 1. 4 gas lines - one .76 m (30"), two .50 m (20") and one .46 m (18"), 2. a 450 Kv transmission line with 2 30.5 m (100') towers and 3. one road (Figure 3). One of the transmission towers was located over the first panel while the other was located over the second panel. Prior to the mining and subsidence of this area, the three smaller and older gas lines were disconnected and abandoned with the gas being rerouted through a new single line outside the area of subsidence (see utilities).

The data that was gathered throughout the subsidence of all three panels and the affected surface structures is summarized below.

Subsidence Response and Rate

The initial response of a surface point to the advancement of a longwall face occurs when the face is between 244 - 205 m (800' - 1000') away from the point. (Figure 4) When this range is reached, the surface begins to bob up and down in what appears to be a response to the cantilever action originating at the shield line. This positive effect or heave is also felt outside the final "0" subsidence line along the sides of, in front of, and behind the panel. This pulsation or wave in front of the face continues until the face is within 15 - 61 m (50' - 200') of the point. At this time continuous subsidence begins and continues until the face has gone some 305 m (1000') past the point (see below).

Maximum rates of subsidence of .06 - .12 m (.2' - .4') per day occur 46 - 122 m (150' - 400') behind the face in high cover and 46 - 91 m (150' - 300') behind the face in low cover. It was found that approximately 85 - 95% of the expected subsidence had occurred after the face had gone 305 m (700' - 1000') past a point which almost mirrors the initial response distances to an oncoming longwall face. In general, these maximum rates of subsidence were found to change with the speed of the face. As it so happened, a required increase in the horse power of the shearer in the middle of the first panel increased the rate of face advance and in so doing also increased the rate of subsidence at the surface.

Further reaction of the surface to face movement was discovered during the monitoring of the .76 m (30") gas line by gas company personnel. It was found that major movement on the surface stopped 48 hours after the face had stopped (i.e. weekend) and restarted only 24 hours after the face had restarted. This timing only applies to those areas that are undergoing continuous critical subsidence and are situated directly over the panel. It is felt that this type of information will prove valuable during the subsidence of critical structures.

Amounts of Subsidence

Subsidence over the front and back barrier pillar areas turned out to be somewhat less than that along the panel edges under similar conditions and distances from center or panel edge. This is in part due probably to the unusual "lenght of panel-width of panel" situation that occurs at the beginning and end of each panel. The direction of the panel's width and length of course change when the panel has advanced as far as its face length. Underground instrumentation in these areas did not show any excessive pressure build up due to a lack of complete caving near or over these pillars, somewhat confirming the above idea.

Probably the most interesting aspect of subsidence is the final and maximum amounts that are experienced in different areas directly over or surrounding the panel.

The maximum amount of subsidence caused by the first panel that occurred in the low cover areas was less than that which occurred in the high cover areas (1 m vs. 1.22 m (40" vs 48")) (Figures 5 and 6) This finding is contrary to the National Coal Board (NCB) predictive methods and is thought to be caused by inadequate caving in these areas causing large brows to extend into the gob. These maximum amounts of subsidence represent 51% and 62% of the 2 m (78") of coal height that was mined on this panel. It is interesting to note that in the second and third panels of this series this phenomenon does not appear to hold true until the unmined or virgin side of the third panel is reached. Amounts of subsidence in these central areas of low cover equal or exceed those of the higher cover areas and therefore, agree with the NCB. (Figure 7) From this it appears that adjacent areas of gob do not appear to provide the support that the "unmined" side does.

As was stated earlier in this report, monitoring is still occurring on most of the original points that still exist. Subsidence that occurred between the second and third year over the first panel still measured .08 - .13 m (3" - 5") in several areas (6% - 8%). Monitoring on a once a year basis will occur on most of the points for one or two more years or until subsidence is reasonably complete.

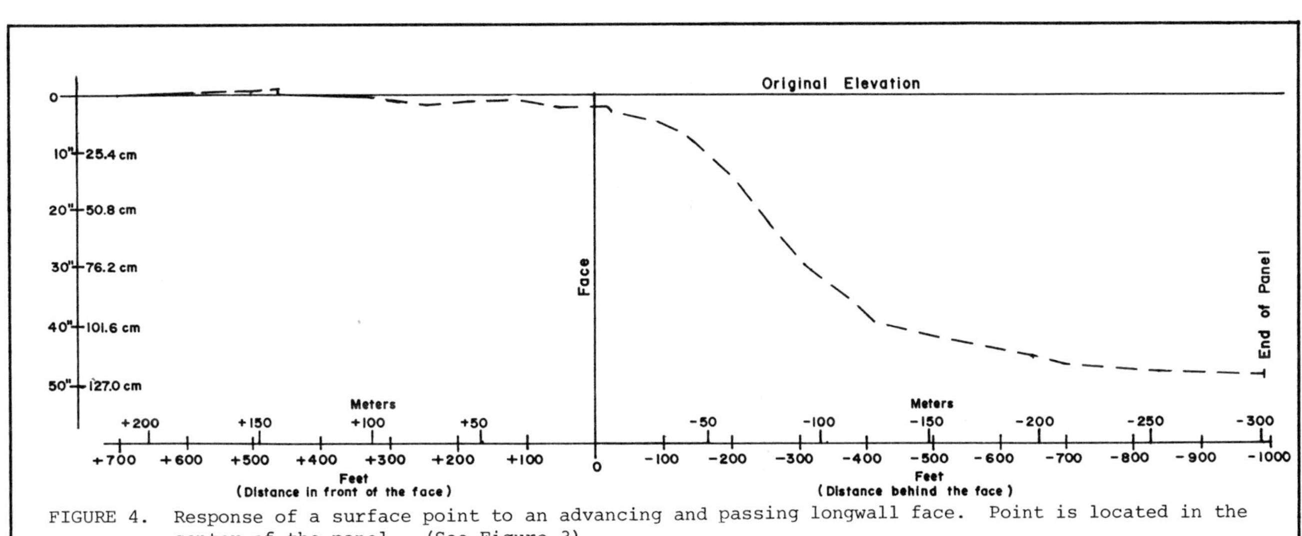

FIGURE 4.   Response of a surface point to an advancing and passing longwall face. Point is located in the center of the panel.   (See Figure 3)

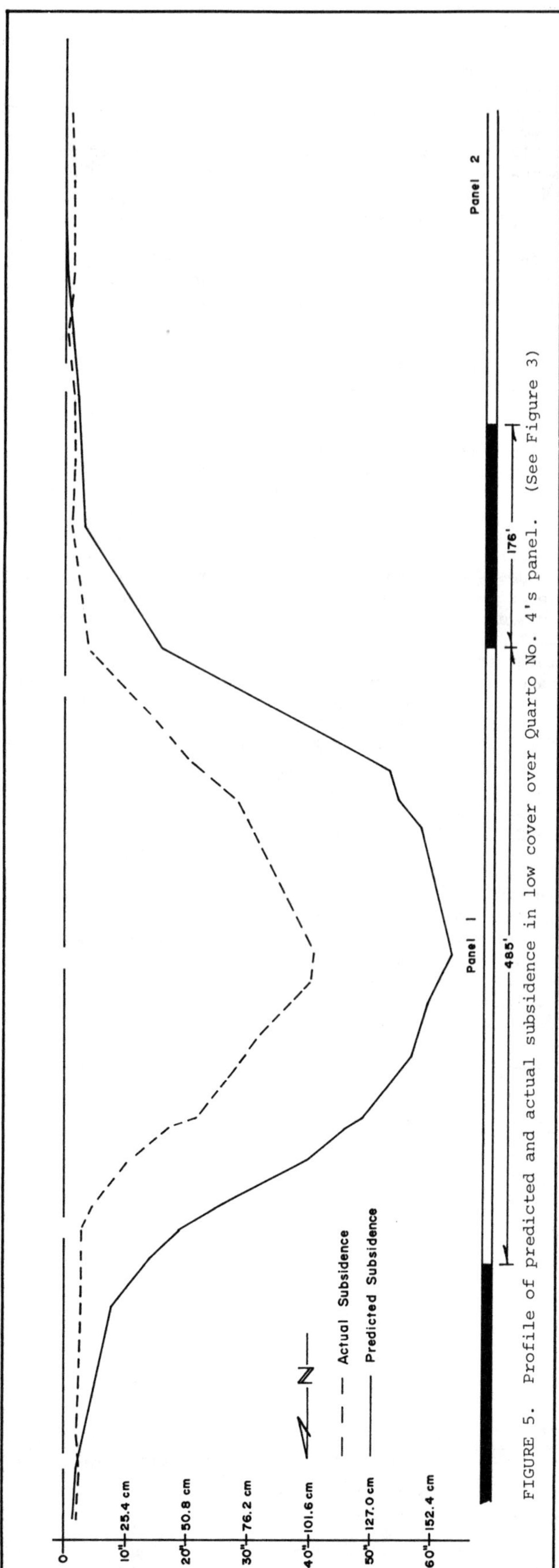

FIGURE 5. Profile of predicted and actual subsidence in low cover over Quarto No. 4's panel. (See Figure 3)

Angle of Draw

The approximate angle of draw was also established during this initial study and was found to range from 22° - 38°. The average, however, landed in the 32° - 34° range. Further work, both in subsidence arrays and on private residences, showed that the standard 35° angle of draw is a good and safe approximation in most cases. Complicating the accurate determination of the angle of draw is the 1. positive heave experienced near the "O" subsidence line, 2. ability to precisely measure very small changes on crude monuments, and 3. spacing of the monuments in the areas of subtle change.

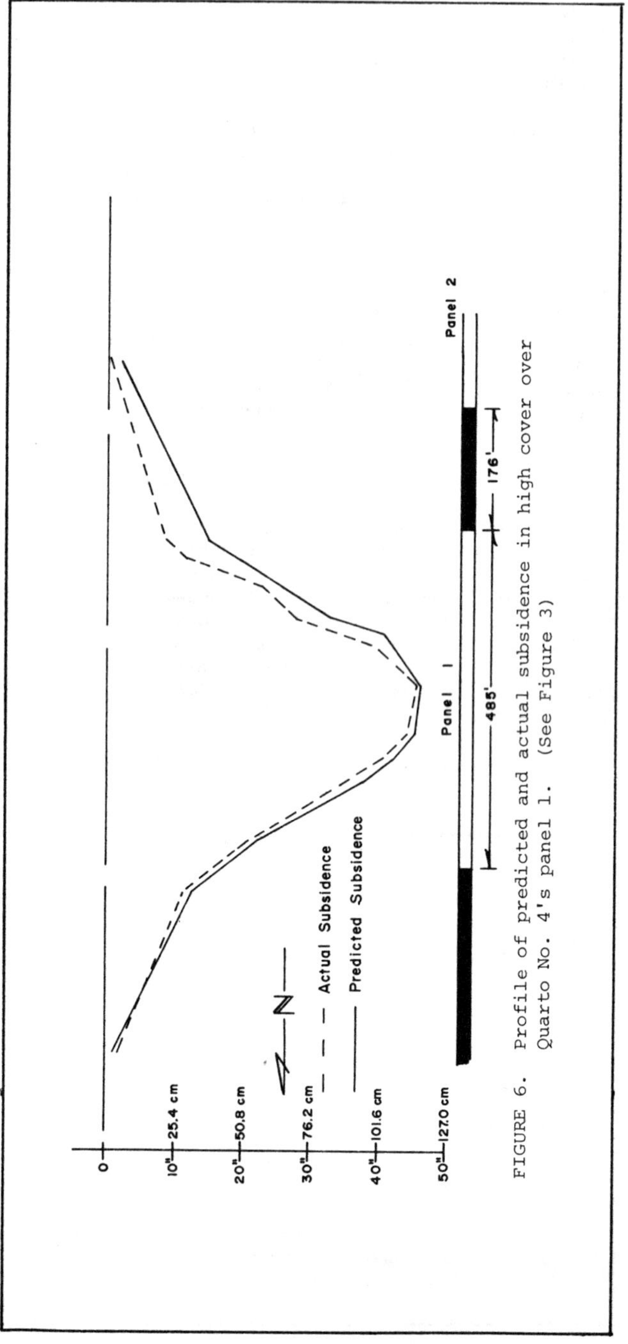

FIGURE 6. Profile of predicted and actual subsidence in high cover over Quarto No. 4's panel 1. (See Figure 3)

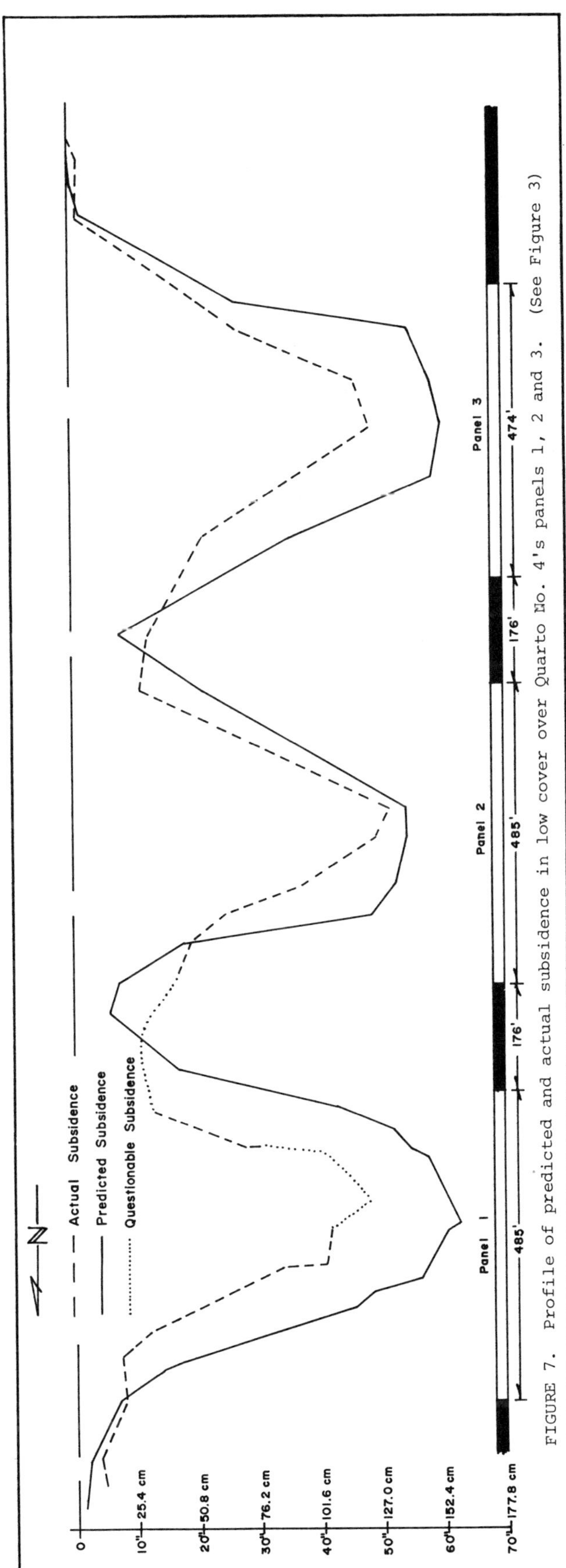

FIGURE 7. Profile of predicted and actual subsidence in low cover over Quarto No. 4's panels 1, 2 and 3. (See Figure 3)

Comparison with NCB

All data that were gathered were compared with predictive amounts that were determined using the NCB predictive emthods. Initially only the high cover data matched the predicted amounts with an accuracy rate of approximately 90%. (Figure 5) It is felt, however, that this data with time would have approached 100% of the NCB amounts but the overlap of the second panel's subsidence profile has complicated the interpretation of numerous points in this area.

As was stated earlier in this report, the initial data from the low cover areas were much lower than what was expected or predicted using NCB methods. (Figure 6) It is felt that in the first panel, the lack of a large amount of overburden did not prevent total caving to occur adjacent to the unmined area. It is also felt that lateral support generated by rapid variations in surface relief was in part responsible for some of this lack of expected subsidence. Data from underground instrumentation show extreme pressure increases in the gates under low cover over the second panel and a portion of the third panel came closer to equaling the amounts produced by the NCB methods. This is in all likelihood due to the fact that the neighboring panel (gob) cannot give as much support as areas that are adjacent to virgin coal.

Other problems were encountered when applying the NCB methods to the areas over the front and back barrier pillars. The panel width, length, and center location were rotated 90° and shifted back and forth to match actual subsidence levels. By using this "hit and miss" method, a usable method was "backed into" for predicting these areas more accurately.

Data from all studies and structures show that the immediate subsidence (first 6 months) is always 10 - 20% less than the NCB amounts. These predictive amounts, however, became more accurate with age as low level secondary subsidence continues to take place. Long term subsidence rates which equaled 6% - 8% for the second year and 1% - 2% during the third year were discussed in an earlier portion of this report. It is felt that if an area is not influenced by a neighboring panel it will be for all intents and purposes stable three years after undermining.

Unstable Ground

During the course of mining the first series of panels, it became evident how critical the effect of subsidence was on unstable or over steepened ground (i.e. ancient and recent slips and slides). During the mining of the first panel, data gathered over the back barrier pillar area showed the initiation and progression of a rotational slump. (Figure 3) This occurrence negated any subsidence monitoring value these points could have contributed to the study. Other instances of downward land movement occurred over this and other panels. The project water well (See Ground Water) was displaced down hill in several segments during subsidence and because of this, down hole pump tests could not be performed. Another slip movement that was thought to be caused by subsidence extended through a rock and gravel base country road. Minor tension cracks developed as well as "stepping" which required remedial work. These experiences with unstable ground dictate that extreme care be taken during the subsidence of critically situated structures even in those areas near the "O" line.

Ground Water

To determine the effect of subsidence on the local ground water, a well was drilled on the ridge along the center line of the first panel (244 m cover (800' cover)). (Figure 3) The well encountered mostly shales and several thin limestone beds non of which were distinct aquifers. As expected, subsidence did affect the ground water, but somewhat differently than had been anticipated. The elevation of the groundwater was not affected until the face had gone 23 m (75') past the point. The water level then began to drop (9 m (30')) as the face advanced 23 - 61 m (75' - 200') past the well and then began to fluctuate as the face advanced to 92 m (300') past the well. This impact on the groundwater coincides with the occurrence and timing of the period of maximum subsidence activity.

After subsidence had occurred at the well and during an attempted pumping test, it was discovered that the well had moved down hill in several segments effectively closing off the opening. The well was therefore redrilled and the bottom portion was grouted to seal any possible fractures that were thought to exist in the bottom horizons. With time, (3 months) the groundwater table returned to near its original elevation. Levels of production which were minimal initially remained essentially the same after the well was subsided and redeveloped. No water quality data was generated during the subsidence of this well.

Another well was drilled in an area of low cover over another panel to monitor the groundwater levels during subsidence and also to determine if groundwater would enter the mine through possible fractures. (Figure 3) The well was developed along a narrow drainage approximately 85 m (280') above the coal and was drilled to a depth of 15 m (50'). The water table in this well as was the case in the initial well dropped during the period of critical face advance. (Figure 8) Levels started dropping when the face passed beneath the well and continued until the face had gone 61 m (200') past. Fluctuation of the water level occurred when the face was advancing from

61 - 92 m (200' - 300') past the well at which time a steady rise in the water level began to take place. (Figure 8)

At about this time the amount of incoming water into the workings began to increase in the area of the gob below this low cover area. The amount continued to increase during the following 6 months but it is now on the decrease. Because of this experience it was determined that the 70 - 76 m (230' - 250') cover range would act as the lower limit of cover (angle of draw extensions included) for longwall mining. It is felt that the general shaley nature of the section will provide for fracture healing with time. (See Geology) However, due to the fact that 1 - 3% of total subsidence occurs in the third year after mining, some of this healing might not start until 3-5 years after mining has occurred.

Utilities

Complicating the mining of Quarto Mining Company's first series of panels was the critical nature of the surface structures that were to be subsided. As was earlier in the report these structures included 4 gas lines and two electric transmission line towers.

Three of the four gas lines (.46 m (18"), .51 m (20"), .46 m (18")) were owned by East Ohio Gas Company who controlled most of the subsidence rights over these initial panels. Since these lines were old (near replacement) and were coupled at all joints, it was determined that they could not be successfully subsided. To avoid subsiding a replacement line, a new single .76 m (30") line was rerouted not only around the initial series of panels, but a future series of panels to the North. Since Quarto did not control all of the mining rights in this Northern area, a cost-share agreement was entered into with East Ohio Gas and the line was rerouted. The original three lines were abandoned and subsided and during subsidence several exposed couplings showed .03 - .05m (1" - 2") of extension.

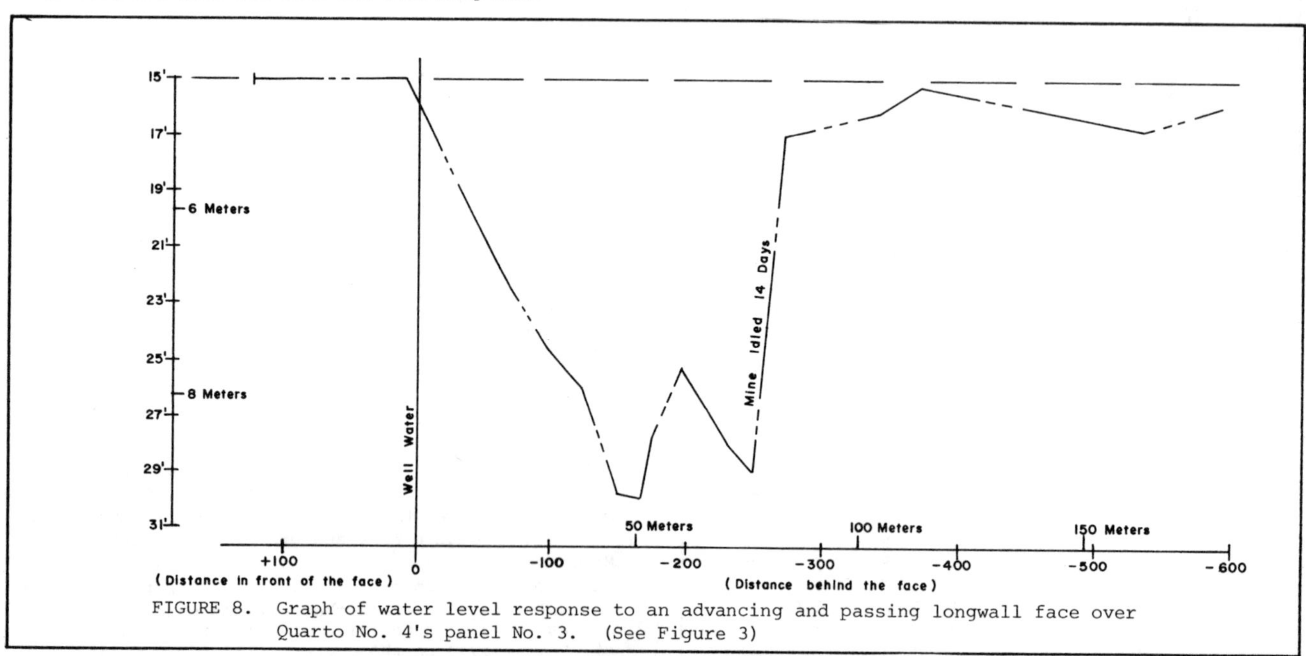

FIGURE 8. Graph of water level response to an advancing and passing longwall face over Quarto No. 4's panel No. 3. (See Figure 3)

The remaining fourth line .75 m (30") was owned by Consolidated Natural Gas Company and even though this line ran parallel and sometimes crossed the 3 East Ohio Gas lines, Consolidated did not control the subsidence rights over these panels. This discrepency between the two gas companies rights of course was due to the timing of Quarto's aquisition of mineral rights in the area. Becasue this line was newer and was a one piece continuous welded pipe it was determined that it could be successfully "subsided."

Prior to the longwall face advancing into the area, Consolidated Gas uncovered the line and under cribbed it at approximately 31 m (100') intervals and in critical spots. During subsidence, elevations at numerous points on the line were taken on a daily basis to determine when "lifting" of the line was necessary. When a reading of .06 m (.2') or more of movement was recorded, the area was jacked or shimmed back up to its original elevation. With this method the line was kept at its original location while the ground subsided below it. A knowledge of where the face was and how fast it was advancing was of utmost importance to the survey crew and because of this, daily communication between the crew and the Quarto Mining Company engineering staff was maintained during critical face movement.

Other data derived from the monitoring of this line showed a good time response to face advance. It was found that subsidence became negligible 48 hours after the face has stopped but began to increase again 12-24 hours after it has restarted. The maximum amount of subsidence that occurred under this line was on the order of 1.27 m (50") or 66% of the mining height, while the maximum subsidence rates of .12 m (.4')per day occurred when the face was 76 m (300') past the face. Data gathered during the subsidence of the gas line involved some error due to the fact that subsidence was occurring between morning measurements and afternoon shimming and leveling. Because of this the total subsidence measured was on the low side.

Using these above methods, the gas line was successfully subsided over the initial panel as well as the remaining 2 panels in this series. The same procedure is currently being used on this same line where it crosses another series of Quarto panels several thousand feet to the north.

The subsidence of the two electric transmission towers was somewhat less involved but of equal importance. The structures were four legged with a 9 m x 9 m (30' x 30') base dimension and were 31 m (100') in height. Since the four legs stradled a subsidence profile, differential subsidence was expected to occur and cause problems. The amount of differential subsidence on this type of structure is critical because: 1. the torque and strain that could develop at the lower portion of the tower could cause damage either to the tower or its foundation and, 2. minor amounts of differential settling become exaggerated when projected to the top of the tower during tilting. Further complicating the subsidence of these towers was the fact that the transmission lines on one of the towers entered and exited at an angle. It was felt that the addition of this lateral force could increase the importance of differential subsidence.

Both towers were subisded without incident with only minor adjustments to the trailing ground wire being required after subsidence was complete. A maximum subsidence of .9 m (36") with .15 m (6") differential subsidence was experienced on the first

tower. The second tower was located near the edge of the second panel and therefore experienced subsidence from two panels. Because of this the maximum subsidence on this tower was somewhat larger at 1.12 m (44") with .18 m (7") of differential subsidence. A discussion on stress and strain on these can be found in Kohli's article on Structural Damage (1980).

No other utilities were subsided during the mining of the initial series of longwall panels at Quarto No. 4 Mine.

APPLICATION

The initial panel of Quarto No. 4's second longwall system to the North was to undermine one residence. As is the case with most of Quarto's Longwall reserve areas, the company owns or controls all of the mining rights under these properties including the right of "lateral and subjacent support." It was determined, however, that the property owner would be approached and compensated in full for any damage that would occur to his house during the mining of this panel. The property which was located on the upper edge of a very steep hill was evaluated using percentages established from the earlier studies. It was found that the residence would experience very minor amounts of subsidence since it was very close to the limits of subsidence. Due to the potential of subsidence causing side hill instability, however, the minor amounts that were predicted became more critical in nature. (See Unstable Ground)

Now that the Quarto Mining Company had a good working knowledge of amounts and timing of subsidence, they needed a method of compensation for subsidence related damage to private residences or structures. Two financial packages or contracts were therefore generated to satisfy this need. Both plans require that the homeowner and Quarto Mining Company obtain independent appraisals of the structure and property and the average of these two is used. In the first option Quarto purchases the structure from the owner for 130% of the appraised average value with the owner holding a 90 day option to repurchase the property after he has been notified that damage causing subsidence is reasonably complete. At this time the "owner" can repurchase the structure at 100% of the appraised value. Quarto is not responsible for any damage repair under this option and during the term of this contract, the owner can live in "his" home rent free.

The second option involves a simple payment scheme of 10% of the average appraised value at the signing of the subsidence contract, a 5% payment 60 days prior to undermining and a 5% payment when damage causing subsidence is reasonably complete. During the life of this contract, the homeowner retains ownership of the structure. The second payment prior to undermining is intended to also act as a notification to the homeowner of the proximity of undermining. If the homeowner selects this option he would receive total of 20% of the appraised value of his house and Quarto Mining Company would pay for all repairs if any in excess of $1000.

After these two options were worked out they were presented to the homeowner whose house was just inside the angle of draw. He selected the purchase-repurchase option and exercised his option to repurchase. Expected amounts of subsidence were small .03 m - .04 m (1" - 1½") but as was stated earlier, the side hill location of this structure made this small amount critical. Subsidence to date under this house ranges from .01 - .04 m (½" - 1½") with no subsidence damage to the foundation or structure.

Quarto Mining is presently negotiating with several other homeowners, but at this writing they have not made their selection. Future longwall development over the next 5 years at both Quarto mines will involve many more houses and Quarto Mining Company's plan is to approach these people prior to longwall gate development into these areas. It is felt that the above approach to homeowner compensation exceeds that which is currently required by OSM regulations. At this writing, however, no decision on the state level has been made concerning this opinion.

The locally derived subsidence data was also applied to the predictive modeling for the subsidence of the .76 m (30") gas line over Quarto No. 4's northern series of panels. The actual amounts of subsidence along this portion of the line were not as critical as when the subsidence was to occur. As was the case with the other gas line section, time delays between leveling and shimming missed some minor amounts of subsidence during critical times. Estimates, however, were still quite accurate and helped the gas company to dispatch their crews when necessary.

SUBSIDENCE INVESTIGATION - QUARTO NO. 7 MINE

In order to evaluate the possible influence of changing stratigraphy on the known subsidence characteristics, it was decided to initiate a study similar to Quarto No. 4's study on Quarto No. 7's properties. Dames and Moore, in conjunction with the DOE, expressed interest in this project and were permitted to implement a study of their own. This study included 103 surface monuments to study subsidence, two .3 m (12") instrumented and monitored bore holes to establish caving characteristics above the upper gob horizon and a reconstruction of an old stone foundation to test the structures reaction to strain during subsidence. The array of points was similar to that of the Dames and Moore study at Old Ben Coal consisting of one row of points running down the center of the panel and two rows running at right angles to the central row.

The condition and position of the Redstone Limestone in the study area differs from that at No. 4 in that it is several feet higher above the coal (2 - 2.4 m (6' - 8') at No. 4 Mine versus 2.4 - 3 m (8' - 10') at No. 7 Mine) and is thinner and more bedded. The coal is also thinner at this location than at the original study area (2 - 2.1 m (6' - 7') at No. 4 versus 1.2 m - 1.3 m (48" - 52") at No. 7).

The study took place during 1979 and 1980 in conjunction with the mining of the initial panel of Quarto No. 7's second longwall system. The panel dimensions at this location were 1280 m (4200') long, 148 m (485') wide, and 1.3 m (52") high. The panel was developed on the 3 entry gate system and had amounts of cover ranging from 183 - 259 m (600' - 800'). The rate of advance on this new face averaged 5 m (16') per day with numerous stops due to labor and equipment problems.

Unlike the Quarto No. 4 results, the amounts obtained from this study almost immediately conformed with amounts derived using the NCB methods. (Figures 9 & 10) It will be interesting to watch these points over the next few years to see if

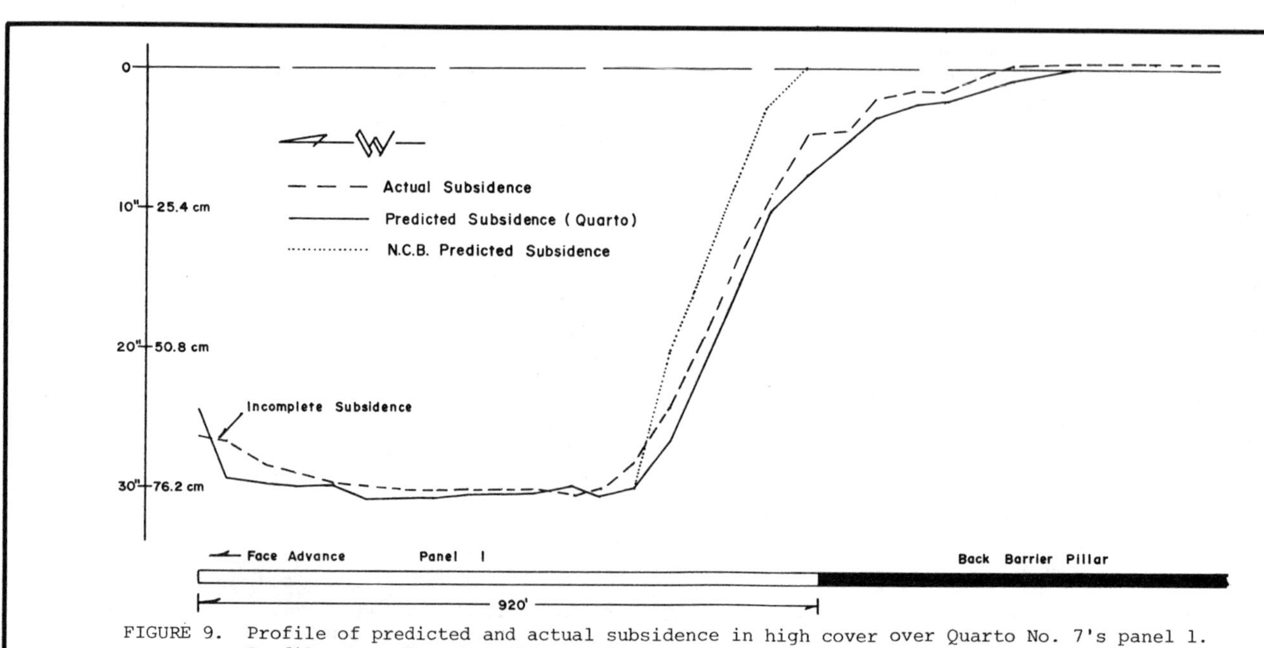

FIGURE 9. Profile of predicted and actual subsidence in high cover over Quarto No. 7's panel 1. Profile shows "panel end" condition.

and when they will exceed the predictive amounts. Data collected from this study was bi-monthly rather than daily and because of this, no detailed analysis of the subsidence rate and response is possible.

The rapid conformity of the results with the predictive amounts is felt to be caused by two factors. The first factor is the uniformity of the amount of cover. The cover over this panel was more consistent and somewhat larger than that at Quarto No. 4. It is believed that this aided in more complete caving over larger areas of the panel. The second factor is the nature of the Redstone Limestone in the area. Since this unit is slightly higher above the coal and more bedded, it does not resist caving to the degree it did at Quarto No. 4 Mine. This factor also aids in more complete and rapid caving in the crical horizons.

Since this is an ongoing project data concerning the angle of draw is inconclusive at this time. Preliminary data show the angle ranging from 23° to 37° under different terrain conditions. This will be investigated further when the total data package is analysed.

The two bore holes were drilled to evaluate caving and tilting of the strata at different horizons and at different locations. One was drilled over the center of the panel while the other was drilled over the gates. But due to installation problems, neither hole was instrumented to its total critical depth. The hole over the center of the panel was drilled to within 15 m (50') of the coal but could only be instrumented (cased) to 61 m (200') above the coal. The hole over the gates was drilled to within 15 m (50') of the coal and could only be

instrumented to 26 m (86') above the coal. Because of this some of the critical lower most horizons could not be monitored during subsidence. A detailed analysis of this portion of this study will be released by Dames and Moore at a later date.

The monuments of this study, like those of the No. 4 study, will be monitored for the next 2 - 3 years. The results will be evaluated and then used to further refine predictive methods for ongoing structure subsidence.

CONCLUSION

North American Coal's initial and subsequent studies have permitted an accurate comparison with existing methods of subsidence prediction. Results of these comparisons have shown that the National Coal Board methods are fairly accurate when applied to local conditions, and with minor adjustments are now successfully being used.

BIBLIOGRAPHY

Adler, L., and Sun, M.C., 1968, "Ground Control in Bedded Formations," Virginia Polytechnic Institute, Research Division Bulletin 28.

Cochran, D. R., 1971, "Mine Subsidence--Extent and Cost of Control in a Selected Area," U.S. Bureau of Mines Information Circular 8507.

Crane, W.R., 1931, "Essential Features Influencing Subsidence and Ground Movement," U.S. Bureau of Mines, Information Circular 6501.

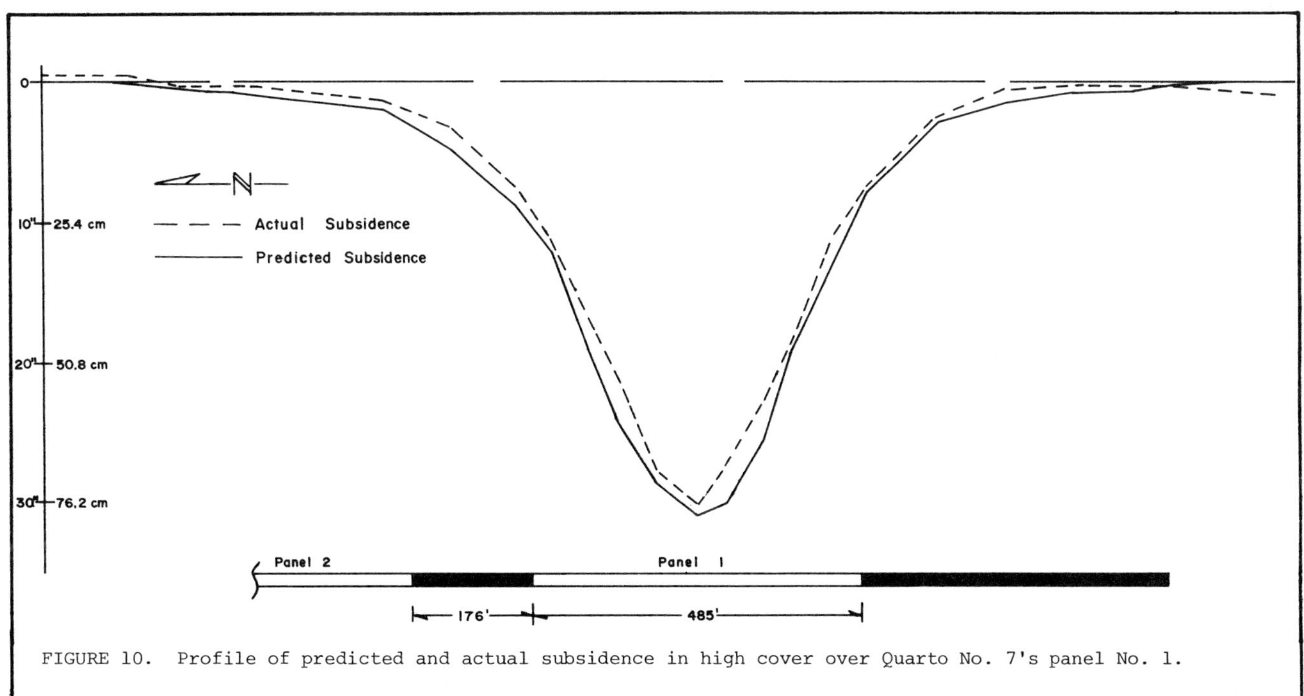

FIGURE 10. Profile of predicted and actual subsidence in high cover over Quarto No. 7's panel No. 1.

King, P.K., and Gentry, D.W., 1979, "Development
of Subsidence and Horizontal Strain Models for
Longwall Mining at the York Canyon Mine,"
Transactions SME-AIME, Annual Meeting, Preprint
79-85, 3 pp.

Kohli, K.K., and Peng, S.S., 1980, "Surface Subsidence Due to Underground Longwall Mining in
the Northern Appalachian Coal Field," Transactions SME-AIME Annual Meeting, Preprint
80-53, 8 pp.

McClain, W.C., 1966, "Surface Subsidence Associated with Longwall Mining," TRANS. AIME,
Vol. 235, pp. 231-235.

National Coal Board (U.K.) 1975, "Subsidence
Engineers Handbook," Mining Department.

Peng, S.S., Kohli, K.K. and Cheng, S.L., 1980,
"Surface Subsidence and Structural Damages
Due to Underground Longwall Coal Mining - A
Case Study," Proceedings of 21st U.S. Nat-
ional Symposium on Rock Mechanics, University
of Missouri, Rolla.

Roscoe, M.S., 1978, "Strata Control by North
American Coal Corporation in the Pittsburgh
No. 8 Coal Seam, Ohio," Stability in Coal
Mining, Brauner, C.O., Dorling, I.P.F., Ed.,
San Francisco, Miller-Freeman, pp. 270-289.

Wade, Lewis V., and Conroy, Peter J., 1977, "Rock
Mechanics Study of a Longwall Panel," Transactions SME-AIME Annual Meeting, Preprint
77-I-391.

Chapter 12

SURFACE SUBSIDENCE DUE TO UNDERGROUND LONGWALL
MINING IN THE NORTHERN APPALACHIAN COAL FIELDS

K.K. Kohli

S.S. Peng

Department of Mining Engineering
West Virginia University
Morgantown, West Virginia

R.E. Thill

US Bureau of Mines
Twin Cities Research Center
Minneapolis, Minnesota

Introduction

Since the adoption of the Surface Mining Reclamation and Control Act of 1977, which mandates that surface subsidence be an intergral part of the underground coal mine design, there has been a sudden rise of demand on guidelines for minimizing surface subsidence and its associated structural damages. Unfortunately there is no sufficient amount of systematic subsidence data to fill such a gap, although considerable research has been devoted to the subsidence-induced damages of surface structures in the past.

However, it was said that many surface subsidence surveys have been performed and monitored by coal companies during the past twenty years, and that it will be a great source of information for subsidence engineering design if they can be collected and assembled together.

An effort was therefore initiated in the Department of Mining Engineering, West Virginia University, under contract with the U. S. Bureau of Mines in December 1978 to contact each individual coal company in the Northern Appalachian Coal Field for any surface subsidence survey data. After 10 months' survey, the data collected include: 17 longwall panels and 2 room and pillar sections. The data in each panel and section cover the complete set of subsidence survey raw data, underground mine layout maps, surface topo maps, geological loggings showing the stratigraphic sequences and thickness of the overburden, and if available, locations and types of surface structures. On-site visits were followed to study the surface terraines and structures. Interviews were conducted with the owners of the surface structures in most cases about, if any, the history and characteristics of damages.

Most subsidence surveys started with the first panel and ended with the second or sometimes, but rarely, the third panel. The first panel may be located in a virgin area but frequently it was adjacent to (on the tailentry side) a mined-out room and pillar section. It must be noted that all of the subsidence data is restricted only to the amount of vertical subsidence and that the horizontal displacement measurement was completely ignored. Primarily due to the fact that vertical subsidence is the most striking feature in a subsidence trough, which is mistakenly considered to be the only critical contributor to surface structural damages, and that a horizontal displacement survey is much more complicated and time-consuming. Another shortcoming of some of the data is that the time interval between subsidence surveys was not well-planned to coordinate with the face locations. In some cases subsidence surveys were conducted after the face had passed the surface monuments. In spite of these shortcomings, the amount of data obtained by the coal companies is impressive and extremely useful.

Subsidence monuments in most cases are wood stakes driven at least 10 inches into the ground. Railroad spikes driven all the way and reset with the surface of the pavements were used for subsidence surveys of roadways. The amount of subsidence in each survey was determined to an accuracy of at least 0.01 ft.

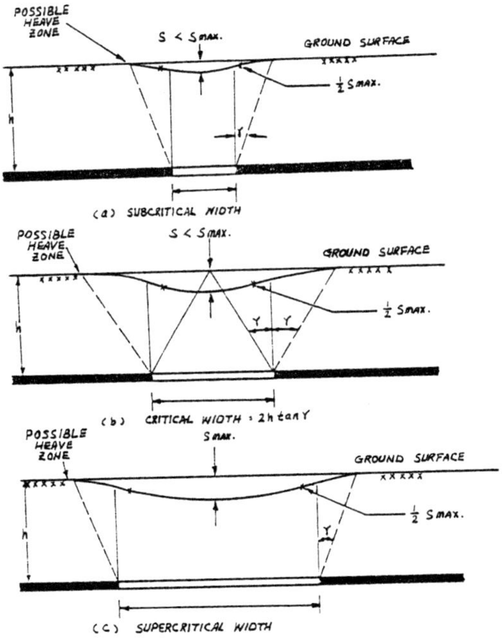

FIG. 1 EFFECT OF WIDTH OF GOB ON SUBSIDENCE

A comprehensive in-depth analysis of the data is underway. This paper presents the preliminary results of the longwall subsidence data. It includes angle of draw, subsidence profiles, subsidence factors as affected by the seam depths, subsidence development curves, and the effects of time, surface topography and multi-panel mining.

Table 1. Angle of Draw

Case No.	Panel Width W, (ft.)	Depth h, (ft.)	$\frac{W}{h}$	Angle of Draw (deg.)
1	480	600	0.80	21
2	480	540	0.88	25½
3	480	600	0.80	30½
4	480	760	0.63	23½
5*	480	760	0.63	23½
6	430	665	0.65	26
7*	430	665	0.65	34½
8	430	700	0.61	23½
9*	430	700	0.61	35
10	410	750	0.55	23
11	410	750	0.55	24

* One side of these panels were adjacent to a gob area mined by room-and-pillar method.

Angle of Draw

The angle of draw is the angle between a vertical line from the edge of workings and a line to the point at which subsidence tails out to zero. Any surface structure lying outside the angle of draw is not affected by mining.

Observed angles of draw are shown in Table 1. It ranges from 21 to 35 degrees. A majority of the angle of draw lie between 21 and 26 degrees. In fact, with only one exception (Case No. 3), those cases whose angles of draw exceed 28 degrees were for panels with one side adjacent to a gob area previously mined by room-and-pillar method. It was the angle of draw measured in this side that exceed 28 degrees (see also Figure 3).

Subsidence Factor

The width and depth of an underground panel are intimately related as far as subsidence is concerned because together they determine the critical width. A critical width is the width of a panel which causes the maximum possible subsidence at one point (usually at the center of the panel) on the surface (Figure 1).

Experience has shown that for a flat coal seam at a fixed depth, if an underground long-wall panel is as wide as the critical width with respect to a surface point, the maximum possible subsidence occurs at this point only. If the panel is wider than the critical width, an area in the mid-span will experience the maximum possible subsidence. The panel width is then a super-critical width. On the other hand, if the panel is narrower than the critical width, the maximum subsidence occurs at the mid-span but is less than the maximum possible subsidence

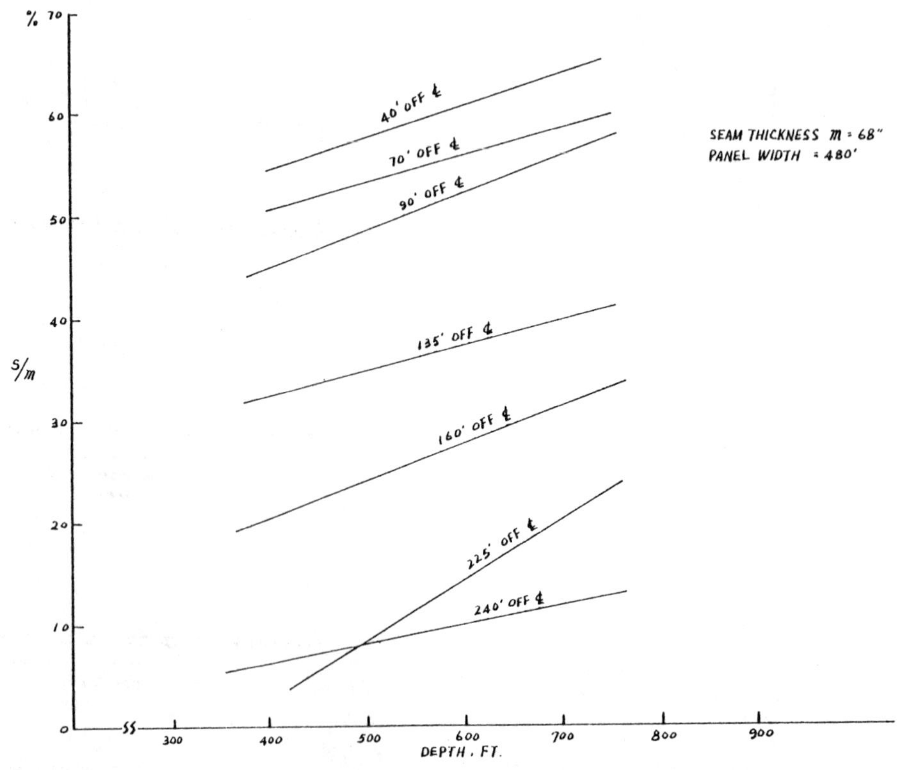

FIG. 2 DEPTH EFFECT

of the seam. The panel width is then called a subcritical width.

The subsidence data indicate that the subsidence factor ranges from 0.22 to 0.7 and the subsidence factor increases with seam depth. The panel width of the cases studied ranged from 450 ft. to 600 ft. and depth from 300 ft. to 1000 ft. Table 2 shows the observed maximum possible subsidence, subsidence factor and width-to-depth ratio for seven cases. The panel width in all the cases is either critical or supercritical because the point of half maximum subsidence in each of the subsidence profiles falls within the gob. Figure 2 shows graphically the depth effect on the subsidence factor for one case where subsidence factor vs depth has been plotted for various points located in different parts of the panel. Regardless of the locations, the subsidence factor increases with depth. Similar trend has been found in other cases. This phenomena is contrary to the NCB's (1) investigation but not rare as found by Wardell and Webster(2), Zwartendyk (3) and Gray and Slaver (4).

The increase in subsidence factor with depth can be explained as follows:

Longwall mining induces caving at the immediate and intermediate roof strata. The caving process propogates upward to a horizon located at a distance of approximately 35 to 50 times the mining height over the coal seam. The deformation above this horizon observed by Dahl and Von Schonfeldt (5) is continuous, i.e., no appreciable breakage with little or no bed separation occurs. The main roof strata settle down in more or less continuous pieces on the gob without any appreciable increase in volume. At greater depth this main roof strata will have more weight to exert on the gob which will compress the broken rock more as compared to shallower depth, thus causing more subsidence at the surface.

The subsidence factor observed for the Appalachian Coal Field is considerably smaller compared to the NCB's (1) prediction. Difference in underground panel layouts and geology between the U. K. and the Appalachian Coal Field is believed to be the most important factors.

Table 2. Subsidence Factor

Case No.	Mining Height m (ft.)	Depth h (ft.)	Width W (ft.)	$\frac{W}{h}$	S_{max} (ft.)	$\frac{S_{max}}{m}$
*1	5.6	391	480	1.23	3.3	.59
*2	5.6	579	480	0.83	3.5	.63
*3	5.6	778	480	0.62	3.9	.70
+4	5.3	792	410	0.52	1.2	.22
+5	5.3	893	410	0.46	1.3	.25
+6	5.3	984	410	0.42	1.5	.29
#7	6.0	650	430	0.66	2.5	.42
#8	6.0	750	430	0.57	2.6	.44

* data from first mine

\+ data from second mine

\# data from third mine

Subsidence Profile

Subsidence profiles drawn from the observed data can be approximated by the following profile function (Brauner, p. 11, eq. 3):

$$S = \frac{1}{2} S_{max} \left[1 - \tanh \left(\frac{2X}{B} \right) \right] \qquad (1)$$

where S is the subsidence for the point of interest located at an horizontal distance, X, from the point of half-maximum subsidence. B is one half of the critical width, and S_{max} is the maximum possible subsidence. Note that $B + d = \frac{W}{2}$ where W is the panel width and d is the horizontal distance from the point of half-maximum subsidence to the edge of the panel.

If an average value of d = 0.1h (see Table 3) is used in Equation 1, the predicted subsidence profile for all the cases are found to match the observed profile closely (Figure 3) with the following exceptions: (1) For some profile, it is not symmetrical about the center line. One

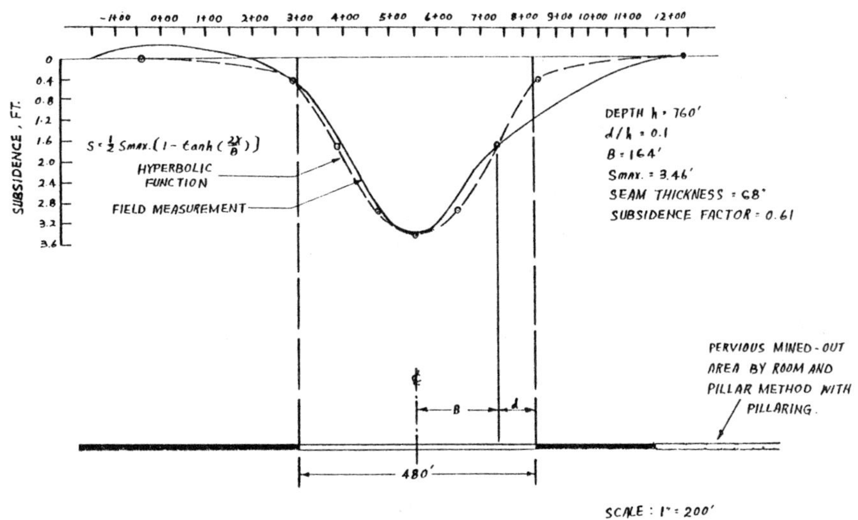

FIG. 3 SUBSIDENCE PROFILE , OBSERVED VS. HYPERBOLIC PROFILE FUNCTION

Table 3. Distance of Half Maximum
Subsidence Point From the Edge

Case No.	W (ft.)	h (ft.)	$\frac{W}{h}$	d (ft.)	$\frac{d}{h}$
1	440	600	0.63	50	.092
2	480	760	0.63	90	.118
3	430	700	0.61	50	.071
4	430	670	0.64	100	.149
5	430	700	0.61	100	.142
6	430	750	0.57	70	.093
7	430	660	0.65	90	.136
8	480	950	0.51	90	.094
9	480	900	0.53	80	.088
Average					0.108

W = panel width

h = panel depth

d = horizontal distance from the panel edge to
the point of half-maximum subsidence

side of the profile which lies toward the side
where previous mining (not necessarily longwall
mining) has already been done does not match
with the profile on the other side. This variation seems to be due to the subsidence effect
from the mined out area causing the observed
profile to be flatter than predicted from the
hyperbolic function. (2) In some case, the
predicted and the observed profiles do not match
for the portion starting immediately above the
rib to the point of zero subsidence. The profile
drawn from the observed data shows a slight heave
before approaching a zero point.

Heaving beyond the ribs of the panel being
mined is due to bending of the strata above the
panel using the ribs for cantilever action. This
further strengthens the idea that main roof
strata settles down in one section without
appreciable breakage as explained in the subsidence factor section.

Subsidence profile can be plotted along the
longitudinal or the transverse cross-section of
the panel. A curve depicting subsidence on the
section drawn parallel to the direction of
advance of the longwall panel is longitudinal
and at right-angles to the direction of advance
is transverse.

In order to draw a subsidence profile, the
locations and amounts of subsidence for at least
five positions are needed. The positions are
(1) The position of zero subsidence. These are
the end points of the subsidence trough. e.g.
Points 1 and 2 in Figure 4. These two points
can be located by drawing the angles of draw
from the edges of a longwall panel.
(2) The position where the maximum possible
subsidence occurs. e.g. Point 3 in Figure 4.
This point is usually located at the center of
the panel.
(3) The positions of half maximum subsidence
e.g. Points 4 and 5 in Figure 4. These points
are located at a distance d from the edges of
the panel.

The positions of the zero subsidence and
maximum subsidence can be easily located. The
position of the half maximum subsidence point,
d, is found by a profile function (Eq. 1) where
the value of d is found empirically from several
subsidence data. Table 3 shows the values of d
for several cases with different panel depths
and widths.

Subsidence Development Curve and Time Dependent Effect

The subsidence development curve depicts the
changing amount of subsidence of a surface point
in relation to the position of a longwall face
passing through the critical width of that point.
From the subsidence development curve, prediction
can be made as to when any point on the surface
is like to subside first, when the likelihood of
subsidence is most acute and finally when the
subsidence will be complete.

Figure 5 represents the subsidence development
curve for one case, where the vertical axis is
subsidence in terms of subsidence factor, also
subsidence in terms of maximum subsidence and
the horizontal axis denotes the face location
with respect to surface point, P. The curve
indicates the subsidence at point P as a function
of face locations which is advancing to the
right. Subsidence at P begins when the face is
at point A about 400 ft. (.5 h) ahead of P and
increases linearly as the face advances. It
reaches 3% of the maximum subsidence when the
face is directly below P. The subsidence rate

FIG. 4   TYPICAL SUBSIDENCE PROFILE
FOR CRITICAL WIDTH OF EXCAVATION

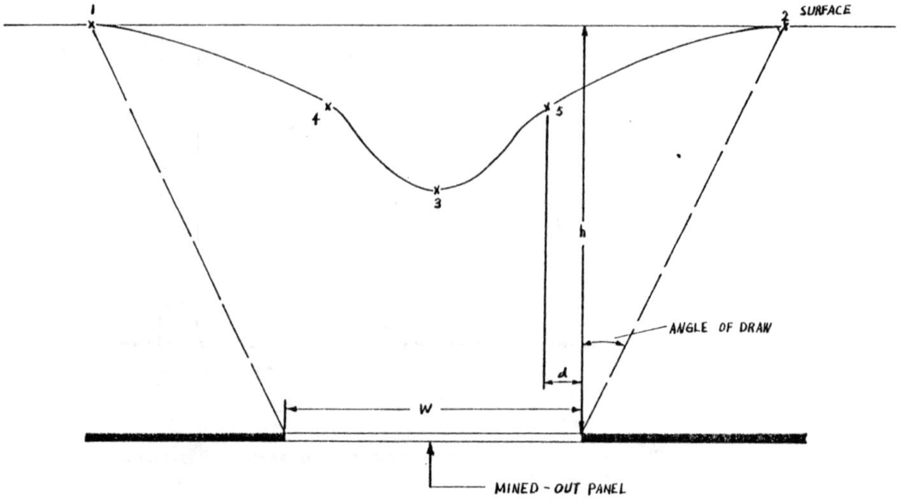

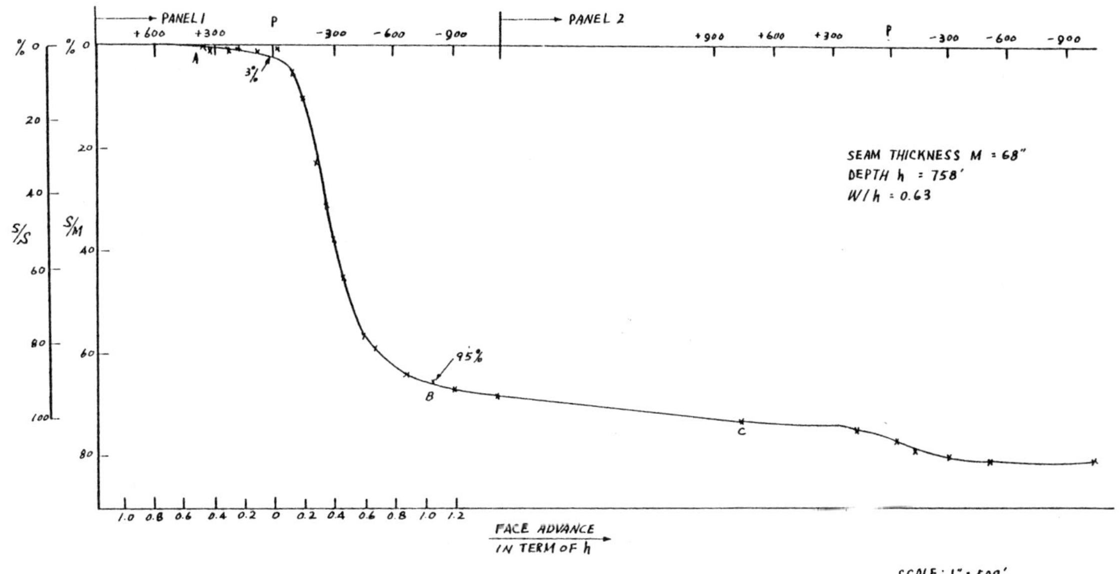

FIG. 5, SUBSIDENCE DEVELOPMENT CURVE OR TIME - DEPENDENT CURVE

is maximum (which ranges from 0.5 ft. to 1.25 ft. per 100 ft. of face advance) almost as soon as the face passes beyond P. The subsidence reaches 95% when the face is at Point B about 800 ft. (1.0 h) beyond P.

At Point B the active subsidence is complete; the remaining 5% (from Point B to C) is due to gradual compaction of the subsided ground and is known as the residual subsidence or time dependent subsidence.

Table 4 lists the data for several cases which show when the first movement started and when the subsidence was complete in each case. The distance ranges for the first movement varied from 1.0 h to 1.6 h. The maximum rate of subsidence in the majority of the cases was found to occur when the point was between 200 ft. and 400 ft. behind the face.

Table 4. Subsidence Development Data

Case No.	Depth (ft)	D_1 (ft)	$\dfrac{D_1}{h}$	% of S_p at P	D_2 (ft)	$\dfrac{D_2}{h}$	% of S_p at B	S_t (%)
1	758	400	.5h	3	800	1.1h	95	5
2	373	300	.8h	6	600	1.6h	87	13
3	700	400	.6h	10	750	1.1h	90	10
4	749	400	.5h	3	750	1.0h	91	9
5	792	300	.4h	10	800	1.0h	92	8

D_1 denotes the horizontal distance between the point (Point A) where first movement occurs and the surface point, P

D_2 denotes the horizontal distance between the point (Point B) where the active subsidence ceases and the surface point, P

S_p denotes maximum subsidence at point P when the subsidence is complete

S_t denotes time dependent subsidence

Field measurements for all the cases also have indicated that the time dependent subsidence is less than 13% of the maximum subsidence (Table 4) and that it generally does not have any significant effect on the surface structures.

Surface Topography Effect

The majority of the mines where these data were collected are located in medium to steep hills of the Northern Appalachian Coal Field. The surface slope varies from 0 to 35 percent.

No significant effect is found due to change in surface topography as shown in Figure 6 which is a graphical presentation of topographic effect on subsidence for the majority of the cases. Subsidence monitoring points shown in this figure are located at the center of each panel and thus have the maximum possible subsidence. There is no trend in the observed data thereby suggesting that surface topography does not have a significant effect on the subsidence. Similar trend has been found for other cases where the subsidence monitoring points are located away from the center of the panel.

Dahl and Choi (7) have also observed that surface topography does not have a significant effect on subsidence in the Appalachian Coal Field. But Gentry (8) found that it does in the western coal field. However, if the surface topography is steep, the observed effect on subsidence may be due to the depth instead of surface topography. It is believed that the effect of surface topography can be greatly altered by the type and degree of vegetation on the surface. An area with heavy vegetation can be rather stable even if the terrain is steep whereas the surface in a desert will be disturbed by slight underground movement.

Interpanel Effect

Interpanel effect may be divided into two parts. Firstly, the effect of old panel on the new panel. Field measurements have indicated that mining in an adjacent new panel has a strong influence on the maximum possible subsidence in the gob of the old panel. But the

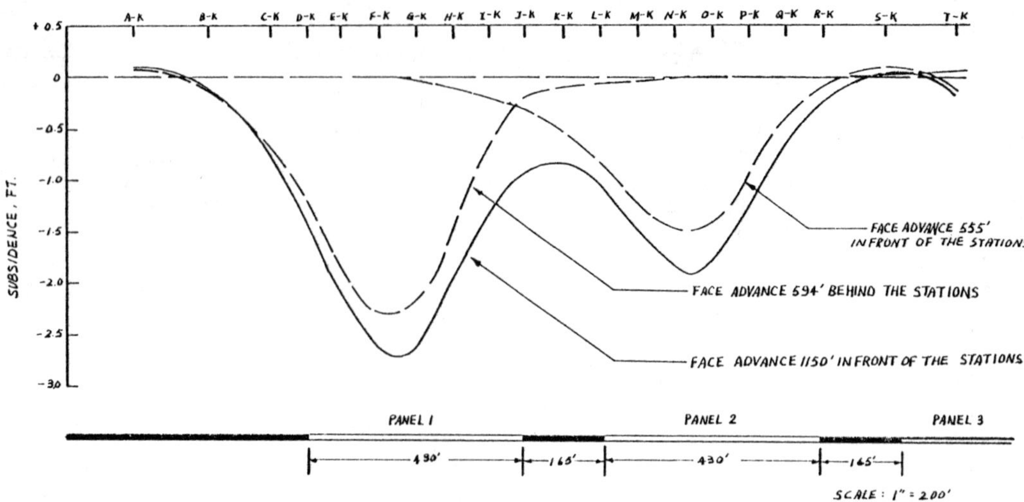

FIG. 7 INTERPANEL EFFECT (PANEL 1 OVER PANEL 2)

influence of old panel on the new panel is not significant. Figure 7 shownns the transverse subsidence profiles of two adjacent panels and indicates that the maximum possible subsidence in old panel (#1 panel) increased by almost 15% due to the influence of mining in new panel (#2 panel). Figure 8 shows transverse subsidence profiles of two adjacent panels which indicate that the influence of old panel (#1 panel) over the new panel (#2 panel) is not significant. Field measurements have revealed that panel #3 also influences the subsidence in panel #1 but the data available is only for a few cases. The interpanel influence on subsidence for various cases studied ranged from 15 to 33% of maximum possible subsidence. Because of this influence on maximum possible subsidence due to adjacent panels, the subsi-

dence factor will also increase. The subsidence factor accumulated as a result of the interpanel effect is called the maximum subsidence factor (a_{max}) whose value will depend on the combined influence of all the adjacent panels.

Summary

To the best of our knowledge, the vast amount of subsidence data collected in this research constitutes the first effort ever attempted in this country. The results of the analyses can therefore be reasonably assumed that they are applicable to most subsidence cases in the Northern Appalachian Coal Field. The results of preliminary analysis indicate that:

1. The angle of draw ranges from 21 to 30 degrees with a few exceptions.
2. The subsidence factor ranges from 0.22 to 0.7. It increases with seam depth.
3. The subsidence profiles can be approximated and predicted by a simple profile function in most cases.
4. The time dependent or residual subsidence is less than 13% of the total subsidence.
5. Surface topography or surface slope does not seem to have any effect on surface subsidence, and
6. In multipanel mining, interpanel effect will add from 15 to 33% to the maximum possible subsidence obtained in single panel mining.

References

1. U. K. National Coal Board, The Subsidence Engineering's Handbook, 1975, 111 pages.

2. Wardell, K., and N. E. Webster, "Surface Observation and Strata Movement Underground", Colliery Engineering, V. 34, 1957, pp. 329–336.

3. Zwartendyk, J., "Economic Aspects of Surface Subsidence Resulting From Underground Mineral Exploitation", Ph.D. Thesis, Pennsylvania State University, March 1971 NITS No. PB 207 512.

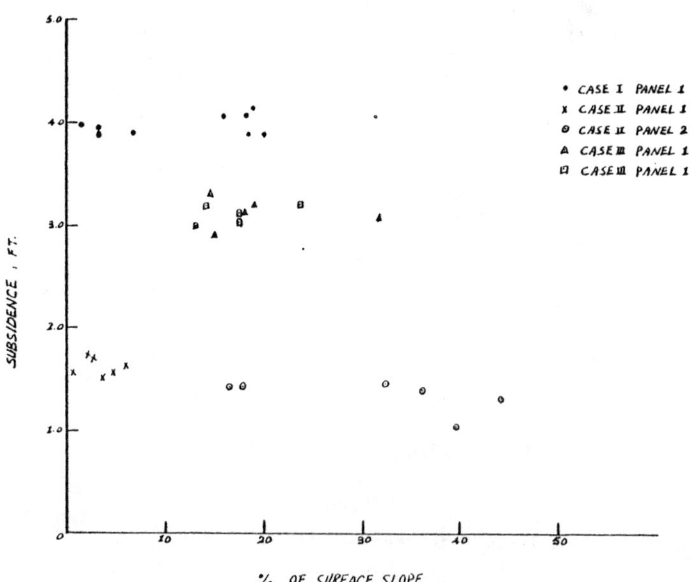

FIG. 6 SUBSIDENCE VS. SURFACE SLOPE

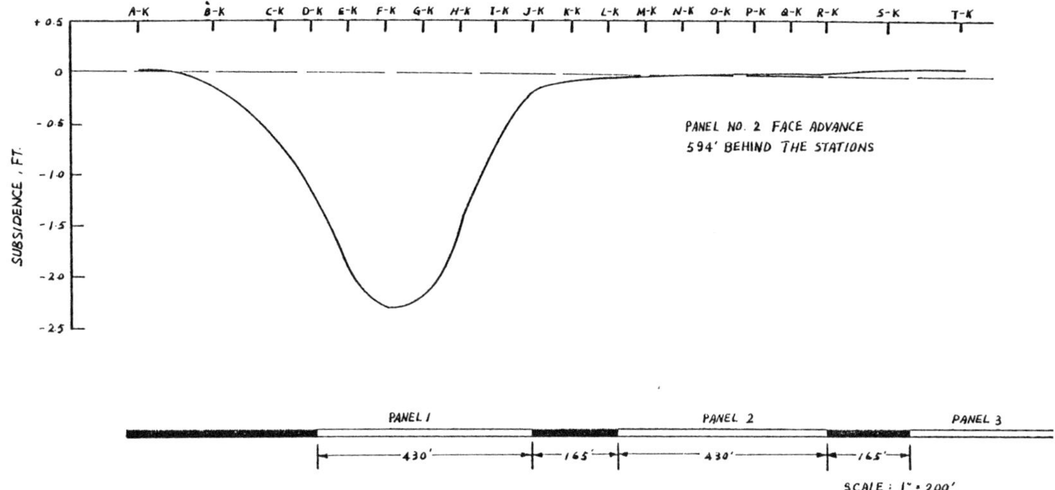

FIG. 8 EFFECT OF PANEL NO. 1 OVER PANEL NO. 2

4. Gray, R. E., and H. A. Slaver, Discussion of
    "State of Predictive Art in Subsidence
    Engineering," J. Soil Mechanics and Founda-
    tion Division, Proceeding, ASCE, Vol. 97,
    No. SM1, January 1971, pp. 258-260.

5. Dahl, H. D., and H. A. Von Schonfeldt, "Rock
    Mechanics Elements of Coal Mine Design,"
    17th U.S. Symposium on Rock Mechanics,
    Snowbird, Utah, 1976.

6. Brauner, G., "Subsidence Due to Underground
    Mining", U.S.B.M. IC 8571, 1973.

7. Dahl, H. G., and D. S. Choi, "Some Case
    Studies of Mine Subsidence and Its Mathemat-
    ical Modelling", Proc. 15th U. S. Sym. on
    Rock Mechanics, Custer State Park, SD, 1973,
    pp. 1-22.

8. Gentry, D. W., "Rock Mechanics Instrumenta-
    tion Program for Longwall Support at York
    Mine, New Mexico," Interim Technical Report,
    July 1976, 613 pages.

Chapter 13

MEASUREMENT OF THE SUPPORT RESISTANCE OF SHORTWALL CHOCKS AND ITS APPLICATIONS[†]

Duk-Won Park, Ph.D.
Assistant Professor

Syd S. Peng, Ph.D.
Professor and Chairman

Department of Mining Engineering
College of Mineral and Energy Resources
West Virginia University
Morgantown, WV 26506

ABSTRACT

For an adequate design of the shortwall face support, it is necessary to understand fully the support-roof interaction. A series of studies has been carried out at a shortwall panel to develop the methods of monitoring support resistance and determining load density of the chocks. Various aspects of applications, for improvement of chock design and understanding of roof conditions, are discussed.

INTRODUCTION

Shortwall mining method is a relatively new technique in the United States coal mining industry. In comparison with longwall mining method, its development and application are not encouraging[1,2]. One of the major causes of unsuccessful operation comes from ground control problems. The lack of understanding of the support-roof interaction resulted in improper chock design. A series of studies has been carried out at the shortwall panel in Valley Camp No. 3 Coal Mine near Triadelphia, WV.

The methods of monitoring support resistance and determining load density of the chocks were developed, and their applications are discussed in this paper.

Shortwall Mining at Valley Camp No. 3 Mine

Valley Camp No. 3 Mine is located in the northern panhandle of West Virginia. The thickness of the coal seam (Pittsburgh Seam) ranges from 5 to 6 feet. The shortwall panel, located under rugged and hilly topography, has an overburden whose thickness ranges from 800 to 950 feet. The typical stratigraphic columns are shown in Fig. 1. The first 12 to 15 feet immediately above the coal seam is shale with irregular coal partings and clay veins.

The shortwall panel was 150 feet wide by 2,655 feet long. The panel employed retreating shortwall mining method with advancing tail entry (Fig. 2). In the head entry side there were 6 rows of chain pillars, while the tail entry was constructed by setting cribs in a regular interval as the face advanced. As shown

[†Editor's note: This paper was presented at the 20th U.S. Symposium on Rock Mechanics, held in Austin, Texas, June 4-6, 1979. Permission to reproduce here has been obtained from the authors and Dr. Ken Gray, U. of Texas at Austin, Petroleum Engineering Dept.]

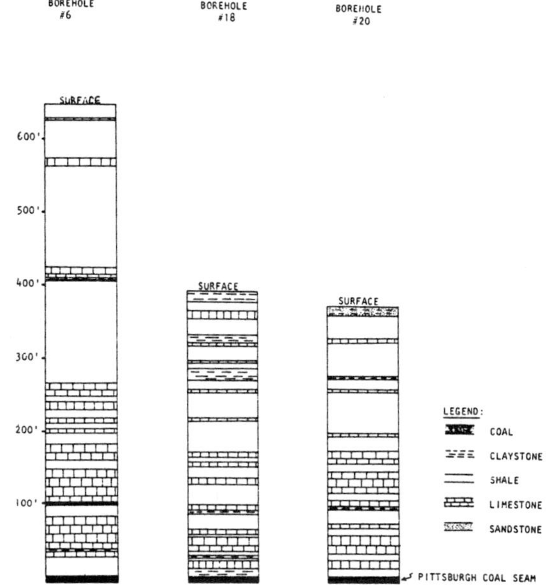

Fig. 1 – Typical stratigraphic columns near the shortwall panel.

in Fig. 2, roof falls occurred frequently at the crosscut areas. The roof falls resulted in dome shaped voids, which measured 1-20 feet wide and 1-15 feet high. In the panel roof falls were made along the clay veins. The roof falls were very severe along the final working face. They covered about half of the working face, and they extended as high as 20 feet into the roof. A simplified shortwall face layout is shown in Fig. 3. Coal was mined by a Joy 12 CM continuous miner, and a loader and a shuttle car were used for face haulage. The miner cuts a web width of approximately 9 feet.

Along the face, 39 four-legged Gullic-Dobson shortwall chocks with 500-ton capacity were laid at 4 feet center. The setting pressures ranged from 1500 to 3000 psi and yield pressures were 7100 psi for the legs and 8300 psi for the front canopy. The shortwall chocks differ from the longwall chocks by having an extensible canopy in front of the front canopy for immediate roof support near the face line area.

The sequences of chock advance are shown in Fig. 4. Before the miner started to make a new cut, the ram jack of the chock was in fully extended position which

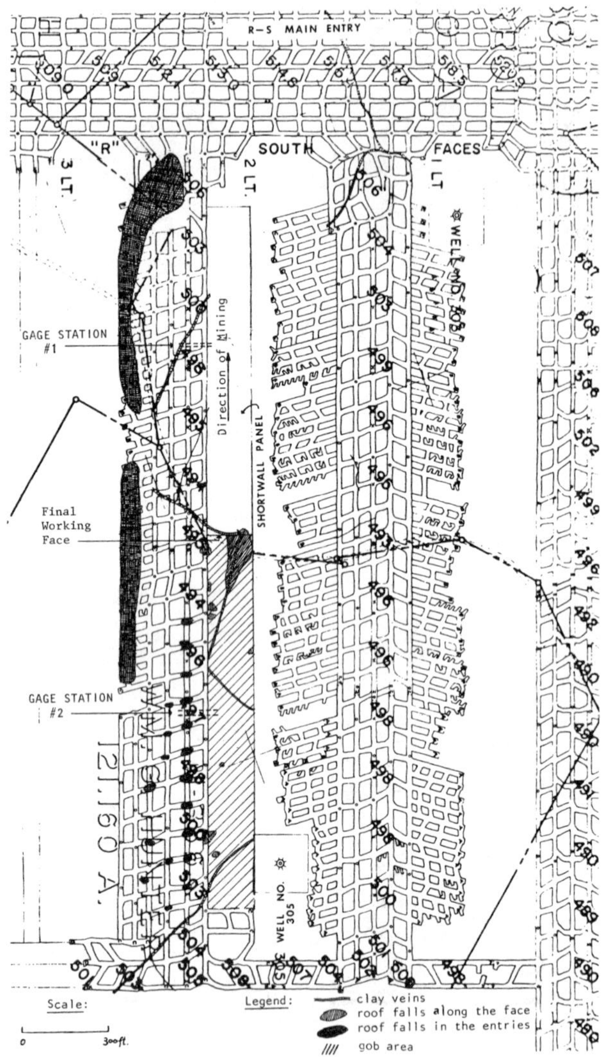

Fig. 2 - Shortwall panel layout with clay veins and
　　　　roof falls.

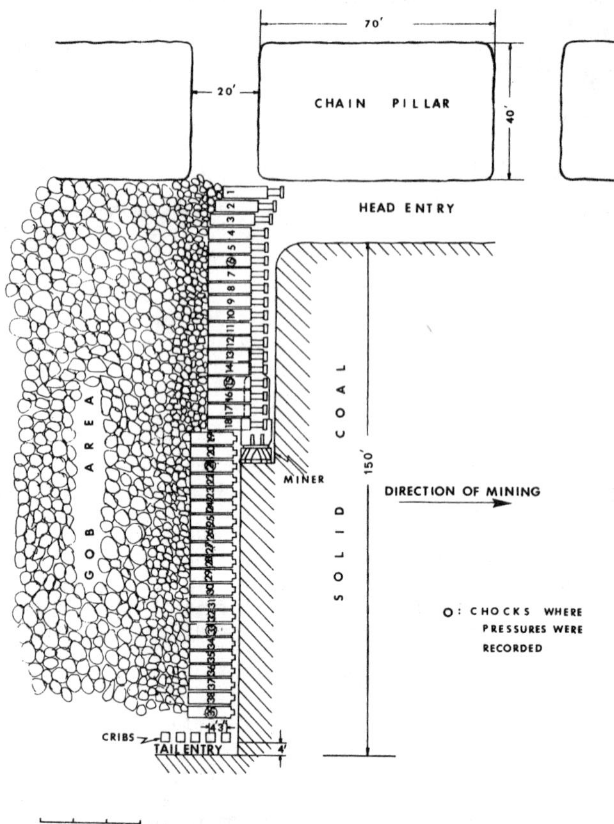

Fig. 3 - Shortwall face layout.

is 4 feet long (Stage #1). When the miner cut beyond
two chocks past the chock to be advanced, the chock
was advanced to the position immediately behind the
spill plate. Simultaneously, the extensible canopy
was fully extended to the 4-foot length (Stage #2).
Stage #3 involved extension of the ram jack and
advance of the spill plate. In Stage #4 the chock
was advanced to the spill plate which was then
advanced to Stage #5.

For convenience of analysis, the shortwall mining
activities are divided into 3 periods: Period 1
covers Stages #1 and #2; Period 2 covers Stage #3; and
Period 3 covers Stages #4 and #5.

SUPPORT RESISTANCE OF THE CHOCKS

Face powered supports for shortwall or longwall
panels are designed to support the immediate roof,
which behaves like a cantilever beam with a soft
support clamped at the front abutment. The location
of the maximum abutment pressure was determined to be
10 feet ahead of the face line. Chock pressures were
recorded by Weksler pressure recorders where

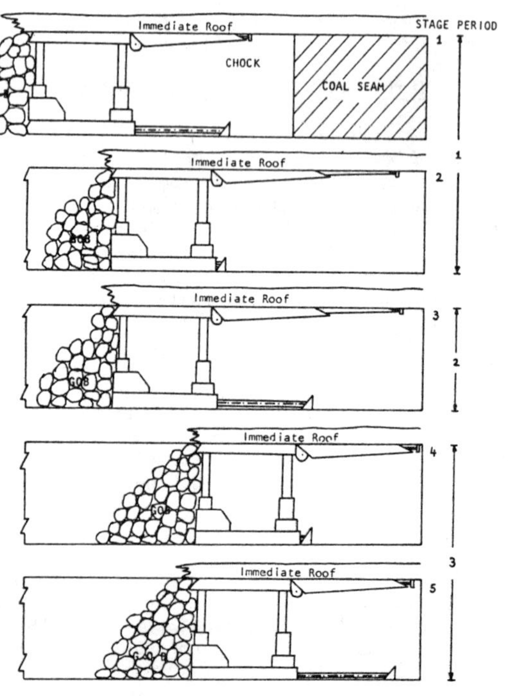

Fig. 4 - Sequence of chock advance.

mechanically wound clocks were attached. Chocks #6, 15, 21, 33 & 39 were selected for monitoring pressure variations (Fig. 3). In each chock, 3 separate line pressures were monitored, i.e., front and rear legs, and front canopy. Fig. 5 shows a typical pressure variation chart as recorded by the pressure recorders installed at Chock #21. The chart with 24-hour per revolution serves better use of establishing pressure variation trends associated with each stage of mining activities while the 7 days' chart was used to confirm the reproducibility of each pressure variation in a larger period.

The letters with arrows in Fig. 5 indicated the corresponding events of the face operations which can be summarized as follows:

A. The chock at this point was advanced and reset in preparation for the new cut, which was early in Stage #1. The setting pressure for the front leg was approximately 2600 psi and dropped rapidly to 800 psi responding to the crushing of roof steps. Thereafter it increased slowly but steadily. The trend of variation of pressures applied in the canopy was also similar to that of the front legs. Pressure in the rear leg remained extremely low.

B. At this point a new cut started at the head entry.

C. The new cut had been made to Chock #20. A surge in pressure resistance for the front canopy and front legs is shown.

D. After midnight shift, at this point, the cutting had resumed and reached Chock #21.

E.,F.,G. The rest of the chocks in the panel were individually lowered, advanced and reset as the cutting progressed toward the tail entry. Pressures in the front legs and front canopy stepped up more and more.

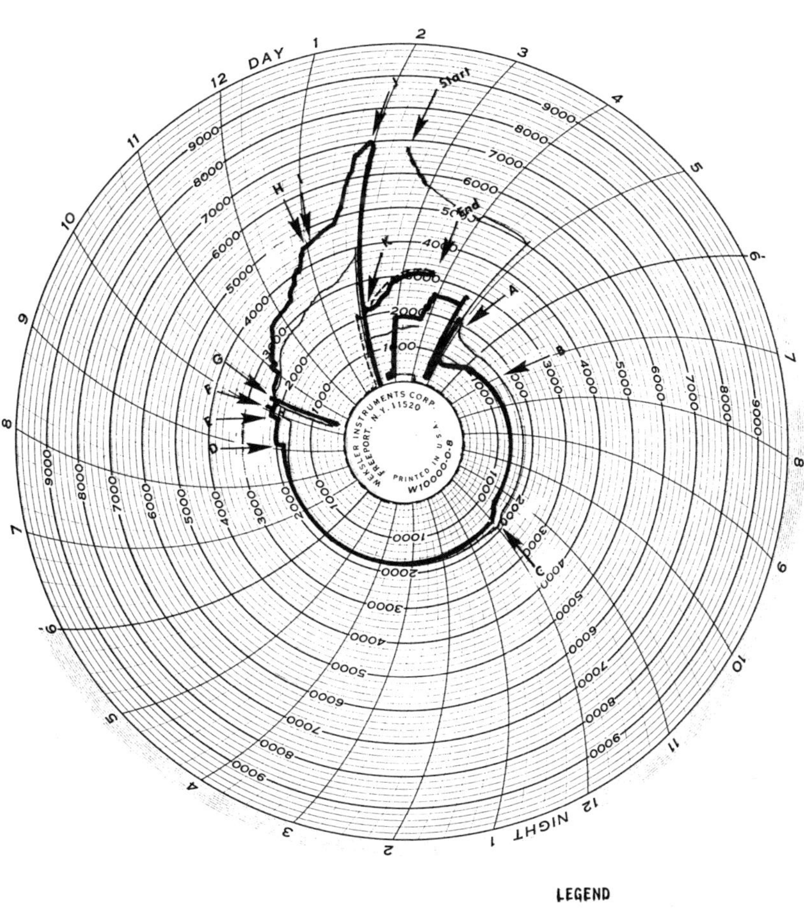

Fig. 5 - Typical pressure variation recorded by a pressure recorder set for 24-hours per revolution.

H. At this point the cutting had completed the full face.

I. The spill plate was advanced and the chock entered into Stage #3 or Period 2.

J. The advance of the chocks into Stage #4 started from the center to the sides of the face. As it proceeded and approached Chock #21, the pressures increased to a very high range.

K. At this point Chock #21 was reset. The pressure in the rear legs sustained for the first time during the day, because the immediate roof near the gob edge was intact at this moment.

However, it was noticed that the top surface of the chock canopies seldom contacted the immediate roof line completely, because the roof line was not smooth enough. A systematic underground observation showed that the point of contact at the front canopy was generally restricted to the front half when the extensible canopy was retracted. Once the extensible canopy was extended, the major contact point moved to the tip of the extensible canopy.

Based on these observations, the free body diagrams for support-roof interaction are shown in Fig. 6. The hydraulic pressures measured at the front (FL) and rear (RL) legs represent the total concentrated force at each point but those measured at the capsule (Pc) are not a direct indication of the forces applied at the front canopy; rather it reduces by an amount proportional to the relative length of the level arms.

$$Ex = \frac{18}{113} \, Pc = 0.159 \, Pc \qquad (1)$$

$$AME = \frac{18}{55} \, Pc = 0.327 \, Pc \qquad (2)$$

where,

Pc = measured pressure at the capsule (psi)
Ex = resultant force at the tip of the extensible canopy (point a)(lbs.)

A. EXTENDED CONDITION

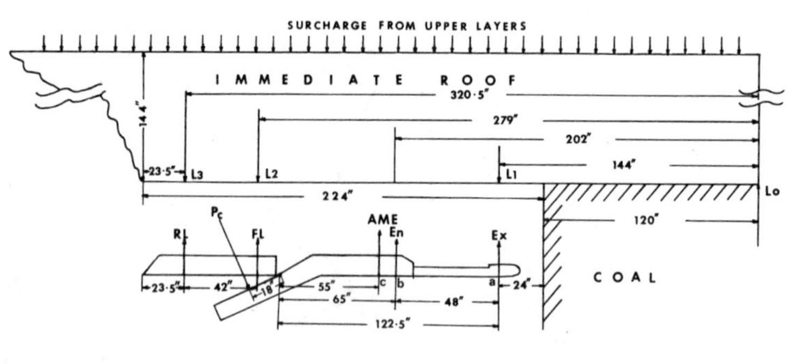

SCALE:

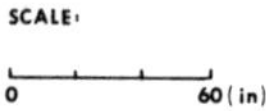

B. UNEXTENDED CONDITION

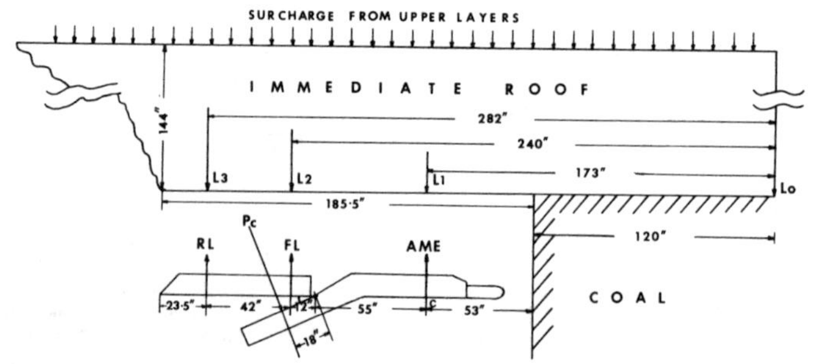

Fig. 6 - Schematic diagram for support-roof interaction.

AME = resultant force at point c (lbs.)

The total support resistance of the chock is the summation of the forces at the front and rear legs and the load Ex resulted by Pc, i.e.,

$$Tex = 2A(Pr + Pf) + 0.159\ APc \qquad (3)$$

$$Tun = 2A(Pr + Pf) + 0.327\ APc \qquad (4)$$

where,

   Tex = total support resistance for extended
         condition (lbs.)
   Tun = total support resistance for unextended
         condition (lbs.)
    Pr = measured pressure at the rear legs (psi)
    Pf = measured pressure at the front legs (psi)
     A = cross-sectional area of the hydraulic
         cylinders (in.^{2})

The total resultant force applied on the chock can be found by taking the moment about the clamped point Lo. The algebraic sum of the moments must be zero for an equilibrium condition, which results in the relationship:

$$Lex = \frac{641\ Pr + 558\ Pf + 22.9\ Pc}{2\ (Pr + Pf) + .159\ Pc} \qquad (5)$$

$$Lun = \frac{564\ Pr + 480\ Pf + 56.6\ Pc}{2\ (Pr + Pf) + .329\ Pc} \qquad (6)$$

where,

   Lex = distance of the resultant force from the
         front abutment when extensible canopy is
         fully extended (in.)

   Lun = distance of the resultant force from the
         front abutment when extensible canopy is
         fully retracted (in.)

The total support resistance for each period of each chock monitored was calculated by using Eqs. 3 and 4. It was then divided by the supporting area of each chock to obtain load density (LD)(Fig. 7). The average load density for the 3 periods is defined as the mean load density (MLD).

As shown in Fig. 7, load density and mean load density vary with the periods in which the chock is operated. Load density is higher for Period 1 than Periods 2 and 3. The data also indicated that load density and mean load density vary with the location of the chock along the panel face. A lower load density is recorded at the chocks near the head entry (Chocks #6 through #15), while a higher load density is recorded for the chocks near the tail entry. The nonsymmetrical distribution of the load density is apparently related to the panel layout (retreating method with advancing tail entry).

The load density for after weekend condition is the highest of all measured, especially those near the tail entry. This could be attributed to the fact that over the weekend the cantilever beams deflect and drop more weight on the support. Conversely, the load

density under the bad roof condition was considerably lower, because the canopy lost full contact with the immediate roof.

The tip loads at the extensible canopy were also determined for each period and "after weekend" conditions (Fig. 8). With the exception of those near the head entry and tail entry, the tip loads are much higher for Period 1 than Periods 2 and 3. The tip load after the chock has been advanced (Periods 2 and 3) never regained the top values achieved in Period 1. The maximum recorded tip load was 17.1 tons for Chock #15 in Period 1, which was far below the rated tip load of 25 tons.

Using Eqs. 5 and 6, the locations of the total resultant force for each period of the chocks were determined (Fig. 9). Again it varies not only with the locations of the chocks but also with the periods of operation. The average location of the total resultant force in Period 1 was about 10.4 feet from the face, while it moved toward the gob at 12.9 feet from the face for Periods 2 and 3. Under the bad roof condition the location of the resultant force is closer to the face (10.4 feet) whereas over the weekend, it moved further back toward the gob (14.9 feet from face). The location of the total resultant force is an indication of the length of the cantilever beam. The longer the beam is, the farther it is from the face. The cantilever beam is shorter in Period 1 when the immediate roof causes the location of the total resultant force to move toward the face. Similarly, when a bad roof or roof fall is encountered, the

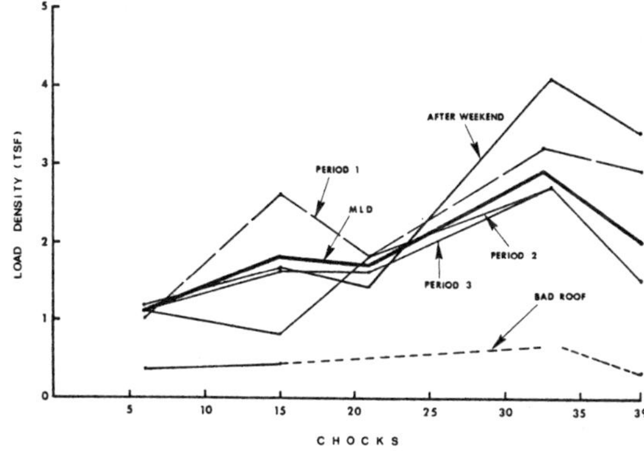

Fig. 7 - Load density (LD) and mean load density
         (MLD) on chocks.

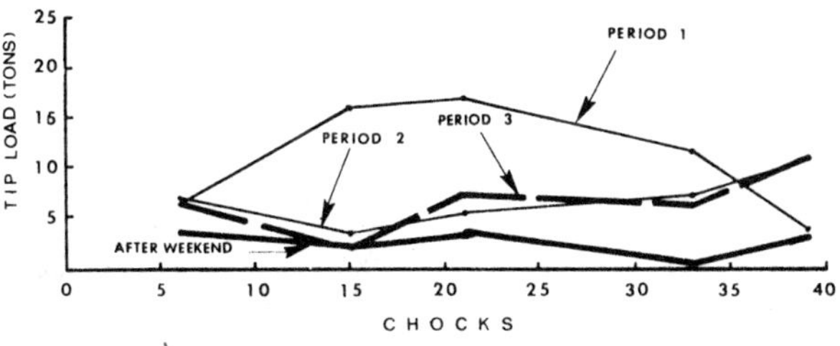

Fig. 8 - Measured tip loads at the extensible canopy.

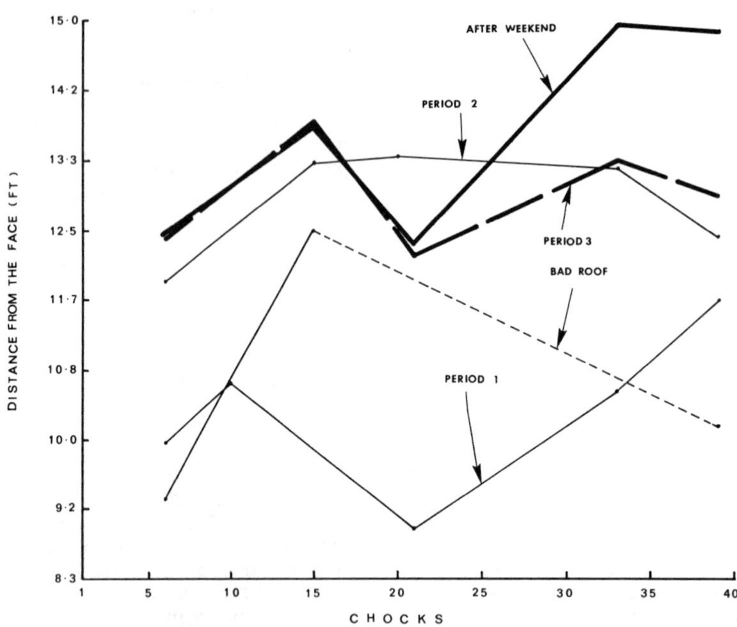

Fig. 9 - Locations of total resultant forces.

effective length of the roof beam becomes shortened because of the high degree of fractures. On the other hand, the standstill waiting time over the weekend allows the immediate roof to deteriorate all the way to the main roof which deflects and applies more load on the chock. This accounts for the fact that the location of resultant force for the after weekend period is the farthest from the face.

The location of the resultant force is also dependent on the integrity of the immediate roof. When the immediate roof is weak, the roof breaks immediately behind the chocks. Further deterioration at the end of the roof beam causes the rear legs to extend after the chocks are advanced and set. This way the pressure in the rear leg will decrease and the location of the resultant force will move forward.

APPLICATIONS

The data for determining support resistance can be easily acquired in a continuous manner with simple equipment. Analyses of the mode and magnitude of support resistance for the chocks reveal the following:

1. Understanding the overall picture of ground response to the chocks: The length and thickness of the immediate roof which rests on the chocks can be estimated.

2. Estimation of required chock capacity: From the continuous monitoring of load density, its complete range including the maximum value can be defined. For example, Fig. 7 indicates that the normal ranges of load density were 1–3 tsf, and that the maximum load density was 4.1 tsf which corresponds to a total load of approximately 200 tons per chock. A 500-ton chock as used was obviously overdesigned.

3. Balance of the support resistance: In order to make the chocks stable, it is required to make the applied load equal for every leg. Uneven load will cause the chocks to tilt or misalign. By tilting, the toe of the chocks tends to punch into the floor where floor strength is low. Also, the uneven load will cause stress concentration, which might exceed the design capacity of a particular part. The uneven load distribution can be determined by calculating the location of the resultant force. The optimum location would be in the middle between the rear and front legs. For instance, the locations of the resultant force at the short-wall panel in Valley Camp No. 3 Mine were mostly closer to the face. To cope with this problem, chock setting time had to be longer than was usually practiced. This way, the portion of the immediate roof about the rear leg can be fully consolidated.

4. Setting pressure: The effect of setting pressure was not fully studied during the study period. However, two setting pressures (2300 and 2900 psi) were used in this panel. The preliminary results indicated that a higher load density was reached for the higher setting pressure. Further study for this matter is necessary for determining the optimum level of setting pressure.

5. Predicting bad roofs: Whenever a bad roof was encountered, the chock pressure dropped and the location of the total resultant forces moved closer toward the working face. From this principal, a bad roof can be predicted and remedial measures can be taken before the weak roof falls.

CONCLUSIONS

A method of determining support resistance is
developed based on pressure recording from the shortwall chocks and its applications are discussed. The
method is relatively simple and economical so that
continuous monitoring system can be made in order to
achieve the optimum operating conditions and predict
bad roofs.

ACKNOWLEDGEMENTS

The results presented in this paper were based on
the data obtained at Valley Camp No. 3 Coal Mine. The
program was funded by the U.S. Bureau of Mines under
Contract No. JO 155125. The authors wish to express
their sincere gratitude to all those personnel in each
of the above mentioned organizations who made contributions to the project's success.

REFERENCES

1. Katen, Kenneth P., 1979, "Analysis of United
      States Shortwall Mining Practice," Mining
      Congress Journal, Jan., pp. 62-69.

2. Peng, S. Syd and Park, Duk-Won, 1977, "Shortwall
      Mining in the US: A Record of Failure and
      Success," Coal Mining and Processing, Dec.,
      pp. 54-59.

Chapter 14

DUST CONTROL IN LONGWALL SHEARER FACES
THROUGH AIR MOVEMENT BY WATER SPRAYS

Natesa I. Jayaraman and Fred N. Kissell

U. S. Bureau of Mines, Pittsburgh Research Center
Pittsburgh, Pennsylvania

INTRODUCTION

This paper describes a recent Bureau of Mines research project that resulted in a simple procedure for reducing the dust exposure of longwall shearer operators. This project involved the development and testing of a water spray and passive shield system to hold dust against the face and away from the shearer operators. This system is known as the Shearer-Clearer.

DEVELOPMENT AND TESTING OF THE SHEARER-CLEARER SYSTEM

Respirable dust conditions have improved immeasurably in the past decade. However, a 1978 study (Mundell et al, 1979) shows the difficulty in bringing double drum shearer operations into compliance with Federal regulations. The study found only 30 pct of these operations to be in compliance with the standard. This non-compliance of longwall shearer operations is due primarily to the design of the system which places workers downwind of dust sources. Of critical importance in coal mine ventilation is the splitting of airstreams to ensure a fresh source of air at every working location. In this study, the airflow split approach has been applied to shearer dust control.

RESEARCH IN A FULL-SCALE LONGWALL TEST FACILITY

It has been noticed that when a double-drum shearer cuts against the face airflow, from tailgate to headgate, a plume of dust laden air leads the shearer. The air entrainment caused by conventional water sprays, which are directed towards the cutting area, tends to push the dust laden air as far as 4.57 meters (15 ft) upwind of the headgate drum. This contaminated air mixes with the clean intake air, upwind of the shearer and travels down the walkway to the shearer operators. Thus, the shearer operators must inhale dust laden air due to rollback. Figure 1 illustrates the effect of such conventional water sprays on face airflow patterns.

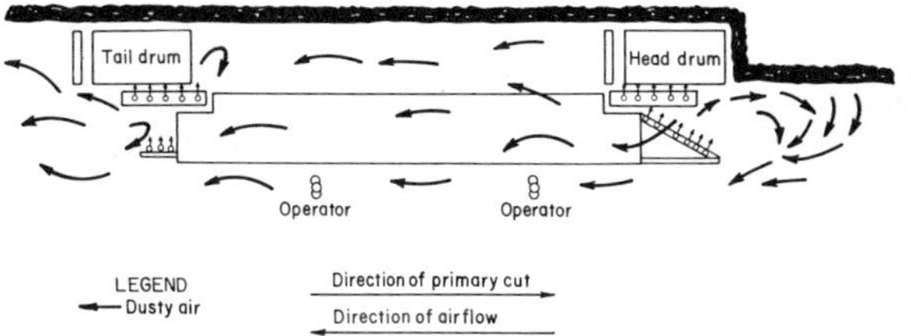

Fig. 1. Dust plume being pushed upstream by a conventional water spray system.

Such patterns have also been observed when tracer

gas measurements and visual smoke tests were conducted on a full-scale model of a longwall face.

This full-scale model was constructed by Foster-Miller Associates under Bureau of Mines contract J0387222. The facility is 38.4 m (126 ft) long and represents a 2.13 m (7 ft) high coal seam. It contains a full-scale wooden model of an Eickhoff EDW 300 double-drum shearer with revolving drums. The roof supports simulate Klöckner shields. To study airflow patterns in this model, tracer gas was released at the drums to simulate the generation of respirable dust. High density smoke was also released at the drums to aid in visualizing the airflow patterns and the effect of water sprays in changing airflow directions. Figure 2 shows the interior of the longwall gallery when smoke was released at the headgate drum in a test with conventional water sprays.

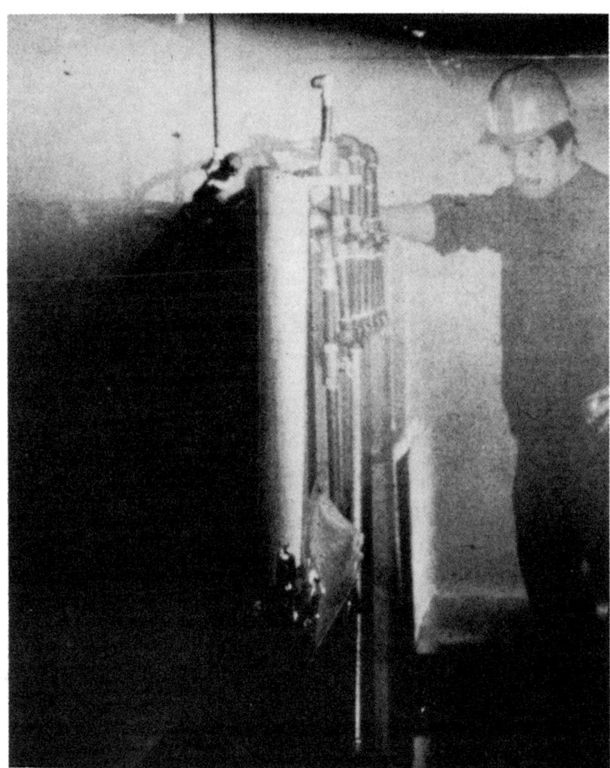

Fig. 2. Foster-Miller longwall test gallery, with smoke released at the lead drum during a test of conventional water sprays.

The smoke was pushed against the face airflow and onto the walkway where the shearer operator was exposed to this smoke rollback. The face air velocity was approximately 1.52 meters per second (300 fpm). Rollback was found to occur with most of the conventional spray systems tested.

Attempts were made to split the relatively clean intake air upstream of the shearer into two streams: A dusty split held close to the coal face and a clean one moving down the walkway. Factors that inhibited splitting of the airstream were:

1. Conventionally located water sprays.

2. The obstruction to airflow caused by the shearer body itself.

These factors caused the dusty air from the coal face to mix with the clean intake air, thus exposing the shearer operator to face dust.

The solution, therefore, was to contain the face dust on the face side of the shearer, leaving the walkway and shearer operators in relatively clean intake air. Figure 3 illustrates this concept.

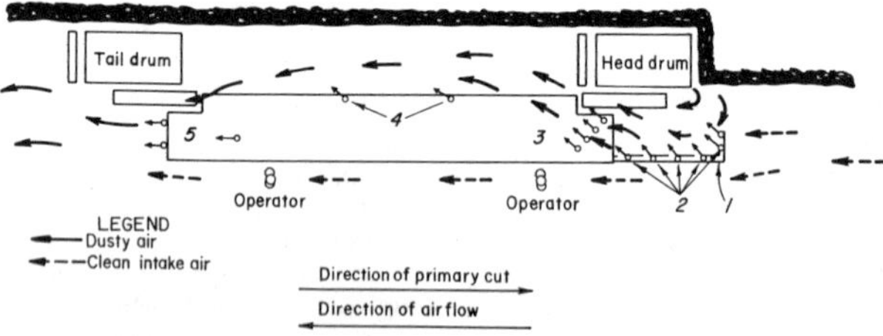

Fig. 3. Components of the Shearer-Clearer illustrating how the system splits the primary airflow and forces the dusty split toward the face.

The hardware is inexpensive and can be installed in a single shift. The hardware consists of a "splitter arm" and water sprays mounted on this arm, as well as on the shearer body. The water sprays are directed in such a way as to move the air in the desired direction. Figure 3 shows the splitter arm (1) extending from the edge of the shearer body to .46 m (18 in) beyond the cutting edge of the drum. Conveyor belting hangs from the splitter arm and extends down to the pan line. Several sprays (2) are also mounted on the splitter arm to cause the required air entrainment. The splitter arm, together with water sprays, splits the intake air into two streams. The dusty air is pushed over the intake end of the shearer body and the relatively clean intake air passes along the walkway.

A set of "crossover" sprays moves the dusty air over the shearer body and towards the face. The location of the bank of crossover sprays is such that it does not obscure the vision of the shearer operator. It should be noted that the dusty air passing over the sprays on the splitter arm and also on the body of the shearer tends to be cleaned in addition to being diverted by the sprays.

The velocity of the dusty air split moving against the face is maintained by the two sprays (4) mounted on the face side of the machine. The three sprays (5) at the end of the shearer body moves this dusty air over and around the drum. These sprays also act to suppress the dust as the dusty air passes over them.

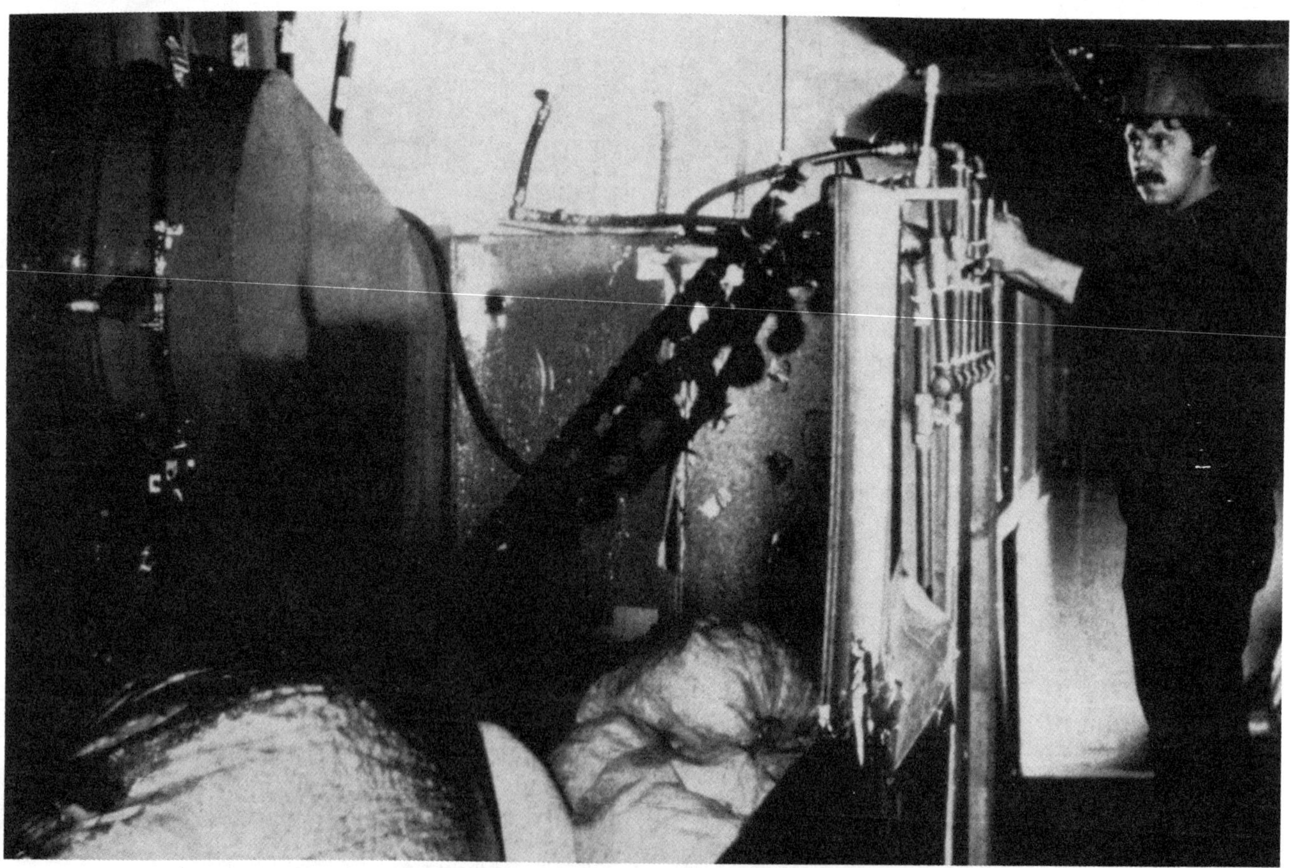

Fig. 4. Foster-Miller longwall test gallery with smoke released at the lead drum during a test of the Shearer-Clearer system.

The location of sprays for the Shearer-Clearer, their orientation and the types of nozzles, were optimized through a series of visual smoke tests and tracer gas concentration measurements. Methane was released from manifolds at the headgate drum and tailgate drum, respectively, and concentrations were determined at the two operator locations. Figure 4 shows the Shearer-Clearer in operation, with smoke released at the headgate drum.

Comparison with Figure 2 shows the walkway and the shearer operator in clean air.

The Shearer-Clearer system was compared with two basic spray configurations: (1) Drum sprays only, and (2) conventional sprays directed at the drum cutting the top coal. The air velocity along the face was 1.52-1.78 m/s (300-350 fpm). Table 1 shows the results of testing with tracer gas.

TABLE 1. - <u>Results of testing in the longwall facility</u>

	Lead Drum Oper. CH_4 Concen. (PPM)	Tail Drum Oper CH_4 Concen. (PPM)
Drum Sprays Only	185	250
Conventional Sprays	1250	780
Shearer-Clearer System	10	10

The operating pressure at the spray nozzles was 965kPa (140 psi) for the Shearer-Clearer and 690kPa (100 psi) for the drum sprays.

UNDERGROUND TEST

A preliminary underground test of the Shearer-Clearer system was conducted in a mine in northern West Virginia which works the Pittsburgh Seam. The longwall face was 175 m (575-ft) long and the seam 1.98 meters (78 in) high. An Eickhoff Model 300 double-drum shearer with 1.52 m (60 in) diameter drums was cutting the face. The face ventilation air was traveling from headgate to tailgate at an average velocity of 1.68 m/s (330 fpm). The cutting direction was from tailgate to headgate against the airflow, and the "clean up" pass was from headgate to tailgate with the airflow.

A steel framework splitter was attached to the intake air end of the shearer body, level with the top of the shearer. A conveyor belt was suspended from the framework and extended the entire length of the splitter frame (80 in).

Figure 5 shows the Shearer-Clearer system used for underground testing.

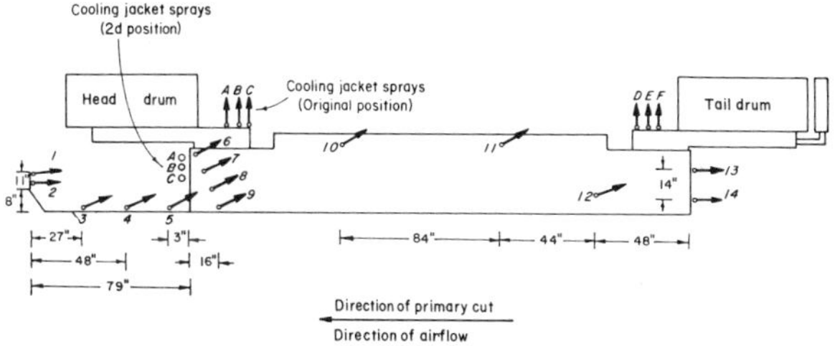

Fig. 5. Shearer-Clearer water spray system. Sprays 1 and 2 were Senior Conflo Co. Venturis. Sprays 3 through 14 were Spraying Systems Co. BD-20-2.

Sprays 1 and 2 located on the splitter frame were angled slightly downward at the intake air end of the shearer. Sprays 3 through 9 were angled at 30° with respect to the length of the shearer; sprays 10 to 12 were at the same angle. Sprays 13 and 14 suspended .30 meters (1 ft) from the top of the return air end of the shearer were pointed almost parallel to the face. Sprays 5 to 12 were mounted .30 m (1 ft) above the top of the shearer, using semi-rigid hose.

The mine-installed conventional water spray system (Figure 6) consisted of four Senior Conflo venturi sprays.

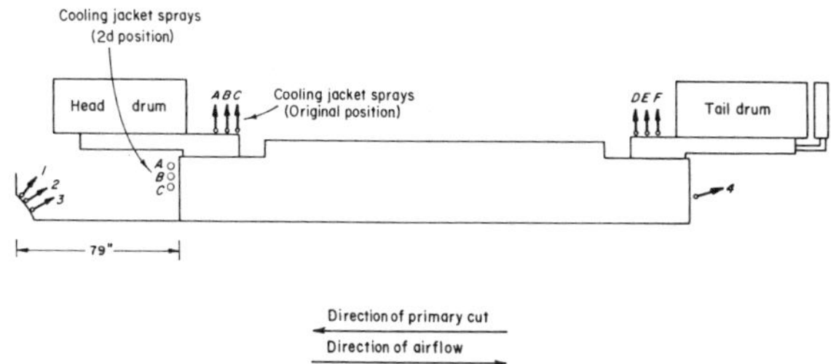

Fig. 6. Conventional water spray system. All sprays were Senior Conflo Co. Venturis.

Sprays 1, 2 and 3 were located on the splitter arm at the intake air end of the shearer. Spray No. 4 was suspended .30m (1 ft) from the top of the return air end of the shearer and directed at the drum.

Both the Shearer-Clearer and the conventional spray systems used 24 sprays on each drum to wet the coal while cutting. Both systems had cooling jacket water sprays (A through F) secured to the face side of the ranging arms.

SAMPLING PROCEDURE AND RESULTS

The respirable dust data presented here were obtained using gravimetric samplers worn and carried by three individuals, two stationed at the shearer operator positions and the third stationed 7.62 m (25 ft) upwind of the shearer. All three moved with the shearer as it cut. Each individual carried four samplers. The dust levels reported are an average of these four sets of measurements.

Sampling began only after the shearer had completely "sumped in" and continued only as long as the shearer continued to cut. Consequently, these data cannot be viewed as average 8-hour shift measurements, but as a worst-case used only to compare the effectiveness of two different spraying systems. Because 8-hour shift concentrations were not recorded, dust levels are reported as a "sample filter loading rate" in µg/min. This is the rate at which the sample filter gains weight as it collects dust; it can be related to an expected 8-hour average.

In the present study, the filter loading rate measured 7.62 m (25 ft) upwind of the shearer (typically about 5 µg/min) was used as a baseline or "intake" concentration, and subtracted from the loading rate measured at the operator positions. This ensured that changes in dust levels reflected only changes occurring at the shearer, not at the headgate or pan line.

On March 13, 1980, gravimetric sampling, as described above, was carried out. The shearer made four cuts from tail to head. The conventional spray system was on for the first and third cut (Figure 6) while the Shearer-Clearer was on for the second and fourth cuts (Figure 5).

The following filter loading rates were obtained with the conventional system on during cuts 1 and 3 when the intake values were subtracted: 45 µg/min for the headgate shearer operator; 22 µg/min for the tailgate shearer operator. Sampling time was 44 minutes, and the average tramming speed (while cutting) was .09 m/s (17 fpm). During cuts 2 and 4, with the Shearer-Clearer system operating, the rates were: Headgate shearer operator, 20.9 µg/min; tailgate shearer operator 12.5 µg/min. The average tramming speed was .09 m/s (17 fpm) for the two cuts and sampling time was 50 minutes.

Following the review of the March 13th data, several improvements were made to the Shearer-Clearer system. Sprays (1), (2), and (5) were installed on the splitter, sprays A, B, and C were relocated from the ranging arm to the intake air end of the shearer (aimed straight down toward the panline), and the splitter assembly was spring-mounted. These modifications were permanent; that is, no valving arrangements were made which would allow a return to either of the spray systems used on March 13. Consequently, the conventional system (now to be called the modified conventional system) was also improved, and future data reflect this.

Following the sampling procedure previously described, data collection during 2 tailgate to headgate cuts began again on March 17, 1980.

With the Shearer-Clearer water spray system fully operational, sampling during the first cut furnished the following concentrations: Headgate drum operator, 4.5 µg/min; tailgate drum operator, 5.7 µg/min. Approximately two minutes into sampling, a roof fall completely buried the splitter assembly, damaging it and bending it down at approximately a 20-degree angle. Sampling continued for a total of 36 minutes in spite of the damage, but the sprays on the splitter could not be properly oriented. The tramming speed while cutting averaged .06 m/s (11 fpm).

Data for the second cut were collected with the modified conventional spray system operating. The following filter loading rates were measured: Headgate drum operator, 9.3 µg/min; tailgate drum operator, 9.7 µg/min. Average tramming speed while cutting was .10 m/s (19 fpm) and total sampling time was 17 minutes.

These results may be summarized as follows:

TABLE 2. – Underground test results

Date System Tested	Sample Filter Loading Rate	
	Headgate Drum Operator	Tailgate Drum Operator
3-13-80 Conventional (original system used by mine)	45.0 µg/min	22.0 µg/min
3-13-80 Shearer-Clearer	20.9 µg/min	12.5 µg/min
3-17-80 Modified Conventional	9.3 µg/min	9.7 µg/min
3-17-80 Modified Shearer-Clearer (final system tested)	4.5 µg/min	5.7 µg/min

On the following day, March 18, bad roof conditions required that most of the shields be advanced over the shearer as it cut, instead of during the cleanup pass.

Dust from shield movement drifted down over the sampling packages, making it impossible to ascertain how much dust was coming from the shearer. This brought all testing to a halt. More underground testing is currently being planned. Meanwhile, other studies have been conducted in the test facility. A system was designed for shearers that cut in the same direction as the airflow, and another version suitable for use where the clearance between supports and shearer is low was devised.

Appendix A presents details of the Shearer-Clearer system including nozzle layout, nozzle types, and splitter arm and spray manifold designs.

SUMMARY

In summary, a new type of dust control system for longwall shearers has been developed. The system partitions the airflow around the shearer into a clean split and a contaminated split. The dust cloud is confined to the vicinity of the coal face, while the shearer operators remain in the clean split on the gob side of the machine. The Shearer-Clearer operates on the principle that each water spray moves air like a small fan and can be positioned to direct the dusty air towards the face. The hardware is inexpensive and can be installed in a single shift. It consists of several strategically mounted water sprays and one or more passive barriers.

REFERENCES

Mundell, R. L., Jankowski, R. A., Ondrey, R. S., and Tomb, T. F. Respirable Dust Control on Longwall Mining Operations in the United States, Second International Mine Ventilation Congress, Reno, NV, Nov. 4-8, 1979.

APPENDIX A

Figures A-1 through A-4 present preliminary details of a Shearer-Clearer installation designed for longwalls where the clearance between shearer and supports is more than .3 m (1 ft). A low clearance version has been devised and will be described in a subsequent publication.

Figure A-1 contains details of the nozzle layout including nozzle types, installation locations and directions.

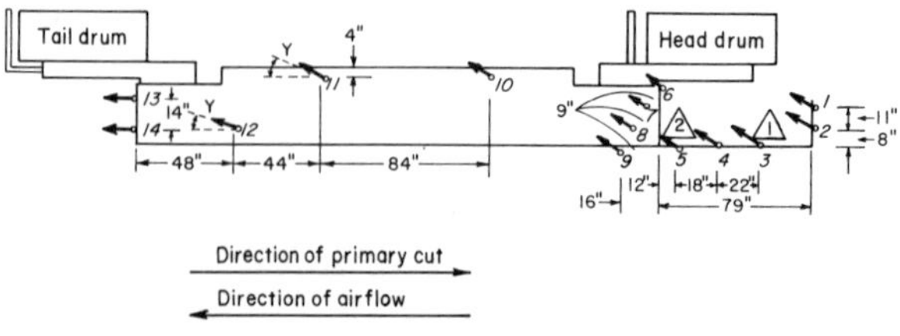

Direction of primary cut
Direction of airflow

Fig. A-1. Shearer-Clearer nozzle layout.

Nozzle descriptions and relevant notes are as follows:

NOZZLE DESCRIPTION

Nozzle No.	TYPE [3]	ANGLE TO PLANE		[4]
		X	Y	
1	GG 3014	45° DIP	45°	
2	GG 3014	30° DIP	10°	
3	BD-20-2	0°	30°	
4	BD-20-2	0°	30°	
5	GG 3009	0°	30°	
6	BD-20-2	0°	20°	
7	BD-20-2	0°	30°	
8	BD-20-2	0°	30°	
9	GG 3009	0°	30°	
10	BD-20-2	0°	30°	
11	BD-20-2	0°	30°	
12	BD-20-2	0°	20°	
13	BD-20-2	0°	0°	
14	BD-20-2	0°	10°	

[1] THE DIMENSIONS AND NOZZLE ANGLES SHOWN ARE FOR A SPECIFIC MACHINE INSTALLATION AND SHOULD ONLY BE USED AS A GUIDE.

[2] THE HEIGHT OF NOZZLES 1-12 IS APPROXIMATELY MIDWAY BETWEEN THE TOP OF THE SHEARER BODY AND THE UNDERSIDE OF THE ROOF SUPPORT CANOPIES. NOZZLES 13 & 14 ARE MOUNTED 8 in. BELOW THE TOP OF THE SHEARER ON THE TAIL.

[3] ALL NOZZLES LISTED HERE ARE MANUFACTURED BY SPRAYING SYSTEMS, INC.[1]

[4] Y – ANGLE FORMED BETWEEN A LINE PARALLEL WITH THE FRONT EDGE OF THE SHEARER AND THE NOZZLE CENTERLINE (MEASURED CLOCKWISE).
X – ANGLE FORMED BETWEEN A PLANE PARALLEL WITH THE SHEARER TOP AND THE NOZZLE CENTERLINE.

[1]Reference to specific manufacturers does not imply endorsement by the Bureau of Mines.

Figure A-2 shows an artist's conception of the splitter arm installation on the lead drum end of the shearer.

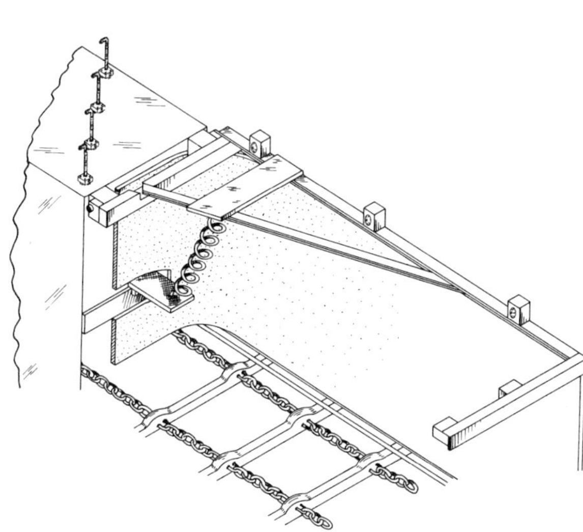

Fig. A-2. Splitter arm.

It extends approximately .46 m (1.5 ft) beyond the leading edge of the lead drum and down to the pan line. The entire splitter arm is spring-loaded and pivots on the end of the shearer to absorb any falls or face rolls. Spray nozzles 1-5 are shown on the splitter arm and nozzles 6-9 are shown mounted on the shearer body.

Figure A-3 shows the flexible mounting manifolds used for spray nozzles 6-12 (refer to Figure A-1).

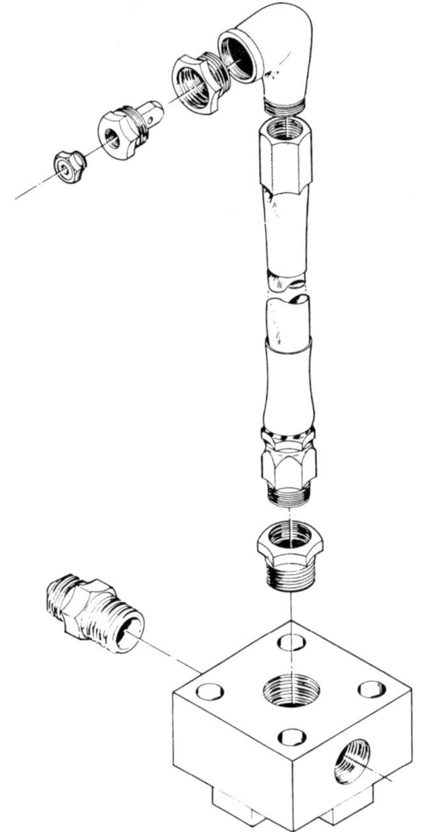

Fig. A-3. Flexible nozzle manifolds.

The centerline of the nozzle is located midway
between the top of the shearer body and the underside
of the roof support canopies. The elevation keeps the
spray nozzle free from rubble and keeps the spray cone
from impinging onto the shearer or roof supports
prematurely.

Figure A-4 shows the spray manifold used for spray
nozzles 1-5 on the splitter arm.

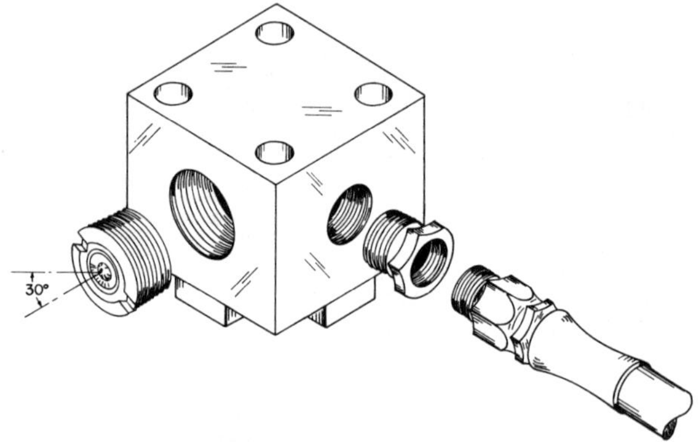

Fig. A-4. Spray manifold (with required angle).

The nozzle is screwed into an adapter with the
required mounting angle machined into it. The manifold can be welded or bolted into place.

Chapter 15

SOME ECONOMIC AND SAFETY BENEFITS OF ENVIRONMENTAL
MONITORING OF COAL MINES

Albert E. Ketler

President of Minetics Corporation
Pittsburgh, Pennsylvania

1.0   Introduction -

The introduction of computerized monitoring
and control (M/C) equipment into the U.S. coal
mining industry may prove to be of truly rev-
olutionary proportions in the 80's. Many of you
are aware that the British have developed and
have been using such sophisticated M/C equipment
for performing production and safety related
tasks in their coal mines for nearly a decade.

The benefits of M/C systems in the mining
industry are usually separated into those which
improve productivity, and hence make money for
the company, and those which improve safety,
and hence cost money.

The economic benefits of M/C are startling
for the production related functions such as for
monitoring and controlling conveyor belts, de-
watering pumps, electrical power distribution,
and the like. It doesn't take much analysis to
show that hundreds and even thousands of tons
of coal don't get mined every time a belt shuts
down or when power is out due to circuit break-
ers tripping for any number of reasons. A flood,
caused by a malfunctioning pump which goes un-
detected for lack of feedback monitoring of
pumping parameters, can destroy literally mil-
lions of dollars worth of mining equipment. The
M/C system would be justified or not based on
whether or not the productivity benefits balanced
the cost, within a timespan commensurate with
the return on capital philosophy of the company.

The safety related benefits are easily under-
estimated or ignored completely, and the purpose
of this paper is to under-score some of these
benefits, as well as to review the application
of M/C to the coal mining industry. The companies involved are Transmitton Ltd of England
and Minetics Corp of Pittsburgh, Pennsylvania.
Transmitton is probably the most experienced
manufacturer of M/C equipment in the world,
having some 200 installations to its credit.
Minetics is the marketing and manufacturing arm
for Transmitton in the U.S.A., and is responsible
for all aspects of distribution and service.
Plans are underway for manufacturing of the
Transmitton hardware for U.S. sales to be trans-
planted into U.S. subcontractors by 1981. Thus,
our coal mining industry will benefit from
British technology, but without the delays and
additional costs usually involved in international shipments.

But what about cost benefits of M/C applied to
safety? What is a non-explosion worth? How
about a non-fire? What would that be worth?
And how about a non-bearing overheat resulting in
a non-shutdown of a conveyor, a fan or a pump?

Those are safety related functions which could
have some pretty important influences on the
profit and loss statement at the end of the
fiscal year.

And injuries and fatalities? If you have a
serious fire or explosion, possibly involving
injuries or fatalities, the first thing that
happens is...MSHA shuts you down for investigation. Then, after a month or so, if all
looks o.k., you can start up again. Naturally,
your people have to be re-assigned, filling in
the gaps around the missing individuals.
You'll have to hire a few people, too. Of
course, they will all need training, so you
better hire someone to handle that. Replacing
the damaged equipment is no picnic, with
inflation, long deliveries, and all. A year
later, after coping with law suits, fines,
capital outlays, loss of production, and
horrible indigestion, you may find that a M/C
system capable of surveying your mine every
few seconds just might be one of your higher
priorities.

A mine-wide M/C system can make your life a
lot easier and a lot richer. You've got your
control console there in the mine office, with
its video display showing all the equipment is
working just fine. A note on the bottom of
the screen reminds you that the belt on conveyor No 24 is starting a tear, but its not
serious yet, you'll have someone fix it during
the maintenance shift tonight. Then there's
the circuit breaker on South 6 West which has
to be readjusted; it has tripped twice today
already, once taking out the whole 8 sections.
Luckily, you had the means to reset the power
remotely. Sure saved a lot of trouble today.
Walking over to the console you key in your
password "BIGFOOT", and up jumps a menu of
commands at your disposal. You decide on
"Methane", and up comes a screen full of data
showing that all sensors are indicating low
levels. And, you continue, bringing up "Air
Velocity", "Smoke", "Temperature", saving each
display on paper using the high speed printer
by simply pressing "P".

Continuing your surveillance, you note that
the shift is over and now you can find out how
you did that day. You key in commands to list
all production parameters...total weight of coal
produced, by section, by hour. Print it.
Total hours of machine operation, by machine,
by shift. Then, you review all of today's
data...graphically. The computer prints up a
3 color chart, right on the video display. You
see the trends of every measured parameter in
your mine for the past half hour, the past hour,
the shift, day and week. The data is all

Fig. 1. Mine Operating System (MINOS) manufactured by Transmitton Ltd. for the National Coal Board of Great Britain. MINOS is a proprietary monitoring and control concept wherein all software is certified by the National Coal Board.

stored in the computer's memory, and you can analyze it in any of a hundred ways...to get the information you need to do your job.

But how is all this actually done? What sort of equipment is needed to perform all these data gathering and analysis functions? Let's get down to some basics and review the hardware.

2.0   Environmental Parameters -

The main parameters which are important for monitoring include air velocity, methane, carbon monoxide, humidity, air pressure, and smoke. Devices are commercially available for monitoring these parameters on a continuous basis. There are many sensors available, some better than others in certain respects. It is not our intention here to make an exhaustive analysis of every sensor available, but merely to assure you that practical devices do exist which are compatible with the Transmitton M/C system. Sensor technology is improving continuously, and we can expect more and better sensors in the years ahead.

The telemetry package is a mature system, with literally hundreds of installations in hundreds of mines, some in operation for nearly a decade. Transmitton, one of the largest suppliers of M/C equipment for coal mines, offers telemetry packages ranging in size from a single unit outstation with a few sensors to an elaborate mine-wide system consisting of literally hundreds of outstations and with elaborate control panels designed for efficient human/machine interfacing and extensive analysis and documentation.

In order for environmental data to be analyzed effectively to uncover trends of increasing or decreasing values that could become hazardous at some later time, it is essential that the sensors be capable of providing analog outputs, and that the telemetry system be capable of transferring these analog values to the surface for analysis, comparison, and documentation. Simple on/off detectors indicating only if the parameter is below or above a particular value

are of little value for detecting small changes in parameters which could be analyzed to provide early warning of worsening conditions. The object is to head off a shutdown situation by correcting conditions before they reach alarm levels. This is an important way in which economic benefits can significantly accrue.

3.0   The Monitoring and Control System -

Figure 1 shows an elaborate control console for a Minos (Mine Operating System) which is a concept developed by the British National Coal Board for performing production and environmental monitoring in British coal mines. This above ground unit communicates by a telemetry system to underground outstations which gather information throughout the mine using a variety of sensors and detectors. Among the environmental parameters which can be measured are air velocity, methane concentration, humidity, air pressure, carbon monoxide, smoke, and the like.

Figure 2 shows a typical underground outstation with a front panel removed exposing the plug-in electronic circuit cards which perform the data acquisition, analysis, and transmission functions. The outstation is also capable of receiving commands from the central control station and executing them by turning on or off motors or actuators. The power supply which converts mine power to 5 volts dc for powering the outstation is shown in its explosion-proof enclosure on the left. At the top of the outstation are a number of glands for admitting the numerous cables for connecting the various sensors, switches, and relay closures which represent the electrical analogs of the measured environmental parameters.

An important safety feature of any M/C system, particularly where control is involved, is the feedback of measured data confirming the implementation of the control command. Thus, if a ventilation fan is turned on, there should be increases in air velocity, vibration, temperature changes, and so on which confirm that the fan did indeed turn on as commanded.

An open-loop command with no confirming feedback would constitute a hazard in itself, in that the failure of the fan to respond to the command and indeed turn on and increase ventilation, would lead to false security and environmental hazards which could lead to explosions or asphyxiations. It is therefore imperative that remote controls be supplemented with fast response monitoring feedback. Likewise, a monitoring system without a remote control capability leads to another set of frustrations. Should methane build up in a particular section, the only alternative may be to remove the personnel, since dispatching a person to a remote fan location to implement a control may not be feasible within the prescribed time limitations. Further, should a fire develop, rapid changes to the ventilation pattern to bring fresh air to evacuate underground crews could be impossible in the time frame of rapidly deteriorating conditions, without remote control.

Losses to equipment and plant facilities due to fires or explosions can be minor compared with the losses from death or injuries to personnel. Claims and law suits, continuing for years and amounting to horrendous expenses for payoffs to beneficiaries, medical expenses, diversion of management to court appearances, continuous legal counsel, and the like,can amount to fantastic sums. Indeed, safety is a mainline economic factor which should be given a high level of consideration. M/C systems can enhance safety as well as productivity, thereby providing a dual means of payback recovery of the cost of such systems. Indeed, M/C should be considered a capital item just as important as conveyors, mining machines, roof bolters, and any other piece of directly productive equipment used in the mining operations.

4.0   <u>The Telemetry Link</u> -

The last 15 years has brought significant advances in the field of data transmission and

Fig. 2. Typical Transmitton outstation. Front panel is open showing plug-in circuit cards and wiring details. Glands on top are for the numerous sensor and control cable connections.

remote control. The resulting systems are reliable, low-powered, flexible, and secure. As with all new product developments, many variations and techniques have been tried and discarded along the way. For example, the British mining industry has shown that direct, multi-conductor, hard wire systems are uneconomical due to the long lines involved and also the high cost and maintenance of large wire bundles. Further, frequency based communications (FSK) has been shown to be unreliable and inflexible, compared with the digital DC system used in the Transmitton equipment. Radio communications are ideal for above-ground applications, but were found to be prone to interference problems and required high cost antennas due to the restrictive signal propagation in mine tunnels.

4.1　Time Division Systems -

The technique found to provide answers to most problems is the transmission of digital signals on a time-sharing basis, termed Time Division Multiplex (TDM). Practical systems have a wide variety of forms, but the essential principal is the sharing of the transmission line between all the signals from all of the sensors in remote data-collecting stations on a one-at-a-time basis. Time division multiplexing effectively reduces the number of wires from perhaps thousands to only two or four wires, depending upon whether half or full duplex communications are used. In half or semi-duplex, communications are in one direction at a time, mainly from the sensor to the data-collection and analysis center. Since data is transmitted only on command from the communications traffic control unit at the control console, semi-duplex communications on a single pair of wires is slowed substantially in order to time phase the command and data signals. On the other hand, a full duplex system utilizing two pairs of wires can speed things up substantially by allowing full-time communications in both directions.

The four-wire communication cable is distributed around the mine forming one or more transmission highways. Data collection points consist of outstations having a capacity of a dozen or more analog inputs or as many as 28 binary inputs, depending upon the requirements. As many control channels are also available for starting motors or actuating devices for remotely controlling actuators, motors, fans, pumps, and the like.

Each outstation can handle a group of sensors in any general location in the mine. The communication controller polls each outstation in sequence, whereupon, the state of each data bit is transmitted from the outstation to the data collection central station three times. The three transmissions are compared electronically, bit by bit, to assure that the data has not been contaminated by line noise. Any discrepancies result in the data being rejected and a new sampling is automatically requested by the communication control unit. Each outstation is programmed with a unique address, and there is one section of the central station (whether hardware or software driven) dedicated to each outstation address. Each outstation polled replies in turn and the data is filed in an appropriate

address in the control console located on the surface. The data can then be directly displayed or passed on to a higher level computer for analysis, processing, and documentation.

The polling of very large telemetry systems can be speeded up substantially by dividing the mine into two or more sub-systems, each with its own four-conductor transmission highway and communication controller. This also adds a large measure of security, because the entire mine is not tied to one cable link. A rock-fall tearing or pinching one cable will not effect the telemetry on other communication highways leading to other sections of the mine.

5.0　Outstation Design -

Environmental monitoring is only one aspect of monitoring control as applied to coal mines. The same telemetry and command system used to monitor the environmental sensors can also be used for other purposes. In addition, there are other specialized outstations which are extremely valuable in the operation of extensive coal mining operations. To illustrate the safety features and flexibility of outstation design, two types of outstations are described below: a pump outstation and an environmental outstation.

5.1　Pump Outstation -

Any float switch can be used to activate a pump motor, on when it's high and off when it's low. But what happens if the pump fails to pump for any of a dozen reasons such as loss of power, loss of prime, clogged inlet, broken pipe, burned-out motor, and so forth? The pump outstation is designed to accept sensor data to determine the conditions surrounding the pump operation, to assure that it's working when it's supposed to be working and not when it's not.

Linear transducers or limit switches are frequently employed for monitoring:

a)　High water level in sump
b)　Low water level in sump
c)　Actual water level (analog)
d)　Output flow
e)　Pump primed
f)　Suction clear
g)　Pump motor running
h)　Motor bearing overheat
i)　Various environmental conditions such as methane, smoke, etc.
j)　Miscellaneous electrical or mechanical parameter such as voltage, current, vibration, etc.
k)　Clock

In essence, this outstation is designed to start the pump at high level, stop it at low level. Before providing the start command, however, the interlocks will be checked to see that the priming is complete and the suction state is satisfactory. Once the pump motor is started, the monitoring system then insures that flow is established and that any other dangerous conditions do not occur. The pump is then stopped at low level. All of these conditions are

automatically dealt with at local control interlocks, while the alarms and levels are continuously transmitted to the central control
station. Figure 3 shows typical pump display.

Water in many mines is of such high quantity
that multiple pumps are employed. For these
installations the control circuits are arranged
to transfer the pumping to the stand-by units
whenever a lack of flow is detected on the main
unit. Additional level switches can be incorporated to indicate extra high and extra low
levels, while even more complex arrangements
can be employed to arrange for priming actions
to be controlled together with control of the
pipeline valves for routing the water.

```
13/11/79  10:43
                LIST OF OUTSTATION CO3 STATUS              ISOLATED

NAME: CO3 PUMP     KIND: PUMP          TYPE: 03      STATE
     CHANNEL           STATE                 COMMAND

01  EXTRA HIGH       N               START 1      N
02  LOCAL STOP       N               STOP  1      N
03  COMMON FLT       N               START 2      N
04  REMOTE FLW       N               STOP  2      N
05  OVERHEAT 1       N               START 3      N
06  FAIL     1       N               STOP  3      N
07  ELEC FLT 1       N               SPARE        N
08  FLOW     1       N               RESET        N
09  OVERHEAT 2       N
10  FAIL     2       N
11  ELEC FLT 2       N
12  FLOW     2       N
13  OVERHEAT 3       N
14  FAIL     3       N
15  ELEC FLT 3       N
16  FLOW     3       N
17  ANALOG LSB       N
18  ..........       N
19  ..........       N
20  ..........       N
21  ..........       N
22  ..........       N
23  ..........       N
24  ANALOG MSB       N
25  MUX 0            N
26  MUX 1            N
27  ..........       N
28  EXTRA LOW        N
    ANALOG 1             O
    ANALOG 2             O
    ANALOG 3             O
LIST COMPLETE
```

Fig. 3. Typical pump display.

Local displays of status are provided on the
front panel of the outstation, along with remote
start/stop switches for local control. Clock
inputs can be utilized for minimizing peak
power demand through the use of priority running
times during non-peak power periods. The idea
here is to show that a totally integrated
sensor/control capability of the remote out-

stations can provide human-like monitoring and
control of the important mining functions. In
addition to pumping, similar outstations are
available for monitoring and controlling conveyor belts, power distribution systems, and
even elevators, all automatically and with the
highest degree of security. Figure 4 shows list
of commands available behind **the** password.

5.2   Environmental Outstation -

The environmental outstation is tailored to
interface with 8 analog tranducers or sensors
of a wide variety of types. Naturally, digital
inputs can be accommodated, as for connecting
existing fire detection systems and alarms and
communicating this information to the surface
for evaluation and documentation.

Each analog sensor is connected to an outstation data channel using an input conditioning
circuit especially tailored for the particular
sensor unit. The following digital signals are
derived:

a) Analog warning level
b) Analog alarm level
c) Loss of analog input.

These signals plus the actual analog values
are transmitted to the interface on 8 separate
replies or multiplex levels. The outstation
design contains local displays of status using
LED's founded on the front panel. Local control outputs are also available which are derived from the incoming analog signals. The
modular concept of the outstation is shown in
Figure 5. It should be noted that all Transmitton outstations are designed with full
safety interlocks and monitors to provide a self-
contained unit which can continue functioning
locally even in the event of failure of the
telemetry link. This is important to assure
continuing safe operation in the event of a
damaged communication cable.

6.0   Data Processing, Analysis, and Documentation -

```
PASSWORD PROTECTION COMMANDS

These are command protected by the PASSWORD. The operator
would type P (CONTROL P) followed by an 8 character password.

Correct operation of the password gives the following
instruction set on the screen:

1 - Change Date and Time
2 - Change Password
3 - Define Transducer
4 - Define Type of Outstation
5 - Define Outstation
6 - Define Analog
7 - Define Graph
8 - Define Analog on Display
9 - Save Files

x - Exit Password
```

Fig. 4. Commands available.

```
          Type 3 (Define a Transducer)
               —
          INPUT OPTION (1 CHARACTER)
          TYPE 'I' TO INSERT,'C' TO CHANGE,'D' TO DELETE,
          'L' TO LIST OR 'X' TO EXIT.

Type I
     —
          INPUT TRANSDUCER TYPE (E.G. BM1A)

LABEL 'A'
          TYPE IN FULL SCALE AS 3 DIGITS AND DECIMAL POINT
          (E.G. 1.23)

          3.00
          DO YOU WANT ALARM AS ANALOGUE INCREASES (TYPE Y OR N)

          Y
          —
          TYPE IN OFFSET AS A % FULL SCALE (E.G. 25)

          20
          —
          TYPE IN UNITS (E.G. % METHANE)

          % METHANE (ALLOWED 10 characters)
          ——————————
LABEL 'B'
          CONTINUE ? (By typing Y the computer will go back to
                      the input option request; by typing N the
                      computer will go back to the original
                      password instruction set).

TYPE C
     —
          INPUT TRANSDUCER TYPE (E.G. BM1A)

          BM1A (This would take the computer back to the
          ——   question made at Label 'A').

Type D
     —
          INPUT TRANSDUCER TYPE (E.G. BM1A)

          BM1A (Deletes appropriate files or says DID NOT EXIST,
          ——   then the computer will go to the question made
               at Label 'B').
```

Fig. 5. An example of the computer question and answer facility.

Data telemetered to the surface contains a complete picture of the mine environment as well as the status of monitored underground equipment. This data can be displayed instantly by command from the human operator on the surface. In order to make this information more easily understood, various graphic displays can be commanded which will automatically summarize in chart form all of the data of a particular type which occurred over a period of time ranging from hours to days. Thus, subtle trends can be discerned, and unusual or dangerous conditions can be quickly identified. Data can be stored, manipulated, analyzed, and even preserved in hard copy if desired.

6.1 <u>Operational Facilities Offered by an Auxiliary Micro-Computer</u> -

Although an auxiliary computer is not essential to the operation of the monitoring/control system, it does provide a very valuable adjunct to the overall mine management capabilities afforded by this powerful M/C concept. Once the

```
     TYPE L
          —
          LIST OF TRANSDUCERS

          NAME          FULL SCALE        OFFSET        ALARM

          BM1A          3.00% METHANE       20%          HIGH
          BA2           5.00% METR/SEC      20%          LOW
          BBBB          1.23% C-0           20%          HIGH
          LIST COMPLETE

          CONTINUE ?

          (Similar to Label 'B')
```

Fig. 6. Typical VDU display.

```
13/11/79  10:41
LIST OF OUTSTATION KIND PUMP  TYPE 02

        CHANNEL     AL   RA
01   EXTRA HIGH      Y
02   LOCAL STOP      Y
03   COMMON FLT      N
04   REMOTE FLW      N
05   OVERHEAT 1      Y
06   FAIL     1      Y
07   ELEC FLT 1      Y
08   FLOW     1      N    1
09   OVERHEAT 2      Y
10   FAIL     2      Y
11   ELEC FLT 2      Y
12   FLOW     2      N    2
13   ..........      N
14   ..........      N
15   ..........      N
16   ..........      N
17   ANALOG LSB      N
18   ..........      N
19   ..........      N
20   ..........      N
21   ..........      N
22   ..........      N
23   ..........      N
24   ANALOG MSB      N
25   MUX 0           N
26   MUX 1           N
27   ..........      Y
28   EXTRA LOW
NUMBER OF PUMPS
LIST COMPLETE
```

Fig. 7. Inputs and definition options.

data becomes available to mine management the interest in analysis and documentation is a natural, almost instinctive outgrowth. Some additional uses for such analysis are noted below.

6.1.1 Conveyor and Pump Software -

These software packages provide summary information on the performance of every conveyor or pump, simply by keyboard command. Such information is useful for determining total running times for each motor, total volume or weight produced, total down time, etc. Such information is useful for management planning and for uncovering any inefficiencies which might be correctable through adjustments, repair or replacement. Figure 6 shows a typical display as would appear on the visual display unit (VDU) for a particular pump outstation number 30C.

6.1.2 Analog Analysis -

These software packages enable any incoming analog value to be displayed continuously in alphanumeric form or to be stored in a graphical format. Absolute values of alarm and warning levels can be set by the operator for each analog function. Also, rate of rise or rate of fall alarms can be programmed to call attention to rapidly changing conditions. This is important for detecting problems well in advance of alarm, possibly avoiding shutdown conditions.

If a new analog sensor is added to the system, the operator merely informs the computer of its

```
13/11/79   10:40

                    LIST OF OUT-STATIONS

    O/S      NAME        TYPE    ANL    FULL SCALE          AL    RATE

    CO3   CO3 PUMP        03
    CO7   CO7 PUMP        02
    CO8   CO8 PUMP        02
    DO1   DO1 ENVRMNTL    01      1    BA2  5.00 MTRS/SEC    60%   22%
    DO1   DO1 ENVRMNTL    01      2
    DO1   DO1 ENVRMNTL    01      3
    DO1   DO1 ENVRMNTL    01      4
    DO1   DO1 ENVRMNTL    01      5
    DO1   DO1 ENVRMNTL    01      6
    DO1   DO1 ENVRMNTL    01      7
    DO1   DO1 ENVRMNTL    01      8
    DO2   DO2 ENVRMNTL    01      1    BM1  3.00 % METHANE   30%   08%
    DO2   DO2 ENVRMNTL    01      2    BM1  3.00 % METHANE   40%   05%
    DO2   DO2 ENVRMNTL    01      3    BM1  3.00 % METHANE   50%   20%
    DO2   DO2 ENVRMNTL    01      4    BM1  3.00 % METHANE   70%   40%
    DO2   DO2 ENVRMNTL    01      5
    DO2   DO2 ENVRMNTL    01      6
    DO2   DO2 ENVRMNTL    01      7
    DO2   DO2 ENVRMNTL    01      8
    DO3   DO3 ENVRMNTL    01      1    BA2  5.00 MTRS/SEC    10%   10%
    DO3   DO3 ENVRMNTL    01      2    BA2  5.00 MTRS/SEC    20%   20%
    DO3   DO3 ENVRMNTL    01      3    BA2  5.00 MTRS/SEC    30%   40%
    DO3   DO3 ENVRMNTL    01      4    BA2  5.00 MTRS/SEC    70%   80%
    DO3   DO3 ENVRMNTL    01      5
    DO3   DO3 ENVRMNTL    01      6
    DO3   DO3 ENVRMNTL    01      7
    DO3   DO3 ENVRMNTL    01      8
    LIST COMPLETE
```

Fig. 8. Operator/computer interface systems.

type, full-scale value, alarm settings, etc. For maximum security, these additions or deletions can only be implemented using a restrictive access password sequence, normally reserved for cognizant supervisory personnel. All information on the screen at any time can be reduced to hard copy using a keyboard request to the on-line printer.

Figure 7 shows inputs and definition options available to the operator and other authorized personnel.

Figure 8 gives an example of the operator/computer interface sequences when adding information to the computer.

Figures 9, 10, and 11 indicate the types of management information that can be made available on command. These illustrations are relegated to environmental parameters, however, similar types of analysis can be used for virtually every outstation usage imaginable, including:

a) Atmospheric and ventilation monitoring
b) Pump control and complete water management
c) Auxiliary ventilation fan monitoring
d) Automatic control of ventilation doors
e) Main ventilation fan monitoring and control
f) Electrical distribution M/C
g) Emergency procedures (in the event of pre-selected conditions arising, the computer can be programmed to display an emergency procedure to be followed.)
h) Calculation of the CO/O_2 ratio to indicate spontaneous combustion.
i) Conveyor belt M/C
j) Monitoring of miscellaneous transportation, car loading, elevators, and the myriad motor and actuator controlled equipment underground.

The eventual scope of the use of monitoring and control systems in coal mining operations is only limited by the ingenuity of the mining industry to conceive of sensors and actuators to reliably measure and control, to achieve the end objective of improving safety and increasing productivity. Experimental installations of monitoring of longwalls and continuous mining operations are in the offing which will further enhance mining operations in the years ahead.

7.0 <u>Conclusion</u> -

M/C is a management tool. It can help you manage the production, and it can help you manage the environment on which your people rely for their next breath of air. It is a tool that can help you save money and make money. It can also help make your mine a safer and more pleasant place to work for all those people you employ and depend on. M/C can help you do your job, but better. And, by doing your job better, won't that make your life a little bit easier? Feeling good, knowing things are under control, the mine is humming, no indigestion, just visions of good times ahead for the weekend. Your boss will have all those reports on his desk, printed up with no mistakes, first thing Monday morning. Might be a good time to ask for that raise!

```
13/11/79  10:38

           LIST OF GRAPHS DISPLAYABLE


     DO2 ENVR ANL 1   8 HOUR

     DO2 ENVR ANL 2   8 HOUR

     DO2 ENVR ANL 3  24 HOUR

     DO2 ENVR ANL 4   7 DAYS

     DO3 ENVR ANL 1   8 HOUR

     DO3 ENVR ANL 2  24 HOUR

     DO3 ENVR ANL 3   7 DAYS

     DO3 ENVR ANL 4   7 DAYS

     DO1 ENVR ANL 1   8 HOUR

LIST COMPLETE
```

Fig. 9.

```
              13/11/79  10:39


         LIST OF ENVIRONMENTAL TRANSDUCER TYPE


O/S   NAME   ANL  TYPE   CURRENT VALUE    ALARM

DO2   ENVR    1   BM1    0.00 % METHANE    (00)

DO2   ENVR    2   BM1    0.00 % METHANE    (00)

DO2   ENVR    3   BM1    0.00 % METHANE    (00)

DO2   ENVR    4   BM1    0.00 % METHANE    (00)

      LIST COMPLETE
```

Fig. 10.

13 11 79 10:47
3 HOUR D02 ENVR 1 BM1 GRAPH
0   0.1   0.2   0.3   0.4   0.5   0.6   0.7   0.3   0.9   1
3.00   %METHANE

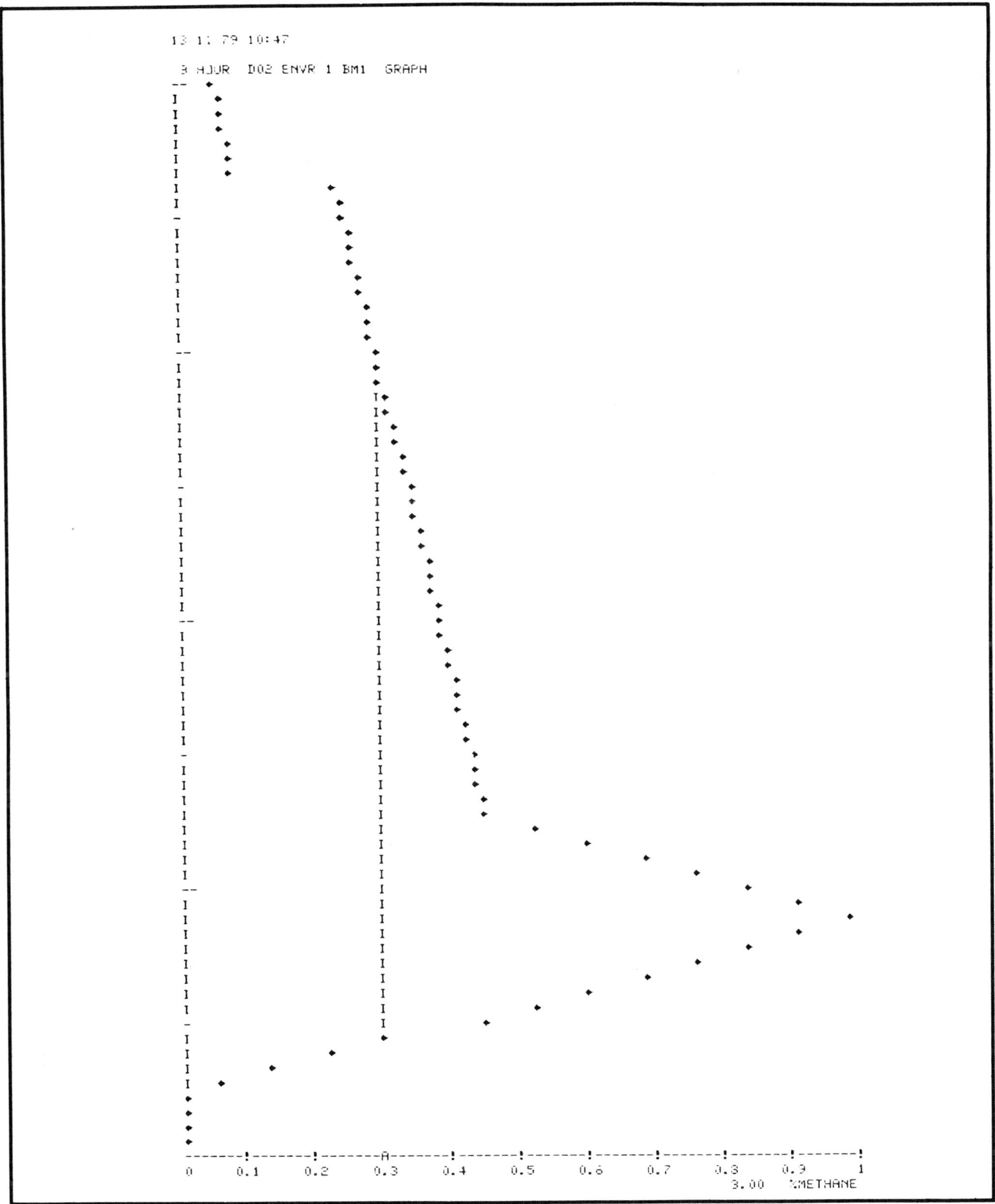

Fig. 11.

III. Unit Operations

John P. Baugues
Mountain, Inc.

Chapter 16

U. S. LONGWALL TECHNOLOGY AND REGULATIONS

K. Thirumalai

Manager Rock Mechanics and Mining Technology
Rockwell Hanford Operations - Rockwell International
Richland, Washington
(previously MSHA New Technology Coordinator)

D. P. Schlick

Vice President Engineering
Jim Walter Resources Inc.
Brookwood, Alabama
(previously MSHA Director of Technical Support)

SUMMARY

Longwall Technology has a potential of resulting in safety and productivity improvements of U.S. underground coal mining similar to those experienced by the introduction of continuous miner systems. The study presents a comprehensive national evaluation of longwall technology to assess the potential of longwalls to improve health and safety and coal productivity. The study was conducted by the authors during the spring of 1980 as part of MSHA New Technology Coordination activities and with a view to examine the factors that may limit the acceptance of longwall technology in the U.S. coal mining operations.

The study points out areas where adjustments need to be made in MSHA regulations and enforcement to help longwalls. A two-entry longwall system appears to offer a practical method of improving underground coal productivity and safety. Applying regulations that are primarily designed for room and pillar methods to longwall mining, in some instances, reduces the effectiveness and productivity benefits feasible by new technology application.

Controlling dust is the single predominant problem in the design and operation of longwalls in the United States. The results of this study points out a need for a coordinated program between government and industries on dust control technologies for examining acceptable technological options to control dust in longwalls without restraining the production capability of the system.

INTRODUCTION

The results of the study is based on a national survey and evaluation of all longwall operations in the U.S.A. Candidate mines, operators and equipment manufacturers were selected for a detailed analysis. In all, a sample of about 24 longwall sections, nine major coal operators and four equipment manufacturers were examined in this study.

Discussions were made with individuals who had a key roll in formulating, developing and establishing the operating policy for longwalls. Technical and operational discussions included scores of miners, mine supervisors and engineers working in longwall sections. A preliminary finding of the problems and potentials related to longwall mining was a first stage. The first stage findings were carried into a second stage discussion with representatives from the National Coal Association and the American Mining Congress. The results of the second stage were carried into a third stage discussion with union representatives. The findings of these evaluations were discussed with the researchers conducting R&D programs in longwall mining. The recommendations resulting from these discussions are presented in this paper.

A REVIEW OF U. S. LONGWALL MINING

Longwall systems with self-advancing powered supports were first introduced in U. S. coal mining in 1966. New sections were added at the rate of about four new sections each year during the period 1966-1972; six new sections each year during the period 1972-1975; eight new sections each year during the period 1975-1977 and about 12 new sections each year from 1977-1979. The rate of growth of the number of longwall sections has doubled within the past five years.

In the beginning of 1980, about 90 longwall sections were in operation in eight Coal Mine Health and Safety Districts in the United States. MSHA Coal Mine Health and Safety District 2 (Pittsburgh, Pa.) leads the list of longwall shearer sections with about 20 shearers in operation and MSHA Coal Mine Health and Safety District 5 (Norton, Va.) leads the list of longwall plow sections with about 12 plows in operation. Eastern Associated Coal Company leads the list of coal operators with the highest number of longwall sections and is followed by Consolidation Coal Co.

U. S. longwall operations are predominantly retreat operations using three or more entries on either side of the longwall panel. Two advancing longwall sections were in operation during 1979. Less than 10 percent of the total longwall sections operate with double or single entries.

U. S. coal operators have invested close to one billion dollars in longwall systems. Only a minor fraction of hardware used for longwall systems is manufactured within the United States. German and British manufacturers dominate the U. S. longwall market. Present equipment costs for a typical longwall section with a 600 foot panel at 6 foot seam height is estimated at about 7.5 million dollars. Several U. S. coal operators are able to pay off the large investment capital to purchase new longwall equipment in about five years.

U. S. coal operators favor the use of shearers for cutting, and shields or hybrid shield systems for supporting longwalls. Shearers overcome the limitations of plows to cut hard coal strata, and shields and hybrid shield systems overcome some of the supporting and other constraints associated with chock supports. Supporting with shields costs about $7,000 per linear foot of longwall face at the present time.

New technological improvements are underway in German and British longwall systems. Some of these improvements are currently being planned for application in the U.S. longwall market. These improvements are primarily in the area of increasing the supporting capability, improving mobility, operating with remote controls and increasing productivity of the cutting system. German and British manufacturers are also incorporating dust control systems and accessories for dust entrapment while cutting.

Longwall technology is in the upward curve of technological growth. Mining equipment experts predict that during the period 1980-1985, radical improvements will be made and new design innovations will be available in longwall systems. Mine technology forecasters predict that longwall technology will make a significant impact on U. S. coal mining within the next decade, which, in many ways, will be comparable to the introduction of continuous miner technology during 1950-1960.

Accidents and Injuries

A review of accident and injury data in longwall operations in the Unites States shows some significant advantages of the technology in terms of reducing mine fatalities and serious injuries.

TABLE 1

A RELATIVE INDEX COMPARISON OF LONGWALL
INJURIES WITH CONTINUOUS MINERS

Type of Injury	Relative Index
Fatality	0
Total disability	0.6
Disabling injury	1.09
Non disabling injury	1.24

Table 1 compares the relative index of longwall mining injuries with continuous miner operations. The relative index represents a ratio of injuries experienced in longwall sections to corresponding injuries for producing equivalent amounts of coal by continuous mines. For example, the total disability relative index of 0.6 represents a 40 percent reduction of total disability in longwalls when compared with continuous miner operation.

These evaluations are based on 1978 and 1979 mine accident and injury data which represent the first two years of quality information collected from operators since the beginning of MSHA's Part 50 reporting procedures. No fatalities were recorded in longwall sections during the two years. Longwalls have a potential to significantly reduce fatalities near the working face and reduce the number of serious injuries (permanent impairment) by at least 40 percent.

Longwalls also appear to significantly reduce disabling injuries (injuries which have one or more lost work days) in electrical injuries and roof fall injuries which are among the leading causes of underground injuries in today's continuous miner operations.

TABLE 2

A RELATIVE INDEX OF LONGWALL
DISABLING INJURY ACCIDENTS

Accidents	Relative Index
Electrical	0.35
Roof fall	0.05
Handling materials	1.22
Hydraulic sources	1.30
Machinery	1.33
Fall of materials	1.50

Table 2 compares the relative index of longwall disabling injuries for the various injury categories. Disabling injuries associated with electrical and roof fall accidents shows a marked decrease in longwall operations. Machinery and material related accidents show an increase in longwall operations.

A comparative analysis of the disabling injury data indicates that longwalls appear to have the potential to achieve approximately 90 percent reduction in roof fall injuries and approximately 60 percent reduction in electrical injuries.

A relative comparison of the longwall disabling injury data for other injury categories indicates that longwalls produce about 20 percent to 50 percent more injuries related to handling of materials, hydraulic repairs and failures, machinery, and fall of materials when compared with continuous miner systems. These injury categories generally depend on the amount of training received by a miner and the extent of the miner's experience on the job.

Coal Production

A comparative analysis of U. S. coal mining production data clearly shows the relative advantages of longwall systems to improve coal production. In 1978, for example, longwalls produced about 22 million tons, which represent about 10 percent of the total underground production in the U.S.A.

The national average production in 1979 is estimated at about 800 tons per working shift for longwall sections and about 225 tons per working shift for continuous miner sections. Several longwall sections in the United States operate with about the same number of miners in face operations each shift as in a continuous miner section.

About 75 percent of underground production in 1979 is from continuous miners. Estimates made in the study indicate that every 10 percent replacement of continuous miner production with longwalls has a potential to increase total underground productivity by 20 percent. Longwall technology, therefore, has a good promise to improve productivity with increased health and safety.

IMPACT OF REGULATIONS ON LONGWALL TECHNOLOGY

Longwalls did not have significant impact on U. S. coal mining, and were not a primary concern at the time of formulating MSHA regulations for coal mine health and safety subsequent to the Federal Coal Mine Health and Safety Act of 1969 (P.L. 91-173). Some provisions were made while making the regulations to accommodate longwalls. Applying the regulations that are primarily designed to safeguard health and safety of room-and-pillar methods to longwall mining, in some instances, lead to interpretations and reduce the advantages and production capabilities available with the new technology.

Longwall and MSHA Expertise

At the time when these investigations were made, MSHA, coal operators, operator associations and labor were, in general, not fully familiar with all options available with longwall methods and have inadequate communications among themselves on longwalls. U. S. coal operators are more familiar with retreating longwalls having three or more entries than advancing systems or retreating systems with less than three entries. Operators appear to lean on a choice of a longwall system that will have the least problems with MSHA regulations. Several operators appear to have a tendency to design a layout for longwall systems that will make a maximum utilization of continuous miners available at the mine sections.

It will be a step in the right direction for MSHA to take a lead role in helping with the longwall technology. MSHA enforcement and technical personnel, at the time when these investigations were made, were not in a position to guide operators on longwalls during the plan approval stages or help with the choice of a longwall layout to increase health and safety wihout reducing productivity or possible regulatory interferences to production. A good approach to develop this expertise will be to select a small team of technical and program specialists from MSHA, USBM and industries to familiarize themselves with the cross section of problems in U. S. longwall operation and equipment developments in the United Kingdom, Germany, Poland and Canada. The team can then form the nucleus for coordinating longwall technology and provide expert guidance on longwall operations in the U. S. in conjunction with MSHA.

Longwall Respirable Dust Control

Respirable dust is the single most critical health problem in U. S. longwall operations. Development and application of an effective dust control is the key for the future growth of longwall technology in U. S. coal mining. The answer to the question on whether technology is available to control respirable dust in longwall sections operating at full production levels, is as yet, not available. Over 50 percent of longwall sections are currently out of compliance, at any given time, with the respirable dust standard which includes about 2/3 of the population of double

drum shearers and 1/3 of the population of longwall plows. Table 3 shows a comparison of average coal production and dust compliance data for longwall sections.

TABLE 3

COMPARISON OF COAL PRODUCTION AND COMPLIANCE DATA

	Average Production Tons/per shift	Compliance High Risk Average	Section Average
Plows	493	77	100
Single Drum Shearers	430	66	88
Double Drum Shearers	593	33	75
Overall	547	48	83

Table 4 shows a sample data of consecutive measurements of respirable dust and production data. It is obvious that with the use of available dust control technologies in a longwall section, the dust levels reduce only if production is correspondingly reduced.

TABLE 4

A SAMPLE PRODUCTION AND RESPIRABLE DUST DATA OF A TYPICAL DOUBLE DRUM SECTION WITH WATER SPRAYS

Day	Production Tons/shift	Average Respirable Dust mg/m^3
1	690	3.6
2	540	3.0
3	520	2.5
4	490	2.5
5	440	2.3
6	460	2.2
7	470	2.8
8	450	2.6
9	420	2.5
10	370	2.3
11	350	2.3
12	350	2.3
13	350	2.3

Several dust control techniques have been developed in recent years by the Bureau of Mines and industry programs. These techniques include water spray systems and additives to water sprays, dust collectors, face ventilation and augmentation, water infusion of coal strata and control of machine operating and cutting parameters. These individual techniques have shown to reduce respirable dust levels. A collective evaluation of a systems application of these techniques to bring the dust level to compliance at full production levels has not yet been made. In the absence of results from such a systems study, MSHA has to depend on a trial and error approach by the oeprators to comply with dust standards.

MSHA should evaluate the application of alternate approaches and should jointly launch a major short term program to examine the systems approach and collective applicability of proven dust control techniques to bring longwalls into compliance with respirable dust standards in conjunction with industry, the Bureau of Mines and the Department of Energy. The systems study should provide specific technical guidelines for compliance. These evaluations should be carried out in at least two or three operating longwall sections selected from east and west of the Mississippi. The system study should emphasize testing available dust control technology, including water infusion, water sprays, face ventilation, and dust collection systems to achieve compliance in operating longwall sections. The technical evaluations emerging from this study should specifically identify the capability of existing technologies to control dust in longwall operations at full production level and establish the existence or non-existence of technology to control dust in longwall operations at full production levels.

Application of Air Helmet Technology

Air helmets provide a column of clean air in front of the miners face. A modified miners helmet with a face protection shield pumps filtered clean air between the miners face and the transparent protection shield. The helmet was first developed in the U. K. for industrial application and has been tested in several longwall sections in the United States. The effectiveness of these helmets to protect the miner from breathing high respirable dust concentrations is unquestionable. Miners working in longwall sections and mine operators have expressed a positive note and enthusiasm for using the helmet in longwall sections. The use of this technology to control dust, raises the following issues:
1. Would the use of the technology, if permitted, discourage operators from developing and using other appropriate technologies to control dust at the mine face?
2. Would the acceptance and approval of the technology violate the intent of the mandated health provisions of the mine Safety and Health Amendments Act of 1977?
These issues cannot adequately be answered without an exhaustive evaluation of the effectiveness of available technologies to control dust and analyses of the legal implications of approving its use for all longwall sections. The technology is a feasible approach to comply with the dust standards in longwall sections without constraining the production capability of longwall systems. A unified effort by miners, mine operators and MSHA may be necessary to take appropriate measures for application of this technology to insure the quality of respirable environment in longwall sections.

Ventilation and Entry Design

Ventilation requirements (Part 75-Subpart D) necessitates a design of retreating longwalls, and in most cases, with four entries on either side of the panel in order to comply with all MSHA regulations without exceptions. In some instances, operators have applied petitions for modification of the

application of the regulatory provisions and have successfully mined retreating longwalls with two entries and advancing lonwalls with single entry under specific ground conditions.

Longwalls require a large capital investment and long-term production plans and, as such, operators do not want to afford the risk to design longwalls based on exceptions. A design of longwalls with four entries on either side tends to increase the share of coal produced by continuous miners in longwall sections to about 40 percent. In other words, irrespective of the rate of production, mines having retreating longwalls in the U.S.A., use continuous miners for about 40 percent of their steady-state production. This, in turn, reduces the overall health and safety and productivity benefits of using the longwall technology.

Reducing the number of entries has a potential to reduce the exposure of miners to mine hazards during the development, reduce ground control problems adjacent to the longwall panels, reduce cost of longwall development and reduce the delays to bring longwalls to production. Formulating a new set of MSHA regulations for longwalls will be counterproductive and will increase regulatory workload for MSHA and operators on longwall sections. An approach to modify existing regulations, where appropriate, with an objective to reduce the number of entries required for longwall mining, without reducing the protection for the miner from mine hazards, will have a positive impact on longwall technology. These modifications could be made within existing regulations and can be tailored to apply only for longwall operations. As far as possible, a systems approach should be considered in evaluating the impact of the individual regulatory requirements as applied to longwalls.

Roof Control - Need for Guidelines

Controlling the roof is a critical factor in the design of longwall entries and the panels. At present, MSHA considers longwall mining as a modification of the open ended method of pillar extraction and approves the support system for longwalls on an individual basis. Maintaining tailgates has been a consistent problem in retreating longwalls. Adequate guidelines are not available for operators to design appropriate tailgate supports or plan on procurement of necessary materials in advance.

Establishing mandatory regulations for roof control in longwalls will be counterproductive, but providing a minimal set of criteria and guidelines for entry and tailgate support in longwalls having less than three entries will help in the design of longwalls and expedite approval of longwall systems. Sufficient number of longwall sections have been completed to date to provide the experience and information necessary for establishing minimal guidelines for supporting double entry longwalls.

Longwall and Enforcement Procedures

Need for Uniform Guidelines: Applying current regulations to longwall operations tend to lead to interpretations and differences in enforcement procedures between individual field areas. In some instances, different enforcement procedures are used for comparable longwall operations within the same MSHA district and between district and district. MSHA should examine existing guidelines for longwall enforcement procedures including: Ventilation of entries and procedures for change in ventilation; mining longwall panels with gas wells: moving longwall equipment; maintenance of head and tailgate entries and belt haulage, and insure uniformity and clarity.

Longwalls and Near Term Research Needs

Though there are no criticisms on the validity of the various research projects on longwalls conducted by the Bureau of Mines and the Department of Energy, operators, in general, have expressed some concern on the direction of R&D programs which seem to elude near term application to several urgent problems in longwall mining operations. The following areas related to operational and production bottlenecks in longwalls will have immediate application in operating mines:
1. Cost effective method and procedures to detect and seal abandoned oil wells in longwall panels.
2. Prevention of coal blockage on face belt haulage.
3. Methods and systems to reduce delays and manpower required to move longwall equipment from one section to another.
4. Development of effective methods and procedures to produce controlled breaking of massive hard roofs behind longwall supports.

CONCLUSION

Using longwall technology has a potential of resulting in safety and productivity improvements similar to those experienced by the introduction of continuous miner systems. Domestic mining equipment manufacturers do not, at the present time, have any appreciable track record in the longwall equipment market. Longwall technology in the U. S. can be viewed as imported technology with imported systems. Since the regulations, strata conditions and mining practices are, in part, different from overseas the approach for application of longwall technology in the United States may need to be different from overseas practice. MSHA, coal operators, operator associations and labor are not fully familiar with all options available in using longwall technology. MSHA may need to develop the technical expertise necessary to help operators plan longwalls to improve safety as well as productivity.

Controlling dust appears to be the largest problem in the design and operation of longwalls in the United States. The results of the evaluation emphasize a need for a coordinated research and development program

between the Department of Energy, the Bureau of Mines and MSHA to demonstrate available technologies to control dust in longwalls without restraining production. A decision must be made on whether technology is available to control dust at full production levels of a longwall operation and on using miner protection devices until other technologies are developed.

Present capital expenditures required to install new longwall systems range between 6 and 10 million dollars. Because of the large investment risk, and the lead time necessary to bring longwall panels into production, it is difficult for operators to plan longwalls based on exceptions to MSHA regulations. MSHA should consider establishing specific criteria and guidelines applied to longwalls with less than three entries, without 101 (c) procedures, if the health and safety of the system shows to be equal or more than other methods.

MSHA is in the process of implementing several of the technical recommendations made in the study. It is apparent that the effectiveness of this technology application will not be realized unless MSHA, operators, and coal mine research and development programs work together and make a concerted effort.

Chapter 17

AN OVERVIEW OF LONGWALL UNIT OPERATIONS
IMPACT ON PRODUCTION

William Laird
Sr. Vice President-Special Projects
Gates Engineering Company
Pittsburgh, Pennsylvania

The predominent method of coal mining in the United States for years has been room-and-pillar mining. The reason for this has been the advantages of room-and-pillar method which is a relatively flexible system and its capital investment. The room-and-pillar method permits some selectivity in mining, especially where gas wells, faults and possible coal quality may be a factor. The mobility of continuous and conventional mining equipment used to develop entries may also be used for retreat mining of rooms and pillars.

Longwall mining of coal takes place along a long straight face, or wall, blocked out between two butt or panel entries. The long faces or walls are normally projected in lengths of 61 to 183 m (200 to 600 ft) in this country.

The longwall method of mining had its beginning in the United States during the 1800's. During that period, pack wall and wood posts were used for roof support. The longwall face was mined by miners using picks to undercut the coal. To keep the coal face from turning over, short wood posts called sprags were set in a slant position as the coal face was undercut to the depth of two to three feet. When the undercutting cycle was completed, the sprags were removed to permit the coal to fall from the roof pressure.

The longwall mining of coal did not take a firm hold during the 1800's, and room-and-pillar prevailed. In the early 1900's, steel jacks were introduced, and along with it, the modified caving theory. In March of 1950, a retreat longwall mining system, utilizing a plow, manually operated props, wood cribs and chock releases, was installed in one of Eastern Associated Coal Corp.'s mines.

In early 1960, a plow cutter loader with hydraulically operated self-advancing frame type roof supports was installed in Eastern Associated Coal Corp.'s Keystone Mine. The success of this installation encouraged other coal companies to purchase longwall equipment. There were, however, longwall mining failures that were attributed to the low resistance roof support. These failures brought on the demand for higher resistance roof supports.

In July 1966, the first heavy-duty roof supports were installed in Island Creek Coal Company's Beatrice Mine. In August 1966, the second installation of heavy-duty roof supports was installed in Eastern Associated Coal Corp.'s Kopperston No. 1 Mine. These installations compiled an excellent record. The roof control was consistant. Regular caving was established, and periodic weight was contained. The success of this system at these mines was communicated to other mining companies, and it wasn't long until interest in longwall mining was re-established in the mining industry.

In 1970, there were 27 longwall units operating in the United States. It was during this period that coal companies accepted longwall as a fully-engineered total mining system. Today, we have 128 operating longwall mining installations. This is a growth of over 100 operating longwall installations in 10 years, or a growth of 474%.

The first shield support installation in the United States was made by Consolidation Coal Company on its longwall face in its Shoemaker Mine. This longwall operation started operating on April 14, 1975.

Today, there are 71 longwall shield-type roof support installations in the United States, of this number, there are 41-2 leg; 29-4 leg; and 1-6 leg shields. In the last two years, there has developed considerable interest in the 4-leg shield. Low coal mine operators are seriously considering the 6-leg shield.

Listed below are the companies manufacturing shields and the number they have sold to mining companies in the United States:

Dowty Mining International	22
Klockner-Becorit	17
Hemscheidt	12
Mining Progress (Westfalia)	9
Thyssen	6
Joy Manufacturing Co.	4
Gullick Dobson	1
	71

The yield load of shield roof supports operating in the United States ranges from 350 to 400 tons for 2-leg shields, and 570 to 610 tons for 4-leg shields.

There are 48 chock and 9 frame-type roof support installations operating in coal mines in this country. The yield load of 4-leg chock roof supports operating in this country ranges from 280 to 720 tons.

The overall trend appears to be in the direction of increased roof support capabilities. This is evident of the fact that of the 128 operating longwall faces in the United States, 71 are shields, which have been purchased within the last five or so years. This growth of longwall indicates that longwall is getting firmly entrenched in the United States.

We conducted a survey recently to find out how many plow-loaders longwall installations were operating in the United States and learned that there are 21. The plows were all purchased from Westfalia Mining Progress. There were 12 hook plow systems, one "conversion" system (Anbau and eight Gleithobel). These plows operate with cutting speed ranging from 30 to 91 m/min (100 to 300 fpm).

Our survey also showed that the cutting depth of the plows ranged from 51 to 203 mm (2 to 8 in.). The overall average cut, however, was 96.5 mm (3.8 in.).

We also learned from our survey that there are 107 shearer loaders mining coal from longwall faces. The

shearer manufacturers and the number operating in the
United State's coal mines are listed below:

Anderson Mavor	31
Eickhoff	58
Sagem Sirus	6
Sagem DTS	5
British Jeffrey Diamond	1
Joy Manufacturing Co.	6
	107

The webb or depth of cut of the shearer loader
ranged from 0.6 to 0.8 m (24 to 33 in.). The majority
of the shearers operating are cutting a depth of
0.76 m (30 in.). We also learned that there are 14
single-drum ranging arm; six single-drum fixed arm;
and the balance of 87 double-drum ranging arm shearers
cutting coal from longwall mining installations in
this country.

The shearer loader haul system available today is
the rack and pinion-type, or the chain pinion system.
They are reconized by such names as Joy rack,
Eicotrack, Eicomatek haulage unit, et cetera, for
propelling the shearer-loader along the longwall face
conveyor. These methods are much safer than the
original pull chain method.

We found from our recent survey that most of the
longwall face and stage-loader conveyors were being
supplied to the coal industry by Westfalia, Eickhoff,
Dowty Meco and Huwood Irwin. We also learned that the
majority of conveyor chains in use were three-strand
18 mm (0.708 inch). The single-strand 30 mm (1.18
inch) chain was the next most used conveyor chain. It
is our understanding that for medium-to-thick coal
seams, the trend is to use double center 26 mm (1.02
inch) and 30 mm (1.18 inch) conveyor chains.

Coal produced from longwall faces is being conveyed
by 0.9 x 1.1 m (36 x 42 in.) panel belts to mine car
loading stations or main-entry belt conveyor systems.

Longwall mining projection data obtained from our
survey show longwall panel lengths ranged from 274 to
1829 m (900 to 6000 ft). The overall average length
of longwall panels was found to be 1107 m (3633 ft).
Panel heading centers ranged from 18 to 30 m (60 to
100 ft). Panel breakthrough centers ranged from 12 to
32 m (40 to 105 ft). The coal seam thickness ranged
from 0.9 to 2.9 m (38 to 116 in.). The average coal
seam height mined by the longwall method was found to
be 1.5 m (61 in.). The longwall face length ranged
from 61 to 183 m (200 to 600 ft). The average longwall face length was found to be 142 m (466 ft).

The lowest average daily production reported for a
longwall having a face length of 116 m (380 ft) and a
mining height of 1 m (40 in.) was 619 t (682 tons).
The coal seam was reported to be practically level.

The highest average daily production reported for
a longwall having a face length of 110 m (360 ft) and
a mining height ranging from 1.1 to 1.2 m (42- to 48-
in.) was 1814 t (2000 tons). The coal seam ranged
from level to having inclinations of up to 4.5%.

The lowest average daily production reported for a
longwall operation in the Pittsburgh seam of coal
having a face length of 168 m (550 ft) and a mining
height of 2.2 m (86 in.) was 1361 t (1500 tons). The
coal seam was reported to be level.

The highest average daily production reported for a
longwall operating in the Pittsburgh seam of coal
having a face length of 183 m (600 ft) and a mining
height of 1.7 to 2.3 m (66- to 90 in.) was 5443 t
(6000 tons). The coal seam was reported to have
inclination of 1 to 2%.

The maximum coal production in any one day from a
longwall having a face length of 137 m (450 ft) and
a coal seam height of 2.1 to 3.3 m (84-130 in.) was
reported to be 15 604 t (17,200 tons). The seam
inclination was reported to be about 8%.

The overall average daily production from all the
longwall reported was 3384 t (3730 tons). The coal
seam thickness ranged from 1 to 2.4 m (40 to 96 in.).

Special scoop-type longwall roof support trans-
porters, special rail cars and hoists have been
designed, built, sold and in operation to facilitate
and expedite longwall moves.

Longwall moves and setup time are taking from two
and one-half weeks to six weeks to complete.

Dust control for longwall mining has been a serious
problem. Methods of controlling the dust include the
installation of cowls, venturi and other types of
dust collectors and water sprays strategically located
on the shearer-loading machine. Spray systems have
been developed to control dust by directing water at
the source of dust, the cutter bit. Bit lacing, bit
penetration and cutter drum speed all have their
affect on the quantity of dust produced during mining
operations.

Plows have water sprays along face conveyor line
and on the plow itself to control dust. The plow,
however, does not generate anything near the quantity
of dust generated by the shearer loader.

In spite of all the work that has been done and is
still underway, dust control problems are having their
effect on longwall mining production.

The longwall units in our coal mines today are
producing from 45 to 80% of all the coal mined per
day. When a mine is depending upon high production
units, any interruption of production from these units
will have a serious impact on the overall mine production because the tonnage is lost and can rarely,
if ever, be made up. A coal company depending upon
a number of high production units must plan their
production to prevent or minimize mine production
interruptions because of the effect it will have on
the company's profits, not to say anything about the
failure to fulfill their obligation to its customers
and possible litigation.

In conclusion, it appears that longwall mining is
here to stay and enjoy a steady growth. At this time,
there isn't any continuous miner unit producing
thousands of tons of coal per day. Our mining people
are well indoctrinated with longwall mining technology.
They, as well as longwall manufacturers, have over the
past ten years, become more innovated and participated
in technological transfer with mining people in the
United States and other nations of the world. The
exchange of technology has brought about today's fully-
engineered longwall mining system.

The cutter loaders, articulated face haulage, and
the hydraulic self-advancing shield roof supports

technology makes it possible for mining people to
select the longwall equipment to fit the mining conditions in their mine or mines.

Longwall mining is by far safer than the room-and-
pillar method because the miners are protected at all
times from roof falls by steel roof supports during
mining operations.

Longwall retreat mining is dependent upon the speed
with which butt and main entries are developed. Today,
we are primarily depending upon continuous miners to
fulfill this task. There have been many times when
the development did not get completed in time for
setting up the longwall equipment. Rapid-entry developing equipment is needed. The automated extraction
system machine may, in the distant future, be able to
fulfill this need. Serious consideration should be
given to the application of tunnel boring machine
systems for rapid-entry development.

Chapter 18

LONGWALL DEVELOPMENT PROBLEMS

Robert A. Stansbury
Superintendent - No. 9 Mine
Gary Coal District
United States Steel Corporation
Gary, West Virginia

Keeping development ahead of retreat mining is a problem which occurs nearly everywhere longwall mining is used in the United States. As improved equipment technology has allowed the application of longwall mining to more difficult mining conditions, and with higher retreat rates, the problem has become more critical. Many factors determine development speed, including roof and floor conditions, mining heights, seam heights, and the number of entries. In an effort to maximize the rate of development and production from development sections, U. S. Steel's Gary District has used five entry, three entry, four entry and is planning two entry development panels at its different mines.

With the introduction of various types of shield supports, it is now possible to utilize longwall mining in higher coal over 3 meters (10 feet), lower coal under one meter, in difficult roof conditions, and with soft floor. Improved shearers of 300 KW (400 HP) and more make cutting heights over 3 meters (10 feet) possible, and off pan shearers and improved Gleithobel plows make cutting less than 1 meter (40 inches) seam height possible.

Longwall mining under these conditions puts an added strain on the ability to develop panels as fast as they are retreated. Poor roof conditions require a substantial amount of additional roof support to ensure that roof control in the head and tail entries will not hinder longwall retreat. The additional measures of longer bolts, resin bolts, point anchor resin bolts, roof mats, roof trusses, steel or wood cross bars, and roof grouting all place an additional burden on man power, installation time, and cost on the mining cycle. Additional roof bolting on development can cut the potential mining rate in half, and any roof falls cause significant delays to mining. Timbers and cribs can restrict ventilation and require more entries. Soft floor conditions can greatly impede development through tramming difficulties with shuttle cars and roof bolters, time required for cleaning and repairing roadways, and maintenance difficulties on equipment. High coal sometimes requires rib support to be installed during the mining cycle, further cutting the mining rate. Mining low seams requires that roof or floor be mined in the head and tail entries as well as the track or tractor roadway entry to provide clearance of the longwall equipment. Mining this rock slows down the mining rate because of the extra material being mined and the extra wear and tear on equipment. To a large degree these conditions effecting development speed cannot be controlled.

The number of entries is one factor which can be controlled, however. For the fastest advance rate and highest ratio of retreat tonnage to development tonnage, the ideal number of entries would be one. However, logistics and regulatory impediment make it practically impossible to use single entry develop-ment in the United States. The factors influencing the choice of number of entries are; roof conditions, ventilation requirements, logistic problems and over-all percentage extraction desired. If roof support is not impeding production and the lower percentage extraction is tolerable, productivity may be increased by using four or five entry panels instead of three to minimize logistics problems, to the point where development rate can be as fast with more entries as it is with three. However, if roof support is the determining factor in production capabilities due to poor roof conditions, three or even two entry panels may improve development rates over four or five entries in spite of restricting face haulage potential and making face ventilation more cumbersome. Ventilation problems due to high methane liberation, long panel lengths, low seam heights, or restricting the air flow with cribs can make four or five entry development mandatory for proper ventilation.

Due to the variety of mining conditions at the different mines in our Gary District, we have made several variations in our mining plans for longwall development. Longwall development began at our No. 9 Mine in 1970 with five entries. This was our standard for continuous miner pillar sections because of optimization of face haulage, ventilation, and roof control. Due to seam heights as low as 900 mm (36 inches), the average rate of advance was 120 meters (390 feet) per month. In order to improve the percentage recovery and the advance rate we changed to three entries with the belt and track in separate entries for the third panel. Using this plan for seven panels with a seam height of 1.2 to 1.3 m (47 to 51 inches), the average advance rate was 135 meters (440 feet) per month. Easier belt maintenance and regulatory considerations required the avoidance of having trolley wire in the intake escapeway, thus the mining plan was changed to use three entries with the belt and track together in the head entry. For the next two panels with 1.2 to 1.4 m (47 to 55 inches) seam heights, the advance rate improved to 150 meters (500 feet) per month. The next three panels had a seam height of only 0.9 to 1.2 m (36 to 49 inches) and the advance rate dropped to 125 meters (410 feet) per month. Using belt and track both in the head entry has the disadvantage of requiring the entry width to be at least 6 to 6.7 m (20 to 22 feet) wide. Due to the width problem and added ventilation requirements of longer panels and lower coal, we will be changing to four entries with two returns in the future. When we installed a second longwall at No. 9 Mine in 1976, we close a 150 m (500 feet) face width as opposed to 110 m (360 feet) for the original longwall in order to cut the required development by 26% for the same amount of longwall tonnage.

In 1976, we began longwall development at No. 50 Mine using three entries with the belt and track in separate entries. Bad roof conditions slowed the advance rate to 100 m (330 feet) per month on the

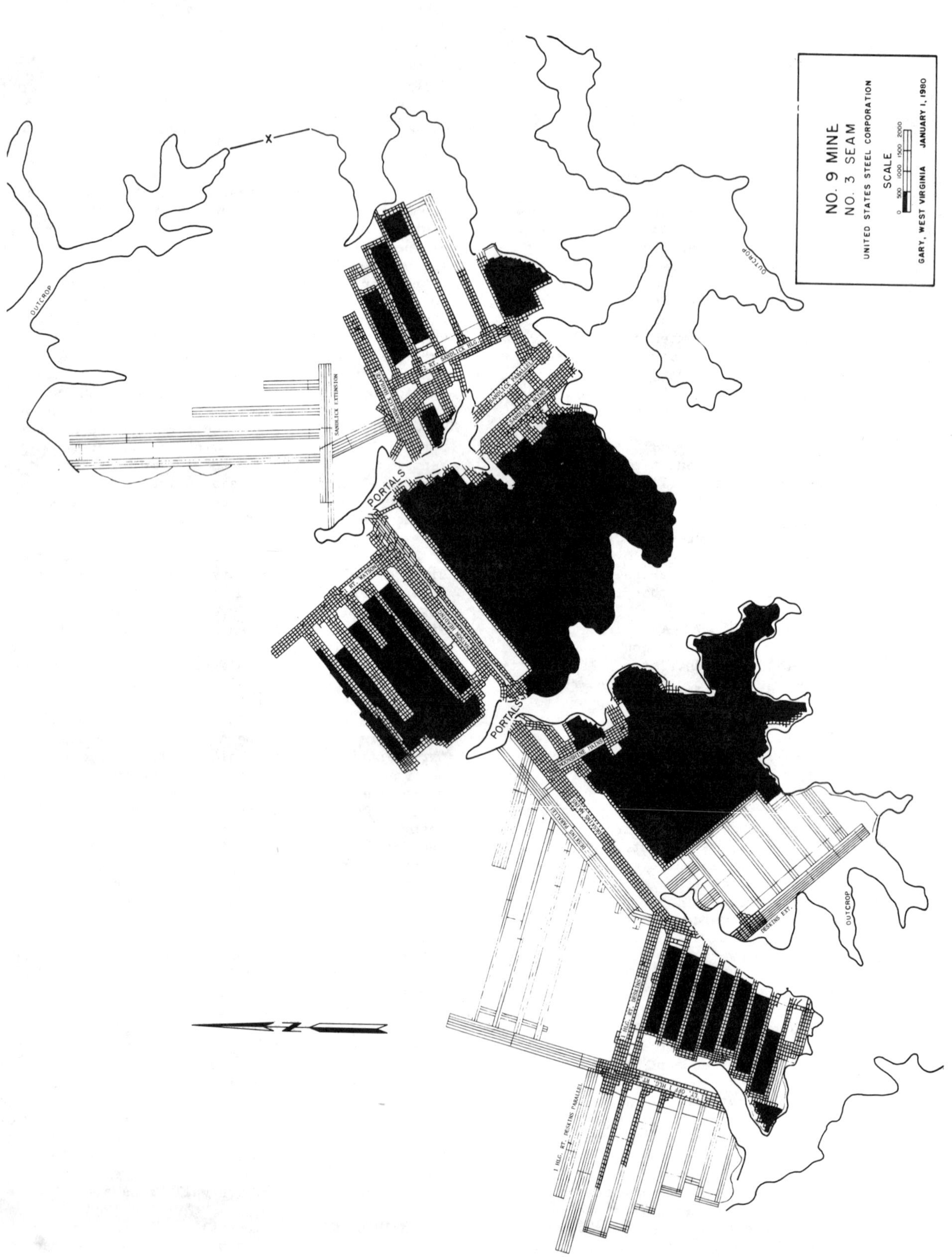

NO. 9 MINE
NO. 3 SEAM
UNITED STATES STEEL CORPORATION
SCALE
0 500 1000 1500 2000
GARY, WEST VIRGINIA   JANUARY 1, 1980
OUTCROP
OUTCROP
OUTCROP
PORTALS
PORTALS
X
N

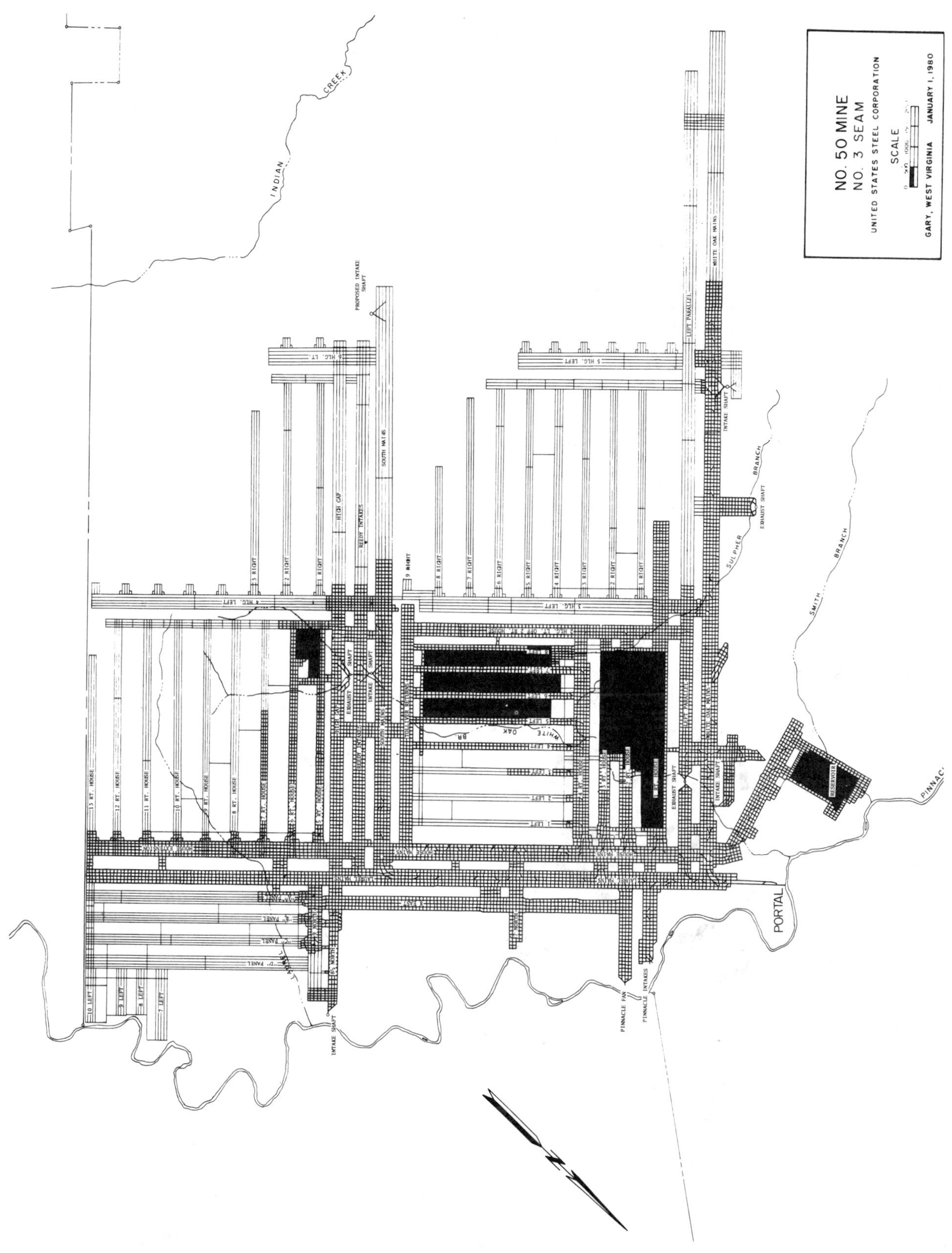
NO. 50 MINE
NO. 3 SEAM
UNITED STATES STEEL CORPORATION
SCALE
GARY, WEST VIRGINIA   JANUARY 1, 1980
INDIAN CREEK
PROPOSED INTAKE SHAFT
SOUTH MAINS
HIGH GAP
REEDY INTAKES
PORTAL
PINNACLE FAN
PINNACLE INTAKES
INTAKE SHAFT
EXHAUST SHAFT
WHITE OAK MAINS
LEFT PARALLEL
SULPHER BRANCH
SMITH BRANCH
WHITE OAK

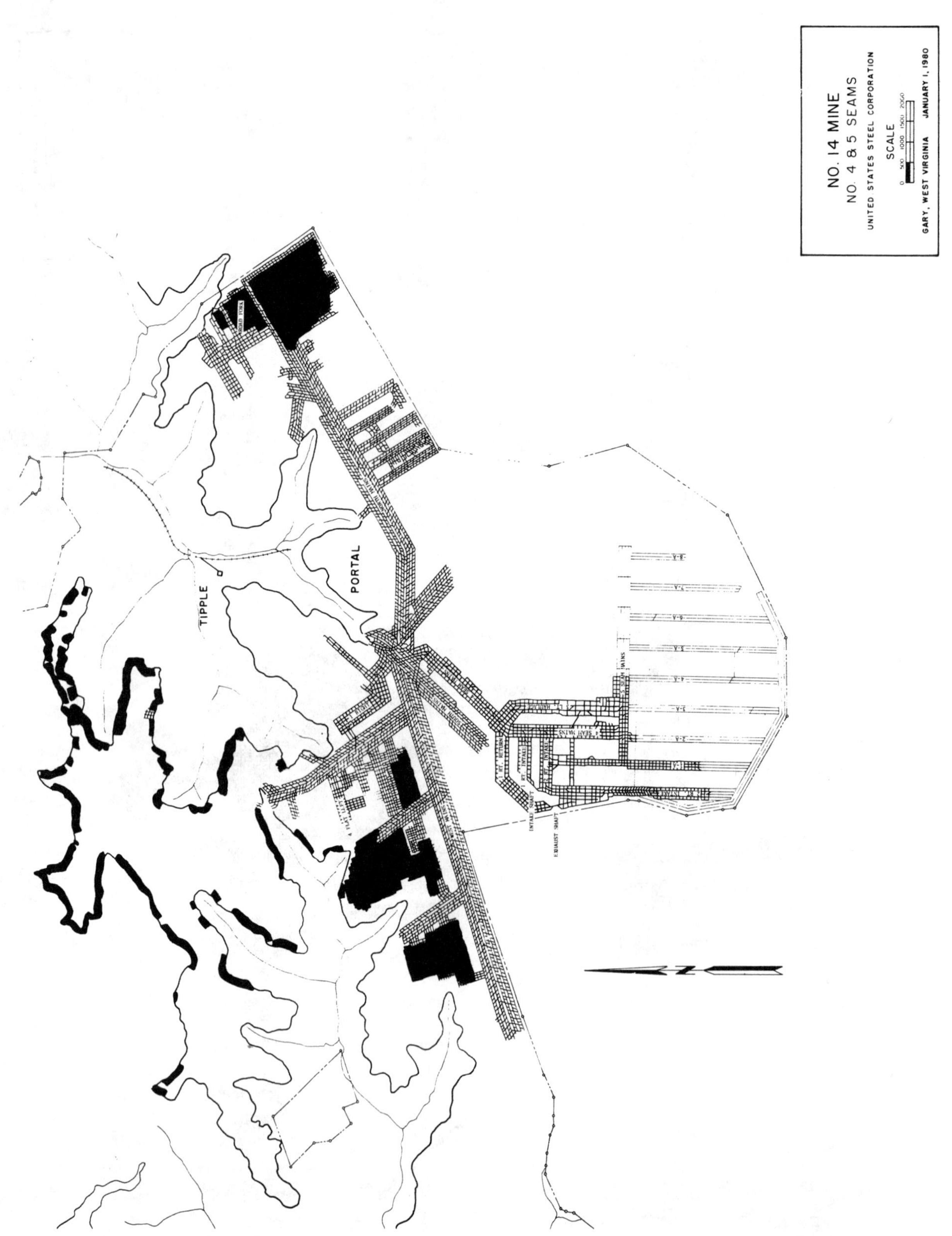
NO. 14 MINE
NO. 4 & 5 SEAMS
UNITED STATES STEEL CORPORATION
SCALE
GARY, WEST VIRGINIA   JANUARY 1, 1980
TIPPLE
PORTAL

first panel but by the fourth and fifth panel, the
advance rate with a seam height of 1.4 to 1.5 m (55
to 60 inches) was 190 m (620 feet) per month. Due to
longer panels [1700 m (5500 feet)] and high methane
liberation, the development which began in 1977 for
the second longwall unit at No. 50 has been four
entry panels with the face width increased from 145 m
(475 feet) on the first longwall to 165 m (550 feet)
on the second to maintain about the same percentage
recovery. Development rates with four entries and
a seam height of 1.2 to 1.5 m (47 to 60 inches) was
120 m (440 feet) per month on the first panel, but
has improved to 190 m (620 feet) per month by the
third and fourth panels. This shows that in relatively good roof conditions, four entries can advance at
the same rate as three with a resultant 33% increase
in productivity.

At our No. 14 Mine roof support problems in the 2.1
to 2.4 m (7 to 8 feet) thick Pocahontas No. 4 Seam
restrict final mining to the longwall method. Panel
development began as four entry on the first panel at
a rate of only 60 m (200 feet) per month. The high
cost of developing these four entries and the fact
that minimizing the number of entries to minimize
disturbance of the strata would improve roof control
led to the change to three entries for the second
panel. Plans are currently being reviewed by the
W. Va. Department of Mines and M. S. H. A. for two
entry panels to be developed starting with the third
panel.

In conclusion, there are many factors determining
the rate of advance of longwall development panels.
But by careful evaluation of all the factors as they
apply in each specific case, a longwall development
plan can be devised to suit circumstances involved.
At U. S. Steel's Gary District, we have used many
different plans to keep the mining plan suited to
the conditions.

Chapter 19

MOVEMENT OF LONGWALLS AT SHOEMAKER MINE

Edward C. Mack, P.E.

Mine Engineer

Ronald G. Stovash

General Superintendent

Consolidation Coal Company
Shoemaker Mine
Moundsville, West Virginia

INTRODUCTION

Consolidation Coal Company's Shoemaker Mine is located in West Virginia's Northern panhandle near Wheeling, adjacent to the Ohio River. Steam coal is mined from the Pittsburgh #8 seam which averages 65 in. in thickness. The 10 in. of draw rock directly above the coal is taken on advancement, but remains when longwalled. The top consists of up to 12 in. of lower quality coal, and immediately over that, up to 10 ft of multi layered shales, coal, and bone. The cover ranges from 300 to 800 ft of shales and limestone.

Two longwalls operate in the mine. The first longwall started in April 1975 and is retreating its eighth panel. This wall was the first to use shield supports in the United States. The second longwall started in September 1979 and is currently retreating its first 7,000 ft panel.

PLANNING FOR A LONGWALL MOVE

The key to a quick, successful longwall move is proper planning. Since techniques and procedures have previously been established, the planning now involves organizing and accumulating the materials and equipment for the move, and where possible, improving procedures.

The main phases of a move are: (1) Preparation of the new face; (2) Establishment of ventilation controls and preparing the moving route; (3) Gathering the supplies and equipment needed for the move; (4) Wire meshing; (5) Face conveyor removal and face roof support; (6) Shield recovery; and (7) Finish installing the new face.

A good set of planning notes should be maintained to be constantly checked and updated in order to make sure that plans are being followed according to schedule. A formal form of this would be a PERT chart or a Critical Path.

As with any major project, a team effort provides a system of checks and new ideas. At Shoemaker, the General Superintendent is ultimately responsible for the mine, and therefore oversees all direct and indirect aspects of the move. More directly responsible is the Supervisor-Longwalls and the Coordinator for that particular longwall. Others having various responsibilities include the Superintendent, Master Mechanic, and Mine Foreman.

PREPARATION OF THE NEW FACE

The lead time is known for new and rebuilt parts required for the next face. These parts are placed on order well in advance so that they will be available when needed. Such parts include the panline (rebuild or replace sections), shearer (use a spare if available), stage loader, haulage chain, hoses, cable, etc.

When mining in the development section is finished, the equipment is moved to another location. The longwall transformer is brought in to supply power during the setup and for the operating longwall. Though not immediately needed, the high pressure pumps are also brought in.

A battery charging station is established and a scoop car is put into service. Sloughage, dirt, etc. on the face and adjacent areas is then cleaned up by the scoop. From the preparation of the new face to finishing the longwall installation, the scoop is used to transport the parts, equipment, and materials. After the face is ready, face conveyor assembly begins.

The type of face conveyor used consists of a single center strand 30 mm chain. Before assembling the panline, the bottom single strand of conveyor chain is laid from the headgate to the tailgate at a distance away from the face where the panline will be placed, with flights installed on each side of the chain couplers only (if available, the top chain is laid along the face and put on the panline later). The reason for the flights is that they are needed to prevent the chain couplers from turning over as they pass through the drives. However, the flights are not able to be put into the flight guides as the panline is installed over top of it. Too many flights between the pan and the bottom will cause excessive drag which will prevent the conveyor chain from moving when the drive motors are started, putting the motors into stall.

The tailgate drive is then placed into position at the tailgate entry and the pan sections are installed over the bottom face conveyor chain. The pan sections are assembled outside, three together, with the middle section firmly attached to the other two. The bolts on the ends are loose so that the three-piece sections may be assembled with less difficulty to the adjoining pan sections.

As the pan sections are installed, the top conveyor chain is put in place along with the flights which are inserted into the conveyor flight guides. Upon completion of the panline, the headgate drive may be left off so that the shearer can be installed at a later date. If the shearer is not readily available, the headgate drive is installed to run the face conveyor in order to bring the bottom chain to the top so that the remaining flights may be installed.

The stage loader is assembled, including the power control boxes, accessories, and electric cables to the transformer. All the cables, water and dust hoses are placed in the trough or cable hangers down the length of the panline.

From a track laid up to the headgate, the shearer is brought in on a special dolly equipped with a section of eickotrack. The dolly is pushed up to the face conveyor, the trailing cable is connected to the shearer, and it is trammed onto the panline.

The headgate drive is installed and the face conveyor is made operational so that all the flights may be installed (if not previously done). Finally, the stage loader is connected to the headgate drive frame.

VENTILATION CONTROLS AND MOVING ROUTE

Before the start of the move, a plan is made on a map of the area that shows in detail the present ventilation controls, the controls needing changed for the move, and the various routes of equipment. A plan is also made that shows the final ventilation after the move is completed and for longwall mining the next panel.

Well before the move, the control changes are made that will not affect the mine ventilation. Material is delivered for stoppings that need to be built after mining ceases. With proper planning, no time is lost during the move due to ventilation.

A switch is placed and a track is laid in the crosscut where the longwall will finish. This will be used to tram the shearer onto a dolly for removal.

GATHERING SUPPLIES AND EQUIPMENT

The supplies and machinery needed for face recovery are ordered and stored at the site. Such items include the scoop cars, roof bolter, dollies, bolts, wire mesh, wire rope, steel beams, oak plank, tie wires, wire fasteners, crib blocks, wedges, posts, etc.

The scoop car charging stations are established and power is made available to the roof bolter. This machinery is placed near the longwall ending point and is checked to insure reliable operation.

Everything needed for the move is made readily available so that no time is wasted due to a lack of any item.

WIRE MESHING

The purpose of wire meshing the top is to hold back the fractured rock during shield removal. A 3/4 in.

wire rope used in conjunction with the mesh provides additional strength and support.

When the longwall face is 35 ft from the recovery point, the roof stone is removed along with the coal. The height mined becomes increased from $5\frac{1}{2}$ ft to $6\frac{1}{2}$ ft. This extra height allows more working room under the shields and provides adequate clearance for the scoop car.

At 25 ft from the stopping point, the first 3/4 in. wire rope is spanned from the headgate to the tailgate. It is attached at the headgate to a crib built for such a purpose, and attached at the tailgate to one of several cribs previously built. This first rope serves mainly as anchorage for the wire mesh.

The 12 ft ends of the 25 ft long wire mesh rolls are attached to the wire rope with prelooped tie wires by the tie wire fastener---a device that hooks the two loops and twists the wires when pulled. This method quickly produces a tight connection.

The rope and wire mesh is then placed over the shield tips and is draped back and hooked to the shields so that a pass with the shearer may be made (see Figure 1).

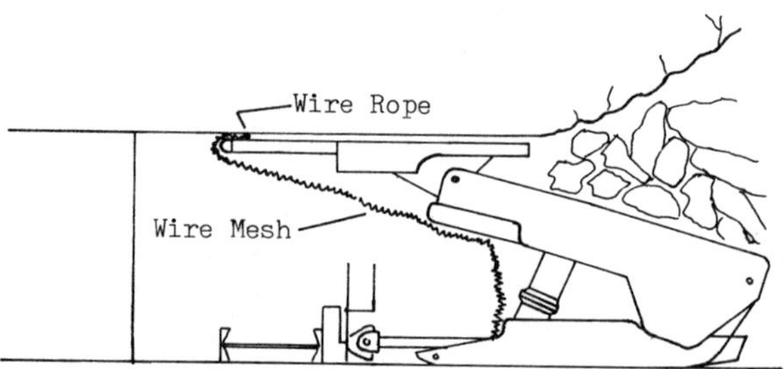

Figure 1. Profile of longwall face, wire meshing.

On the headgate, another crib is built outby the recovery point and a wire rope is attached to it and is stretched and attached to the one previously built to provide an anchorage point for the remaining cables that will span the face. The same procedure is used on the tailgate, though sometimes the spanning rope is attached to other cribs previously built.

After a pass with the shearer is made, the mesh is unhooked. Another cable is attached with tie wires 30 in. from the first cable. Two men then hold the mesh out to prevent snagging on the shield tip while one man lowers and advances the shield. Since this is a slow process, the next cable is simultaneously strung out and attached. The wire mesh is then hooked back to the shields and the procedure is repeated with every pass of the shearer up to 5 ft from the stopping point.

The shields are not advanced on the last two 30 in. cuts in order to provide room for the scoop and recovery operation. The panline is disconnected and advanced with ram jack extensions so that the shearer is able to make the last pass.

Oak plank is delivered to each shield by placing them on the face conveyor, running it in reverse, and unloading them where needed. The plank is placed over the shields and under the wire at an angle to rest on two shields and span the 5 ft from the shield tip to the face (see Figure 2).

In addition to the wire rope and mesh, 4 in. steel H-beams are installed over the shields at the headgate and tailgate (see Figure 3).

FACE CONVEYOR REMOVAL AND FACE ROOF SUPPORT

Upon completion of mining, the stage loader is disconnected from the panline and is pulled down the belt entry to be recovered later.

As soon as the plank are delivered, the headgate and tailgate drives are disconnected and are removed by the scoop cars. The track previously laid for the shearer recovery is extended as far as possible to the panline. The shearer is then trammed onto the special dolly and is taken outside for rebuild.

The panline is pulled away from the face to provide clearance for the roof bolters. The roof is then bolted by two bolters from both the head and tailgates (see Figure 2 for bolt location).

While the roof is being bolted, the lighting fixtures and wiring on the shields are removed. At the same time, the panline is removed three sections at a time by cutting the bolts with a torch. The conveyor chain is also cut (if worn out) and a scoop removes the panline and conveyor chain to a storage area so that they may be recovered later.

SHIELD RECOVERY

After the panline is removed and the top completely bolted, the hosing is disconnected on the two tailgate long tip entry shields. Individually, each shield is lowered, pulled out with a cable attached to the scoop, loaded into the scoop's bucket, and set on a dolly to be taken and installed on the new face. The previously installed steel beams and cribs support the top.

The hydraulic hoses are disconnected from the next two shields. Each one is lowered and turned parallel to the face by the headgate scoop using a wire rope, sheave wheel, and at times a waterjack to support the sheave. These are the recovery shields used to support the top behind the shields as they are recovered (see Figure 2).

The next ten shields are hydraulically isolated, and through captive pressure they will remain up.

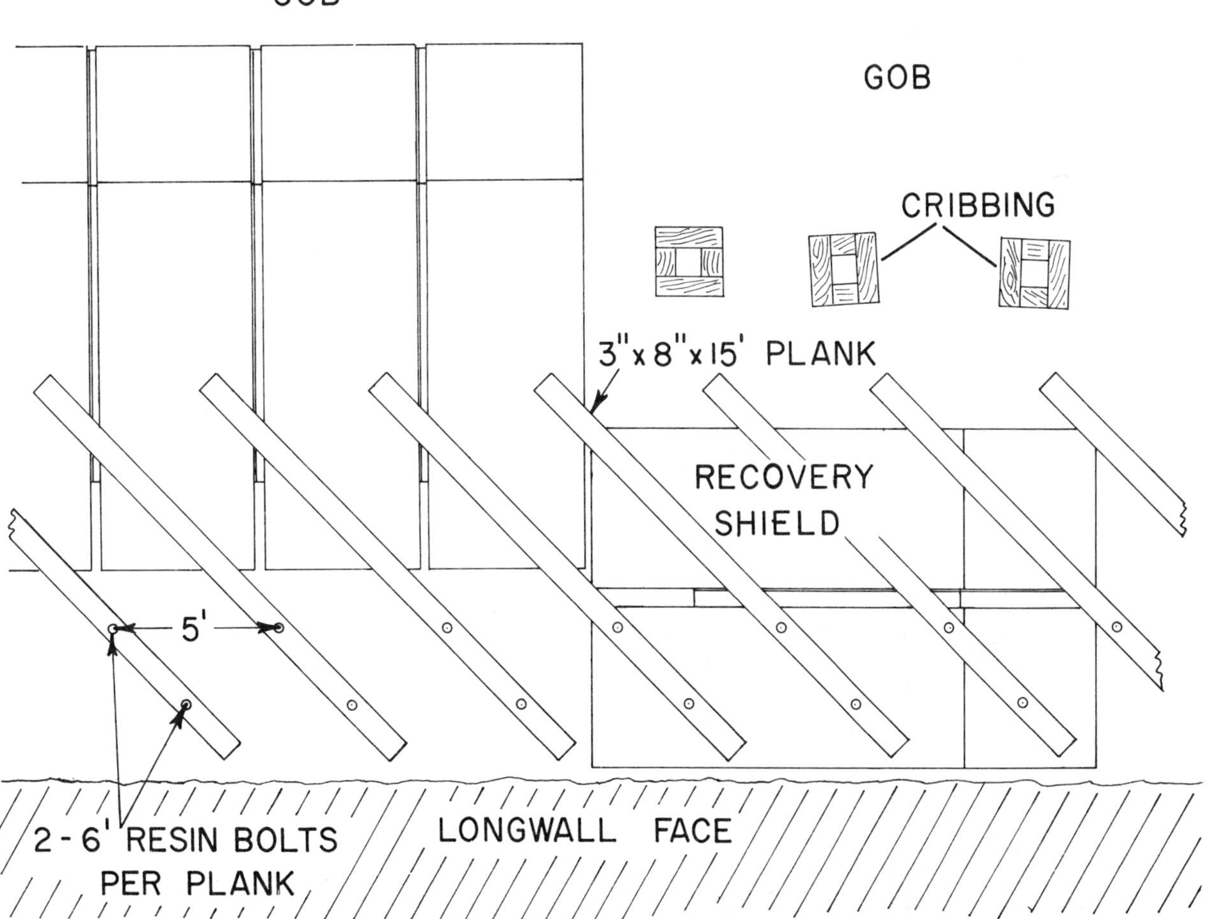

FIGURE 2

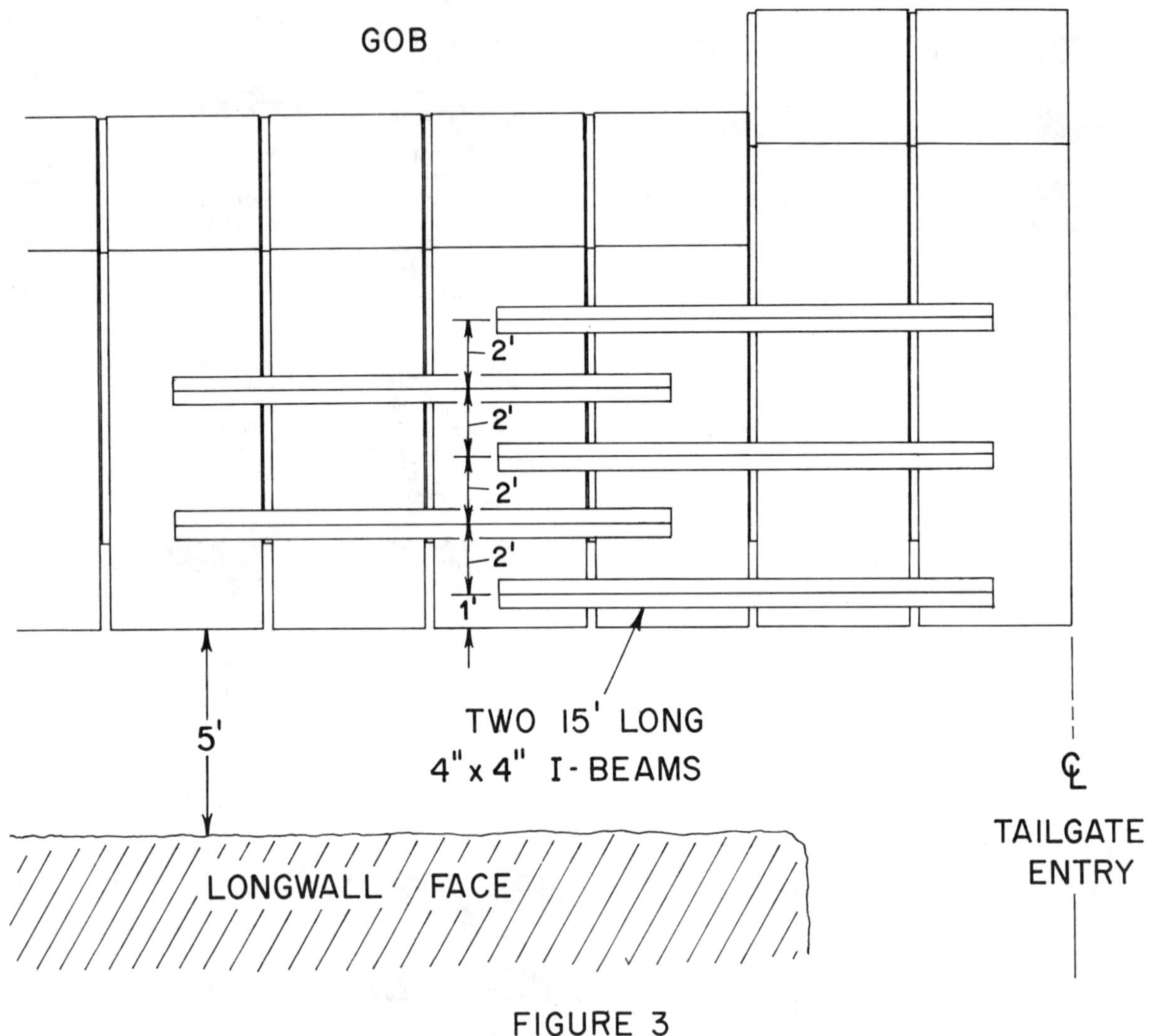

FIGURE 3

Hydraulic hoses are installed from the eleventh or last pressurized shield to the two recovery shields. These shields will then be able to be lowered and raised as needed.

The shield being recovered has its hydraulic hoses disconnected and is lowered. The two recovery shields are individually lowered, pulled forward by the scoop, and are pressurized against the top. The shield being recovered is then pulled out with a wire rope attached to the scoop. The rope is placed around a sheave at the face so that the shield can be pulled straight out.

The shield is then turned, pulled down a few yards, loaded onto the scoop, and taken to the dolly. It is then transported to the next face and installed. Meanwhile, a couple of cribs (using 24 in. crib blocks) are set in the void of the shield pulled for the purpose of holding up the sagging wire mesh so that the rock does not break through and override the next shield. The cribbings also help to keep weight off of the next shield so that it may be pulled out easier (see Figure 2).

The procedure is repeated until all the shields are recovered.

FINISH NEW FACE

As the shields are being installed on the new face, they are hosed, connected to the panline, and have the lighting installed. When the last shield on the headgate is installed, all systems are checked, and the longwall is ready to begin mining.

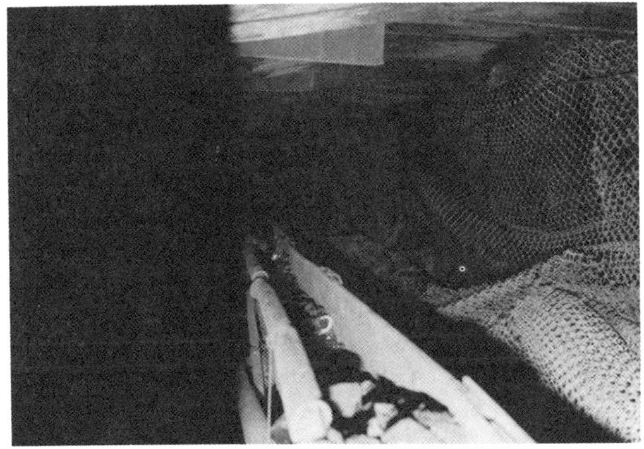

Picture 1. Wire mesh with wire rope attached, ready for shields to be advanced.

Picture 4. Roof bolting at longwall face.

Picture 2. Wire mesh pulled back, ready for next pass of shearer.

Picture 5. Shield being recovered pulled toward face.

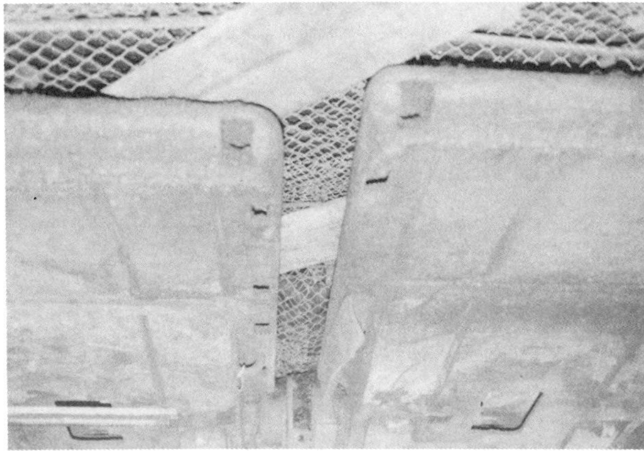

Picture 3. Shield tips showing plank, wire rope, and wire mesh.

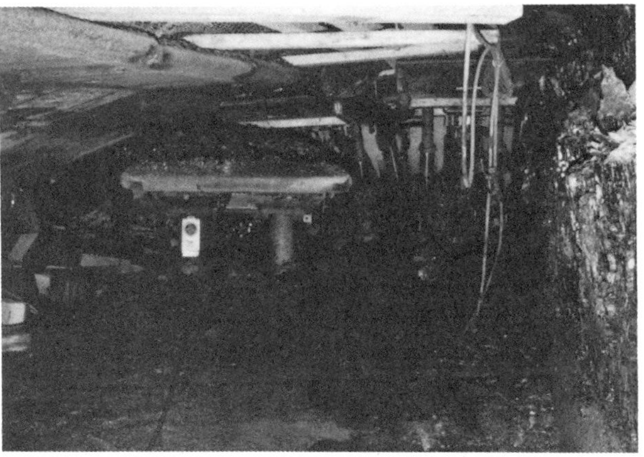

Picture 6. Shield being recovered turned and pulled away from recovery shields.

Picture 7.   Building cribs after shield has been removed.

Picture 8.   Effectiveness of cribbing, wire mesh, and wire rope shown.

Chapter 20

POLYURETHANE AIDS ROOF CONTROL IN LONGWALL MINING AT ISLAND CREEK COAL COMPANY

B. Rao Pothini
Hilmar A. von Schonfeldt

Mining Research Group Head
Manager Coal Research
Occidental Research Corp.
2100 S. E. Main St.
Irvine, CA 92714

ABSTRACT

One of the key elements essential to a productive and uninterrupted longwall mining operation is roof control. In several of the Island Creek Coal Company's mining sections, polyurethane has been infused into strata to control severe roof failures in the face and gate entry areas. To date about 50,000 kgs (110,000 lbs) of the chemical have been consumed in a number of applications. Roof-to-floor convergence and differential roof sag have been monitored to determine the need for polyurethane infusion and also to evaluate roof stability following the infusion. Stratascopic observations of the roof formed the basis for confirming the need and also for estimating the required quantities of polyurethane. Both chemical infusion and roof control monitoring techniques have been applied successfully under very adverse roof conditions; substantial production benefits have been realized. Operator acceptance of the system has been very good.

INTRODUCTION

Successful roof control next to equipment reliability is of utmost importance in longwall mining because it ensures the continuous and smooth operation. An uninterrupted operation is of particular importance because the longwall system is capital intensive and has a high productivity potential. It is also less flexible in adapting to the varying geological conditions than other more established mining techniques. Roof control affects mine safety, mining productivity and costs to a significant degree in all underground mining operations. Recent advances in roof control technology such as the fully-grouted resin bolt in gate entries and shield supports in the face have reduced accidents. However, roof control related injuries remain the largest source of mine accidents emphasizing the need for further improvement (MESA, 1977). Some of the improvements in roof control have increased unit mining costs by either reducing productivity or increasing material costs, or both. For example, the fully-grouted resin bolts require more time for installation at a higher material cost than conventional bolts of comparable length. Roof control operations are relatively more labor intensive, often accounting for nearly 25% of direct face labor (Pothini, et al, 1976). As indicated in Figure 1, roof control cost, including labor and materials, is the highest single cost item in the unit mining costs. It is evident, from the foregoing discussion, that further advances in roof control are necessary to improve safety and increase productivity in an attempt to slow the rise in mining costs.

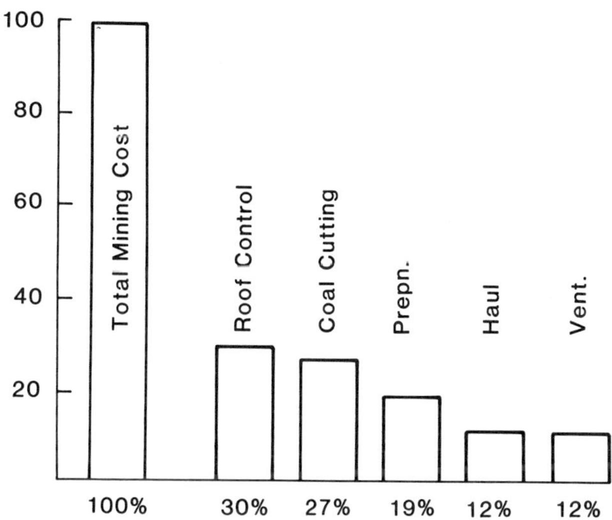

Figure 1
Functional Distribution of Mining Costs in a Coal Mine

PREDICTION OF ROOF CONDITIONS

A program to evaluate roof stability has been undertaken in Island Creek Coal Company's deep mines. The purpose of this work was to optimize roof control practices recognizing that the principal objective of roof control is the prevention of roof falls. All roof falls occurring in a mine over a 10 year period were mapped and analyzed (Pothini and von Schonfeldt, 1978). The study revealed that roof falls occur primarily in four-way intersections within close proximity of the mining face (Figures 2 and 3).

Any attempt to prevent all roof falls by installing sufficient support would be impractical, inefficient and perhaps impossible. A more prudent approach would be to prevent most of the roof falls through routine support design and to predict the remaining extraordinary cases with the help of an appropriate monitoring technique. Once predicted, either an impending fall could be stabilized through additional supports or permitted to fall in isolated and noncritical areas. The referenced roof fall analysis permitted selective monitoring of a limited area of the mine for a limited length of time which proved to be an important factor in the entire roof fall prediction process.

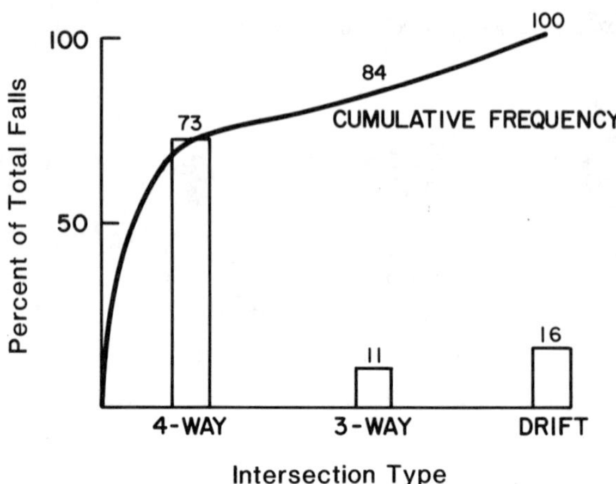

Figure 2
Roof Fall History of a Mine

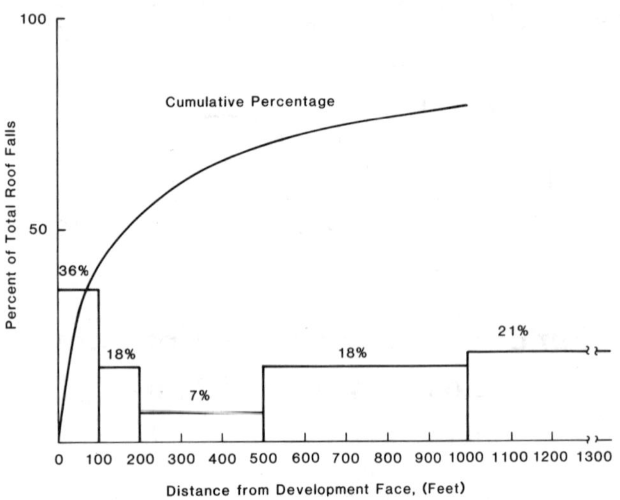

Figure 3

Distribution of Roof Falls vs Distance From Mining Face

With the help of numerous field measurements it was discovered that the deformation behavior of the rock layers surrounding the mine opening indicated instability and impending roof fall. The parameters monitored were roof-to-floor convergence and differential roof sag. The magnitudes of cumulative deformations, rates of change in deformation per unit time or velocity and the acceleration of deformations together indicated the stability or lack of it at a given location. Instability was indicated when cumulative convergence and convergence rate exceeded certain predetermined values. Any instability so detected was verified through visual inspections with the help of a borescope (Figure 4).

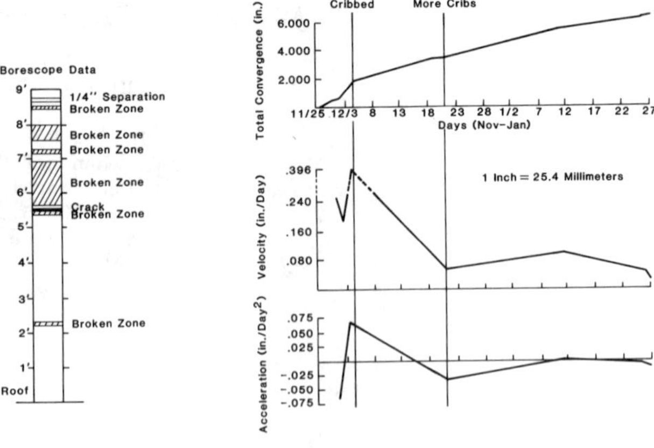

Figure 4
**Convergence and Borescope Data for an
Unstable Area in a Development Unit**

Numerous falls have been prevented since this technique was first developed. The tools used were simple and are described in more detail elsewhere (Pothini et al, 1980). To prevent a roof fall various types of auxiliary supports in addition to the existing roof bolts were installed. The severity of the deformation in the roof dictated the type of supports. Often the functional use of the area also restricted the selection of additional supports. Supplementary supports normally included longer roof bolts, wooden posts, wooden beams, steel columns, steel beams and cribs. For extreme cases where the conventional supports were either inadequate by virtue of their limitations or impractical by virtue of their bulkiness a new technique has been developed. The principle of this technique was to "glue" the severely broken roof rock together with polyurethane foam.

ROOF STABILIZATION WITH POLYURETHANE

The technique was originally developed for the European coal mining industry. The method has since found wide application especially around longwall systems in Germany. The method is based on placing polyurethane into the friable strata through a predetermined number of drill holes. The chemical is forced into the fissures of the broken rock from the borehole and upon setting re-cements the broken material into a compound mass. The placement of polyurethane is accomplished by direct infusion of its two chemical components into the holes through a mixing element (Bergwerksverband GmbH, 1976).

The volume increase associated with the foaming action has a beneficial effect on the consolidation process. The expanding material causes a pressure forcing the chemical out into cracks and fissures. This is in contrast to most other chemicals considered for chemical stabilization which generally shrink in volume upon curing.

The viscosity and setting time of the material is temperature dependent. It is therefore important to specify the range of rock and ambient mine temperature at which the material is to be applied before ordering the material.

The chemicals are available in fast setting and in slow setting formulation. The fast setting system has a setting time of approximately two minutes after the components are mixed. The short curing time makes this system well suited for applications in areas with large cracks. The material will set in the fractures quickly thereby blocking passage to further flow.

A slow setting resin on the other hand takes 15 minutes to cure. It has the potential of penetrating further into the broken mass. In large fractures, however, it may lead to considerable waste caused by the material dripping through open cracks and bolt holes. It would also cause considerable delay during the infusion process which is frequently done in a "stop-and-go" operation to allow the polyurethane to set in the larger fractures first.

The chemicals are supplied in 18.9 liter (5 gallon) containers for ease of transportation.

INFUSION EQUIPMENT

The infusion equipment consists of a specially designed pumping unit, packers, a mixer to mix the chemical components and two lines of high-pressure hoses from the pump to the infusion hole.

The pumping unit is commercially available. It is of tubular, compact construction and has two gear pumps, one for each chemical component (Figure 5). The pumps are driven by a compressed air motor. The motor operated at an optimum air pressure of 689 kPa (100 psi), using 0.014 m^3/s (30 SCFM) of air. Both chemical components are pumped separately to the hole in equal volumes and are mixed at the mouth of the hole in a labyrinth mixer.

Figure 5
Polyurethane Pumping Unit

The pumping unit used in the present work was equipped with a 9.79 MPa (4260 psi) pressure gauge to monitor the infusion pressure at all times. The pump stalled out at a pressure of approximately 14.69 MPa (2130 psi). High-pressure hydraulic hoses with a pressure rating of 68.95 MPa (10,000 psi), 19 mm (3/4 in.) and 13 mm (1/2 in.) in diameter respectively were used to pump the two chemical components[1] to the infusion hole. The difference in hose diameter was necessary because of the differences in viscosities and flow rates of the two chemicals.

INFUSION PROCEDURE

A minimum of three men were required for the infusion test: a pump operator, a chemical feeder and a man at the infusion hole, who acted as the supervisor for the operation.

An effective and fast communication system between the crew members is essential. The supervisor controlled the entire process by stopping and starting the pumping at his discretion to prevent spills. Rags were stuffed into the cracks as soon as polyurethane appeared on the surface to prevent the material from dripping on the mine floor and equipment. Small amounts of polyurethane were frequently extruded through roof bolt holes and fissures (Figure 6). This indicated good lateral and vertical penetration of the rock mass. The pumping operation was terminated when the pumps stalled out. Stalling of the infusion pump indicated that all cracks were filled and the material was set. The process was then repeated at the next hole.

Figure 6
Polyurethane Extruding Through Open Fractures in Roof

[1]The chemical components known as Riklok B-3 Binder, Component A, and Roklok B-3 Binder, Component B, are available from Mobay Chemical Company, Pittsburgh, Pennsylvania.

PRE-INFUSION INVESTIGATIONS

The West German longwall operations consume 10 Mkg (22 million lbs) of polyurethane per year for longwall face and entry stabilization (Giesse, 1978). However, prior to this work polyurethane had not been used in Island Creek Coal Company's mines or any mines in the U.S.A. on a significant scale as a roof support means. The chemical is expensive at $3.04 a kg ($1.38 per lb) (1981 price). It is imperative, therefore, that any application be well justified. This is accomplished through the following pre-infusion investigations:

1. Visual inspection of the proposed test area for obviously large deformations and extension cracks.

2. Roof to floor convergence and differential roof sag monitoring to detect deformations not obvious to the naked eye.

3. Borescoping to verify the mangitude and extent of fractures and bed separations in the roof. This is found most useful in also estimating the quantities of the chemicals needed to fill all the voids in the roof. A borescope is an optical device that can be inserted into a drill hole allowing direct inspection of the borehole wall.

The foregoing is easily accomplished in the gate entries. However, in an active longwall face it is not practical to develop a convergence history or even to do borescoping. In such cases the pre-infusion investigation has to be limited to visual inspections only.

SITE PREPARATION

Site preparation involves setting auxiliary supports, drilling infusion holes, installation of packers and installing a communication system.

Auxiliary Supports

Polyurethane is infused into broken rock at pressures of up to 14.69 MPa (2130 psi). Upon curing it expands 3 to 5 times in volume (Bergwerksverband, 1976). The infusion and curing pressure will generally induce a downward movement of the immediate roof. It is therefore essential to install auxiliary supports at a proper density prior to infusion.

It is important that this procedure be followed to avoid major difficulties. The chemical merely consolidates the fragmented rock into an "aggregate" mass or slab but does not tie the broken material back into the solid rock with the original strength. The entire unsupported chemically grouted mass could fall without warning (Glaesmann, 1977).

The recommended auxiliary support density used in West Germany is 8 tons/m^2 of exposed roof for each meter (0.243 tons per ft^2 per ft.) of coal seam thickness (Glaesman, 1978).

The types of auxiliary supports used depend on the location. Hydraulic props of 40-ton (44-short ton) capacity with steel I-beams or link bars are preferred over the passive support of cribs and posts.

Infusion Hole Drilling

The infusion holes were 44 mm (1.75 in.) in diameter and drilled dry to prevent any moisture from interfering with the chemical curing process. The depth of these holes depended on the condition of the roof and the locations. In the entries approximately 1.5 m (5 ft.) deep holes were found to be adequate. An idealized drilling pattern for a typical headgate entry is illustrated in Figure 7. Spacing of these holes is dictated by the degree of fracturing. It is necessary to ensure filling of all voids between holes. The larger the fractures, the farther the holes can be spaced apart. Our own experience indicated that a 3 m (10 ft.) spacing between rows of 3 holes in the 6 m (20 ft.) wide entries with fast setting chemical and 4.6 m (15 ft.) spacing in the longwall face areas with slow setting resin is optimum.

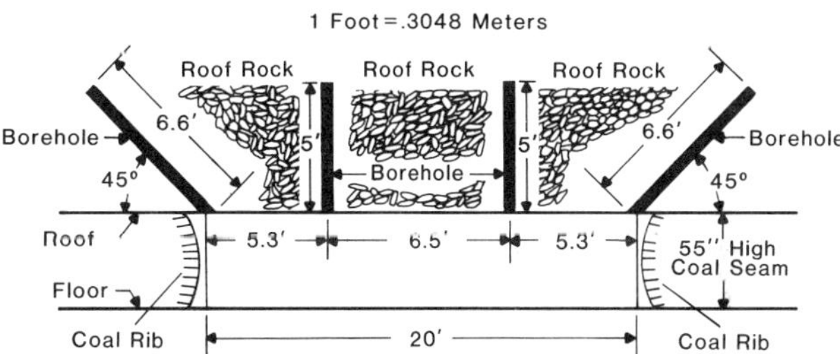

Figure 7A
Idealized Drilling Pattern, Vertical Section

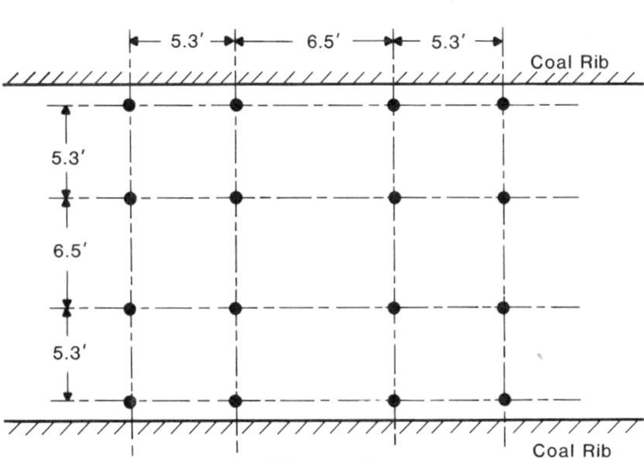

Figure 7B
Idealized Drilling Pattern, Plan View

Packer Installation

Each infusion hole was sealed off to within 0.2 to 0.3 m (8 to 12 in.) from the end with packers.

The packer consisted of a hollow metal rod with a nonreturnable ball type check valve on one end. A brass adapter was provided on the outbye end for a 10 mm (3/8 in.) staplok fitting. A 20 cm (8 in.) long soft rubber element between two nuts on the metal rod provided the packer seal. As the rod was turned clockwise in the hole with the aid of a box wrench on the brass adapter, the soft rubber extruded against the hole wall to seal the hole. The packers had to be extremely tight to prevent leakage at high infusion pressures. A 2 m (6.5 ft.) handle on the socket wrench is needed for adequate tightening by one man. Figure 8 shows a packer installed in an infusion hole.

Communication

The site preparation also included a set-up of an independent telephone system for communications between the infusion pump and infusion hole. To minimize traffic, to afford additional safety and to improve efficiency, the infusion pump and the chemical containers were centrally located remote from the individual infusion holes.

Air Supply to Pump

The infusion pumps were driven by a motor operating at 758 kPa (110 psi) with compressed air. The main compressed air line was tapped to provide the air for the motor through a 20 mm (3/4 in.) diameter flexible hose line.

Transportation

Site preparation included transporting several tons of chemical in 18.9 1 (5 gallon) cans, packers, hydraulic hoses and the pump.

CASE HISTORIES

Island Creek Coal Company has refined the technique of polyurethane infusion and used over 50,000 kgs (110,000 lbs) of the chemical during the last 3 years for roof consolidation. It has prevented falls in critical head entries and face areas of several longwall mines. The technique has found acceptance of the supervisory and union personnel. Details of 3 case histories will be discussed in the following.

Figure 8
Packer Installed in Drill Hole in Roof

Case History No. 1

Infusion of the Headgate Entry of a Longwall.

The very first test conducted in Island Creek's mines and perhaps the first of this scale in this country was in a 61 m (200 ft.) long portion of the headgate entry of a longwall in southwest Virginia (von Schonfeldt, H., and Pothini, B.R., 1979).

The longwall system was located in an older part of the mine. The longwall panels in this area are generally developed through a four-entry system. Roof control problems in the head and tailgate entry of this panel under consideration resulted in low mining productivity for almost the entire history of the panel. The age of the entry was 5.5 years which was believed to be one reason for the bad roof conditions. Heavy auxiliary support in the form of cribs had already been installed in many locations of the entry.

The roof in the test area was fractured and broken. In the most critical section the roof had already sagged below the 1.2 m (47 in.) minimum clearance required for the coal conveying equipment. A roof fall had already occurred in this entry in the intersection adjoining the proposed test site.

In the judgement of Mine Management an entry that had deteriorated this much could not be held open for the coal plow to mine past (Dulaney, 1978). Logistics would not permit leaving cribs and collars in place as the plow advanced closer. Although roof bolts, 2.4 to 3 m (8 to 10 ft.) long could have been used and angled into the ribs, the chances of successfully holding the roof were considered rather remote. It was therefore obvious that mining through the gate entry area after polyurethane consolidation, without a roof fall and without serious impediment to productivity would provide a sound basis for considering such a test a success.

Visual inspection of the area formed the basis for the selection of this test site. A program of monitoring the convergence and sag was instituted to evaluate the entry stability before, during and after the infusion.

Much of the entry deformation occurred prior to installation of the convergence stations. But because unstable roof areas continue to converge at higher rates than stable ones, it is possible to identify bad top even if the complete convergence history is not known. Early stabilization of hazardous roof is generally more economical because less chemical will be required to fill the still small fractures. It might also be possible to stabilize the area in such cases by cribbing, timbering or rebolting.

In addition to roof-to-floor convergence, roof sag measurements were made at various locations to determine whether the deformation occurred mainly in the roof or the floor.

Most sag and convergence data indicated no unusually large deformations during the pre-infusion period. This is not too surprising in view of the entry age plus the presence of the heavy supplementary support under the weakened roof.

Eight 44 mm (1.75 inch) diameter holes were drilled to a 3 m (10 ft.) depth in the roof for the purpose of borescope inspection. The fractures, broken zones and the separations of roof layers observed in each of the holes are shown in the composite map of Figure 9. This vertical section through the entry roof gives an indication of the roof damage along the axis of the entry. It was used to estimate the number of infusion holes and the density of auxiliary support required.

An important correlation existed between the borescope data and the residual convergence data. Comparison of the convergence and borescope data suggested that the more highly fractured roof continued to deform more than lesser damaged areas. The highest residual convergence for a period of 3.5 months was over 23 mm (0.9 in.) (Figure 10). The less severe fractured areas corresponded to lower residual convergence totals of less than 2.3 mm (0.09 in.) for the same time span. It was apparent therefore, that convergence rates at any stage of the life of an entry reflect its state of stability.

In the test area, the average normal seam height was approximately 1.4 m (4.6 ft.) and therefore the required support density was about 11.2 tons/square meter (1.14 tons/ft^2) of roof area. The proposed test area comprised 61 m (200 ft.) of entry at an actual width of about 6 m (20 ft.) for a total of 365 m^2 (4,000 ft^2) roof area. Therefore, a total support capacity of 3600 tons (4,000 short tons) was required.

Hand-cranked props with a 25-ton (27.5 short tons) yield load and an 8-ton (8.8 short ton) setting load were used in the test to achieve the required support density.

The support system consisted of 10.2 cm (4 in.) high by 3.65 m (12 ft.) long I beams supported by two props. The props were set on 1.8 m (6 ft.) centers with the beams installed end to end (Figure 11). In addition most of the previously installed cribs, collars and wooden posts were left in place.

Figure 12 shows the 60 infusion holes drilled to a depth of 1.5 to 2.0 m (5 to 6.6 ft.). A total of over 9800 kgs (21,560 lbs) of polyurethane was infused. The consumption rate per hole varied between nothing in a hole with no cracks to 2125 kgs (4675 lbs) in the heavily fractured area.

Advanced deterioration of the entry had opened up large gaps and cracks in the rock; furthermore, numerous open holes existed stemming from the use of conventional roof bolts. Therefore the fast setting chemical system was used to minimize material waste.

Post Infusion Monitoring

Following completion of the infusion tests, roof to floor convergence and roof sag in the test area were monitored until the arrival of the mining front. The objective was to determine whether the area had been sufficiently stabilized to allow the longwall face to pass without premature caving.

It is significant to note that the total post infusion convergence of 16.6 mm (0.652 in.) in the heavily glued and fractured area was only half as much as the 37.7 mm (1.483 in.) of post infusion convergence in the lightly glued area. The roof stiffness appeared to increase with the amount of chemical grout infused.

The purpose of the roof sag measurements was to detect time dependent deterioration of the chemical bond, if any. Prior to the chemical consolidation all sag stations exhibited differential displacements

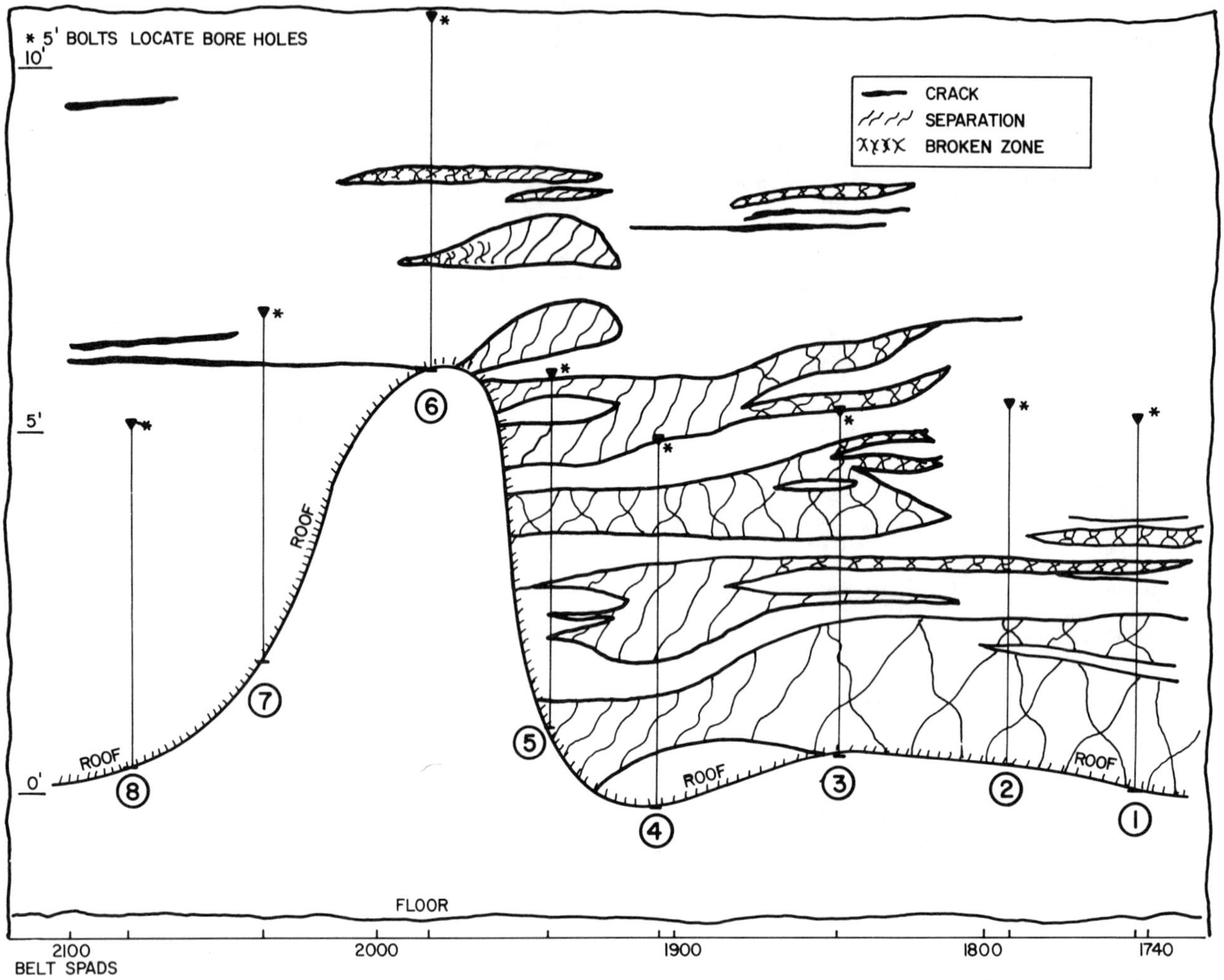

Figure 9

Vertical Section of Roof in Headgate Entry Depicting Fractures

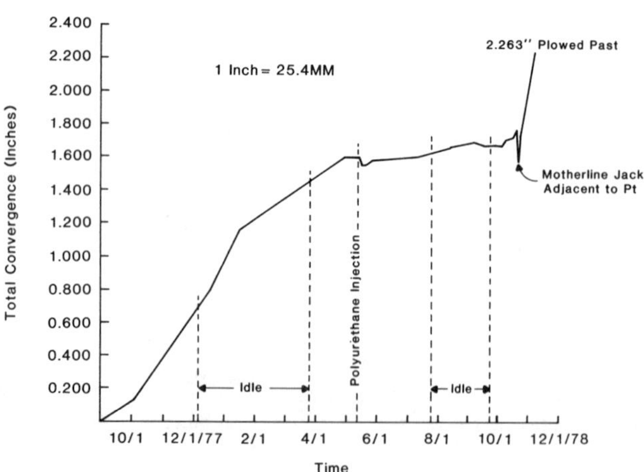

Figure 10
Longwall Headgate Entry, Point H15

between the immediate roof and the strata 3 m (10 ft.) above. During the infusion process sudden displacements were observed both in the immediate roof and the 3 m (10 ft.) deep horizon (Figure 13). This was a confirmation of the penetration of the chemical into the strata up to 3 m (10 ft.) deep in the roof. It had been hoped to bond together the lower 1.5 m (5 ft.) of the roof strata; the penetration to 3 m (10 ft.) depth provided additional assurance for the ability to form a well consolidated continuous roof beam.

After completion of the infusion process the differential sag was reduced to insignificant proportions of the total movement. For instance, the differential sag was 40% of the total movement prior to infusion. Following the infusion and until the mining front passed the same sag station the differential sag was only 7% of the total sag. Thus it was obvious that the integrity of the bond was maintained for the 6 month period between infusion and mining.

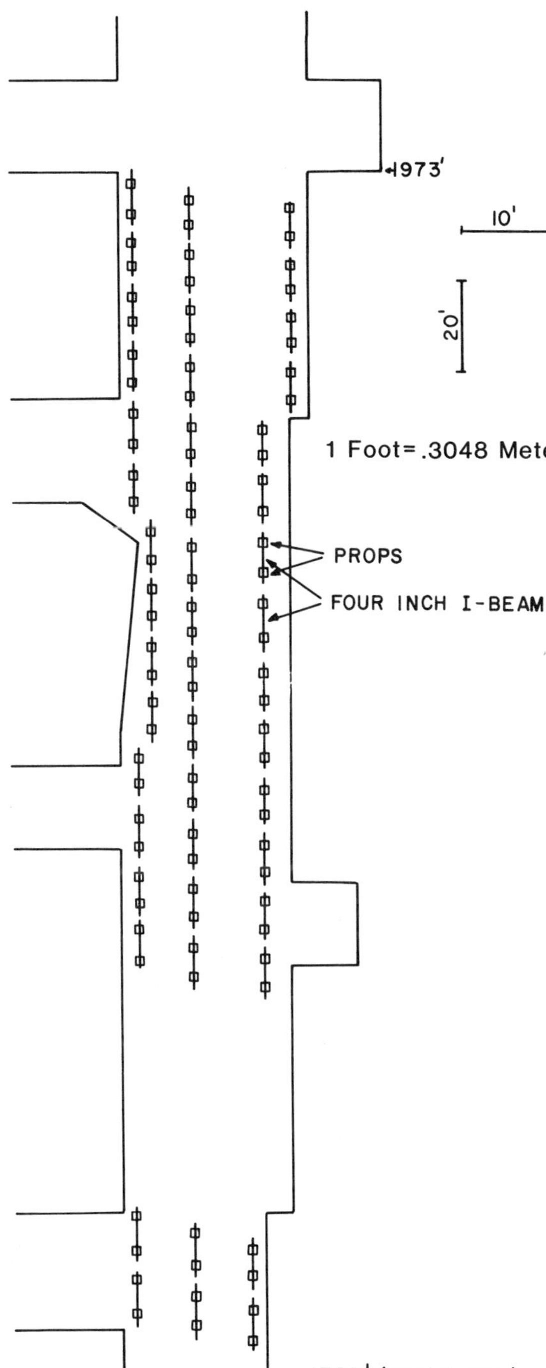

**Figure 11
Polyurethane Injection Area
with I-Beams and Props**

Roof Caving Characteristics

The stratigraphic properties of the rock in the mine roof determine its caving characteristics behind the longwall support. Thinly laminated and fractured strata cave in right behind the supports. Massive rock on the other hand tends to hang to significant distances behind the face supports.

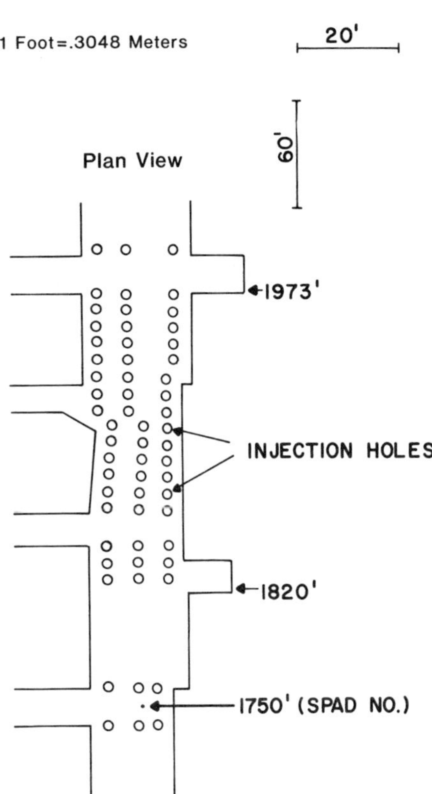

**Figure 12
Polyurethane Injection Holes Headgate Entry**

The purpose of the chemical consolidation was to create a massive uniform beam from the fragmented rock mass. Consequently the grouted roof should behave in its caving characteristics similar to that of a massive rock. Such was indeed the case. The roof overhang in the infused area varied between 2.1 m (7 ft.) and 12.2 m (40 ft.) with an average of 6.4 m (21 ft.). In the untreated roof areas the overhang distance behind the face supports varied between 1.5 m (5 ft.) and 7.6 m (25 ft.) for an average of 3.7 m (12 ft.). Thus, the chemical consolidation increased the span of the overhang by over 62%. In the extreme case of the 12.2 m (40 ft.) overhang the roof actually converged down to the floor without actually breaking. No excessive pressure on the face supports was experienced due to the overhang. It is speculated that the reason for this lies in the relative flexibility of the artificial roof beam.

The data regarding the roof span before caving represent estimates and not precise measurements for obvious safety reasons. Operating personnel working in the area from the start of the demonstration test confirmed that the span of the overhang was significantly greater in the infused area than in the untreated area.

Longwall Productivity in Test Area

The infusion of polyurethane was done at a significant cost in both labor and materials. Its purpose was not only to avoid a roof fall but also to minimize losses in productivity while mining through the test area. A comparison of the longwall productivity for

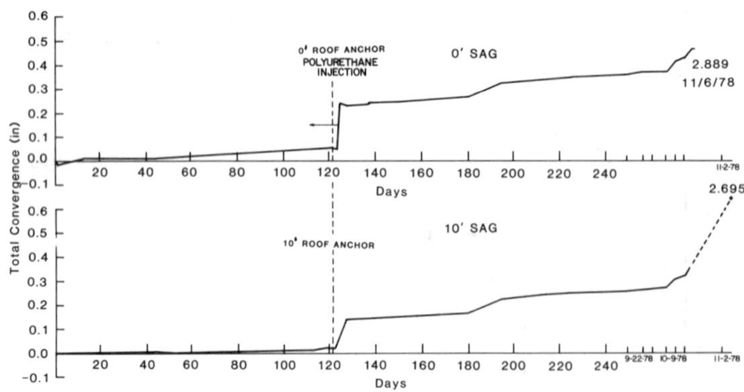

Figure 13
Convergence Vs Time For Immidiate Roof (0' SAG)
and Upper Strata (10' SAG) SAG Station No 4

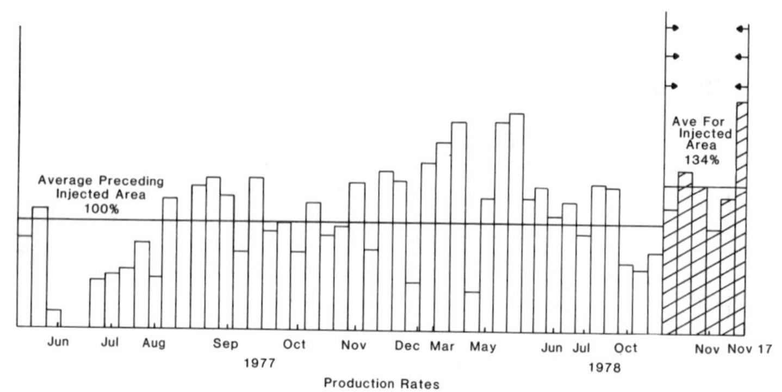

Figure 14
Longwall Productivity Before and During Mining in Test Area

the entire history of the present longwall panel is presented in Figure 14. The productivity of the longwall was 34% higher through the test area. The productivity improvement may not be that significant in itself because of the relatively short duration of the test compared to total panel life. It is significant, however, that no loss in longwall productivity was experienced while mining through the test area in spite of the difficult roof conditions.

The demonstration test therefore is regarded as successful in achieving both objectives of preventing a sizable roof fall and preventing a drop in the longwall productivity.

Response of Operating Personnel

When the demonstration test was planned and carried out most of the operating personnel were uncertain of its effectiveness. In their experience it had never been possible to mine through an area like the present one without sizable roof falls and extensive cleanup work.

As the longwall face advanced into the area and the span of the overhang of the treated roof in the headgate entry increased the attitude of the operating personnel became more optimistic. The face then moved through the heavily fragmented zone without requiring significant scaling of loose rock fragments from the roof. The span of the overhang was significantly greater than in the untreated area indicating a well cemented roof beam. Even before the mining in the test area was completed all operating people from the

mine superintendent to the hourly worker in the face area were satisfied with the improved roof control that this chemical consolidation system provided.

After completion of the demonstration test the opinion of all those associated with the longwall appeared to be unanimous that without the polyurethane infusion, mining through the test area without a major roof fall would have been impossible.

A pertinent concern was that the roof behind the longwall support caved without audible noise prior to failure. This is unlike their previous experience with untreated roof. The infused roof just collapsed suddenly in large blocks. The lack of audible warning before failure was certainly considered a disadvantage regarding mine safety.

This type of failure was anticipated, however, when the project was planned. For this reason the auxiliary support was left in place and checked at regular intervals until the arrival of the mining front.

Case History No. 2

Roof Stabilization in the Longwall Face Area with Inclined Holes.

The second case study involved a longwall at another mine in the same geographical area. The roof had failed in shear at the face and caved ahead of the longwall chock supports at two different locations covering many chocks. At one location the cave was 9 m (30 ft.) high and there was no contact between the longwall support canopies and roof. Such a situation was considered extremely dangerous because rocks falling from such a height could easily damage the face equipment and cause injury to workers in the area. The conventional procedures for advancing the longwall through such areas would be to fill the gap between the top of the canopies and the roof with wooden crib blocks. This work again would be extremely dangerous because it had to be carried out under unsupported roof. In view of the type of jointed cohensionless sandy roof strata encountered it was considered likely that the roof in the area would continue to fail in a similar manner. This implied that roof conditions could only worsen when mining was resumed.

It was concluded that polyurethane infusion was the safest and most economical way to increase the cohesion between the massive blocks of roof rock in the face.

A physical inspection of the area indicated that the fractures in the rock over the longwall face were small and extended over considerable distances. A slow setting chemical with 15 minutes curing time was procured, therefore, to allow better penetration.

The infusion holes were drilled at a rising angle in the roof rock above the face to a depth of 2 to 5 m (6.6 to 16.5 ft.) followed by packer installation. The pump station was set up in the headgate. Two high-pressure hoses were used to move the polyurethane components from the headgate to the infusion holes. A total of 3000 kgs (6600 lbs) of foam was infused into the area through the 19 holes shown in Figure 15.

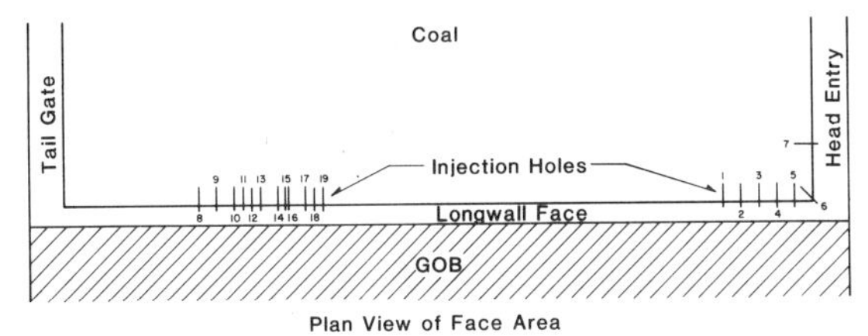

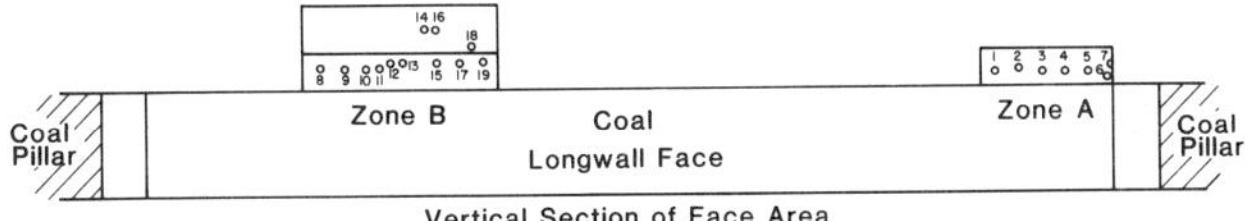

Figure 15
Polyurethane Injection Holes Face Area

Post infusion monitoring was limited to visual observations of roof behavior while longwall mining resumed. The face was advanced past the area without any roof control problems and the test was therefore considered successful.

Case History No. 3

Infusion of Stalls in a Longwall Face to Control Severe Roof Failures.

The last case involved massive roof caving over and in front of the chock supports. The caving extended upto 15 m (50 ft.) at places. This was considered one of the most difficult roof conditions on

a longwall face in the mines in this division of Island Creek Coal Company. Attempts to crib on top of the canopies and blast away any loose boulders of rock in front of the canopies further aggravated the problem of caving. As an extreme measure to ensure success multiple infusions were made. The first step involved drilling slant holes at a rising angle into the roof ahead of the face to a depth of 5 m (16.5 ft.). In addition to infusing these holes a number of 3 m (10 ft.) wide by 3 m (10 ft.) deep stalls were excavated ahead of the face in the infused area. Figure 16 illustrates the approximate position of these stalls in the face area. The stalls were in turn bolted with 2 m (6.6 ft.) long bolts on 1 m (3.3 ft.) spacing with steel mats to control the draw rock. The stalls were supported with props on 1 m (3.3 ft.) spacing and the roof in stalls was infused again. Figure 16 also gives a close up view of these stalls, the supports and the infusion holes.

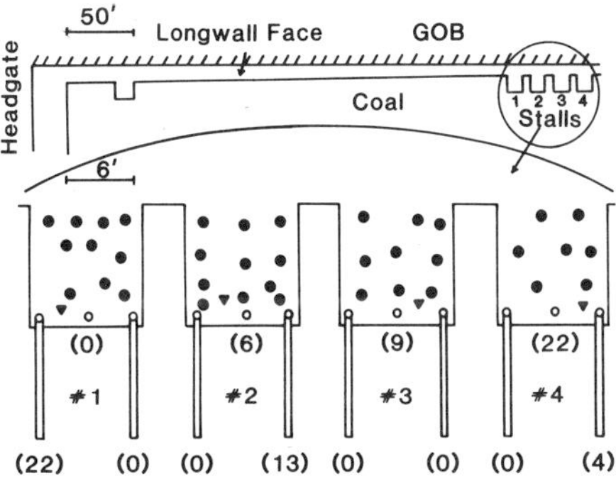

Figure 16
Stalls and Polyurethane
Injection Holes-Face Area

The combination of these measures was successful in permitting the longwall face to advance past these difficult roof conditions.

CONCLUSIONS

1. Polyurethane has been proven as a viable supplementary stabilization method to control extremely adverse roof conditions both in the face and gate entries of longwall faces in deep coal mines in south western Virginia.

2. The mine operators acceptance of the chemical system for roof stabilization has been excellent.

3. It has been proven that chemical
    stabilization of the roof can
    arrest the decline in longwall
    productivity even through difficult
    ground conditions.

SUGGESTIONS FOR FURTHER RESEARCH

1. Polyurethane infusion techniques to
    control draw rock in the longwall
    face and entry areas should be
    developed.

2. The logistics of transporting the
    chemicals need to be worked out to
    facilitate easier infusion in tail-
    gate entry areas of longwalls.

3. More cost effective means of cleaning
    the hydraulic hoses of the chemicals
    following the infusion needs to be
    developed.

ACKNOWLEDGEMENTS

The contribution to the success of this work by
the U.S. Department of Energy who funded part of this
work under Contract ET-78-C-01-3271 is greatfully
acknowledged. Special thanks are due to Mr. Otto
Glaesmann, a Consultant from Steinkohlenbergbauverein,
Essen, West Germany.

REFERENCES

Bergwerksverband, GmbH, 1976, The Use of Polyurethane
    For The Consolidation of Coal and Adjacent Strata,
    351 Frillendorfer Str., 4300 Essen 13, W. Germany.

Dulaney, R.L., 1978, Island Creek Coal Co. Corporate
    Offices, Lexington, Ky. Private communication.

Giesse, A., 1978, Mobay Chemical Company, Pittsburgh,
Pa. Private communication.

Glaesmann, Otto, 1977, Consultant from Steinkohlen-
    bergbauverein, Essen/W. Germany. Private communi-
    cation.

Glaesmann, Otto, 1978, cited above, private communica-
    tion.

MESA, 1977, Safety Reviews, November, 1977, Table No.
    8, U.S. Department of the Interior.

Pothini, B.R., Thaler, J.H., and Finlay, W.L., 1976,
    Rock Mechanics Considerations in Mine Design for
    a Deep, Bedded Deposit Under the Influence of High
    Tectonic Forces, Proceedings 17th U.S. Symposium
    on Rock Mechanics, Snowbird, Utah.

Pothini, B.R., von Schonfeldt, H., 1978, Roof Fall
    Prediction at Island Creek Coal Co., Proc. First
    International Symposium on Stability in Coal
    Mining, April 5-7, 1978, Vancouver, B.C. Canada -
    in print.

Pothini, B.R., Turyn, J., and von Schonfeldt, H.,
    1980, Underground Monitoring as an Operating Tool
    to Prevent Roof Falls, Proc. Eleventh Annual
    Institute on Coal Mining Health Safety & Research,
    VPI, Blacksburgh, Va.

von Schonfeldt, H., and Pothini, B.R., 1979 Consoli-
    dation of Bad Top at the V.P. Division of Island
    Creek Coal Company, Report submitted to DOE,
    Pittsburgh.

IV. Case Studies

William H. Mullins
Freeman United Coal Mining Company

Chapter 21

FIFTEEN YEARS OF CONSISTENT LONGWALL PRODUCTION AT
BETHLEHEM'S CAMBRIA DIVISION, EBENSBURG, PENNSYLVANIA

Edmund J. Korber, Manager, Cambria Division, Bethlehem Mines Corporation
Frank A. Burns, General Superintendent, Cambria Division, Bethlehem Mines Corporation
Donald E. Raab, Assistant Chief Engineer, Cambria Division, Bethlehem Mines Corporation

During the early 1960s, the advent of self-advancing longwall roof supports triggered serious consideration by Bethlehem management to introduce the technique of longwall mining at one of our central Pennsylvania coal mines. Although longwall mining was initially justified as a means to cope with the severe roof conditions at our Mine 32 (Watson, B., 1971), the result of the endeavor has been a consistent and reliable level of coal production, first from Mine 32 and then later from Mine 33, over a period of fifteen years. Longwall production from the Cambria Division has been the mainstay of the low volatile coal required by Bethlehem Steel for steel production. Let us review the events that led to this accomplishment, as well as where we stand today.

LOCATION, HISTORY AND GEOLOGY

Bethlehem Mines Corporation, a subsidiary of Bethlehem Steel Corporation, operates coal mines at six locations throughout the northeastern United States. One of these operations, the Cambria Division, is located near the borough of Ebensburg in central Pennsylvania approximately 121 kilometers (75 miles) east of Pittsburgh. The three mines presently operating in the Ebensburg area provide a high grade low volatile metallurgical coal for use in the steel making process.

One of the three operating mines, Mine 78, is a new mine that was opened by Bethlehem within the last five years. A second mine, Mine 32, is the oldest operating mine of the Cambria Division, having been opened by the Monroe Coal Mining Company in 1916. The first coal was shipped to market from Mine 32 in March, 1918. On October 14, 1948, Mine 32 was purchased by Bethlehem. The third operating mine, Mine 33, was put into operation by Bethlehem in September, 1964. The coal property surrounding Mine 33 was acquired in 1948 by leases from the Clearfield Bituminous Coal Corporation. The coal product produced from the three Cambria Division mines cleans to a sulfur content of 0.95%, an ash content of 6.8%, a reflectance of 1.5, and a volatile matter of about 19% when cleaned at a specific gravity of 1.42.

Geologically, the Cambria Division is located within the Allegheny Mountains section of the Appalachian Plateau Province (Iannacchione, A.T., and Puglio, D.G., 1979). The area is characterized by gentle anticlines and synclines, with most of the relief being caused by a dendritic pattern of stream erosion. The coal seams mined are found in the Kittanning and Freeport Formations, which are part of the Allegheny Group of the Pennsylvania system. Mine 32 produces coal from the Lower Kittanning coal group called the "B" seam. The seam is 1.07 meters (42 inches) thick. A 31 to 46 centimeter (12 to 18 inch) layer of bone coal is also mined, making the

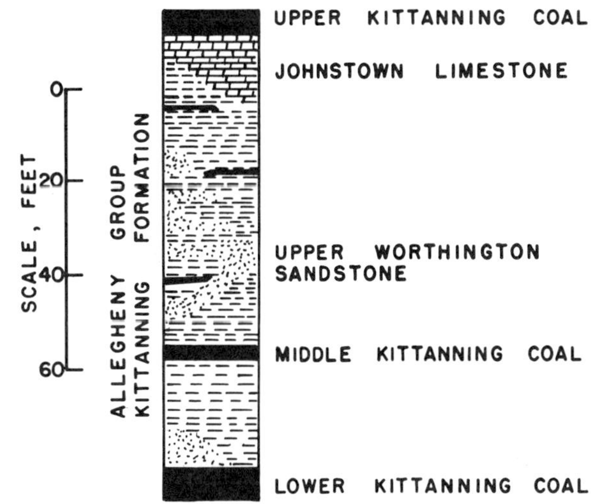

Stratigraphic column of Kittanning coal beds and adjacent strata.

total mining height about 1.4 to 1.5 meters (54 to 60 inches). Mine 33 produces coal from the "B" seam as well as the Upper Kittanning coal group known as the C-Prime seam. The C-Prime seam is approximately 1.4 to 1.5 meters (54 to 60 inches) thick.

Mine Nos. 32 and 33 are contained within the Johnstown syncline, the Ebensburg anticline, and the Wilmore syncline. Present workings are located about halfway down the eastern limb of the Ebensburg anticline with main entries following the axis of the anticline in the S 30° W direction. The dip of the coal seam varies from 1% to 10% and the average depth of cover in the longwall areas varies from 500 to 1,000 feet. Roof rock consists primarily of shales and laminated sandstones with only certain areas having sandstone directly over the coal seam. The floor is a fairly hard fire clay.

MINING CONDITIONS

Adverse roof conditions are common at Mine Nos. 32 and 33, particularly in development sections. Extreme roof pressures cause roof "cutters", pressure falls, and running caves which often course through several hundred feet of entry, zigzagging through crosscuts into adjacent entries. Numerous studies over the past 20 years have been conducted in an attempt to determine the cause of our adverse roof conditions, and to formulate effective strategies for dealing with these conditions. Attempts to improve roof conditions have ranged from epoxy infusion of the roof strata in 1959, to installation of yieldable arches and developing main headings by the pressure arch theory in the early 1960s, to forming "caving chambers" by intentionally caving certain entries for pressure relief in the late 1960s and early 1970s.

As a part of the continuing attempt to resolve the problem of running pressure falls at Mine Nos. 32 and 33, a conceptual study was conducted by a British mining consulting firm, K. Wardell and Partners, during 1969. The study concluded that ". . . the conventional assumption of a vertical or gravitational stress component only . . . is invalid at Mine Nos. 32 and 33 . . . (and that) another parameter(s) must be operative. The most probable missing parameter is the existence of relatively high horizontal stress components." (Wardell, K. and Partners, 1969). A subsequent study conducted for the U. S. Bureau of Mines by the Excavation Engineering and Earth Mechanics Institute of the Colorado School of Mines confirmed this postulation during the mid 1970s. An instrumentation technique which utilizes stress-relief rosettes revealed that horizontal stress concentration at Mine Nos. 32 and 33 is often twice the vertical stress field from overburden (Maxwell, B., et al., 1977).

As a result of our experience and the studies we have conducted, present mining methods are designed to relieve horizontal stresses where possible. Blocks of coal that are to be mined by continuous miner units are developed with 3-entry systems closely paralleling previously established gobs. Although main heading development and longwall development continue to be problems, coal production by the longwall method is effective and provides an improved means to cope with horizontal pressures.

EARLY LONGWALLS

Bethlehem's first longwall panel was started at Mine 32 in July, 1965. Although numerous mechanical problems and adverse physical conditions were encountered in the first panel, it was believed that the longwall had achieved the initial goal of providing an improved means to mine coal in poor roof conditions. The panel was completed in April, 1966, after producing 149,000 metric tons (164,000 short tons) of coal on a two-shift a day schedule at an average production rate of 789 metric tons (870 short tons) a day (Watson, B., 1971).

Subsequent panels were mined at drastically improved production rates and many of the mechanical problems and problems with the configuration of the longwall itself were worked out. The second panel started in June, 1966, and was completed in December of that same year with an average production rate of 1,769 metric tons (1,950 short tons) a day. Peak production rates for this panel reached 2,268 metric tons (2,500 short tons) a day. The third longwall panel at Mine 32 operated from January until May, 1967, averaging a production rate of 2,450 metric tons (2,700 short tons) a day. Later longwalls at Mine 32 did not prove to be so successful. Extremely bad physical conditions as well as excessive water problems were encountered. These conditions coupled with the general deterioration of the longwall equipment itself, particularly the powered supports, resulted in the eventual termination of longwall mining at Mine 32 in 1972. The initial results at Mine 32, however, provided sufficient encouragement as well as knowledge of

longwall mining to justify capital appropriations and mining plans to establish longwalls at Mine 33, a new mine operating in the same seam of coal as Mine 32.

Longwall mining at Mine 33 began in March, 1968. As the mine expanded in both the "B" seam and C-Prime seam, so did the use of the longwall mining technique. Today, three longwalls are in operation at Mine 33, two in the "B" seam and one in the C-Prime seam. Coal production from these three longwalls has averaged over 4,436 metric tons (5,000 short tons) a day for the past five years including all moves from one face to the next and delays due to development requirements. Two world production records were set at Mine 33 by the longwall units. In 1970, 6,604 metric tons (7,280 short tons) were produced in a 24 hour period (Jones, D.C., 1971), and in 1972, 8,192 metric tons (9,030 short tons) were produced in a 24 hour period. Recent world records have been set by longwalls operating in the higher seams of coal which require fewer passes across the face to achieve high tonnages than do the low seam longwalls.

PANEL CONFIGURATION

The first longwall panels at Mine 32 varied between 152 meters (500 feet) and 183 meters (600 feet) in width and 488 meters (1,600 feet) to 732 meters (2,400 feet) in length, depending on conditions in the area that was selected for the longwall site. Later, panel dimensions were standardized at 178 meters (585 feet) in width and up to 1,524 meters (5,000 feet) in length. Initially, the face transportation system consisted of a 76 cm (30 inch) wide armored face conveyor, a 46 meter (150 foot) long 76 cm (30 inch) wide stage loader, a 76 cm (30 inch) wide cross belt, and a 91 cm (36 inch) wide panel belt. The panel belt for the first longwall panel was not located in the entry next to the longwall block itself, but rather in a parallel entry. The cross belt was needed to transfer the coal from the stage loader to the panel belt in the other entry.

Upon completion of the first panel, the longwall equipment was turned 180 degrees and began retreating the block of coal immediately adjacent the first panel. The panel belt that was used to mine the first longwall panel was also used to retreat the second panel. For the second panel, however, the belt was located next to the longwall panel itself, thereby eliminating the need for the cross belt. This technique had the advantage of a simplified coal transportation system in that two longwall panels could be mined off one panel belt and one loading point.

The tail entries for the next longwall panel were developed by a continuous miner unit operating just ahead of the longwall face in operation. This was known as the "advance-retreat" method of longwalling. Due to ventilation, drainage, and mine layout restrictions, however, the advance-retreat method of longwalling, as well as turning the longwall 180 degrees to mine the next panel were not used again after the early 1970s. All longwall panels are now mined in the S 30° W direction, paralleling the main entry development. The butt cleat direction is S 20° W and the face cleat direction is S 70° E. Experience has proven the S 30° W direction of mining to be best from a roof control and drainage point of view, as well as being conducive to a systematic

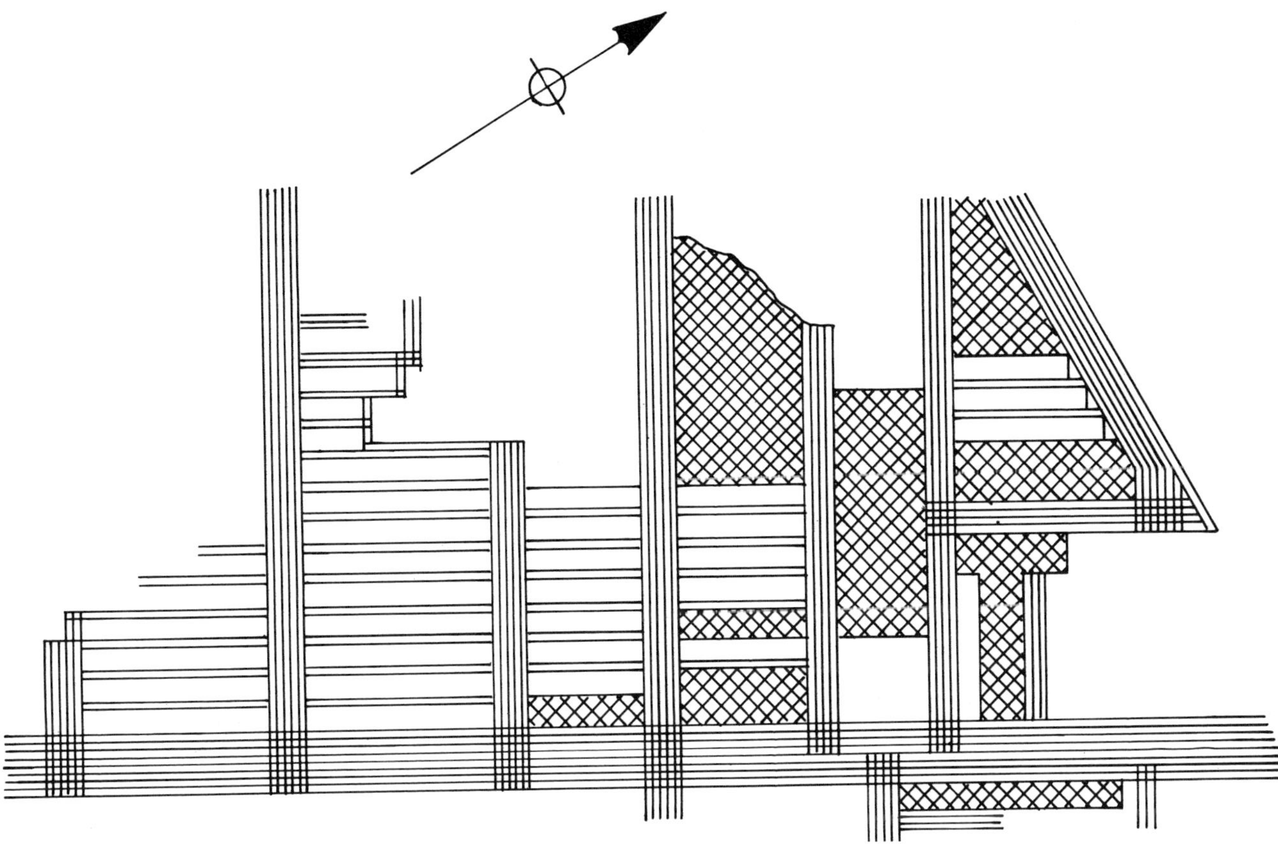

Longwall panels of Mine Nos. 32 and 33.

method of mine development.

Longwall panels are developed by continuous miner units driving three-entry butt headings. The conveyor belt is located in the outside entry next to the panel being developed. When development is complete, the conveyor belt is left in place for use by the longwall as soon as production of the panel begins.

VENTILATION

Although several plans have been approved for ventilation of longwall faces at Mine 33, the most commonly used plan brings the fresh air across the longwall face from the tailgate to the headgate side. Approximately 566 cubic meters per minute (20,000 CFM) is supplied to the tailgate for ventilation of the longwall face. As the air moves along the longwall face, dust and gases are diluted and directed into the return airway on the headgate side of the longwall panel. Some of the air moving along the longwall face is permitted to flow through the chocks and into the gob rock in order to prevent any gases from moving out onto the longwall face. This air also moves to the return entry on the headgate side of the panel.

In addition to the 566 cubic meters per minute (20,000 CFM) of fresh air delivered to the tailgate for face ventilation, approximately 142 cubic meters per minute (5,000 CFM) of fresh air is directed

through the center entry paralleling the panel belt, moving through a crosscut and then into the belt entry at a point just outby the longwall face. This air, along with the belt air, ventilates the headgate machinery, and then passes into the return airway.

This method of longwall face ventilation has been found to be very satisfactory in terms of methane and dust control. In the "B" seam, methane gas is generally not liberated in large quantities from the coal seam itself, but is liberated in large quantities from the roof rock as it caves. The ventilating current moving through the gob rock carries these gases into the return entry. Any dust generated by the panel belt is also directed into the return airway, and does not contaminate the face area itself. The dust generated by the shearer passes down the longwall face, meets the air which ventilates the panel belt, and proceeds directly into the return. The shearer is a single drum bidirectional type, with the drum located on the "down wind" side of the machine. This design permits the shearer operator to work "up wind" of the cutting drum at all times, thereby reducing his exposure to high concentrations of airborne dust.

METHANE CONTROL

As was mentioned earlier, most of the methane gas that is made in connection with mining the Lower Kittanning coal seam is liberated from the roof rock as it caves, rather than from the coal seam as new

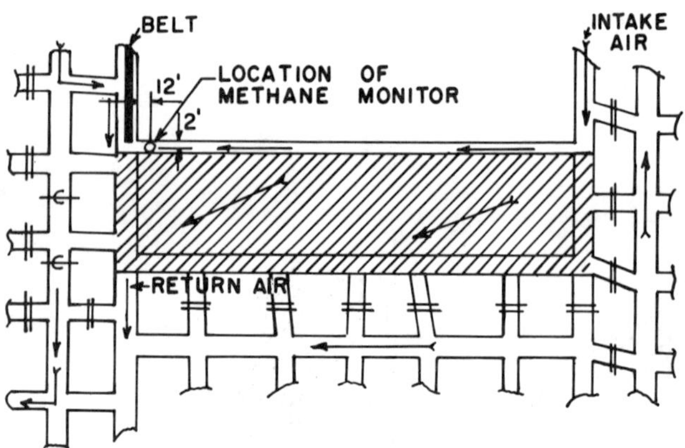

Typical longwall face ventilation plan.

faces are exposed. Consequently, efforts to control methane concentrations center around controlling gas in the gob and in the bleeder and return entries, more so than at the face itself. In the early days of longwall mining at Bethlehem's Cambria Division, production rates were often reduced in order to prevent excessive methane concentrations from accumulating in bleeder and return entries. Since the quantity of gas entering the mine is a function of how fast the roof rock caves, gas liberation could be controlled by reducing the number of passes the shearer takes over a given period of time, thereby reducing the rate at which the roof caves and the rate at which methane gas is liberated.

Improvements in methane control were realized with the introduction of methane boreholes as a means of collecting gas from the gob. Holes from the surface are drilled to the coal seam being mined. Although most of the hole is cased, the bottom portion, which will eventually fall within the zone of broken and caved rock, is left raw. Once the longwall passes the methane borehole, methane gas (which is almost one-half as dense as air) is liberated from the caving rock and is vented directly to the outside atmosphere by way of the borehole. Exhaust fans are installed on the surface at the borehole and are used to assist in this process.

Gas concentrations vented to the atmosphere at the methane borehole vary between 25% and 80% to 90% methane gas. The volume of gas that is exhausted usually varies between 7 and 113 cubic meters per minute (250 and 4,000 CFM). Once the methane concentration at the borehole drops below 25%, the exhaust fan is de-energized, and the borehole is permitted to continue to exhaust gas by "free-flow". Sometime after completion of the longwall panel, the methane concentration and quantity usually drops off considerably. At this point the hole is considered to be of no further value and is grouted to the surface.

Due to a number of factors, including new surface mining regulations, drilling boreholes is becoming prohibitively expensive. Consequently, Bethlehem is now exploring the potential of using underground horizontal boreholes drilled over the coal seam as a means to degasify gobs. A test of this technique is presently in progress at Mine 33 in conjunction with the U. S. Bureau of Mines.

EQUIPMENT

The major equipment which comprises a longwall face are the shearer, the face supports, the face conveyor, the stage loader, the controls and pumps, and the panel belt. An evolutionary process at Bethlehem's Cambria Division has been the means by which specifications for Mine 33's longwall equipment have been developed to best cope with "B" seam mining conditions.

Shearer. The first shearer used in 1965 at Mine 32 was a British Jeffrey Diamond 150 horsepower bidirectional machine with a ranging head, a 137 cm x 61 cm (45 inch by 24 inch) drum, point attach bits, and rope haulage. Later, at Mine 33, the shearer horsepower was increased to 200, and the drum size was reduced to 112 cm x 76 cm (44 inches by 30 inches) to improve coal quality.

Experience with the early shearers provided valuable information in specifying later shearers. It became obvious that increased horsepower and speed, as well as a powered loading cowl and rope haulage were desirable features. The present shearer used at Mine 33 is the Anderson Mavor A. B. Sixteen Mark IV bidirectional shearer with a single drum, a maximum speed of 12 meters per minute (40 feet a minute), a 270 horsepower motor, a 137 cm (54 inch) helical vane drum with a cowl, an 84 cm (33 inch) web, rope haulage and hydraulic drive. A 600 kva transformer is used to reduce the primary voltage of 4,160V AC to a 460V AC machine utilization voltage. The shearer and all other AC electrical equipment are powered by 460V AC motors.

Supports. The chocks that were purchased for the first longwall were Dowty six-leg 181 metric ton (200 ton) chocks with articulated bases and beams, and hydraulic adjacent controls. The chocks were spaced on 112 cm (44 inch) centers across the longwall face. Two 26.5 liter per minute (7 gallon a minute) ram pumps with unloaders and a 757 liter (200 gallon) tank were used to furnish the hydraulic power.

Although adjacent controls were a constant source of trouble with the original chocks, these problems were later resolved. Adjacent controls are specified when new support systems are purchased. The original support system provided evidence that large roof and floor contact areas were desirable, although the width of the chocks had to be minimized in order to avoid excessive areas of unsupported roof when a chock is released. In order to facilitate high speed shearing, later chocks were specified to be capable of "one web back" operation in order to provide an obstruction-free travel way. An increase in the capacity of the hydraulic system was also needed.

Today, Huwood Irwin 254 metric ton (280 ton) chocks are used. These chocks have four legs with double-acting cylinders and a range of 81 cm to 123 cm (32 inches to 48½ inches). They have a setting load of 71 metric tons (78.5 tons) and a yielding load of 254 metric tons (280 tons), a base pressure of 22.5 kg/cm^2 (320 psi) and a top pressure of 9.63 kg/cm^2 (137 psi). The chocks are 81 cm (32 inches) wide and are placed on 107 cm (42 inch) centers across the longwall face. Dowty Dowval valves with adjacent controls are used to control the chock functions. Three Gullick T50 110 liter per minute (29 gpm) plunger pumps furnish the 141 kg/cm^2 (2,000 psi) hydraulic pressure for the chocks. A fire resistant water-in-oil emulsion is used for the hydraulic fluid. Staple-

Mine 33 longwall face showing the shearer mining toward the headgate entry.

Loc fittings are used wherever possible in the hydraulic system.

Armored Face Conveyor. The original face conveyor used was a Jeffrey Mark IV. This conveyor was 76 cm (30 inches) wide with a triple strand 18 mm chain running at 64 meters per minute (210 feet a minute). Two 125 horsepower motors, one located at the head end and one located at the tail end, powered the chain. The unit was equipped with MK11A Bretby handling cable trough and chain. Weaknesses in this conveyor chain were improved on later units. The size of the drive motors was increased, and the clutches were redesigned for better synchronization. Deck plates and the top rails on which the shearer rides were strengthened. Ramp plates were also specified for an angle of 37½ degrees, and the chain speed required was to be at least 61 meters per minute (200 feet a minute).

Today's longwalls at Mine 33 are equipped with Huwood heavy armored scraper chains rated at 662 metric tons per hour (370 tons an hour), running at 68 meters per minute (222 feet a minute). Two 150 horsepower motors, again one located at each end,

drive the 18 mm triple strand chain. The conveyor is 76 cm (30 inches) wide, 19 cm (7½ inches) deep, with flights on a one meter spacing. Ramp plate angles were maintained at 37½ degrees.

Stage Loader. Matched with the original BJD shearer and Jeffrey face conveyor was a Jeffrey stage loader fitted with a tripper to transfer coal to a cross belt. The stage loader was 46 meters (150 feet) long and was equipped with a 76 cm (30 inch) wide chain running at 78 meters per minute (255 feet a minute) powered by one 75 horsepower motor.

The Jeffrey stage loader was the focal point of many of the initial problems with the first longwall. The problems were primarily associated with the tripper device and the fact that the chain was not captivated. Succeeding stage loaders were integrated with the tail of the entry belt. The tripper and cross belt were eliminated. Belt take-ups were also installed in order to permit the stage loader and belt tail to advance as one unit.

The stage loaders now in use are furnished by

Huwood. They are 22 meters (72 feet) long and
equipped with a 76 cm (30 inch) wide 18 mm by 650 mm
triple strand captivated chain running at 93 meters
per minute (305 feet a minute).

OPERATING PROCEDURES

Fifteen years of longwall experience has resulted
in efficient longwall operating procedures at
Bethlehem's Mine 33. A standard cycle begins with
the shearer located at the tailgate end of the longwall face. The body of the shearer and the tail drive
section of the face conveyor are both low profile
units. This feature allows the body of the shearer
to ride out over the tail drive into the tailgate
entry without contacting the roof. When the shearer
is in this position, the ram cylinders on the chocks
push the face conveyor (or pan line) and the tail
drive forward 76 cm (30 inches). The chocks are then
pulled up 76 cm (30 inches) into position. The
shearer drum is now properly positioned to begin
mining a 76 cm (30 inch) cut of coal across the 178
meter (585 foot) long longwall face.

The shearer operator begins the pass by position-
ing the cowl behind the shearer cutting drum and
advancing the shearer down the pan line at an average
rate of 6 meters per minute (20 feet a minute).
Following approximately 9 meters (30 feet) behind the
shearer is a crew member called a "snaker". The
snaker activates hydraulic cylinders which push the
pan line up to the newly exposed face. In the process of moving the pan line up, most of the coal
which remained on the bottom in the newly exposed
76 cm (30 inch) cut is bulldozed by ramp plates into
the pan line.

Following behind the snaker is the chock mover, or
chock boy. The chock boy releases the pressure from
the chock canopy, and advances the chocks back up
into position behind the pan line. This step in the
cycle usually results in the initial caving of roof
immediately behind the chocks. When the shearer
reaches the headgate entry, the cycle is half complete.

Since the shearer drum is positioned on the headgate side of the shearer, only the drum itself is
advanced into the headgate entry. The body of the
shearer does not move out into the headgate entry,
so the machine does not interfere with the stage
loader nor the other equipment in this entry. The
snaker pushes the pan line (and thus the shearer) up
76 cm (30 inches), and the headgate man advances the
stage loader and the belt tail 76 cm (30 inches). The
shearer operator turns the cowl on the cutter drum
and begins to tram the shearer back across the face,
mining another 76 cm (30 inch) pass of coal. The
snaker and chock mover follow the same procedures in
moving from the headgate to the tailgate as they did
when the shearer was moving from the tailgate to the
headgate. During the cutting cycle, men in the headgate and tailgate entries advance hydraulic roof jacks
and shovel coal spillage onto the pan line as needed.
A headgate man advances the mule train and shovels up
any dirt associated with that movement.

Although the standard method of advancing the pan
line and chocks as just described is followed when
roof conditions are poor, the "one web back" method
is usually employed when roof conditions are good.
The one web back cycle begins with the chocks in the
back position, rather than pulled up close to the pan

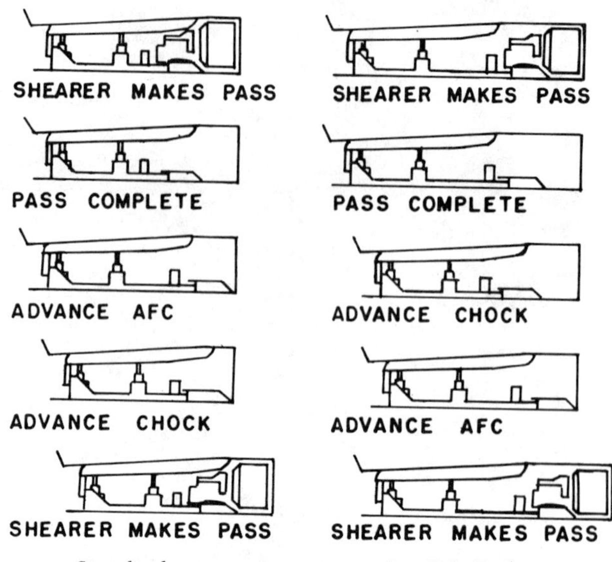

Comparison of two typical longwall methods used at
Mine 33.

line. As the shearer moves down the longwall face,
the chock boy (rather than the snaker) follows 9 to
15 meters (30 to 50 feet) behind, pulling the chocks
up to the pan line. The snaker follows the chock boy,
moving the pan line up to the longwall face. The
advantages of the one web back method are better
ventilation, improved visibility, and the facilitation
of high speed shearing, all of which can be attributed
to the increased working area at the face.

Coal is produced on a three-shift basis at the
Cambria Division. Routine longwall maintenance is
handled during the shift, or between shifts, while
larger jobs are reserved for the weekends. One shift
each day is designated as the shift to straighten out
(or line up) the longwall face. Throughout the course
of the day variances in mining conditions, machine
tolerances, etc., can cause the pan line, and hence
the longwall face itself, to deviate from the straight
condition desired. Since the haulage rope is always
straight across the longwall face, straightening out
the face itself is simply a matter of measuring horizontal distances between the haulage rope and given
points on the pan line, and then moving the pan line
up only the distance required to make these measurements equal. The next pass of the shearer then
straightens out the face.

Efficient operation of the longwall requires a crew
size of ten men plus a foreman. The men are classified as follows:

1	Foreman
1	Shearer operator
1	Snaker
2	Chock movers
1	Tailgate man
2	Headgate men
1	Beltman
2	Mechanics

The longwall foremen report to an assistant mine foreman, who is in charge of all the planning and general

operation of the longwall. The assistant mine foreman reports to the mine foreman who is in charge of all of the underground operations of the mine.

Five continuous-miner units are required to support each longwall at Mine 33. Adverse mining conditions in main, submain, and butt headings often result in slow advance rates for miner units. In order to be sure that development keeps up with the longwalls, miner units are evenly divided between work in main headings and work in the butt entries. This formula has been effective in minimizing delays due to lack of development.

LONGWALL MOVES

As soon as mining of the longwall panel has been completed, all of the longwall equipment is removed from the longwall face and transported to the next face. In the process of making this move, much of the equipment is sent to the outside shop and is rebuilt. The equipment usually rebuilt includes:

1. The shearer
2. The face conveyor head drive
3. The face conveyor tail drive
4. The stage loader
5. The mule train (pumps, motors, starters, etc.)

This equipment is removed from the face as soon as possible after mining has been completed. While the equipment is being rebuilt, the face conveyor and chocks are being dismantled, loaded up, and transported to the new longwall panel. Moving the face conveyor and chocks to the new longwall panel is the most time-consuming phase of the move. The fact that this phase is time consuming, however, provides outside shop crews with the time they need to rebuild equipment. The entire longwall move takes approximately 21 days.

A step-by-step procedure is required to efficiently move a longwall. The procedure that has been developed to dismantle a longwall at Mine 33 is as follows:

1. The shearer mines into the longwall cutout entries.

2. Using the haulage rope as a guide, the pan line is straightened out.

3. The chocks are pulled up to the pan line.

4. The shearer takes a pass, straightening out the coal face.

5. The pan line is advanced 76 cm (30 inches) to the coal face. The chocks are not advanced.

6. The shearer takes another pass. The pan line is not advanced, since it is already fully extended.

7. Without advancing the pan line and shearer forward, another pass with the shearer is made across the longwall face to clean up coal left from the previous pass.

8. The shearer is trammed to a cutout entry and dismantled for transportation to the outside shop.

9. 8 cm x 25 cm x 2.4 meter (3 inch by 10 inch by 8 foot) planks are used to timber the longwall face. One end of each plank is placed on top of a chock, and the other end is secured to the roof by a post placed adjacent the coal face. This provides a 3.4 meter (11 foot) runway for moving chocks once the pan line has been dismantled.

10. The panel belt and belt drive are disassembled and removed.

11. The stage loader and longwall head are disassembled and removed.

12. The mule train is removed, although one hydraulic pump and tank must remain to provide power to the chocks.

13. The tail drive is dismantled and removed through the tailgate entry.

14. The power cable is removed.

15. The telephones and Jabco switches are disconnected.

16. The armored face conveyor chain is taken out.

17. The conveyor pans are removed.

18. The chock rams are pulled back and blocked up.

19. The last chock on the tailgate side is pulled out. The ram of the chock is anchored to a jack by means of a chain, and the chock is pulled out under its own power. After the chock is pulled out into the runway, the hydraulic hoses are disconnected and the chock is chained to a battery powered tractor. The tractor pulls the chock through the runway, down the cutout entry, and into a storage area in preparation for rail transportation to the next longwall face.

20. A minimum of 4 posts are set in the pan line area, and 2 cribs are set in the area from which the first chock was removed.

21. Steps 19 and 20 can be proceeding on the headgate side of the longwall panel or at points in between the cutout entries simultaneously with the removal of chocks on the tailgate side.

22. All chocks are removed as described in Step 19 and the face area is secured with cribs and posts. Cribs are used to replace every other chock removed, and posts are used to secure the rest of the longwall face.

Procedures for setting up a new longwall face vary depending on roof conditions at the face, the rate at which equipment is rebuilt at the outside shop, and other factors. A generalized procedure for setting up the new face is as follows:

A typical longwall move showing timbering at the face and transportation of equipment through a cutout entry.

1. The panel belt and belt drive are set up.

2. The tail drive is installed.

3. Installation of the pan line, armored face conveyor, and chocks begin on the tailgate side and progress toward the headgate.

4. Telephones, Jabco switches, and power cables are installed.

5. The mule train is set up in the headgate entry.

6. The stage loader is installed.

7. The head drive is set up.

8. The shearer is assembled on the pan line.

9. The last few chocks on the headgate side are installed.

LONGWALL SAFETY RECORD

Safety improvement is of paramount importance to Bethlehem management as well as all Bethlehem employees. Not only is longwall mining more efficient and productive than other mining methods, but it is also safer. Over the past five years Mine 33's longwall safety record has been consistently better than the safety record for the rest of the mine.

On a tonnage basis longwall mining obviously outperforms other mining methods, primarily because of the high tonnages produced. In fact, over the past five years (1975 through 1979) Mine 33 longwalls have produced almost seven times (6.71) as many clean tons/lost time accident as have been produced by the rest of the mine. The safety record of the longwalls is also better than other activities at the mine when considered on a man-day basis. Over the same five year period, longwall crews worked almost three times (2.84) as many man-days without a lost time accident than all other employees at the mine. This figure is significant in that it compares safety records on the basis of exposure time, rather than production rates.

Although it is often difficult to pinpoint precisely why one activity is safer than another, the safety advantages of longwall mining over continuous or conventional methods seem to be obvious. The chocks which are used for roof support along the longwall face essentially provide a continuous steel canopy

and means of roof support, which tend to completely
isolate the workmen from newly exposed roof and
caving rock. The longwall technique allows the roof
to cave very soon after it is exposed, whereas continuous mining requires that the roof be bolted and
supported for, in most cases, days, weeks, or months
before it is permitted to cave. Consequently, the
exposure time to a particular bad area is greater for
crews working on continuous miner units than to longwall crews.

Many other functions of a continuous miner unit
which often lead to accidents are not present in the
longwall system. Some of these functions are roof
bolting, rock dusting at the face, shuttle car haulage (which can lead to rutted and uneven runways,
cable problems, dislodged timbers, etc.), cable
handling, installing ventilation canvas, etc. These
factors as well as others are certainly part of what
makes Bethlehem's longwalls a safe means of mining
coal.

CONCLUSIONS AND FUTURE PLANS

Bethlehem Mines Corporation is very pleased with
the success and consistency of the Cambria Division's
longwalls over the past fifteen years. In fact,
success at the Cambria Division was encouragement for
the introduction of longwalls at other Bethlehem
mines in southwest Pennsylvania, West Virginia, and
Kentucky. In light of today's economic conditions
and the subsequent necessity for increased productivity as well as safety, longwall mining will undoubtedly assume a larger role in the long term
planning for Bethlehem's coal mines.

As a part of being considered for a larger role in
long term coal production, various longwall mining
techniques may provide means to recover coal reserves that heretofore have been uneconomical to
exploit. Although conventional retreating longwalls
have greatly improved productivity and safety in
adverse conditions, entry development for these longwalls is still accomplished through the use of continuous miners. Severe conditions result in slow
development rates and high costs. Techniques such as
advancing longwalls or retreating longwalls with advancing tailgates are among the alternatives being
considered to deal with these problems. Based on a
foundation of fifteen years of successful longwall
mining, Bethlehem believes that its operating expertise, coupled with recent technological and engineering advancements, may be the ingredients required to
develop the techniques necessary for the recovery of
difficult reserves.

References

Iannacchione, A.T., and Puglio, D.G., 1979, "Geology
    of the Lower Kittanning Coal Bed and Related
    Mining and Methane Emission Problems in Cambria
    County, PA." Report of Investigations 8354,
    Bureau of Mines, United States Department of
    the Interior.

Jones, D.C., 1971, "Five Years of Longwalling Experi-
    ence Pays Off," Coal Mining and Processing, March,
    pp. 56 - 59.

Maxwell, B., et al, 1977, "Improving Coal Mine Roof
    Stability by Pillar Softening," Study conducted
    by the Excavation Engineering and Earth Mechanics
    Institute of the Colorado School of Mines for the
    U. S. Bureau of Mines, February, pp. 40.

Wardell, K., and Partners, 1969, unpub. "Report on
    Strata Control and Mine Design at Mines 32 and 33,"
    June, pp. 12.

Watson, B., 1971, "Longwall Produces Record Tonnage,"
    Coal Age, March, pp. 64 - 75.

Chapter 22

THE EVOLUTION OF A LONGWALL SYSTEM TO SUIT THE DEEP MINES IN ALABAMA

JOHN F. BRASS

JIM WALTER RESOURCES, INC.
MINE MANAGER
BLUE CREEK NO. 3 MINE
ADGER, ALABAMA

INTRODUCTION

The Jim Walter Corporation is a Florida b a s e d organization with diverse interests. They have operations in many states and overseas, but the h e a r t of the company is in the southeast. The major part of the corporation's business is home building and everything that is related to it. In 1969, the Jim Walter Corporation gained an interest in coal mining through the acquisition of the U. S. Pipe & Foundry Comp a n y which had two slope mines to supply its requirement of metallurgical coal. In the early 1970's it w a s decided to enlarge the coal mining activities and work was started on the Blue Creek project. This involved sinking four new deep mines (originally there were to have been five but it was later decided that the property could be satisfactorily worked by four mines). They are all located in west central Alabama between Tuscaloosa and Birmingham. They have a combined projected output of 7-3/4 million tons of very high grade metallurgical coal per year to be reached by the mid 1980's.

These new mines are all to work the Blue Creek seam which is a very gassy seam and in the area to be mined is found at between 45m (1500 ft.) and 731m (2400 ft.) below the surface.

Longwall is the only practicable method of min i n g to produce the planned outputs in the c o n d i t ions prevailing and each mine is planned to have t h r e e longwalls as the primary production system. Continuous miners will be used just to do the developm e n t work for the longwalls.

No. 3 Mine is the most advanced of t h e mines a n d started the first longwall early in 1979. The second unit started in the middle of 1980 and the third i s scheduled for mid 1981. The first longwall at No. 4 Mine is expected to commence production early in 1982. This paper attempts to relate some of t h e problems that have been encountered and the lessons that have been learned.

SEAM CHARACTERISTICS

The characteristics of the seam vary considerably over the take. The Blue Creek seam itself v a r i e s in thickness, generally between 1.22m (4 ft.) and 2.44 m (8 ft.), and it is exceedingly friable with multiple cleats. The Mary Lee seam, which is harder and has a high sulfur content lies above the Blue Creek a n d is separated from it by a middleman that varies between about 60cm (2 ft.) and 3m (10 ft.) in thickness. The middleman is generally strong, comprising about 60 cm (2 ft.) of hard shaley fireclay and up to 2.44m (8ft.) of strong siltstone. The Mary Lee is generally about 60cm (2 ft.) thick and has 10.7m (35 ft.) of strong sandy shale as a roof. The floor below the Blue Creek varies from a very hard sandy shale to an exceedingly soft shaley fireclay.

Borehole information showed some of the variations in the Blue Creek seam and in the immediate surrounding strata but indicated that it should be possible to work the Blue Creek seam only at No. 3 Mine. T h e first impression was that No. 4 Mine would h a v e to work both seams in most parts of the property a s the middleman appeared to be thin and the Mary Lee roof is exceedingly strong in that area. At Mines No. 5 a n d No. 7 there appeared to be some areas where i t would be necessary to work both seams and others where t h e Blue Creek only could be worked. Although boreholes indicated variations in the seam, the variations found during mining have been far greater and more sudden than anticipated. At No. 3 Mine where the Blue Creek was expected to be between 1.4m (4 ft. 6 in.) a n d 1.67m (5 ft. 6 in.) thick, the thickness has varied be -tween 30cm (12 in.) and 2.75m (9 ft.).

Where the seam has been thin, in general, the floor has been very hard with a compressive strength of u p to about 103,400 KPa (15,000 p.s.i.) Conversely, i n most of the areas where the seam is very thick, t h e floor is exceedingly soft.

At No. 4 Mine, the Blue Creek seam appears to be of much more consistent thickness that at No. 3 Mine and the floor is a good hard one. However, the interval between the Blue Creek and the Mary Lee has been found to vary more and to be generally greater than anticipated.

Effect of Seam Variation

The variations in seam conditions have been at the root of the majority of the problems experienced o n the longwalls at No. 3 Mine and pose some major problems in the consideration of equipment for the other mines.

System of Workings: The basic machine type selected for the longwalls is the double-ended ranging shearer. This is the only type machine capable of satisfactorily coping with variations in seam height and of c u t-ting clear past the tail-drive. This latter requirement is considered an important one, a s i t i s necessary to crib the tail entries. The minimum height that can be extracted using a conveyor-mounted double-ended ranging shearer is about 1.52m (5 ft.) (an off-pan shearer can cut much less, but cannot cope with seam variations) and therefore, we have to cut extra height from either roof or floor when the seam is less than 1.52m (5 ft.). Due to the proximity of the Mary Lee seam above the Blue Creek it is not practicable to cut the extra height out of the top and therefore floor must be taken. Another reason for cutting the floor rather than the roof is that the roof horizon is much more consistent than the floor horizon, so undulations are minimized by cutting floor.

There have been occasions when the coal seam h a s been of such great thickness that the supports would not reach the roof if the entire seam were to be extracted. Attempts to leave coal up to form a r o o f have not been successful so that it is necessary i n these circumstances to leave a coal floor. Due to the very soft nature of the coal, the supports generally have to be advanced on boards in this situation.

Whereas, the entire production from No. 3 Mine i s from the Blue Creek seam only, the majority of development work at No. 4 Mine has involved extracting both the Blue Creek and the Mary Lee seams. T h e b i g

179

question that now must be answered is whether the long
-walls should mine both seams or whether it is p o s-
sible to extract the Blue Creek alone. It is certainly desirable to extract the Blue Creek alone and there
are many areas where it should be practicable proposi-
tion as the middleman is hard and above 1.22m (4 ft.)
thick. There are some places, however, where t h e
middleman is only about 45cm (18 in.) thick and friable and these areas appear likely to present problems.
One possibility being considered is to try and cater
for both situations. The geometry of the shearer i s
one limiting factor, but it should be possible to cut
up to 3m (10 ft.) heights taking both seams, and down
to about 1.6m (63 in.) when cutting the bottom s e a m
only. Where the two seams combined exceed 3m (10ft.),
the middleman is sufficiently thick and strong to permit the Blue Creek only to be worked. The variations
in section are such that it might be necessary to take
one seam in one part of the face and both in another.
The transition from one situation to the other would
probably present some difficulty.

Shield Supports: The two installations of s h i e l d
supports at present in use at No. 3 Mine have a range
of about 1.07m (42 in.) to 2.3m (90 in.). The lower
figure is for ease of transportation as the normal
minimum working height is about 1.47m (58 in.). A s
already noted, there have been occasions when t h e
thickness of the seam has been greater t h a n 2.3m
(90 in.) but the disadvantages in increasing the minimum height might be greater than the benefit t o b e
gained, particularly when it is considered that t h e
very thick coal is in small areas.

The first installation is equipped w i t h 2-leg,
rigid base shield supports, rated at 350 tons, whereas
the second has 4-leg semi-rigid base supports rated at
610 tons. The floor has been sufficiently soft on occasions for problems to be experienced with both types
of support. The problems have been more severe with
the 4-leg supports, which started work in v e r y difficult conditions with thick coal, soft floor and bad
roof. The biggest problems arose due to them not having rigid bases, the net result being that the l e g s
were caused to lean. The bases of these supports have
now been rigidized with a bolt-on bridge and that prob
-lem has been overcome. It is still necessary to lift
both types of supports onto timbers in very soft floor
conditions.

Our first set of shields had single telescopic legs
with mechanical extensions. The mechanical extensions
were found to be a considerable disadvantage in conditions where the working section varied rapidly. Considerable vigilance was required from the operators to
ensure that there were always just sufficient extensions out so that hydraulic travel was available.

The supports for No. 4 Mine require special consideration. If they are to cater for either the B l u e
Creek or the combination of the Blue Creek a n d the
Mary Lee they must have a large range. This could involve different leg angles or possibly other m e a n s.
They will also require a face sprag so that when working the two seams together the middleman can be held
in place as close to the shearer as possible. If this
sprag is readily removable it will improve the clearance over the machine when the Blue Creek a l o n e is
being mined.

Shearer: Three different shearers from three d i f-
ferent manufacturers have so far been used at No. 3
Mine. Two of these are multi-motor machines while the
third is a single motor unit, with a mechanical haulage. Each of these machines uses a chainless haulage
system.

Several lessons have been learned from our experience so far that would influence future machine selection. Although, one of these machines has yet to

cut through an area of thin coal and hard floor, t h e
other two have and both have suffered physical damage
to some degree. The structure of the arms, the ranging jacks, and the jack attachment point o n b o t h
machine and arms are places that are subjected to particularly high stresses. Shearer drum lacing and the
maintenace of shearer drums are important features
that can substantially help in reducing the wear a n d
tear on the machine.

As stated earlier, the shearers that we have used
have had different haulage arrangements. The mechanical haulage is a fairly simple unit and has proven to
be fairly reliable provided it isn't abused. I t i s
not really suitable however in varying conditions as
it is not self-compensating and therefore cannot b e
kept operating at maximum pull. An infinitely variable self-compensating haulage is required in our circumstances for optimum performance. This c o u l d be
either electric or hydraulic although the electric
units that we have used have not been without problems.

Horizon control has proven to be a problem at No. 3
Mine, although the situation is improving as the operators become more experienced. In the varying conditions that we encounter it is not easy to maintain a
horizon that is reasonably consistent and yet m i n i-
mizes the amount of rock cut, is high enough for the
shearer to get through, maximizes the amount of coal
taken and is always within reach of the supports.

Face Conveyor: The face conveyor on the first panel
was a single center strand unit but this has now been
converted to a twin center strand conveyor and our
other conveyors are also of this type. The change was
made partly because of the need to transmit more horse
-power and partly for other reasons. The main one concerned with the variable nature of the seam was t h a t
there was a problem of flight stability that was most
evident when lumps of floor rock went under the m a -
chine. This caused the chain to jerk and the flights
were liable to come out of the top race and would then
be damaged under the shearer. Chain tension was also
critical with the single strand conveyor.

We have also experienced conveyor problems in areas
of very soft floor. In some areas it is sufficiently
soft for the ramp-plates to push the floor up and i n
these cases the conveyor has to be lifted in order to
maintain the desired horizon. Problems were experienced in starting the conveyors in soft floor areas
and consideration is being given to using bottom covered pans. There are some obvious disadvantages t o
bottom covered pans, particularly with a center chain
conveyor and we have been able to substantially improve
the situation on our present conveyors by attention to
our power distribution system so as to give a better
voltage on start-up.

PILLAR SIZES

The development sections for t h e first longwall
panels at No. 3 Mine are three-entry developments with
the entries on 100 foot centers and the cross-cuts at
120 foot centers. This arrangement was satisfactory
for the extraction of the first panel but some problems
have been experienced in the tail-gates of the second
and third panels despite the fact that these entries
were cribbed with double rows of cribs.

Calculations indicate that at 457m (1500 ft.) depth
one single pillar 55m (180 ft.) square, or a double row
of pillars 40m (130 ft.) square is required to accept
the pressures from two longwalls without yielding. A
two-entry development would be ideal as it should b e
capable of relatively rapid advance, it would sterilize
the minimum amount of coal in pillars and it would maximize the ratio of longwall output to continuous miner
output. Unfortunately, it is not permitted within existing legislation. The easiest arrangement to work

from the point of view of the legislation is a three-entry development on 46m (150 ft.) centers. It isn't possible to ventilate a three-entry development, i n the gassy Blue Creek conditions with that s i z e o f pillar and therefore if 40m (130 ft.) square pillars are required the development must be driven as a four-entry section with 40m (130 ft.) pillars on the outsides and an 18m (60 ft.) pillar in the center. This arrangement results in a section that can neither advance rapidly or at high productivity. In addition, it sterilizes a large amount of reserves.

The relative continuous miner and longwall extraction ratios, also recovery of reserves percentages for these two development systems are as follows:

1. Two-Entry System

Longwall Extraction	Cont. Miner Extraction	Ratio L.W. C.M. Output	Recovery Percentage
152.4m^2/m advance or 500 sq. ft./ ft. advance	17.7m^2/m advance or 58 sq. ft./ ft. advance	8.62	77.5

2. Four-Entry System

152.4m^2/m advance or 500 sq. ft./ ft. advance	37.2m^2/m advance or 122 sq. ft./ ft. advance	4.10	69.2

In both the above cases i t has been assumed that the longwall panel is 152m (500 ft.) wide and t h a t the development entries are 6.1m (20 ft.) wide. F o r the sake of simplicity this comparison takes account of the longwall panel development entries only. T h e difference would be much greater if the effect of main road development was included.

It is clear that the development output needed is doubled with the four-entry system and the recovery percentage falls by almost 11 percent. In v i e w of these factors, we are continuing our efforts to devise and gain acceptance of a practicable two-entry develop-ment system. In the meantime, although, we have one four-entry development section with entries on 45.7m (150 ft.), 24.4m (80 ft.) and 45.7m (150 ft.) centers, we are also trying to develop with three-entry sections with both entries and cross-cuts on 36.6m (120 ft.) centers and with improved support systems for both the development and longwall production phases.

HUMAN FACTORS

By no means all the lessons learned in the last two years have related to equipment or systems of working. A mine with a longwall places different demands o n mine management than does a mine that h a s only continuous miners.

Longwall is a completely different system and because there is so much at stake, it must be s o l d to everybody who is even remotely connected with it. They need to be informed why it is to be installed, t h e principles of how it works and what it is expected t o do. We spent a good deal of time, trouble a n d money on this, but feel that our efforts were all worthwhile. There has been great enthusiasm for the system f r o m the outset.

Good management is required to get the best from a longwall, particularly if mining conditions are d i f-ficult. A high standard of discipline is needed as it is important that jobs are done in the prescribed manner. Failure to do some things in the correct manner just one time can cause major problems that can take a long time to work out of the system. A foreman w h o

will do as he is told and inquire if he is in doubt is of much more value than one w h o u s e s h i s initiative and thinks that he knows it all. T e a m work is important and the members of the team should not only have a knowledge of one another's jobs b u t be able and prepared to do them. Conveyors m u s t run continuously. Good maintenance is required a s although longwall equipment is reasonably robust, a failure of just one of a number of compenents c a n stop production completely. A detailed approach i s needed in longwall management, particularly in problem solving. A major job such as a move m u s t b e planned in minute detail if it is to be completed efficiently with the minimum of downtime.

A longwall has the potential to produce large tonnages at high productivity, but the manner in which it is managed merits consideration. With so much at stake, a longwall must receive the best management and maintenance attention so as to ensure that t h e best results are obtained. However, the rest of the mine cannot be neglected. Continuous miner sections developing longwall panels are just as important a s the longwalls themselves and must require appropriate treatment. Further to that, the people who work i n other parts of the mine deserve attention. T h e i r jobs are just as important to the success o f t h e enterprize as the jobs of the longwall crews. T h e y still deserve attention and must not get the impression that the longwall is the only thing that anybody is interested in.

SUMMARY

The conditions encountered on the Jim W a l t e r Resources' longwalls are unlike those met on other longwalls in the United States.

Many problems have had to be faced, but valuable experinece has been gained such that these u n i t s should now be capable of producing the r e s u l t s originally projected for them.

Chapter 23

OPERATING EXPERIENCE IN THICK COAL
LONGWALL MINING, YORK CANYON MINE, RATON, NEW MEXICO

Rodney Lawrence

Mining Engineer, Kaiser Steel Corporation

Tim Hackett

Mining Engineer, US Bureau of Mines

INTRODUCTION

The western USA contains significant coal reserves in seams 10 feet or more in thickness which lie too deep for surface mining. As part of a demonstration of the use of two legged shield supports in the United States, Kaiser Steel Corporation, under contract to the US government used the longwall method to mine a ten foot thick coal seam in New Mexico. Caliper type shield supports and double drum shearers were used in three longwall panels. During the test, technical, production, and safety data were collected to objectively evaluate the performance of the shields in mining the thick coal seam. This paper presents discussions of practical operating experience and summary statistics to aid the mining industry in planning longwall mines in similar conditions.

MINE SETTING AND GEOLOGY

The York Canyon mine is located 16.1 Km (10 miles) south of the New Mexico-Colorado border and 64.4 Km (40 miles) west of Raton, New Mexico (Fig. 1).

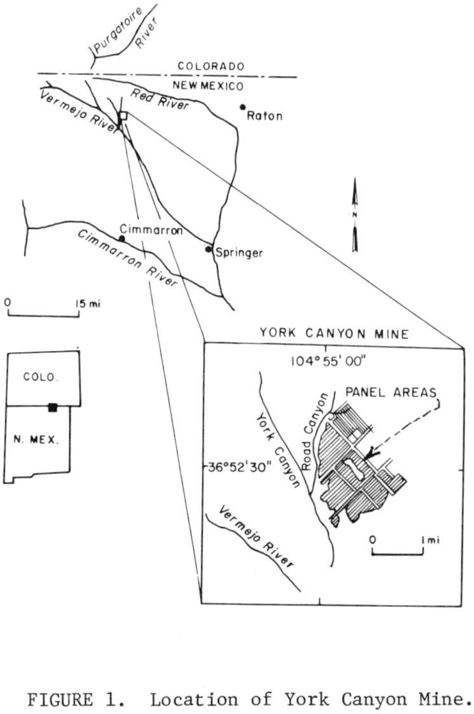

annually from the underground mine. The mine land surface is used for livestock grazing, wildlife habitat, and logging.

Coal is extracted from the York Canyon coal bed in the Raton formation of late Cretaceous to Paleocene age (Pillmore, 1969). Coal thickness varies from 1.22m (4 ft.) to 3.66m (12 ft.) and was 2.3m (7 ft.) to 3.35m (11 ft.) in the three shield demonstration panels. Small scale faults with vertical displacements of up to 3.66m (12 ft.) are common in the mine. Longwall panels are oriented at approximately 45° to coal cleat and roof jointing. Intersecting cleats in the coal face contribute to the formation of large coal lumps which are discussed later in the report. Mine overburden consists of 40-50% shale, 20% sandstone, 10% siltstone, and coal, bony coal and carbonaceous shales (Gentry, 1976). The coal seam was essentially flat lying in the shield demostration panels.

A carbonaceous shale split and rider seam overlie the main coal seam (Fig. 2). The split varies from

FIGURE 1. Location of York Canyon Mine.

The regional topography is irregular and consists of a highly dissected plateau. The mine is developed on the outcrop of the coal seam lying just above the valley floor. Approximately 800,000 mt (900,000 st) of .5% sulfur, 14,000 Btu/lb. coal are shipped

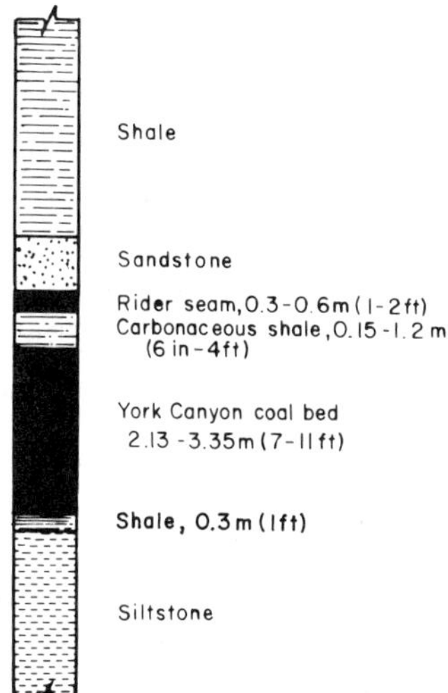

FIGURE 2. Immediate Roof and Floor Strata, Longwall Panels.

15cm (6 in.) to 9.14m (30 ft.) in thickness and both seams are mined where it is less than 45.cm (18 in.) thick. During mining of the third test longwall the shale split fell before the shields could support it. Problems were experienced in holding this shale layer during development of the third longwall and parts of the longwall test panels were subjectively considered as having bad roof (Fig. 3).

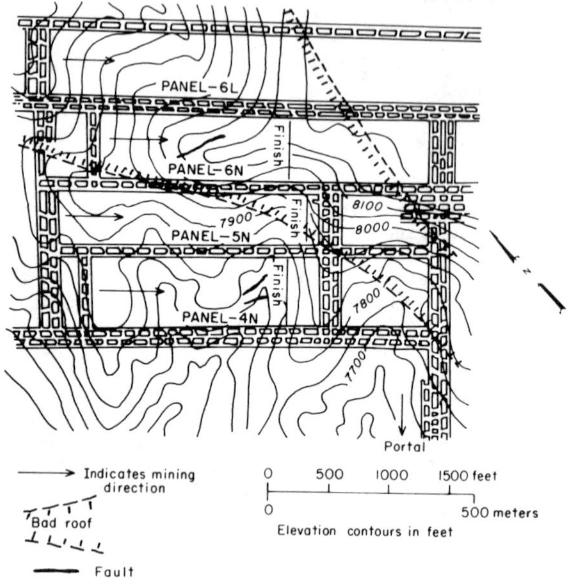

FIGURE 3. Layout of Shield Test Panels.

The floor is composed of a thin shale unit averaging about 0.3m (1 ft.) thick. When wet, it becomes unstable and creates problems, and, therefore, is usually removed with the coal seam. Under the soft shale is a strong siltstone unit generally thicker than 3m (10 ft.) which makes an excellent floor.

Strata strength was determined from plate load bearing and core tests (Gentry, 1976). The maximum bearing capacity of the plate load test apparatus was 3.65×10^7 Pa (5300 psi) and the strength of the thin shale in the floor exceeded the test capacity in most cases. The uniaxial compressive strength of the floor siltstone beneath the shale was approximately 1.24×10^8 Pa (18,000 psi). No strength data were obtained for the carbonaceous shale split above the main seam because it had fallen during mining in entries where plate load tests were conducted or disentegrated during core drilling. A shaley siltstone grading to sandstone overlies the rider seam. Average uniaxial compressive strength of this layer was 6.78×10^7 Pa (9,840 psi) but its properties were highly variable.

Cover depth in the area of the test longwall panels varied from approximately 122m (400 ft.) to 244m (800 ft.), with cover increasing as successive panels were mined (Fig. 3).

LAYOUT AND DESCRIPTION OF TEST PANELS

The longwall test panels were laid out adjacent to each other, separated by double entry systems (Fig. 3). The three test panels are designated 4N (4 North), 5N, and 6N and their dimensions and other characteristics are given in table 1. In the double entry system the conveyor belt carrying coal from the face lies directly next to the panel and ventilation

air travels along the adjacent entry until the last open crosscut. Approximately 850 cubic meters/sec (30,000 cfm) of ventilation air were delivered to the longwall headgate. Air is exhausted from the longwall face through the tail entry, and along the edge of the waste (gob) into the bleeder system.

All three test panels were mined using double drum shearers and caliper shield type supports. Characteristics of the face equipment are given in tables 2 and 3. Components of the shield supports are shown in Figure 4. Three lemniscate shields were used in the headgate and in the tailgate of the third test panel (6N) because entry height was too low for the caliper shields. A lighting system using flourescent luminaires was also used on the 3 test panels.

The operating height of the caliper shields varies from 1.7–3.5m (67–138 in.). Load density varies from 35.2–$51.8mt/m^2$ (3.6–5.3st/ft^2) (mt = metric tons, st = short tons) over this range. A hinge pin (Fig. 4) joining the base and gob shield is moved into the lower position to operate at lower heights. Thinner (2.13m or 7 ft.) coal in the third test panel required operating the shields with the pin in the lower position.

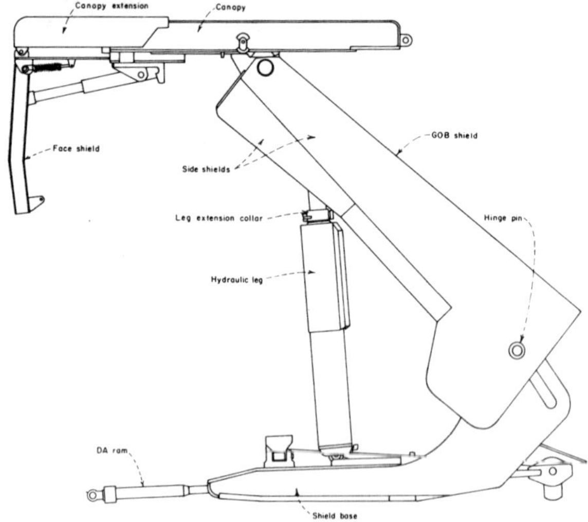

FIGURE 4. Caliper Shield Support

OPERATING EXPERIENCE

Panel 4N

The face height in the first panel mined 4N (see Fig. 3) was 3.05m (10 ft). Pressure on the face was noted just before the first cave after about 25.9m (85 ft.) of advance, and large coal lumps began falling from the face ahead of the shearer (Fig. 5). The lumps were up to .6 x .9 x 1.8m (2 x 3 x 6 ft.) in size and formed where joints in the coal intersected just ahead of the face. To keep large pieces of coal off the conveyor belts, a mine fabricated lump breaker was installed ahead of the stage loader. The face was pinned with resin grouted wooden dowels to try to prevent lump formation, but the time spent drilling holes for the pins was greater than time required to break the lumps.

Lumps adversely affected production primarily by wedging along the face conveyor and at the junction of the face conveyor and stage loader. Considerable downtime was required to break the lumps using a

Lumps created a safety hazard by falling across the conveyor and spill plate into the travelway. During the maintenance shift, face shields on the roof supports were lowered to keep the lumps out of the travelway (Fig. 7). Because of the high face height (3.05m), the size and number of lumps were probably greater than they would have been on a lower face. The lumps tended to fall mostly just ahead of the shearer and could be avoided by the miners. Fall of rib and face, which includes some rib falls not due to lumps, accounted for 10 accidents including one lost time accident.

FIGURE 5. Large Coal Lumps.

double jack or air pick (Fig. 6). One way cutting was required to keep lumps from building up but it did not greatly affect production because the shearing machine can travel without cutting coal to the tailgate almost as fast as it takes to back up and sump into the headgate. Shields were used in the one web back position during most of panel 4N.

FIGURE 7. Face Shield in Lowered Position, Strata Exposed in Face, and Canopies 1.5m From Face After Fall of Caprock.

Conveyor hangups plagued operations during mining of the three panels. Floor heave may have been partially responsible. Buildup of broken shale and mud underneath the conveyor may have also contributed.

Two faults with offsets up to 3.7m (12 ft) were encountered near the end of 4N. In order for the shearer to negotiate faults, the floor rock on the upthrust side had to be drilled and shot before each pass (Fig. 8). The shields upslope of the fault tended to slide downhill when advanced and crowd the shields on the lower side of the fault. Roof bolts and mats were installed in the fault zone as a safety precaution. It was decided that the remaining coal in the panel could be more economically recovered using continuous room and pillar mining and the panel was terminated 30.5m (100 ft) early.

Panel 5N

Roof conditions on the second panel mined, 5N, were good with the exception of minor falls of the shale split or caprock directly above the main coal seam. The caprock would fall after the shearer passed unless the shield canopies were immediately extended. In areas where the seam and roof were fractured or slightly faulted, small pieces of rubble occasionally fell between the shield canopies.

FIGURE 6. Breaking Coal Lumps in Headgate.

TABLE 1 - SHIELD PANEL STATISTICS

	4N	5N	6N
Face Width	168m (550 ft.)	135m (444 ft.)	152m (500 ft.)
Length	479m (1570 ft.)	604m (1980 ft.)	453m (1486 ft.)
Face Height	3.05–3.35m (10–11 ft.)	3.05–3.35m (10–11 ft.)	2.13–3.35m (7–11 ft.)
Raw Tons	331,122mt (365,000st)	313,886mt (346,000st)	257,640mt (284,000st)
Avg. Tps	817mt (901st)	982mt (1082st)	593mt (654st)
Start	5/9/75	7/30/76	8/1/77
Finish	5/14/76	5/12/77	9/29/78

TABLE 2 - SHIELD PANEL EQUIPMENT *

Caliper Shield Supports:	Hemscheidt 320 HSL, 317mt (350st) capacity
Face Conveyor:	Eickhoff EKF-3, 30mm single chain two 111.8kw (150hp), one 93.2kw (125hp) drives
Shearer:	(4N) Anderson Mavor 223.7kw (300hp) water cooled, 152.4cm (60 in.) drums (5N & 6N) Eickhoff EDW 300-L 300kw (400hp) 152.4cm (60 in.) drums
Hydraulic Power Pack:	Two Uracca triplex piston 55.9kw (7.5hp) pumps
Conveyor:	Continental 106.7cm (42 in.) take up capacity 112m (360 ft.)
Stage Loader:	Dowty Meco
Lumpbreaker:	(4N) Mine fabricated (5N & 6N) Klokner Ferromatic SB63

*Reference to specific makes and models of equipment and suppliers is made for identification only and does not imply endorsement by the United States government.

TABLE 3 - CALIPER SHIELD CHARACTERISTICS *

Manufacturer and Type:	Hemscheidt 320 HSL
Capacity:	317mt (350st)
Operating Height Range:	1.7–3.4m (67–132 in.)
Collapsed Height:	1.5m (59 in.)
Width:	1.5m (59 in.) centers
Weight:	9.1mt (10st)

*Reference to specific makes and models of equipment and suppliers is made for identification only and does not imply endorsement by the United States government.

After the first month of production, the face began showing signs of pressure and large coal lumps began forming as on 4N. Numerous conveyor hangups also occurred.

Panel 6N

Panel 6N had extreme ground control problems throughout. These problems resulted from poor roof conditions, operating constraints of the shields, and the close proximity of another active longwall (panel 6L). As on 5N the shale caprock fell after the shearer passed. The one web back system was tried but the caprock fell before shield canopies could be extended. Voids .3–.6m (1–2 ft.) high formed in the roof and broken roof material fell into the travelway. In panel 6N the hydraulic connections were arranged so that each shield was operated from the shield adjacent. The European system of support advance was adopted. In this system the face conveyor is pushed over as soon as possible after the shearer passes and the face supports are then pulled up as tight as possible against the face.

Because coal height had decreased to as low as 2.13m (7 ft.) in panel 6N, the hinge pin in the shields was moved into the lower position. However, fall of the caprock increased face height to at least 3m (10 ft.). Since the gob shield and canopy rotate around the hinge pin, the canopy-face distance increases with face height for caliper shields. Moving the hinge pin to the lower position increased the face-canopy distance and made this problem worse. The shield canopies were left up to 1.5m (5 ft.) from the face leaving the roof between unsupported. Figure 7 shows the caprock and the distance of the canopies from the face. The shields were difficult to advance and some were extended so high they tilted backwards. These conditions probably helped cause several major roof falls up to 2.4m (8 ft.) high on the face and in the gates.

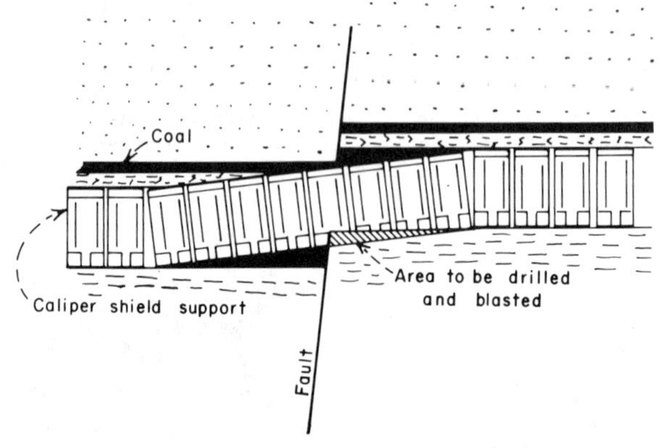

FIGURE 8. Vertical Section Through Longwall Face When Mining Through A Fault.

Before production could resume the roof in the fall areas had to be bolted and the fallen rock removed from the face conveyor. Large pieces of rock on the conveyor had to be drilled and blasted. Smaller pieces were shoveled into the travelway. Several production shifts were required to clean up a major fall.

Soon after starting 6N, the tail entry began to deteriorate as a result of front and side abutment pressures. The tail entry rib spalled and the floor heaved up to 16m (80 ft.) ahead of the face. A single row of cribs was installed down the center-line but conditions continued to worsen. Mechanically anchored bolts 2.1m (7 ft.) long, steel I beams on 1.2m (4 ft.) centers, and wood posts were installed. At approximately 147m (482 ft.) of advance the tailgate roof caved, wedging two shields between the roof and the floor. At the same time, voids .3-.9m (1-3 ft.) high formed in the tail entry roof ahead of the face. Resin grouting of the roof was started and additional beams, cribs, and posts were installed. Although conditions in the tail gate continued to be poor, resin grouting on a continuous basis for the remainder of the panel allowed operations to proceed in that area.

From December 6, 1977 until March 27, 1978 the 6N face was idled by the UMWA contract strike. Much of the face and the tail entry up to 61m (200 ft.) ahead of the face was resin grouted prior to the strike in order to control ground conditions. No significant deterioration of the face or tailgate was noted during this period, but the bleeder entry between 6N and 6L began a rapid and extreme deterioration. Figure 9 shows roof conditions in the bleeder entry. At this time, panel 6L was approximately 152m (500 ft.) behind panel 6N. Overlapping abutment pressures from the two panels may have caused the deteriorating conditions.

FIGURE 9. Roof Conditions in Bleeder Entry Between 6N and 6L.

Mining resumed on March 27, 1978, after the strike was over and steps were initiated to control roof problems on the face. To keep the face height within the operating limits of the shields, a layer of coal .6-.9m (2-3 ft.) thick was left on the floor along the face, and the caprock and rider seam were taken by the shearer in this area (Fig. 10). The shale caprock was left in the gates and ends of the face (Fig. 10) resulting in a ramp upwards from either end of the face towards the middle.

In the main part of the face the roof was competent sandstone and an immediate improvement in roof conditions and production was noted. Two areas of poor roof persisted near each end of the face. A brow or transition zone up to 1.2m (4 ft.) high formed where the roof changed from sandstone to cap rock (Fig. 11). In these areas shields were extended to their maximum height and the canopies were up to

1.5m (5 ft.) from the face. Large piles of rock tended to accumulate in these areas making it difficult to advance the shields. On the tailgate side it was necessary to install bolts and steel mat after every shear.

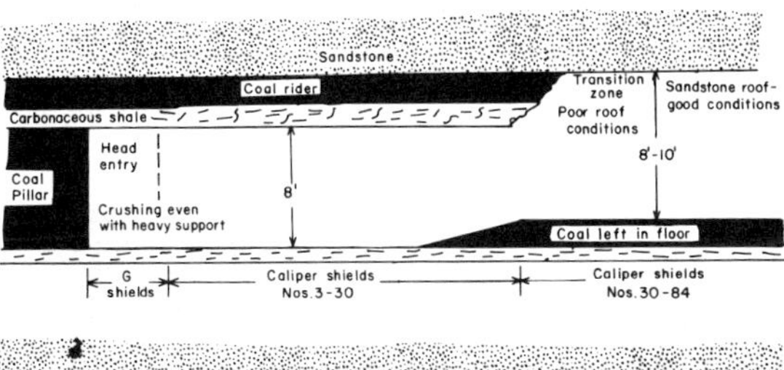

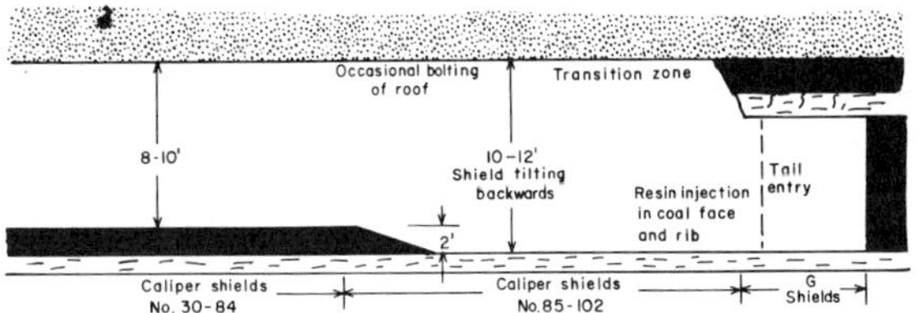

FIGURE 10. Section Through 6N Face After Special Steps Were Taken To Control Poor Roof Conditions.

FIGURE 11. Poor Roof In 6N Transition Zone.

At 320m (1050 ft.) of face advance panel 6L was within 37m (120 ft.) of 6N. The bleeder entries between 6N and 6L became impassable and began taking weight ahead of 6N. Two roof falls up to 1.2m (4 ft.) high occurred in the headgate. Additional cribs, posts, and steel beams were installed and resin injection started on a regular basis. Advancing the shields in 6N became slow and difficult. Realizing that pressure from panel 6L was affecting 6N, the 6L face was kept idle for four weeks to allow 6N to move ahead. However, headgate conditions continued to be poor.

On September 3, 1978 at 444m (1458 ft.) of advance the largest and most difficult fall of the project occurred. The roof fell up to 6m (20 ft.) high between shields 17 and 27 and continued to cave after the shearer passed. In order to advance the face a lip was established, using artificial

support. First the roof ahead of the face was resin
bolted and resin grouted. Next 5cm (2 in.) diameter
holes were drilled horizontally into the face and pipe
set in the holes. The shields were lowered and
advanced under the pipe sticking out of the hole.
Figure 12 shows the face with the pipes installed.
Lagging posts and crib blocks were piled on top of
the pipes to keep broken rock from falling into the
face. With all of this additional support the
shearer was able to cut in the fall area without
roof failure.

FIGURE 12. Establishing A Lip On Panel 6N
Using Pipe.

On September 29, after 453m (1486 ft.) of
advance the headgate squeezed until no more than 3 ft.
of working height was left. At this time the 6N face
was terminated.

PANEL MOVES

After completion of a longwall panel it is
desirable to move the supports as quickly as possible.
Different methods of moving the shields were used
because of different conditions in each panel.

Prior to the move of the 4N panel, the roof was
secured by placing wire mesh and I beams over the
shields during the last 9m (30 ft.) of advance. The
face conveyor, shearer, and stage loader were then
removed. Next railroad rails were laid on the floor
to slide the shields on, a winch was installed at the
headgate, and shields were removed one at a time,
starting at the tailgate. As the shields were
removed, cribs and posts were set in their place to
hold the roof. The shields were then loaded onto
trailers (Fig. 13) and pulled behind battery scoop
cars to the next panel. The move from panel 4N to 5N
is described in US Bureau of Mines IC 8747 (Oitto
et al, 1977).

In the move from 5N to 6N, the rails were not used
on the floor. The last two shields at the tailgate
were turned 90° and used as trailing shields,
supporting the end of the face as shields were
recovered (Figs. 14A and 14B).

The last panel move was much different because
the quick decision to terminate 6N did not allow time
to install wire mesh and I beams over the shields for
the last 9m (30 ft.) of advance. Also the squeeze
in the headgate and heavy support in the tail entry
cut off access to the face. To recover the shields,
two entries were driven through the remaining

335m (1100 ft.) of panel to intercept the face. An
access entry was driven 24m (80 ft.) outby and
parallel to the face, Figure 15. Another entry was
driven from this entry into the face at 3m (10 ft.)
from the headgate. These two entries were supported
with yieldable arches, and the first 30 shields
recovered through them. A continuous miner
was brought in to cut 2.4m (8 ft.) of width along the
remainder of the face and the roof was supported with
bolts and mats. The rest of the shields were
recovered from the two entry system using the last
two shields as in the move from 5N to 6N.

FIGURE 13. Shield On Trailer for Transport.

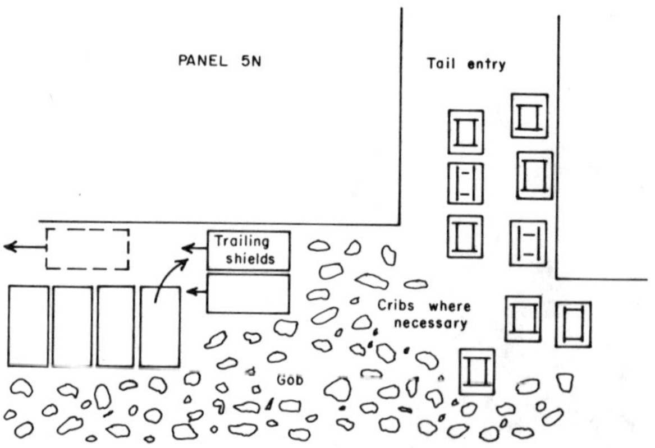

FIGURE 14A. Recovery of Shields From Panel 5N
Using Trailing Shields.

FIGURE 14B. Trailing Shields in Move of Panel 5N.

Roof conditions in the headgate were severe (Fig. 16) but resin grouting in this area cemented the broken roof together adequately to allow the shields to be retrieved without extraordinary problems.

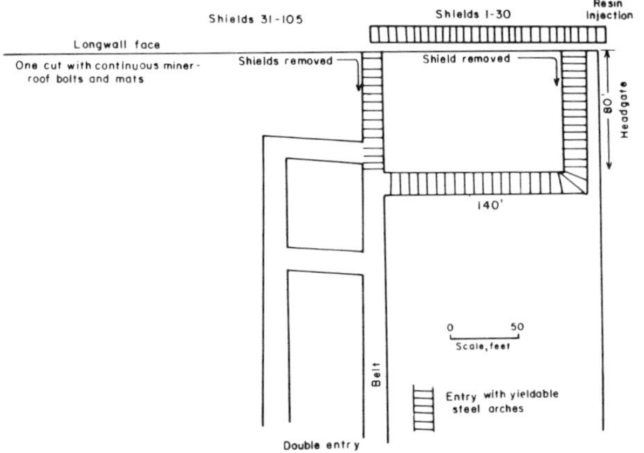

FIGURE 15. Recovery Layout For Panel 6N.

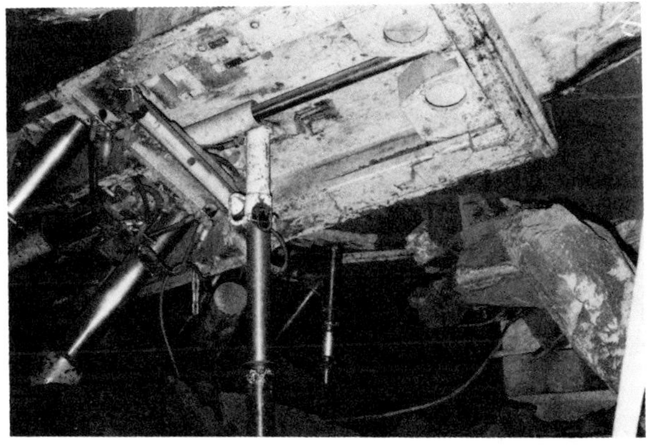

FIGURE 16. Overturned Shield In Headgate Of Panel 6N.

SAFETY

The safety record of panels 4N, 5N, and 6N compares favorably with published safety data for combined USA longwalls (Meritt and Davis, 1977). A total of 114 accidents with 6 lost time accidents occurred. Figure 17 shows the accident record for the demonstration panels compared to national statistics for longwalls. The classification of accident causes was done by the authors and might vary from a classification done by others. The major causes of accidents were roof falls and machinery. Most of the roof-fall accidents occurred during the extremely poor roof conditions in panel 6N. A serious cause of machinery classified accidents was getting caught between a shield base and the face conveyor. Changing the hydraulic hoses to allow operation of each shield from the adjacent support helped reduce this hazard. Utility men and unskilled laborers suffered the greatest number of accidents.

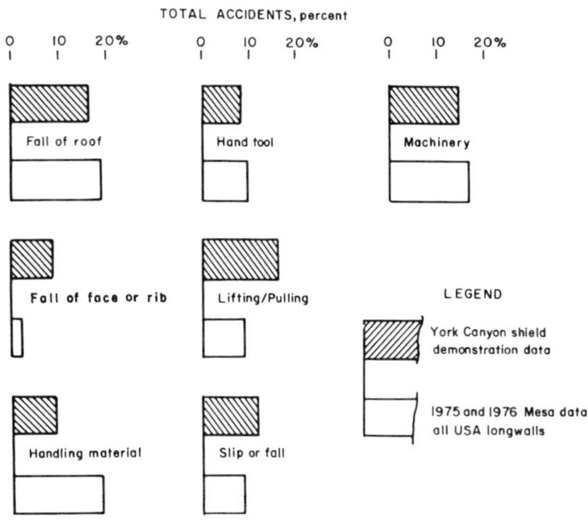

FIGURE 17. Percentage of Total Number of Accidents (114) by Accident Class For Panels 4N, 5N, and 6N.

An experimental lighting system was used on the three panels. The exact contribution of the lights to safety is difficult to assess but they did help to keep the face aligned and generally improved working conditions.

Like most longwalls in the United States, the three shield panels were out of compliance with the 2mg/ℓ respirable dust standard a significant part of the time. Compliance was better on the third panel because production was less, and worst on the second panel because production was best. An innovation now used at York Canyon is the MSA airstream* helmet. The helmets are available to any York Canyon coal miner who wants to wear one.

A complete safety and production record with analyses is contained in Kaiser Steel's final contract report (Lawrence and King, 1980).

*Reference to specific makes and models of equipment and suppliers is made for identification only and does not imply endorsement by the US government.

PRODUCTION

Production for the three shield panels averaged 778mt (858st) per shift. Figure 18 shows the production record for each panel. The effect of moves, the UMWA strike, and poor roof in panel 6N can be seen. The major constraints on production due to the high coal face appeared to be coal lumps and poor roof conditions on panel 6N. Coal lumps form on other faces at York Canyon, but a high face significantly increases their size. Conveyor hangups and other sources of downtime such as shearer maintenance and external conveyor belts were also significant production constraints but occur also on normal height longwalls. An analyses of downtime for panel 4N was made by Jet Propulsion Laboratory under contract to the US Government (Curry et al, 1977).

Shield faces have out produced chock faces at York Canyon. For the period of the first two shield panels chock faces averaged 620mt (684st) per shift compared to 900mt (992st) for the shields. Panel 6L, which used 2 leg lemniscate shields, averaged 1106mt(1219st) per shift and operated in the same bad roof zone as panel 6N without serious roof problems.

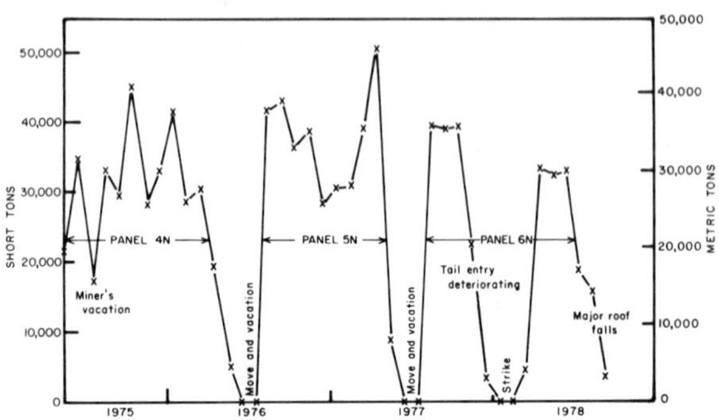

FIGURE 18. Monthly Production For Panels 4N, 5N, and 6N.

CONCLUSIONS

Overall production and safety of the three test panels was satisfactory regardless of the poor conditions in the last panel. The caliper shields performed well when they were correctly adjusted for face height. Use of the shields in the low height range position contributed to roof problems on panel 6N when fall of the caprock increased face height above the intended operating height. The weak shale caprock and the close proximity of an adjacent longwall also contributed to poor conditions. Problems attributable to the high coal faces at York Canyon were coal lumps and the previously mentioned roof problems in 6N. Resin grouting permitted advance of the face and recovery of shields under conditions that would have otherwise been prohibitive. The caliper shield faces out produced chock faces at York Canyon.

ACKNOWLEDGEMENTS

The Caliper shields were purchased by the US Government. Kaiser Steel Corporation used the supports and furnished relevant data under contract to the US Bureau of Mines and, later, Department of Energy. The contribution of all professionals, technicians, and miners who participated is greatly appreciated.

REFERENCES

Curry, K.C., Heimburger, D.A. and Metcalfe, M.A., 1977, "Longwall Mine Availability and Delay Analysis, Phase II; Part I Technical Analysis".

Gentry, D.W., July 1976, "Rock Mechanics Instrumentation Program For Kaiser Steel Corporation's Demonstration of Shield-Type Longwall Supports at York Canyon Mine, Raton, New Mexico", US Department of Energy Report FE 9067-1.

Lawrence, R. and King, R., 1980, "Performance of Shield Longwall Demonstration at Kaiser Steel Corporation's York Canyon Mine", US Department of Energy Report No. DOE/ET/12530-T1.

Meritt, P.C. and David, H., "Longwall Mining: Avenue to Safety, Productivity, and Resource Recovery", Coal Age, p 59-107, January 1977.

Oitto, R., Wisecarver, D.W. and Sikes, W., 1977, "Moving Longwall Shield Supports at the York Canyon Coal Mine, Raton, New Mexico", USBM IC 8747.

Pillmore, C.L., July 1969, "Geology and Coal Deposits of the Raton Coal Field, Colfax County, New Mexico", The Mountain Geologist, v. 6, no. 3, pp 125-142.

Chapter 24

LONGWALL MINING IN ILLINOIS

P. J. Conroy
Dames & Moore, Park Ridge, Illinois
and
E. A. Curth
U.S. Department of Energy, Pittsburgh, Pennsylvania

INTRODUCTION

In June 1962, the first attempt at longwall mining in Illinois was initiated by Old Ben Coal Company in their No. 21 Mine. Longwalling was continued intermittently using various supports until April 1970. During this period, longwall mining techniques were used on separate occasions at six faces within Mine No. 21. However, due to various problems that developed at each location the faces were eventually terminated. The most significant problems concerned the support systems and roof conditions that were present in the mine during each attempt. Between 1965 and 1972, similar conditions were met by Freeman Coal Mining Company during four attempts at longwall mining in their Orient No. 5 and Orient No. 6 mines.

In 1975, the U.S. Bureau of Mines and Old Ben Coal Company entered into a cost-sharing agreement to demonstrate the application of longwall mining in Illinois. The first phase of the demonstration was to conduct a premining investigation to review previous attempts at longwall mining, perform rock mechanics tests, and establish support requirements based on the investigation results. The information obtained during the demonstration at Old Ben's No. 24 has been utilized to optimize subsequent face equipment and panel geometry.

Mining of the first panel began in September 1976 and was successfully completed in May 1977, and two additional panels were mined under the cooperative agreement. Currently, Old Ben Coal Company is the only longwall mining operator in Illinois. In addition to the face in Mine No. 24, five longwall faces are being operated in four other mines.

HISTORY

In 1962 the first mechanical longwall operation in Illinois was installed by Old Ben Coal Company in their Mine No. 21. Roof supports consisted of Rheinstahl-Wanheim four-leg chock sets, each leg with a yield load setting of 4×10^{-1} meganewtons (MN) (45 tons). The face equipment that was used on two faces consisted of an Eickhoff double-strand snaking conveyor and a Joy continuous miner, Model 4, with dual controls for operation from either side. The miner cut a width of 3.7 m (12 ft) off the face. Face No. 1 operated a total of 4 days and retreated 7.3 m (24 ft) while Face No. 2 retreated 45.7 m (150 ft). Lack of roof support capacity was the primary reason for terminating these faces.

After Face No. 2 was terminated, two additional legs were added to the rear of each chock set in order to provide a greater support capacity at the desired roof break line. These modified chocks were used on Face No. 3 along with an Eickhoff EDW 200 double-drum shearer that cut a 63.5-cm (25-inch) wide strip off the coal face. This face operated about 4 months and had a total face retreat of 67.1 m (220 ft) before its termination due to insufficient support capacity (see Figures 1 and 2).

After Face No. 3 a new chock was specified. Face No. 4 was equipped with chock sets with four legs of 8.9×10^{-1}-MN (100-ton) capacity each and the operating pump pressure was increased to set the chocks as near as possible to yield capacity. The main chocks were set at 1.5-m (5-ft) centers along the face. A two-leg "pilot" set was between each main chock set. Each of the pilot sets was positioned one web back from the face conveyor so that as the shearer cut its web, the pilot set could be advanced immediately to provide forward roof support in the newly exposed roof area prior to advancing the conveyor. Although this support system itself provided adequate capacity, stability of the supports became a problem. The chocks shifted position and tended to close the space for the pilot supports making it difficult to move them. For these reasons as well as additional problems with the face conveyor and mining system, this face was terminated in August 1965 with a total face retreat of 178.3 m (585 ft).

Based on the experience gained in longwall mining during Face No. 4, a new Rheinstahl-Wanheim four-leg chock, each leg having a 1.06-MN (120-ton) yield, was built and used at Face No. 5. The new chock had a solid roof bar from the rear leg to the front leg capable of withstanding a 1.78-MN (200-ton) bending moment. The roof bars were sufficiently long to permit the chocks to set one web back from the conveyor. In addition to the redesigned support equipment, an Eickhoff conveyor of heavier design with activated ramp plates was installed. Although operating one web back provided no unusual problems, stability of the chocks became a paramount problem at the face. Between July 1966 and March 1967, the face retreated 326.1 m (1,070 ft). The face was terminated, however, because it was felt that major design changes in the chocks were necessary to provide stability and that design changes in the hydraulic system were necessary to expedite the lowering and advancement of the chocks.

Face No. 6 had a length of 91.4 m (300 ft), which was shorter than any of the previous faces. Total face retreat was 283.2 m (929 ft), although not all of this retreat was at the face length of 91.4 m (300 ft). The support system consisted of Gullick six- and seven-leg chock sets. Except for the center front leg on the seven-leg chock sets, which had a 4.45×10^{-1}-MN (50-ton) capacity, all the other legs were of 7.56×10^{-1}-MN (85-ton) capacity. The seven-leg supports were a conventional Gullick design of a four-leg cluster at the rear of the chock connected by a solid roof

FIGURE 1. OLD BEN LONGWALL FACE NO. 3 SHOWING CRIBBING

FIGURE 2. OLD BEN LONGWALL FACE NO. 3 SHOWING ROOF SEPARATION

canopy with two forward legs connected to the solid canopy by hinged roof bars. Between the two front legs was the seventh leg, which supported a roof bar that could be extended outward and upward to immediately support the newly exposed roof. This face was characterized by massive falls at the head and tailgates and pressure exerted in advance of the face.

In addition to the attempts at longwall mining by Old Ben Coal Company, four attempts were made by Freeman Coal Mining Company. Their first longwall was conducted in their Orient No. 5 Mine in 1965. Supports consisted of Gullick six-leg chocks with a leg capacity of 4.89×10^{-1} MN (55 tons). The face was discontinued after about 54.9 m (180 ft) of retreat. In 1971, Freeman began the first of three longwall mining attempts at their Orient No. 6 Mine using Dowty chocks. Roof control was achieved up to a retreat of 48.8 m (160 ft), when overriding pressures were encountered that broke the roof above the supports and resulted in falls when the supports were moved.

DEMONSTRATION

Although previous attempts at longwall mining were unsuccessful in Illinois, interest in longwall continued. From an operator's standpoint, longwall mining offers the potential for improved productivity, higher resource recovery, better ground control, and generally easier compliance with safety standards. The promotion of the longwall mining technique was a goal of the U.S. Bureau of Mines Research Program under the Advanced Mining Technology Program in order that coal could be mined more safely and efficiently without higher social and environmental cost.

In 1975, the U.S. Bureau of Mines and Old Ben Coal Company entered into a cost-sharing contract to demonstrate that the Herrin No. 6 Coal Mine in Illinois could be mined safely by longwall methods. Three panels were to be mined at Mine No. 24 under this agreement. The initial phase of the demonstration was a premining investigation conducted by Dames & Moore under a subcontract to Old Ben Coal Company. The study consisted of (1) reviewing the previous attempts at longwall and comparing geologic conditions between Mines No. 21 and No. 24; (2) performing in-situ testing; and (3) preparing design recommendations for the support system to be used in the demonstration.

The general geology of the roofs of Mines No. 21 and No. 24 are shown on Figure 3. Although the stratigraphy of the roofs in the two mines are not similar, a review of physical properties, primarily tensile strength, modulus of rupture, and elastic modulus in rupture, indicated that the performance of the roof in Mine No. 24 would be similar to that experienced in Mine No. 21. From the review of performance of supports in Mine No. 21, it was concluded that the supports could be required to provide a support density of 862.5 KN/m^2 (125 psi). In addition, a damage survey

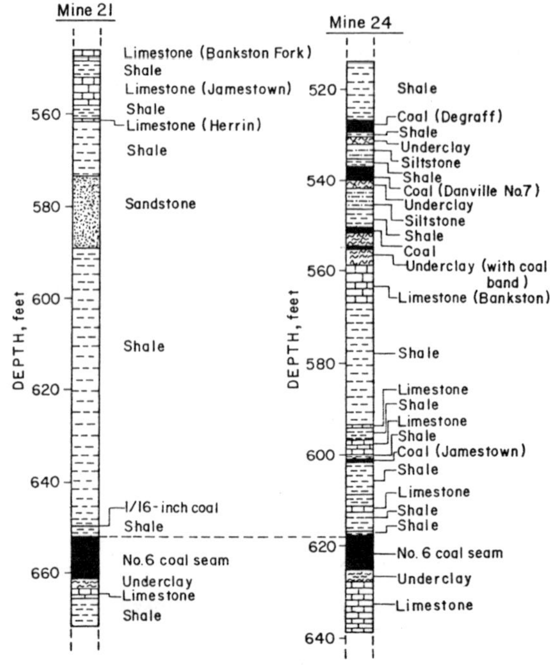

FIGURE 3. BOREHOLE LOGS

performed on the supports used on Face No. 5 in Mine No. 21 indicated that supports could be subjected to a lateral thrust parallel to the face equivalent to the lateral component of the roof load inclined 15° from the vertical.

In-situ testing consisted of measuring the bearing capacity of the roof and floor. The test procedure consisted of jacking various size plates into the roof and floor using the apparatus shown on Figure 4. Roof strength exceeded the equipment capacity of 4.36 MN/m^2 (6,320 psi). The underclay, when wet, had a bearing capacity of 2.2 MN/m^2 (315 psi). Tests were performed on the floor varying the thickness of floor coal. A 15.2-cm (6-inch) thickness of floor coal provided a bearing capacity of approximately 6.9 MN/m^2 (1,000 psi). A 35.6-cm (14-inch) thick floor coal capacitated the loading equipment.

Based on the results of the premining investigation, the following recommendations were made:

1. A minimum of 15.2 cm (6 inches) of floor coal should be maintained to provide sufficient bearing capacity for the supports;

2. Mean load density should not be less than 862.2 KN/m^2 (9 tons/ft^2); and

3. Supports should provide a continuous overhead canopy and stability against lateral thrust.

Following the recommendations of the premining investigations, Old Ben and Bureau of Mines engineers developed specifications for roof shields. A Lemiscate shield manufactured by Thyssen was selected (see Figure 5).

The shield weighs 15.87 metric tons (17.5 tons) and its overall length is 5.0 m (16.5 ft). The powerpack supplies a hydraulic working pressure of 30.0 MN/m^2 (4,350 psi); the shield is designed to yield at 38.0 MN/m^2 (5,500 psi), at which time spring-loaded relief cartridges open. Thus, to maintain a strong thrust against the roof immediately after exposure and to prevent bed separation the working pressure equals 80 percent of the yield pressure.

The shields are specially designed so that extreme power is concentrated in the legs. The two legs together can exert a thrust of 4.61 MN (518 tons) and yield a load of 5.84 MN (656 tons); however, a mechanical disadvantage reduces the force that acts on the canopy hinge. At a 2.1-m (7-ft) mining height, the shield can exert a thrust of 3.59 MN (404 tons) and sustain a load of 4.55 MN (512 tons) at yield.

A minimum of 0.3 m (1 ft) between canopy tip and face must be maintained to provide a safe clearance between the rotating drums and the canopy tip. The shearer drum is 0.6 m (2 ft) in width so that after the shearer has cut by, the roof exposure is widened to 0.9 m (3 ft). This span is reduced to 0.3 m (1 ft) again after the shields are advanced to 0.6 m (2 ft). However, the entire shield can be brought 0.6 m (2 ft) closer to the face and operated in the one-web-back or conventional mode by lowering the canopy extensions, which are 0.6 m (2 ft) long. In the closed position, the shield is rated to yield at a mean load density of 1,197.5 KN/m^2 (12.5 tons/ft^2) before the shearer passes and 958.0 KN/m^2 (10 tons/ft^2) after the shearer has cut by.

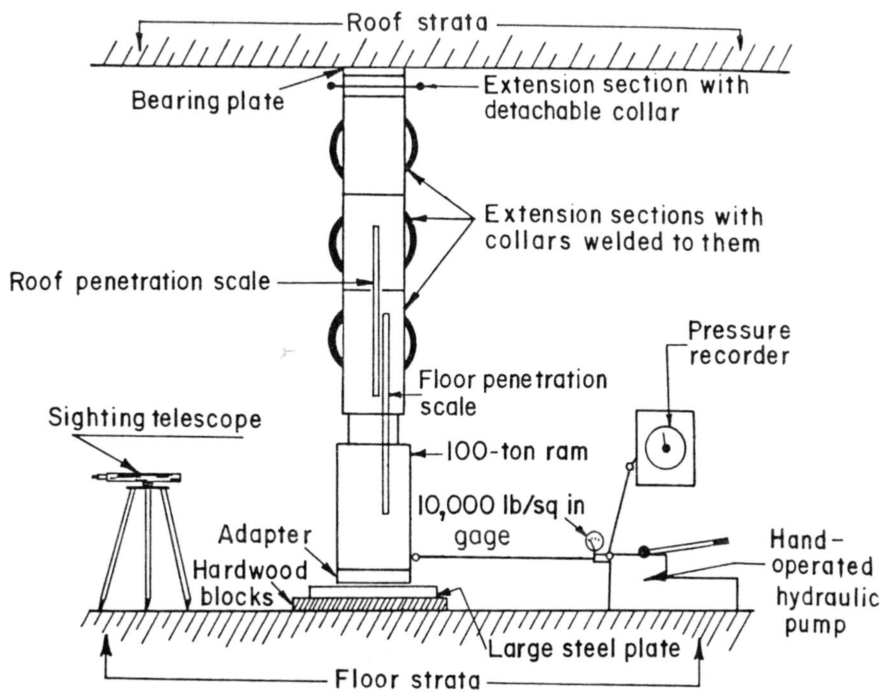

FIGURE 4. BEARING TEST APPARATUS

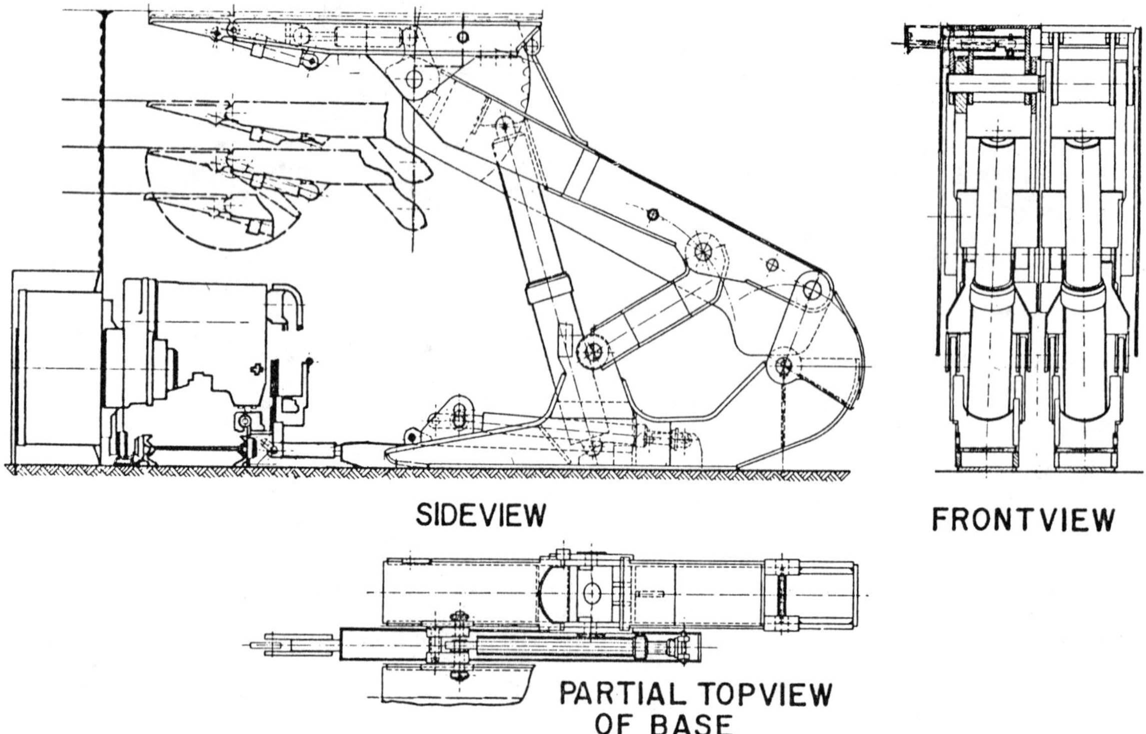

FIGURE 5. ROOF SHIELD 6' / 10'

The canopy is designed to maintain a stable roof contact during the shield advance. This is accomplished by the action of a ram that swings the canopy against the roof and by designing the canopy to have a ratio of front portion to rear portion of not greater than 2. The canopy tip load exceeds 4.45×10^{-2} MN (5 tons) at yield.

The base, including the weight of the shield, exerts a floor pressure of 2.1 MN/m^2 (311 psi) at yield. The advancing shield can overcome uneven or soft floor or obstacles since each half of the base can be lifted individually by its double-acting leg. The base is self-cleaning, in that debris and slack coal can pass into the gob through a sufficient number of open spaces in the base while the shields are advanced.

The tendency of the shield base to dig into the floor is counteracted by mounting the double-acting ram in an inclined position. The ram acts through reverse linkage to raise the base slightly and to exert the full force of the piston end when the shield is pulled up to the conveyor. Also, the shield is designed so that the forward edge of the base is ahead of the projection of the canopy hinge joint.

The following roof support equipment was procured:

1.   A shield line of 95 units on 1.5-m (59-inch) centers to accommodate the conveyor pans, which are 1.5 m (59 inches) long, so that the middle of the pan can be attached to the double-acting ram, thereby minimizing the stress on the pan connections due to eccentric forces. Three of the shields are placed across the headgate and two in the tailgate.

2.   A powerpack consisting of two high-pressure pumps, each capable of operating the full shield line by supplying 0.095 m^3/min (25 gpm) to the dual pressure system; one mixing tank to prepare the 5 percent soluble oil in water emulsion that powers the system; and two filters.

3.   One hundred and fifty hydraulic single props with a range of 2.1 to 3.0 m (7 to 10 ft). Connecting these props to the hydraulic pressure system via hose and quick-connect, pistol-type coupling will extend them to exert a thrust of 2.67×10^{-1} MN (30 tons), if properly blocked. Each prop yields at 39,916.1 kg (44 tons). Bleeding the fluid retracts the prop.

The hydraulic props are placed into the three gate roads and crosscuts less than 24.4 m (80 ft) from the face and remain there until the face approaches within 0.9 to 1.5 m (3 to 5 ft). According to European and U.S. studies, by virtue of early bearing capability, rows of single props control convergence in gate roads better than late bearing cribbing, which relies on rock deformation for building up the support strength. Several props in the supply gate, properly blocked and equipped with hydraulic gauges, provided an early warning system against excessive abutment pressures, which never occurred.

The demonstration also included a ground control and subsidence study. This study was conducted to provide early warning capability by measuring strata movement and pressures during mining so that corrective measures could be taken quickly should an unusual situation arise. The study also evaluated the adequacy of the roof support system and derived design criteria for future longwall mining efforts.

The program included the following tasks:

1. Measuring convergence in gate roads and differential rock and floor strata movement to detect bed separation;

2. Measuring induced stress in coal and floor of the longwall panel and chain pillars;

3. Tracing the loading profile in the roof support system by measuring hydraulic pressures with load indicators, gauges, and recorders; and

4. Determining vertical extent of caving by boreholes from surface to coalbed.

Convergence of roof and floor seldom reached or surpassed 5.1 cm (2 inches) before the face passed. Only in the tailgate did a station show a 10.2-cm (4-inch) convergence at face approach. Measurements of differential roof strata movement did not indicate bed separation between shale and limestone. However, differential floor movement readings showed slight heaving of the underclay, up to 5.1 cm (2 inches), relative to the siltstone base.

Near the staging area, the stress readings showed a gradual rise in stress followed by a sudden drop when the face attained a distance of 45.7 m (150 ft) and the first major roof caving occurred. Afterwards, the stress increased gradually. No recurring pressure fronts were observed.

In midpanel, stressmeters sharply responded to the face approach, one of them indicating a gain of 11,040.0 KN/m^2 (1,600 psi) before destruction.

Stressmeters in the siltstone floor under the longwall panel showed a small increase at face approach, a drop to below setting pressure after the face rolled over them, and a slow gain back to setting pressure when the face moved away.

In the recovery room, stressmeters indicated a sharp rise when the face cut through into the recovery room.

Pressures in the hydraulic system of shields and props were measured to compute the roof loads. Shield loads increased during idle days by approximately 69.0 KN/m^2 (10 psi). A sharp decrease in canopy loads occurred immediately before and following the first major roof falls and simultaneously with the stress relief measured by stressmeters near the staging area.

The successive caving of roof strata over the mined-out area was monitored by Time Domain Reflectometry, as follows: a 2.2-cm (7/8-inch) coaxial cable was grouted into an NQ-size [7.62-cm (3-inch)] borehole to a depth of 204.2 m (670 ft). The borehole intersected the No. 6 coalbed, which is 2.7 m (8.7 ft) in thickness, at a depth of 188.7 m (619 ft). A Tektronix 1502 Time Domain Reflectometry cable tester, which detects cable faults and breaks by radar, was used to monitor the

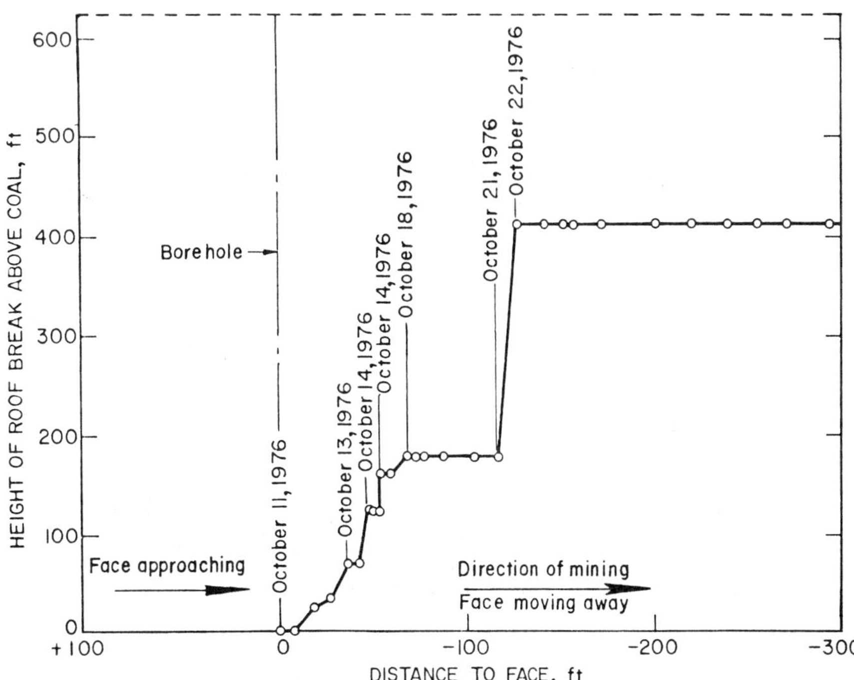

FIGURE 6. STRATA FAILURE IN TERMS OF FACE ADVANCE

condition of the cable. Prior to installation, the cable is crimped at 3.0-m (10-ft) intervals to aid in calibration and measurement. Figure 6 plots the progress of caving in terms of face advance.

General panel geometry is shown on Figure 7. Panels were 140.2 m (460 ft) wide by 530.3 m (1,740 ft) long. The panel entries, in sets of three on 18.3 and 30.5-m (60 and 100-ft) centers, were developed by continuous mining machines and were 4.3 m (14 ft) wide except for the center entry, which was widened to 5.5 m (18 ft) to accommodate the longwall powerpack and the power center along the conveyor belt. The roof of the entries was supported by 1.5-m (5-ft) resin bolts on a 1.5-m (5-ft) grid.

The chain pillar on 18.2-m (60-ft) centers was mined along with the longwall face, adding to its length. This practice is unique to longwall mining and designed to increase recovery.

Panel No. 1 was mined in approximately 34 weeks. A fault was encountered in Panel No. 2. The fault was located approximately 83.8 m (275 ft) east of the headgate and extended south through the panel. Vertical displacement on the fault was approximately 0.9 m (3 ft). The fault was traversed without unusual support problems. Panel No. 2 was mined in approximately 53 weeks.

The three panels mined in the demonstration were successfully completed without problems in roof support.

Recovery of longwall roof supports is always a difficult task and requires circumspect planning and execution. Shield recovery poses an unusual problem because part of the structure is under caved rock. Therefore, during the last ten passes across the face, wire mesh was stretched over the top of the shields. In the first panel, a recovery

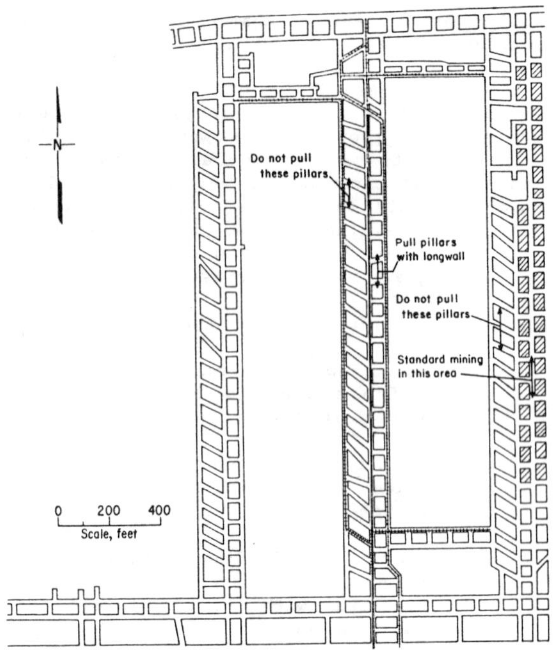

FIGURE 7. PANEL LAYOUT

room was predriven and the longwall advanced into it. In the second panel, a recovery room was driven by a continuous miner along the shield line (Figure 8). The continuous miner loaded the material into the pan line, and it was necessary to withdraw the miner after a run of 2.4 to 3.0 m (8 to 10 ft) so that the roof bolter could be brought in to secure the roof. The roof was secured by installing 1.5-m (5-ft) resin bolts on 1.5-m (5-ft) centers and by bolting wire mesh and 10.2-cm x 3.7-m (4-inch x 12-ft) bridge boards against the

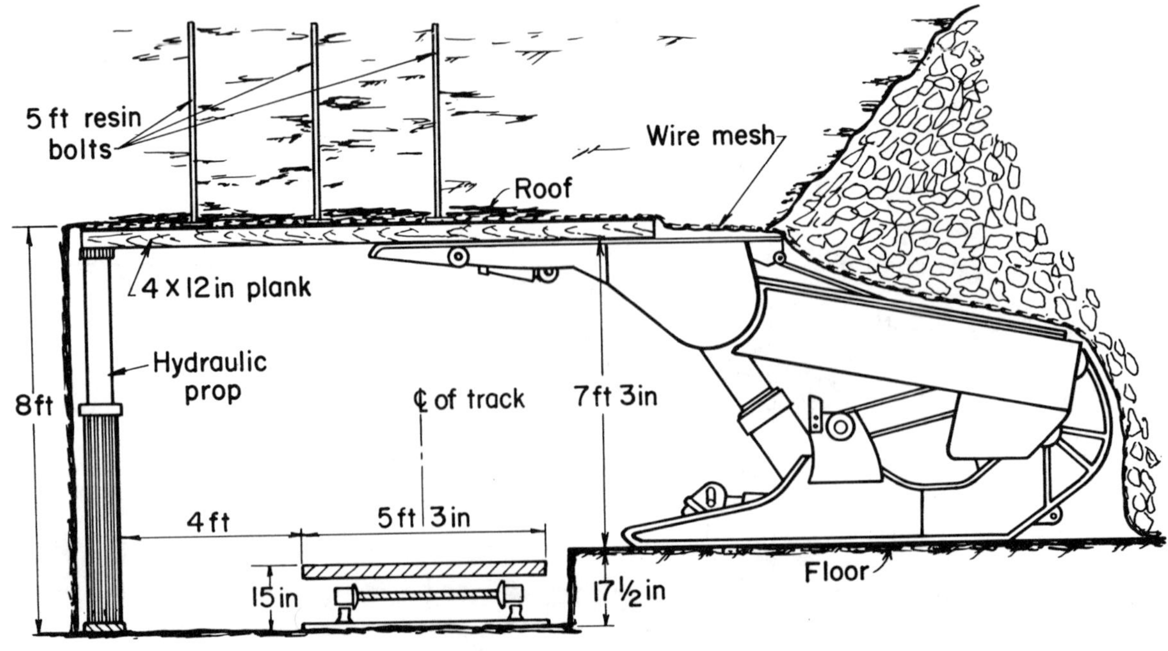

FIGURE 8. RECOVERY AREA

roof. This cut and bolt cycle with associated moves slowed recovery room development to 13 calendar days.

In the third panel, the shearer was used to open up the recovery room. The pan line was advanced from the terminal shield line by means of hydraulic props substituting for advancing rams. A roof bolter, which slides on the pan line, was towed by the shearer and installed roof bolts and wire mesh immediately behind the shearer. After shearer and pan line were removed from the recovery room, the space was further secured by a movable bulkhead formed by two shields moving side by side along the recovery room, while a shield was removed from the shield line and taken off the face (Figure 9). This was an important safety feature. Another precaution taken was to build cribs in place of a removed shield. A third precaution taken was to advance shields from the line of supports under their own power. Under these conditions, shields can be set against the roof quickly should immediate support be needed.

PRESENT PRACTICE

Currently, Old Ben is the only operator using longwall mining techniques in the Illinois Basin. Old Ben is operating five longwall faces, in addition to the previously described panel in Mine No. 24. Two faces are installed in Mine No. 25, and one each in Mines No. 21, No. 26, and No. 27.

The specified support density of the subsequent faces is 862.2 KN/m^2 (9 tons/ft^2), the same as that specified originally for Mine No. 24. Both Thyssen and Hemscheidt shields are being used. The configuration of the shields is shown on Figure 10. As can be seen by comparing Figure 10 with Figure 5, changes in support configuration have been made.

The rear canopy has been eliminated to provide better tip control and a shorter canopy ratio. To aid in recovery, the new shields can collapse to a height of 1.2 m (4 ft) and be removed along the face under the canopies of the operating shields. Canopy extensions, or "flippers", are used on the tips of the shields. The extensions can be set up to 20° above the horizontal. This design permits following uneven roof. Also, with the extension, the shield can be set within 30.5 cm (12 inches) of the face. Each half of the base of the shield operates independently to overcome soft or irregular floor conditions. Floor pressure is 1.6 MN/m^2 (225 psi) at yield. Shield hydraulics are controlled from adjacent shields and hydraulic system pressure is capable of setting the shields at 80 percent of yield.

Headgate support is provided by hydraulic single props with a capacity of 3.56×10^{-1} MN (40 tons) and steel link bars. The steel link bars consist of segments 76.2 cm (30 inches) long that can be rigidly coupled together by means of setting wedges. The 76.2-cm (30-inch) length of the link bar is compatible with web width so a set of bars can be recovered with each shearer cut. The segments can be installed to accommodate roof irregularities up to 7°.

Other face equipment consists of Eickhoff 300-L double-drum shearers with 167.6-cm (66-inch) diameter drums. Face conveyors are Halbach & Braun with 250 hp motors at each end. Illumination on alternating shields is provided by mercury vapor high-intensity lights; however, newer faces are using fluorescent lights.

Present panel layouts are similar to the layouts used in Mine No. 24. The mining practice

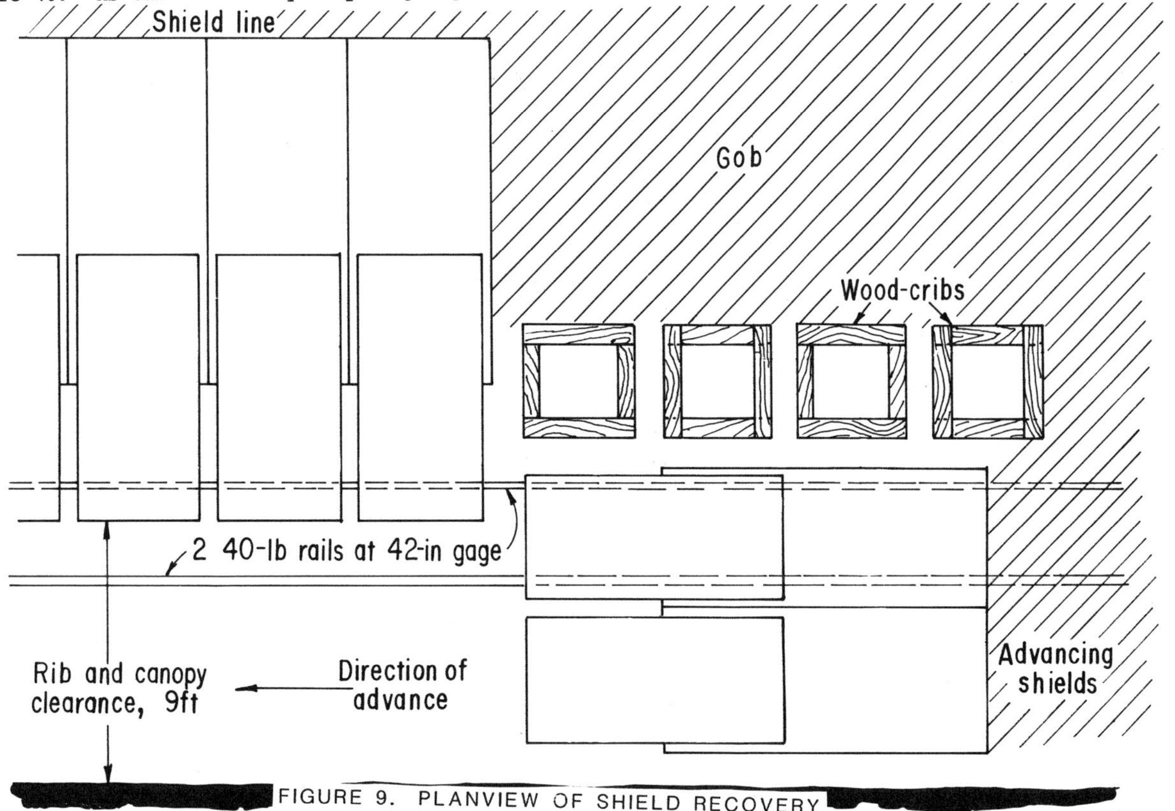

FIGURE 9. PLANVIEW OF SHIELD RECOVERY

followed is to mine to the roof and leave approximately 15.2 cm (6 inches) of floor coal. Chain pillars are not being recovered in the current panels.

Current shield recovery operations consist of predriving two rooms parallel to the face beyond the end of the panel. Three or four crosscut "stubs" are then driven into the panel to provide recovery routes in addition to the headgate. At the completion of the panel, the shearer is advanced two additional cuts while holding the shields in position. This provides a work space of 2.7 to 3.4 m (9 to 11 ft) between the tip of the supports and the face. Two end shields at the tailgate are then brought into the opening and turned 90°. These two shields travel parallel to the face and form a bulkhead as each subsequent shield is removed. Shield removal is accomplished by a scoop powered, rubber-tired low-boy that has an 18,143.7-kg (20-ton) capacity. The low-boy is backed under the bulkhead shields and the next shield to be removed is pulled onto the low-boy. Two scoops work on recovery, and three shields per shift can be recovered.

FUTURE PLANS

Studies have been performed to optimize entry geometry for future panels. The studies have included in-situ stress measurements as well as finite element stress analyses. In-situ stress measurements have indicated an in-situ stress of approximately 20.7 MN/m^2 (3,000 psi) in a north-south direction. Finite element analyses (Figure 11) have indicated that this stress is beneficial to roof stability when the entries are oriented normal to the direction of the stress. One panel is currently being mined in an east-west direction to verify this theory. Finite element analyses have also been performed to optimize entry and pillar geometry. The analyses have indicated that headgate conditions would be improved if the pillar adjacent to the panel was increased to 30.5 m (100 ft) on centers. Longer panels are presently being developed. Future panel lengths will be in the 1,524- to 1,828.8-m (5,000- to 6,000-ft) range.

An automatic shearer is being developed and is expected to be installed in mid 1981. The automatic shearer is being developed by Foster-Miller under a contract with NASA for the U.S. Department of Energy. The shearer will have a sensitized leading drum that will sense roof rock. The trailing drum will be a slave to the leading drum. The shearer will be capable of operating automatically, by radio, or manually.

SUMMARY

The use of shields has been a major factor contributing to adequate roof control in Illinois.

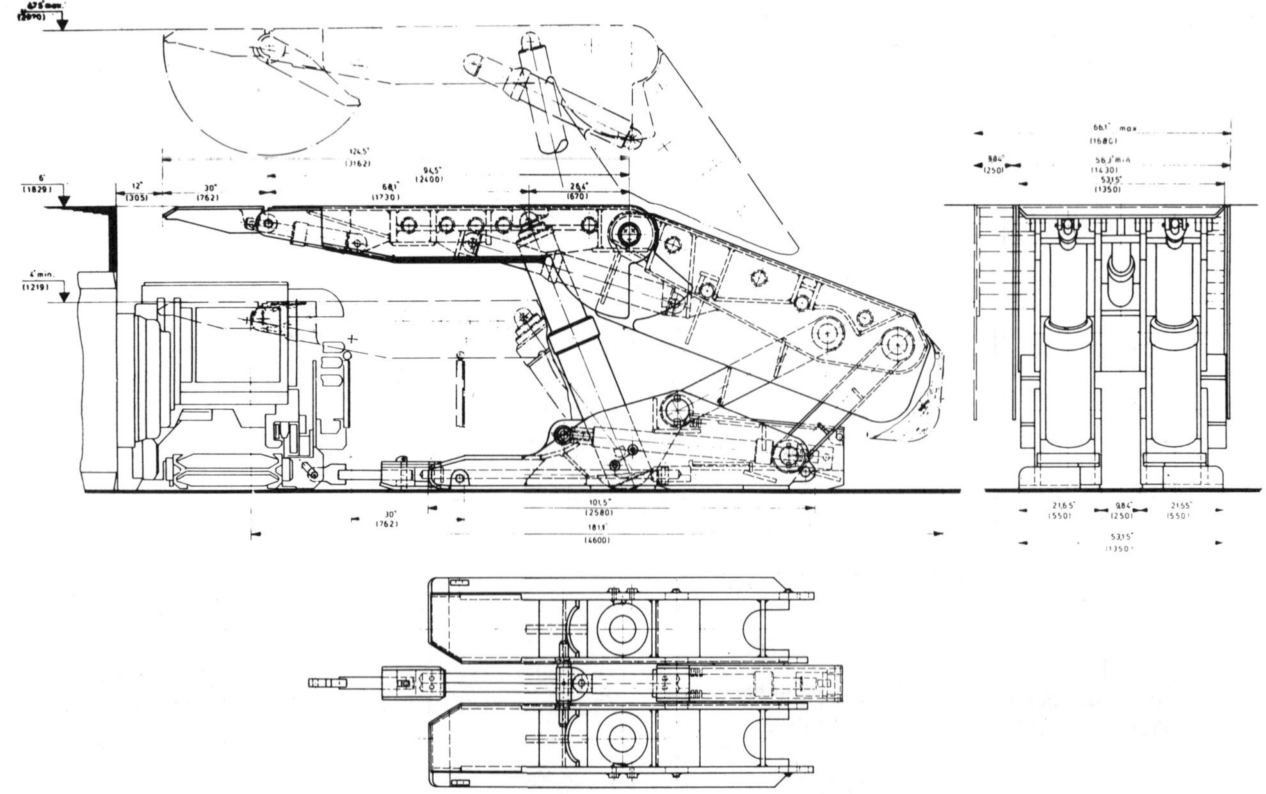

FIGURE 10. ROOF SHIELD 4' / 9.75'

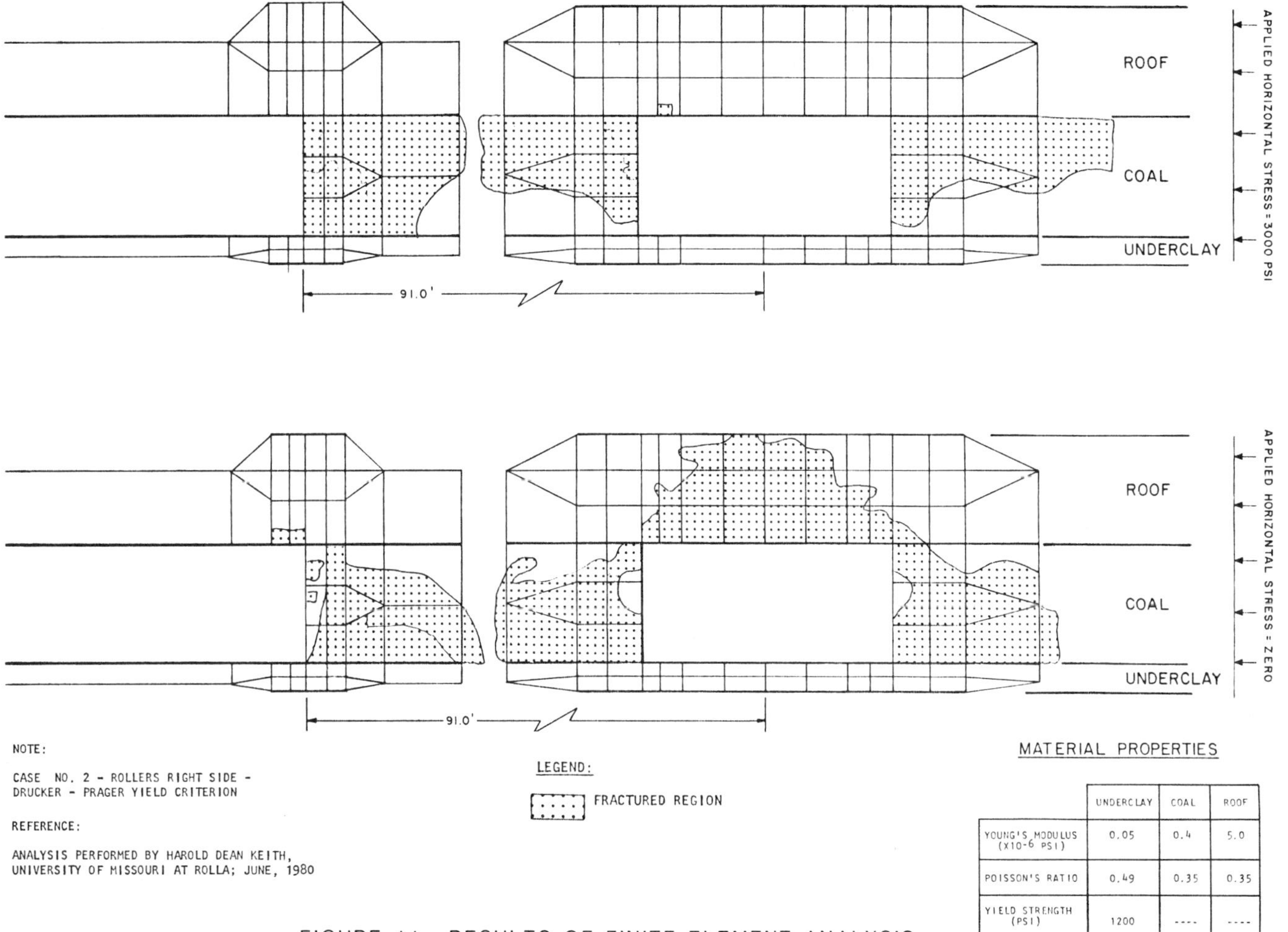

	UNDERCLAY	COAL	ROOF
YOUNG'S MODULUS (X10-6 PSI)	0.05	0.4	5.0
POISSON'S RATIO	0.49	0.35	0.35
YIELD STRENGTH (PSI)	1200	----	----

FIGURE 11. RESULTS OF FINITE ELEMENT ANALYSIS

Short spans of exposed roof, quickly applied support, stability of structure, and full shelter for face crews are some of the significant advantages that led to the adoption of shields. The successful application of shields that were designed to sustain high yield loads resulting in a mean load density of 862.2 KN/m^2 (9 tons/ft^2) has encouraged the growth of longwall mining in the Illinois coal basin from one face to six in less than 4 years. Double-drum shearers of 1.5×10^8 kg/m (400 hp), using 995-volt power, with chainless haulage and high capacity face conveyors and stage loaders are standard face equipment. New coal mines are being developed for longwall mining with panel blocks as large as limiting factors allow. They are projected to have two active longwall faces and as many continuous miner units as are required for development and extraction.

REFERENCES

Curth, E.A., 1978, "Safety Aspects of Longwall Mining in the Illinois Coal Basin", U.S. Bureau of Mines Circular.

Curth, E.A., and Cavinder, M.D., 1977, "Longwall Mining the Herrin No. 6 Coalbed in Southern Illinois", Proceedings of the Illinois Mining Institute.

Harrell, M.V., 1973, "Longwalling at Freeman Coal Mining Company in Illinois", Proceedings of the Illinois Mining Institute.

Moroni, E.T., 1973, "Longwall Experience in the Herrin No. 6 Coal Seam", Proceedings of the Illinois Mining Institute.

Wade, L.V., and Conroy, P.J., 1977, "Rock Mechanics Study of a Longwall Panel", AIME Fall Meeting, St. Louis, Missouri.

Chapter 25

SHORTWALL MINING IN THE US: A RECORD OF FAILURE AND SUCCESS†

Syd S. Peng, Ph.D.
Professor and Chairman

Duk-Won Park, Ph.D.
Assistant Professor

Department of Mining Engineering
College of Mineral and Energy Resources
West Virginia University
Morgantown, WV 26506

INTRODUCTION

More than five years have elapsed since the first shortwall panel was installed in a US mine in 1972. By the latest count, 11 mines have tried the shortwall method on 34 panels.

Potentially, the shortwall method (see Figure 1) offers the combined advantages of longwall and continuous mining. But just like any other mining method, shortwall is not a panacea. It is only applicable under a set of stringent conditions.

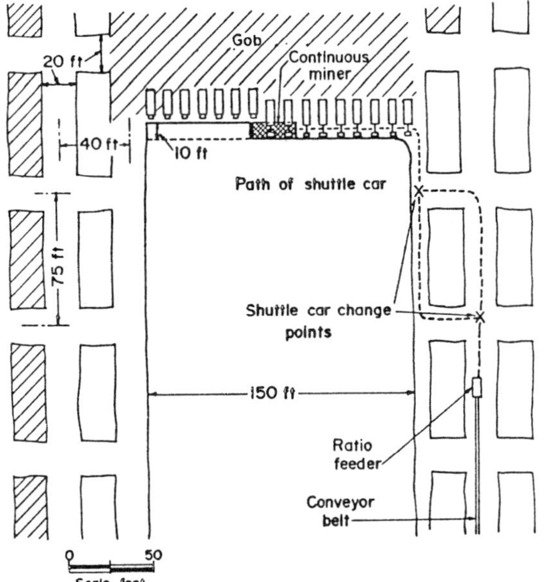

Figure 1. Typical shortwall panel layout with two-shuttle-cars system. It may also be used for one-shuttle car system. However the entry belt is sometimes relocated in the headentry for one-shuttle-car system.

Coal operators at eight of the eleven mines are satisfied with shortwall mining, though there are only five mines presently producing coal from shortwall panels. The most frequently cited advantages for which management decided to adopt shortwall mining are:

____

†Editor's Note: This article was written in 1977 and appeared in Coal Mining and Processing, December, 1977. It is reproduced here with the permission of the authors and Coal Mining and Processing.

A. Shortwall vs continuous mining--(1) higher coal recovery, (2) higher clean coal production, (3) safer and healthier face working environments, e.g., positive roof support and ventilation, and less dust.

B. Shortwall vs longwall mining--(1) less capital cost, (2) higher flexibility, (3) less training period.

In this article, we will discuss the findings of our latest survey of mines employing the shortwall mining method.

COAL SEAMS

All of the shortwall panels are located in the Appalachian coal field (Table I). Four mines (15 panels) have operated in the Pittsburgh seam, and one mine in each of the following seven seams: Upper Freeport, two panels; Lower Kittaning, three panels; Elkhorn No. 3, three panels; Pocahontas No. 4, two panels; Jaw Bold, one panel; Tiller, one panel; and Lower Five Block, one panel. All of the mines except the shortwall stripping on Lower Five Block seam, are underground.

Seam thickness ranges from 40 in. in the Tiller seam to 108 in. in the Pittsburgh seam with an average of 68 in. Bear in mind, however, that shortwall mining is feasible only if the seam is above a minimum thickness--which is the summation of the following parameters: minimum operating height at the drum end of the continuous miner, thickness of the canopy of the face supports, and allowance for floor undulation. For most drum type miners the minimum height is approximately 54 in. Seams with less than the minimum can only be mined by cutting into the roof.

PANEL DEPTH AND LAYOUT

Shortwall panels have been operated in a wide range of depth below surface (Table I). The shallowest panel was the shortwall stripping demonstration project in Julian, WV, where overburden thickness ranges from 30 to 80 ft, while the deepest panels (1000-ft deep) are in the Pittsburgh seam. It seems, therefore, that the success of shortwall mining is not directly related to overburden thickness but depends heavily on the characteristics of the immediate roof.

Panel width runs between 100 and 200 ft. The trend is toward a wider panel. For example, of the five active shortwall panels, one is 150 ft, two are 180 ft, and the other two are 200-ft wide. When the face gets wider, however, the distance between the shuttle cars' loading and unloading points increases. As this distance increases, so does the travel time for each shuttle car trip. As a result, the waiting time of

TABLE I

SEAM CHARACTERISTICS AND PANEL LAYOUT

| | Mine Names and Locations | Coal Seam | | Depth (ft) | Immediate Roof Rock | Floor Rock | Panel Layout (ft) | | | Remarks |
		Name	Thickness (in.)				No. of panel: wd. x lg.	Entry	Chain Pillar	
1	Helen Mine Homer City, PA	Upper Freeport	48 – 84	760	20 ft shale interbedded with sandstone	fireclay	1: 100 x 3400 1: 180 x 3400	20		
2	West Virginia	Pittsburgh	70 – 84	800 – 1000	8" rider coal 20' shale	firm shale	6: 200 x 2000	20	55 x 80	diamond-shaped pillars
3	West Virginia	Pittsburgh	72 – 92	1000	8" rider coal 20' shale	1.5' soft shale	1: 200 x 1400	13½	80 x 80	
4	Delta Mine Stoytown, WV	Lower Kittaning	65 – 89	83 – 220	8' shale 4' rider coal 6' shale	soft fireclay	5: 125 x 700	20	40 x 50	
5	Beth-Elkhorn Jenkins, KY	Elkhorn #3	48 – 60	150 – 800	13' shale 20' sandstone	firm shale	6: 150 x 2400	20	40 x 55	
6	West Virginia	Pocahontas #4	84 – 108	200 – 300	2-3' rider coal 20-25' laminated shale	soft shale	1: 180 x 800 1: 180 x 600	16	48 x 100 48 x 200	staggered pillars
7	WVSM & RA Julian, WV	Lower Five Block	60 – 66	30 – 80	3-6" rider coal 6-9" shale 35-50' shale	rapidly weathering shale	1: 200 x 1200	Coal extraction by under-cutting along outcrop		
8	Tiller Mine Jewell Valley, VA	Tiller	± 40	200	60' sandstone	sandy shale	2: 150 x 3200	20	30 x 50 50 x 50	partial recovery
9	Jewell #12 Ridge Mine, Whitewood, VA	Jaw Bold	± 65	700	20-50' broken shale interbedded with SS	sandy shale	1: 200 x 3000	20	50 x 50	diamond shaped pillars
10	Mathis Mine Bethlehem, PA	Pittsburgh	± 55	300	3' shale with coal parting 26' shale	hard shale	1: 185 x 2500	16 20	60 x 70 40 x 70	
11	West Virginia	Pittsburgh	55 – 72	800 – 900	12-15' shale	extremely hard shale	6: 100 x 400 1: 110 x 1900 1: 150 x 2700	18	40 x 75	

the miner also increases. A system analysis of the
shortwall face operations by E. R. Palowitch of USBM
(paper presented at the 1972 SME Fall Meeting, Birmingham, AL preprint No. 72-AM-356) indicated that the
optimum width with two shuttle cars of five-ton capacity is 150 ft. It is a compromise between productivity, coal recovery, and capital cost for face
powered support (Figure 2).

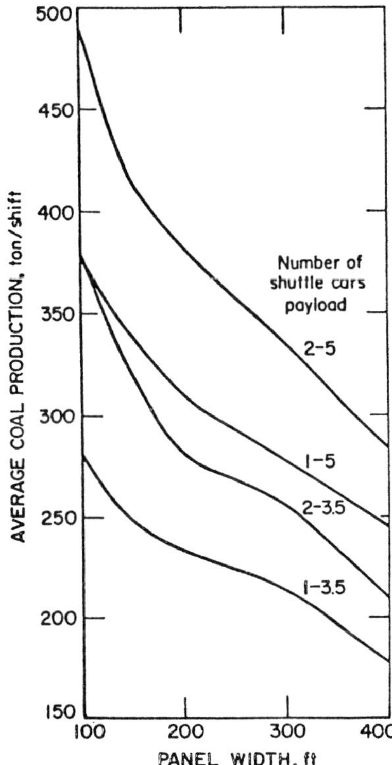

Figure 2. Productivity as a function of shortwall
face width based on the U.S. Bureau of
Mines' system analysis, assuming a scheduled delay of 0.1 min per ft of face.

Panel lengths range from 400 to 3400 ft. Most
companies started with short trial panels and extended
to longer lengths in the later development. For
example, the six trial panels in Mine No. 11 were 400-
ft long; in the last panel it became 2900-ft long. A
shortwall panel length of about 3000 ft is optimum,
considering the number of equipment moves, haulage and
ventilation requirements.

All of the shortwall panels are mined on retreat.
The panels are flanked on both sides by three or four
parallel entries, which leave two or three lines of
chain pillars (Figure 1). Chain pillars are either
uniform in size or nonuniform, with the one immediately next to the tail entry larger than the others.
The nonuniform pillar design is ideal because the tail
entry of a shortwall panel is always immediately next
to the mined-out panels, and excessive roof pressure
is expected on the tail entry side. This is especially true when it comes to the second and third
panel.

But there are exceptions. For example, Mine No. 11
employed a Z-type face layout. It predeveloped four
parallel entries in the head entry side, but no entry
was predeveloped in the tail entry side. The tail

entry was formed by setting a row of cribs 4 ft from
the face end during retreat mining (Figure 3). It
remained open throughout the mining period of the
panel without requiring any special maintenance.
However, this layout was tried in two virgin panels--
and might experience difficulty in the second or third
panel.

Another exception is that of the shortwall stripping
in Julian, WV. The method involves opening a narrow
bench running parallel to and in for about 35 to 50 ft
from the coal outcrop. Development consisted of
driving two 18-ft-wide parallel entries perpendicular
to the bench on 60-ft centers for 200-ft long. Crosscuts were driven on 100-ft centers. The left-hand
entry became the face that continuously advanced along
the coal outcrop.

Most chain pillars are not recovered after panel
retreat mining. But the smaller pillars near the head
entry side in Mine No. 8 and Mine No. 9 were partially
recovered. Pillar recovery operation followed the
advance of the shortwall face but was kept approximately 200 ft behind.

IMMEDIATE ROOF

The immediate roof plays a rather important role in
the success of a shortwall panel because immediately
after a new web cut, the prop-free front is at least
between 14 and 21-ft wide. Hence, strong and self-
supporting roof is required. The strong and massive
immediate roof, however, must cave easily without too
much overhang so that the required capacity of the
support units will not become excessive. Conversely,
a weak immediate roof requires an adequate support as
soon as it is exposed by extraction of underlying coal
to prevent premature failure. This is a very important factor especially for the area near the corner
of the faceline and roofline where the sole support is
offered by the tip of the extensible canopy.

The immediate roof of the Pittsburgh seam, which is
soft shale or soft shale interbedded with coal streaks,
generally satisfies these criteria provided no other
weak structure exists. But weak structures such as
clay veins are found in the Pittsburgh seam. Clay
veins are brecciated claystone and highly slickensided. They fall off easily on and between the face
support units and cause serious roof-control problems.

The premature termination of the shortwall panel in
Mine No. 11 could largely be attributed to the numerous clay veins that criss-crossed the panel. Some of
these were so large compared to the face length that
their fall halted the operation for more than two
weeks.

The immediate roof of the Pocahontas No. 4 seam is
another example. It is an extremely thinly laminated
and highly slickensided shale that extends up to 40
ft in thickness. It falls very easily and breaks
into small pieces. Extremely large roof falls occurred
at the head entry T-junction whenever the faceline
advanced to a point in line with crosscuts of the head
entry chain pillars. Once this happened, it blocked
out the head entry--and was the main reason for discontinuing shortwall mining in that seam.

Floor rock is also important: if it is too soft,
the frequent travels of shuttle cars in the head
entry ahead of the face or even along the face in
front of the powered supports may damage the floor,

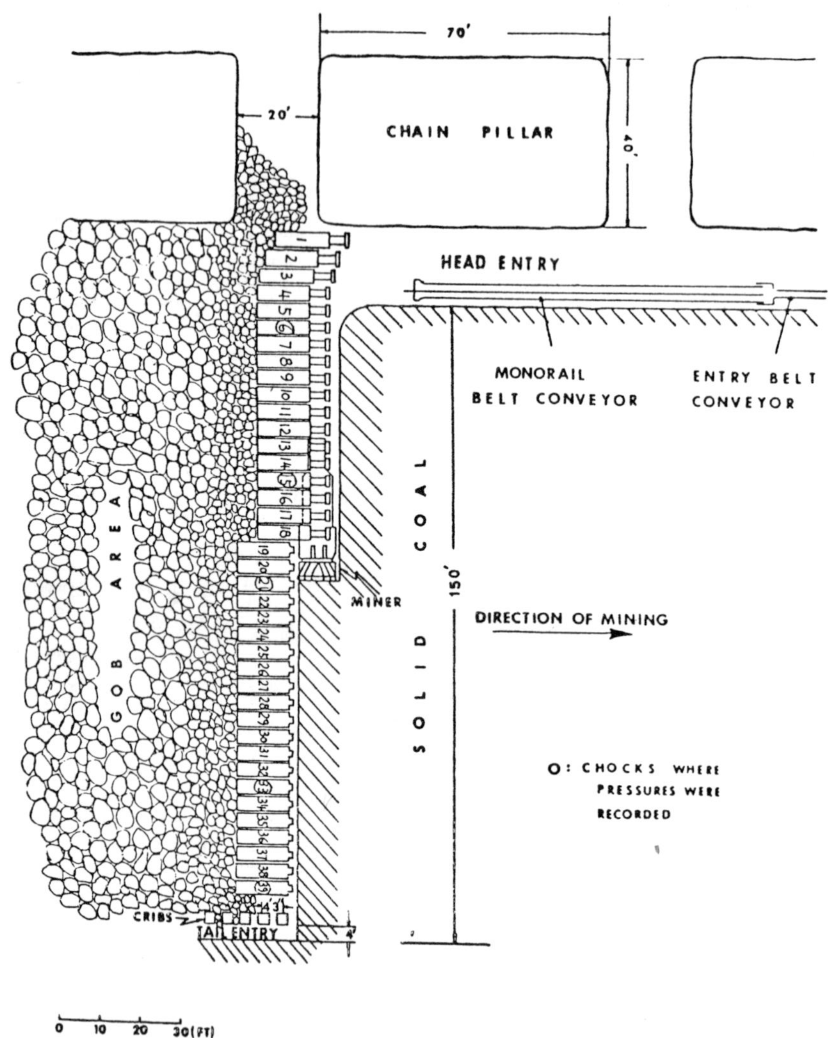

Figure 3.   In this Z-type shortwall panel layout in Mine No. 11, the
head entry was predeveloped whereas the small tail entry
was formed as the face retreated.

which eventually becomes uneven. In addition to
higher maintenance costs for shuttle cars, uneven
floor causes problems in advancing powered supports;
the supports, for instance, will move toward pot holes
during advance and bump into each other, or they cannot be budged because their base tips dig into the
floor. Soft floor problems get compounded by the
presence of water near the face area. Soft floor
problems were found in Mine No. 3 and Mine No. 6.

COAL EXTRACTION AND HAULAGE

Continuous miners used for coal cutting vary from
mines to mines (Table II). They are mainly supplied
by the nation's three largest manufacturers: Jeffrey,
Joy, and Lee-Norse. Both fixed-head and oscillating-
head ripper type miners have been used. But the
oscillating-head miner tends to spill coal over the
dozer beams and causes difficulties in advancing face
supports. Width of cutting can vary from 6 to 11 ft,
though the most popular is approximately 10 ft.

Coal cut by a continuous miner is loaded on a
shuttle car which runs between the miner and the entry
belt conveyor. At the loading end of the belt conveyor, a feeder-breaker is usually installed to
minimize coal spillage during shuttle car's loading.
A coal loader also is frequently used to clean up
spillage along the face: the coal cut by a continuous
miner is dumped in the rear of the floor and picked up
by the loader which then loads on the shuttle car.
This arrangement keeps a clean floor along the face.
A clean floor is essential for uninterrupted operation
of all face equipment.

Since most continuous miners can cut coal at a rate
of 8 to 12 tons per minute--while the capacity of a
shuttle car runs from 3 to 20 tons--a continuous miner
spends most of its shift time waiting for shuttle cars.
Some of this waiting time can be reduced by running
two shuttle cars in sequence, i.e., one on the loaded
trip and the other on the unloaded trip. With two
shuttle cars, the general practice is to put the entry
belt conveyor in the second entry (Figure 1) to create
a separate travel route for each shuttle car and thus
avoid collision.

TABLE II

CONTINUOUS MINERS, COAL HANDLING AND PERFORMANCE EVALUATION

	Mine Names and Locations	Miner	Web Width (ft)	Coal Handling	Duration of Operation					Avg. Prod. Rate (tons/shift)	No. of crew/shift	General Evaluation
					Yr. started	Yr. end	No. of panel finished	No. of panel in progress	No. of panel in planning			
1	Helen Mine Homer City, PA	Jeffrey 120L	10	Mineral→S.C.'s →B.C.'s	1976	operational	1	1	unknown	300	7	Satisfactory
2	West Virginia	Jeffrey 120M	6	Miner→Loader→ S.C.→Sliding Monorail→B.C.	1973	operational	5	1	unknown	400	10	Excellent, it produces more than continuous mining sections
3	West Virginia	Jeffrey 120M	6	Miner→Loader→ S.C.→B.C.	1977	1977	1	0	0	300	7	Discontinued because base tips of the supports tend to sink into the floor &prod. rate is not enough
4	Delta Mine Stoytown, WV	Lee-Norse 106	10	Miner→S.C.'s→ M.B.C.→B.C.	1973	1974	5	0	1	300	9	Satisfactory for decent roof conditions. New panel will start in early 1978
5	Beth-Elkhorn Jenkins, KY	Joy 11CM	10	Miner→S.C.'s→ B.C.	1974	1976	6	0	pending USBM approval	490	10	Satisfactory completion of a demonstration project sponsored by USBM
6	West Virginia	Lee-Norse 45H	8½	Miner→S.C.'s→ M.B.C.→B.C.	1975	1976	2	0	0	300	6	Discontinued because too many roof falls
7	WVSM & RA Julian, WV	Lee-Norse CM 285H	10	Miner→FAC→ truck	1975	1976	1	0	0	200	7	Satisfactory completion of a demonstration project sponsored by EPA
8	Tiller Mine Jewell Valley, VA	Lee-Norse HH 265	11	Miner→S.C.→ B.C.	1977	operational	1	1	1	400	7	So far very successful
9	Jewell Ridge Whitewood, VA	Joy 14 CM	10	Miner→S.C.→ B.C.	1977	operational	0	1	1	300	7	So far very successful
10	Mathis Mine Bethlehem, PA	Joy 12 CM	10	Miner→Loader→ S.C.→Sliding Monorail→B.C.	1977	operational	0	1	12	300	7	So far it's okay
11	West Virginia	Lee-Norse 35Y Joy 12 CM	9½ 8½	Miner→Loader→ S.C.→B.C. Miner→AFC→ Sliding Monorail→B.C.	1973	1977	8	0	0	250	7	Discontinued because of excessive roof falls caused by clay veins

1. S.C.: shuttle or surge car; B.C.: section belt convey; M.B.C.: mobile belt conveyor; AFC: armored face conveyor.
2. crew numbers include a foreman

The trial of extensible belt conveyor for coal
handling between the head entry T-junction and the
entry belt feeder was a failure because the conveyor
broke down frequently. Other attempts at continuous
coal haulage included an armored face conveyor laid
parallel to the face but in front of the dozer beams
or the shortwall supports, very much similar to those
in longwall faces. Coal was loaded onto the conveyor
and conveyed to the head entry T-junction where it was
unloaded on the floor. A loader picked up the coal
continuously and loaded on the feeder of the sliding
monorail. A sliding monorail is a belt conveyor suspended from a rail fixed in the midspan of the head
entry (Figure 4). It can travel on the rail to follow
the loader. The monorail worked fairly well when it
was used in Mine No. 11, but the chain links on the
armored face conveyor broke down so often that it was
removed after approximately 400 ft of face advance. A
mechanical worm, which is a flexible conveyor, had
also been tried as continuous haulage system between
the monorail and the moving continuous miner. It was
unable to follow the miner continuously because it
always bumped into the rib in following the face
curvature at the T-junction.

Figure 4. The feeders end of the sliding monorail as
viewed from the headentry T-junction. The
self-powered feeder is used to move the
monorail.

FACE POWERED SUPPORT

Two types of powered supports are used in shortwall
faces (Table III). The six-leg frame type supports
are solely supplied by Westfalia/Mining Progress, Inc.
The four-leg chock type supports are supplied by
either Hemscheidt America or Joy/Gullick-Dobson.

The frame design provides larger support yield
capacity ranging from 720 to 1050 tons vs 500 to 800
tons for chock supports. However, the setting loads,
which are the initial loads exerted by the hydraulic
legs against the roof, are only a small fraction (from
20% to 40%) of the yield load capacity. This indicates an unsatisfactory state-of-the-art in determining optimum support capacity because in a properly
designed powered support the yield-to-setting-load
ratio should not exceed 1:3. When we conducted an extensive underground investigation, we found that the

Figure 5. Shortwall face at the head entry T-junction
area. A good portion of the roof is unsupported, although a screwjack was set up
along the rib. A continuous miner cannot
make a 90-degree turn at the beginning of
the web cut. Instead, it negotiates a
curvature and enlarges the entry roof span.

shortwall chocks in the Pittsburgh seam experienced a
maximum mean load density of 4 tons per sq ft, which
is equivalent to approximately 200 tons in a shortwall
chock. During the same investigation, we also found
that there were always a few contact points rather
than a full surface contact between the canopy and the
roofline. Since lack of full contact is caused by the
uneven cutting surfaces left by the continuous miner,
some sort of powered valve is needed to ensure a
consistent and full setting load--especially in view
of the fact that the current criterion of setting a
chock is by "seeing" and "hearing" it, e.g., a chock
is properly set when the canopy squeezes into the roof
and makes noises. Furthermore, face supports cannot
be advanced immediately following a continuous miner's
cutting, but must wait until the miner cuts a distance
more than two support widths (usually more than 8 ft
wide) before advancing to support the wide roof undercut by the miner.

For supports with similar capacity, a frame type is
cheaper than a chock, but the roof coverage at the
face for a frame is approximately 55% compared to 70%
for a chock. (In chock supported faces, therefore,
less rock debris falls between the support units and
from the gob.) Since frame supports have a smaller
base and canopy in terms of contact areas with the
floor and roof, a higher pressure tends to localize
at the tip of the base and at the rear edge of the
canopy. A high tip pressure at the base increases the
probability of digging into the floor, and this is
detrimental to support advance. This happened to Mine
No. 3 and was cited as one of the reasons for discontinuing shortwall mining.

PRODUCTIVITY AND PRODUCTION COST

Average production rate for each mine (see Table
II) ranges from 200 to 490 tons per shift or 20 to 57
tons per face man shift. Run-of-mine coal rejects

TABLE III

CHARACTERISTICS OF SHORTWALL FACE POWERED SUPPORTS

	Mine Names and Locations	Manufacturers	Types	Yield Capacity (psi)	Setting Pressure (psi)	Nos. of Support	Support Spacing
1	Helen Mine Homer City, PA	Westfalia	6-leg frame	700	3000	40	59
2	West Virginia	Westfalia Gullick-Dobson	6-leg frame 4-leg frame	? 800	? 3000	24 42	102 57
3	West Virginia	Westfalia	6-leg frame	1050	2500	24	100
4	Delta Mine Stoytown, WV	Hemscheidt	4-leg chock	700	5000	42	51.5
5	Beth-Elkhorn Jenkins, KY	Gullick-Dobson	4-leg chock	500	2400	42	48½
6	West Virginia	Hemscheidt	4-leg chock	600	4125	33	60
7	WVSM & RA Julian, WV	Hemscheidt	4-leg chock	660	5000	42	59
8	Tiller Mine Jewel Valley, VA	Westfalia	6-leg frame	700	?	32	60
9	Jewel Ridge #12 Whitewood, VA	Gullick-Dobson	4-leg chock	700	2400	47	51
10	Mathis Mine Bethlehem, PA	Gullick-Dobson	4-leg chock	650	2500	43	54
11	West Virginia	Gullick-Dobson	4-leg chock	500	2000	35	48

from shortwall panels are less than those from room-and-pillar sections. In the Pittsburgh seam, a one-third reduction in reject was not uncommon. If the roof condition is good, a production of 500–600 tons per shift can easily be attained.

There are many factors that affect the face production. For example, the senior author witnessed the two last production shifts before the summer's two-week vacation in the Mine No. 11. The crew members in each shift were so eager to meet the production rate set by the shift foreman (and then leave for vacation) that one miner's pass which cut approximately 250 tons was completed in less than two hours. Crew experience is also a positive factor. Since there are fewer equipment items in a shortwall face than in a longwall operation, the break-in period is much faster--and as crew experience increases, so does the production rate. This was definitely observed in the Mine No. 11. The average production rate for the first 600 ft of face advance was 100 tons/shift. It increased to 130 tons/shift for the following 300 ft and further increased to 150 tons/shift for the last 300 ft.

In their report entitled "Comparative Cost of Shortwall and Room-and-Pillar Mining Operation" (U.S. Bureau of Mines, IC 8757, October, 1977), Green and Palowitch made a comparative cost study based on the data obtained from the six shortwall demonstration panels in Hendrix No. 22 Mine, near Jenkins (KY), and from the room-and-pillar sections of the same mine. The average production was 975 tons/day (two shifts) from the shortwall section, and 905 tons/day from the

room-and-pillar section. The direct cost for both sections was approximately the same at $4/ton. Obviously a production increase of at least 75 tons/day or 3.75 tons output per man shift (O.M.S.) was needed to recoup the extra capital investment on the powered supports. Though the demonstration shortwall panels showed no significant economic benefit over the room-and-pillar mining, they did increase overall coal recovery and provided improved working conditions in terms of crew's health and safety.

CONCLUSIONS

The foremost requirement for a successful shortwall mining is a moderately strong immediate roof. It should be able to self-support to some extent, but it should also cave immediately behind the gob edges of the powered supports. A successful shortwall operation also requires a relatively hard and even floor. A good roof and floor, however, cannot guarantee a highly productive shortwall panel if continuous haulage cannot be achieved. Sliding monorail works fairly well for handling coal transportation between the head entry T-junction and the entry belt conveyor. But what's needed is a reliable and truly "continuous" face haulage system.

V. Research and Demonstrations

Joseph J. Yancik
National Coal Association
Howard E. Rutherford
Monterey Coal Company

Chapter 26

STEEP SEAM LONGWALL

David W. Wisecarver

Research Supervisor
U.S. Bureau of Mines
Denver Research Center
Denver, Colorado

James K. Greenlee

Director of Underground Mining
Western Associated Coal Corp.
Denver, Colorado

INTRODUCTION

It is estimated conservatively that some 14 billion tonnes of coal reserves in the United States exist in beds considered steeply dipping, i.e. at pitches or slopes in excess of 15° – a slope too steep for practical use of mobile wheel or crawler drive mining equipment. Past attempts to mine these steep reserves are limited owing to production difficulties and high costs, an expected result when mining of slopes must be done conventionally and mining on strike with mobile mining equipment is restricted because of limited cross sectional opening and operating room.

In Europe, these adverse conditions are countered through the use of the longwall mining method. In fact, longwall mining is employed along pitches up to 55°. However, both operating and administrative conditions in foreign steep seam coal mines differ from ours in that (1) the mine does not necessarily have to make a profit and (2) single entry development of retreat mining panels is acceptable practice. In the United States development can be done with a minimum of two entries only, provided belt haulage is not used during development.

The Department of Energy is cooperating with Snowmass Coal Company near Carbondale, Colorado to introduce and demonstrate longwall mining of steeply dipping coal seams in the United States. Plans call for some 2.5 million tonnes of coal to be mined from three panels. Installation of the first face is scheduled for February 1981. Project duration is expected to be about five years during which time the mining operation and progress will be monitored to determine costs, recovery, production, safety requirements, engineering and operational guidelines, and dynamic and static loading of longwall panels and chain pillars. Mining data in these areas will be obtained continually on a regular and routine basis, reported monthly and incorporated in a final report of the demonstration to be released by the Department of Energy.

MINING CONDITIONS

Mining will be in the "A" seam within the Bowie Member of the Williams Fork formation. The "A" seam at the demonstration site averages about 2.1 meters in thickness, strikes at $S10^{\circ}E$ and dips twenty-seven to thirty-three degrees to the west. The seam is underlain by a competent sandstone (Rollins) and overlain by a strong interbedded sandstone-shale. Thus a non-heaving floor and a relatively strong top should maximize entry stability yet permit good caving characteristics behind the longwall supports. Previous longwall experiements and pillar mining have shown good caving characteristics.

FIGURE 1. Snowmass Mine Site

The surface topography is rocky and mountainous. Overburden at the first panel varies from about 152 meters at the main slopes to 533 meters near mid panel and down again to about 107 meters at the starting room. Each successive panel, considering two-entry development, 24.4 meters wide chain pillars and 122 meters face length, goes downslope a distance of about 153 meters. At 30° pitch, each panel then picks up an average 76.2 meters of cover. Thus the third panel will be under 686 meters of cover at the maximum point.

Plans call for the retreat longwall mining of three adjacent panels with successive lengths of 1646 meters, 2012 meters and 2225 meters.

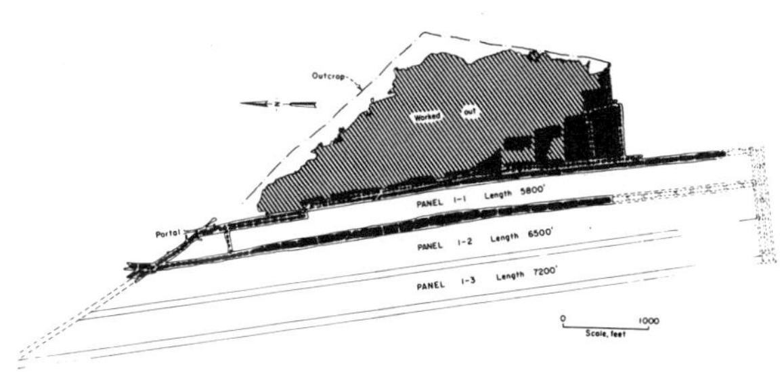

FIGURE 2. Proposed Panels – Steep Seam
Longwall Demonstration

FACE EQUIPMENT

The demonstration at Snowmass will utilize longwall face mining equipment similar to that in other
longwalls but will feature a support advance system
unique to United States longwalls. Yet another
feature will be a safety winch with its wire rope
attached to the shearer to provide back up to the
shearer braking system and chainless haulage.
Following is a general description of the equipment
to be employed along with major specifications.

Supports

The Hemscheidt Troika (Troika - a German term for
group-of-three) system featuring two-leg shield-type
supports will be used. Hook up and operation of the
supports is in sets of three. The supports are not
linked or attached firmly to the face conveyor as in
other setups and thus do not use the face conveyor as
an anchor to enable forward pull of the support.
Instead, a large beam (Fig. 3) is the vehicle
whereby the supports may be advanced.

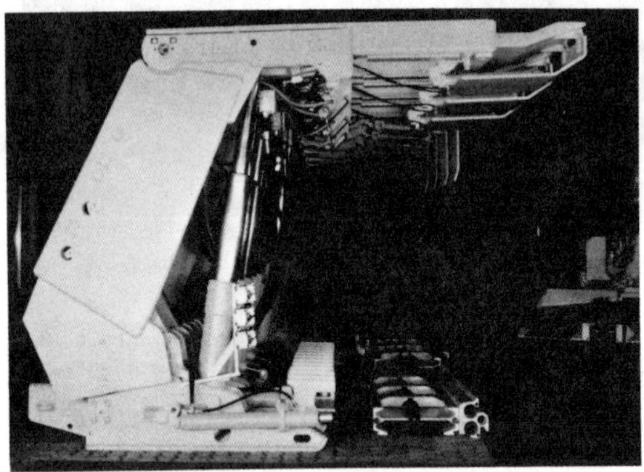

FIGURE 3.   Snowmass Troika Supports Showing
Beam and Conveyor Push Ram.

Support advance is accomplished by pulling against,
or pushing, the Troika beam. The sequence of advance-
ing each set of three supports is center support
first, lower support second and upper support third.
The two outside supports push the beam and the lowered
center support forward. The center support is then
set against the roof and when set, serves to anchor
the beam as the lower support is pulled forward and
set. Now the center and lower support anchor the beam
such as to permit the upper support to be pulled
forward. This procedure is repeated for each Troika.
The system uses adjacent controls. The hydraulics
of each Troika are controlled from the lower support
of the adjacent upslope unit.

Each shield support has a leg capacity (yield) of
159 tonnes per leg, or 318 tonnes per support. Setting
load is 254 tonnes and support load density is
718   kilopascals.   The support has contact advance
capability. Height range is from 1.5 meters closed
to 3.4 meters at maximum. Advancing ram power is
7.25 tonnes thrust on push and 32 tonnes on pull.
System hydraulic power is obtained from a Hauhinco
unit supplying 87 liters/min. @ 32.6   megapascal.

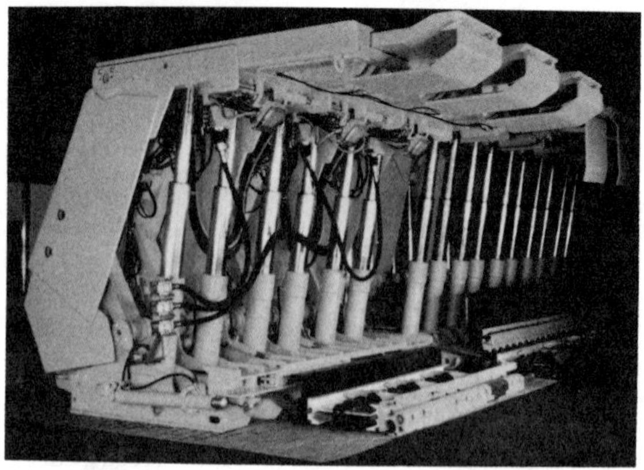

FIGURE 4.   Troika - Oblique View

Face Conveyor

A 635 tonnes/hour, 0.76 meter-wide face conveyor
is supplied by Dowty Meco. The unit has headgate
and tailgate drives at 112 kilo watts each.

A twin outboard chain, in lieu of single or twin
inboard types, was chosen for this conveyor to
provide a more rigid flight bar for better resistance
to tumbling coal and better control of coal as it is
transferred to the stage loader.

A special modification was made to the spill plates
to increase height and provide improved downslope
protection to face miners and longwall support
equipment. The standard 800mm high spill plate
height would leave a 600 - 900mm opening between the
top of the plate and the support canopy at the current
2.1 meter seam height; an opening through which coal
lumps, tumbling at high speed, could pass and wreak
havoc on miners, support legs and hydraulics. This
problem was corrected by increasing the basic spill
plate height to 1400mm and adding an adjustable
height (700mm range) safety rail. This addition nearly
scals off the manway and support side of the face
conveyor from the coal face. Each safety rail is
hinged at both ends and may be swung inward toward
the gob using either upslope or downslope hinges, as
the miner chooses.

The conveyor is fitted with push plates against
which a Troika ram makes contact in advancing the
conveyor. One-hundred twenty-tonne-capacity
connecting links provide additional conveyor pan
coupling strength.

Stage Loader

The stage loader is also supplied by Dowty Meco.
Capacity is 770 tonnes/hr. Chain type is outboard
twin strand with chain diameter at 18mm. A 75 hp.
motor powers the unit. A lump breaker is mounted on
the panline at the pan-stage-loader transfer point.

Shearer

Anderson-Strathclyde makes the 500 hp. double-ended
ranging drum shearer selected for the steep seam
demonstration. The Roll Rack chainless haulage system
will be utilized. Drum size is 1.5 meters x 0.75
meters. Bit blocks were installed with double wrap
per manufacturers recommendation. Pick flushing and
sprays help control respirable dust.

The shearer will be equipped with a lump breaker mounted on the upslope side. Operation will be by remote radio control enabling the operator to stay slightly upslope of the shearer and under support canopies.

The shearer has a 36.2 tonne pull capacity and has been tested at Bretby to verify Roll Rack strength and adequacy of pull power for the 30° slope.

Safety Winch

An automatic safety winch is an additional facility added to this mechanized face. It provides a backup and assist to the shearer haulage system. The possibility of failure of the chainless haulage system resulting in a runaway shearer is remote. However, it was felt that a safety winch is a necessary precaution for this demonstration.

Along with the added safety, the winch can assist the shearer haulage upslope (up to 13.6 tonnes capacity) and too, provide an automatic parking brake in the event of electrical power failure. A braking resistance of 6.8 tonnes may be realized.

Lighting and Communication

Victor lights will be used to illuminate the longwall face. Density will be one light per support with an extra light on every other support. Davis-Derby supplies the communication equipment.

PANEL DEVELOPMENT

Panels will be developed for retreat mining using two entries driven slightly updip in the seam to establish grade at about 1.5 percent and hence overall drainage. Plans call for entries on 29 meter centers with chain pillars 61 meters long by 24 meters wide. A MSHA variance to permit mining entries 61 meters beyond the last open crosscut was granted based on use of auxiliary fans to maintain adequate face ventilation and too, the realization that the variance would permit reducing upslope crosscut mining by one-half. It is the mining of crosscuts which is particularly difficult and hazardous. Three-meter wide crosscuts are mined conventionally with blasted coal pushed and shoveled by hand down to the lower entry.

Both headgate and tailgate entries had to be mined in development of the initial longwall panel. In this case, approximately 1370 meters of an old pillar mining haulageway was rehabilitated for use as a second entry and a permanent air return. (Figure 2.)

Because of the 30° pitch of crosscuts, equipment cannot be transferred back and forth between adjacent entries as may be done in flat lying seams where a single complement of equipment consisting of miner, bolter and shuttle cars may be used in multiple entry development. At Snowmass, each entry must be equipped with a miner and haulage and bolting equipment. F6A Alpine miners are used to cut coal. Rail haulage in the lower of each set of double entries is used to get coal from development faces to main slopes and the main belt. In the lower of these entries coal from the F6A Alpine is loaded onto a monorail belt conveyor from which 4.5 tonne coal cars are loaded. In the upper entry, coal is hauled away from the miner by shuttle car and to an outby crosscut

where coal is dumped into a metal transfer chute which guides coal to cars spotted at the chute discharge. At slopes of 30° and greater, coal will run down the chute freely but where less than 30°, miners must help work it downslope.

Roof bolting consists of 1.2 meter bolts on 1.5 meter (slope distance) centers. Holes are drilled with stopers from under temporarily supported roof. Bolts are tightened and torqued with air powered impact hammers.

Entries are driven approximately 4.25 meters wide. The upslope entry height averages about 3 meters - a combination of the 2.4-meter vertical distance along the rib and a 0.6-meter deep cut into the sandstone floor. The downslope side is about 1.2 meters high, the diminished height resulting from convergence of steeply pitching top to the horizontally mined coal bottom.

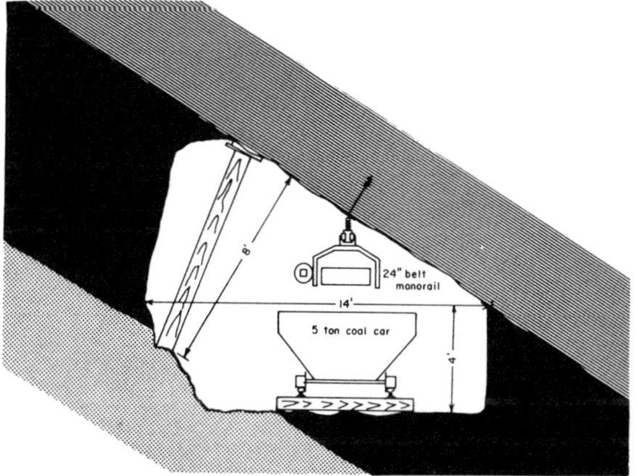

FIGURE 5. Typical Gate Entry Cross Section

A critical dimension in entry development is the 0.6-meter deep cut in the upslope floor of the top headgate entry. It is this cut that determines the position of the stage loader relative to the armoured face conveyor at the transfer point. If the cut is too shallow, the stage loader rides too high and the face conveyor cannot dump into it. If the cut is too deep, the stage loader rides too low and the conveyor can throw coal over it. The problem with too deep a cut can be remedied by blocking up the stage loader but this is time consuming and an unneeded expense. A $\pm$ 250-millimeter tolerance to the basic 0.6-meter depth is allowed.

Another critical dimension in the development of a retreating longwall panel by multiple entries is the width of chain pillars. It follows that using the maximum allowable pillar length along entries reduces the number of steep crosscuts and crosscut footage. It also follows that the narrower the pillar widths, the less crosscut footage to be mined. However, the chain pillars must be strong enough, hence wide enough, to preserve the stability of the lower headgate entry so it may be retained to serve as a tailgate in the next adjacent longwall panel.

An estimate of required widths to best satisfy these steep-seam chain-pillar requirements was obtained from a study done by US Bureau of Mines researchers. Analysis of dynamic and angular pillar loading during mining was made using the finite element modeling technique. The results of this study are illustrated in Figure 6. Per these results, a 25-meter-wide pillar would be required at the 535 meter overburden near the middle of the first panel. The work in this area has been published by the USBM.

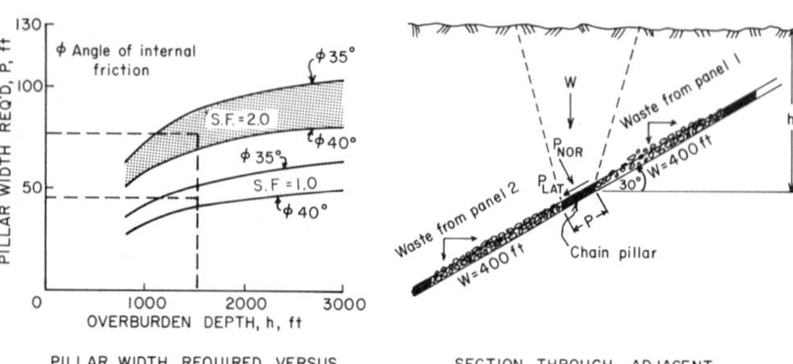

FIGURE 6. Design Curves for Entry Support Pillars for Retreating Longwalls in Steep Coal

A shear zone of severe and unexpected proportion was encountered during development of the first longwall panel. The approximately 36.5-meter-wide zone crosses Panel 1-1 as shown in Figure 7.

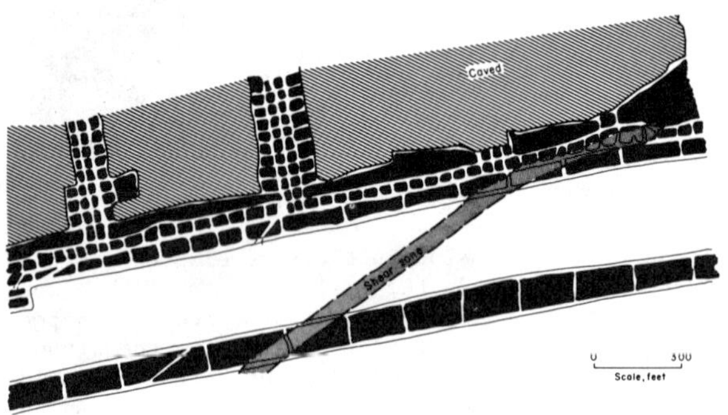

FIGURE 7. Strike of Shear Zone Across Panel 1-1

The zone consisted of a very weak and unconsolidated rock which was quick to fail in unsupported roof spans of two feet and more. Failure was often in the form of "runs" of material where maximum rock size did not exceed 500 millimeters (Figure 8). Advance through the zone in all entries was slow and costly requiring closely spaced supports and tight lagging. On 27 September, 1979, grouting of the shear zone in the upper headgate entry with polyurethane binder was initiated. Use of the polyurethane in this entry was an instant success and was continued in this entry until advance through the shear zone was completed. Earlier, the lower headgate entry had been driven through the same zone without polyurethane grout. Data obtained from this entry enabled engineers to make comparisons of this difficult mining process with and without grout. Briefly, using the polyurethane, (1) the daily advance was nearly doubled, (2) the cost of required artificial support was reduced some two thirds and (3) miner safety was improved.

The savings realized from the reduced support requirements were almost equally offset by the grout and packer cost, but the advance rate was about doubled and resulted in substantial savings in labor costs. The use of polyurethane was considered an unqualified success and the technique will be recommended strongly as an assist in future mining through similar zones.

FIGURE 8. Advancing Upper Headgate Entry Through Shear Zone

As is the case of all retreating longwall panels, it is important to maintain a constant face length throughout panel length to eliminate adding on or taking away supports and pan sections to adapt to changes in face length. At Snowmass that problem will be especially important because the Troika system groups shield supports in sets of three by beams. Adding or removing a shield support makes a set incomplete and eliminates using the beam. Without the beam the supports in the incomplete set must be modified for them to be advanced. Where steep seams vary in pitch, driving of headgate and tailgate entries on parallel lines of sight can result in appreciable changes in panel slope distance (face lengths). Using parallel lines of sight for driving headings, a face length of 122 meters at 30° pitch would be 126 meters at 33° pitch and 118 meters at 27° pitch. Thus care is taken to monitor entry floor elevations so as to determine pitch and up or down slope distance changes in entry center lines to hold a constant face length. Curvature in the headgate entries in panel 1-1 have been made to adapt to pitch changes and maintain a constant panel width.

ROCK MECHANICS PROGRAM

Along with the production, recovery, safety, and cost factors to be realized from this three panel demonstration, a comprehensive research program has been instituted to study the behavior and loading of the underground mine structure during mining along with the surface topography changes resulting from subsidence. A sub-contract has been let to researchers from the Colorado School of Mines who are charged to monitor surface subsidence and identify and quantify stress profiles (1) ahead of the advancing face, (2) in chain pillars and (3) downslope panels. In addition, CSM researchers also have the responsibility of accurately documenting and reporting details of roof falls, floor heaves, chain pillar failures and other ground control problems which might occur during mining. "Other ground control problems" might include the negotiation of the longwall face through fault zones, bad top, and severe variations in coal seam thicknesses.

Subsidence studies will be based on data obtained from level and triangulation surveys of a number of surface reference stations (called subsidence monuments) installed as a network over the three proposed longwall panels. Initially, subsidence monuments will be installed only along three lines normal to the long axis of the panels. Distance between monuments is about 24 meters and distance between parallel lines is about 500 meters. Subsidence information obtained from these perpendicular lines during the mining of the first longwall should determine the location of the maximum subsidence or "trough" along the long axis of the panel. This information will enable researchers to estimate the trough location for panels 1-2 and 1-3. Lines of subsidence monuments will then be installed along these predicted locations to provide best references for obtaining maximum subsidence data during mining of the respective panels.

Monuments immediately affected by the advancing face will be surveyed twice per week and all moving monuments will be surveyed at least once per week. Survey frequency for stabilized monuments will be gradually reduced to cursory check surveys as subsidence ceases.

Survey instruments will include a theodolite and laser distance measuring equipment. Data should be accurate to 30 millimeters.

The underground rock mechanics program is directed towards determining stress field variations resulting from mining, influence of sidehill slopes and rapid increase in depth on rock mass response from longwalling, behavior of strata in the immediate roof above the seam, and design factors for chain pillars on steep seam longwalls.

This information will be obtained from instrumentation installed in chain pillars and panel ribs at selected locations along headgate and tailgate entries.

A support loading study will be included in the underground program. The effects of sidehill loading, a previously caved panel, and width-to-depth ratios on support loading will be accomplished by monitoring hydraulic pressures and pressure differentials in support legs. Data for these determinations will be collected from continuous double pen recorders installed on selected shields.

EQUIPMENT INSTALLATION

The face supports will be installed by lowering from the tailgate with a twin drum winch and setting it up in a headgate to tailgate direction. This sequence will also be used in installing the armoured conveyor.

At this writing, it is not known if the shearer can be taken into the starting room in one piece or if it must be partially disassembled.

OPERATIONAL SEQUENCE

The selection of a face mining or operational sequence for this steep seam longwalling was made after consideration of the need to minimize falling lumps in the open face area, minimize the down pitch shift of support and/or conveyor and minimize the support log time for newly exposed roof.

The system selected is the pseudo one-web-back. In this system, coal is cut down pitch only. At start of the cut, supports are positioned at the panline. As the downslope cut is made, rigid forepoles are extended immediately behind the passing shearer. This supports the new roof immediately. The panline and support position remain the same.

Upon completion of the cut, a cleaning pass is made up pitch. The conveyor is advanced behind the shearer as it advances upslope, followed by retraction of support forepoles and advancing of supports.

Where needed, the support system, along with the face conveyor can be steered upslope. This is done by turning the Troika system a maximum of 5° to the face conveyor and moving the supports upslope on successive cuts until desired position is reached. The face conveyor is steered along with the supports by means of the support-to-conveyor anchor stations.

CONCLUSION

The proposed longwall mining at the Snowmass mine is the beginning of development of steep seam longwall technology in the United States. It is believed that much will be learned from this demonstration about engineering, costs, maintenance, equipment, face mining techniques and safety. At the conclusion of the demonstration, a report detailing the history, findings, and results of the work will be published.

Chapter 27

TWO LIFT LONGWALL MINING METHOD FOR MID-CONTINENT RESOURCES, INC.
28-FOOT THICK SEAM (JOINT DOE/MCR PROJECT)

by

Jasinder S. Jaspal
Mining Engineer
U.S. Department of Energy
Pittsburgh, Pennsylvania

John A. Reeves, Jr.
Manager, Longwall Operations
Mid-Continent Resources, Inc.
Carbondale, Colorado

Bradley J. Bourquin
Chief Engineer
Mid-Continent Resources, Inc.
Carbondale, Colorado

INTRODUCTION

The United States has an abundance of coal reserves in thick seams. Most of these resources are in the West. Surface mining operations in thick seams are highly efficient, particularly in the West, but methods for mining thick seams underground are very limited. The percentage of resource recovery in deep mining of thick seams is often as low as 30 percent. It is estimated that there are 45 billion tons of coal reserves for underground mining in seams greater than 10-ft thick in the western states alone. With such a large available reserve base and low rates of recovery, the need to develop the technology to mine these thick reserves is quite evident.

As the definition of a thick seam varies widely from country to country, a thick seam is defined here as any seam where the height is such that it is not minable in a single pass with available mining equipment.

The development of efficient thick seam mining methods has proved to be difficult because of ground control, ventilation, gas, and a host of other mining problems; hence, evolution of methods for their exploitation has been slow. But, with the rapid development of hydraulically-powered longwall roof supports during the past 25 years and the corresponding advancement in the mechanization of equipment used for coal getting, loading, and transporting techniques for mining thick seams have experienced dramatic improvement.

Thick seams in many countries, which were considered impossible to mine previously, now are worked successfully with the new mining equipment. The most common method of mining thick seams is the multilift (also called slicing) method, which consists essentially of extracting the seam in two or more lifts in descending or ascending order. The current trend favors working the descending mining system, leaving an artificial or natural roof partition between each lift.

In view of the need to extract national resources more efficiently and conservatively, the Department of Energy entered into a cost-sharing agreement with Mid-Continent Resources, Inc. to demonstrate the multilift longwall mining in the 28-ft thick Coal Basin seam.

This paper describes the proposed two lift mining method at Mid-Continent's L.S. Wood No. 3 mine and discusses the changes in mining practices and the special conditions encountered at the mine. It also reviews briefly the geology and the mining background of the mine.

L. S. WOOD NO. 3 MINE

The L.S. Wood No. 3 mine is located on the western slope of the Rocky Mountains near the village of Redstone in Pitkin County, Colo. Geologically, this mine is on the flanks of a plunging, anticlined structure that was formed by laccolithic igneous intrusion. The stresses that produced the anticlinal structure, and the heat from the igneous intrusions were responsible for metamorphosing the coal to its present midvolatile rank.

Mid-Continent mines two seams--the Dutch Creek and the Coal Basin. Mining at the L.S. Wood No. 3 mine is in Coal Basin seam. The mine's portal and lamphouse are at an elevation of 10,000 ft, and workings currently extend to below 3,000 ft of cover. It is one of the highest elevation coal mines of the world, and one of the deepest mines in North America. Although coal was first mined in this area nearly 100 years ago, Mid-Continent started its operations in 1956 after the field had remained idle for nearly 50 years. This mine was developed by driving the main slopes on true dip of 12 to 13 degrees. Entries then were turned left and right on approximately 800 ft centers for room and pillar extraction. As the mining progressed down dip, the depth of cover increased and mining conditions became increasingly difficult. Under 1,000 ft of cover, floor heave became a serious problem, which at places required taking up the bottom three or four times to maintain adequate heights. As mining progressed, the cover approached 2,000 ft. Mountain bumps and outbursts in the development faces were common. This problem was further exacerbated by the fact that Coal Basin seam has a high methane content, and the pressure of this gas in the fracture system in the coal, in conjunction with the overburden pressure, created conditions under which coal faces could not be mined efficiently. As a result, management concluded that room and pillar methods could not work successfully under a cover in excess of 2,000 ft and that alternative mining methods had to be evaluated.

In their quest for a suitable alternative mining method, Mid-Continent engaged the services of the National Coal Board of Great Britain to survey their mining operations. The National Board subsequently recommended the adoption of advancing longwall mining for the following three reasons:

1. The advancing faces would require a minimum amount of narrow heading development;

2. The longwall faces would be stress-relieving because the unconfined area could expand elastically; and

3. Methane emissions would be less sporadic since the slower rate of face advance would allow better drainage.

Accordingly, Mid-Continent decided to install their first longwall face (No. 101) in the Dutch Creek No. 1 mine in May 1975 and augment the ventilation with a methane drainage system. A second advancing longwall of similar fashion was started in their L.S. Wood No. 3 mine in August 1978.

TWO LIFT LONGWALL MINING METHOD

The Department of Energy and Mid-Continent Resources, Inc. entered into a cost-sharing agreement in September 1979 for the demonstration of a multilift longwall mining method; at that time, Mid-Continent had two advancing longwall faces already in operation. The project includes mining seven panels--four upper lift panels and three lower lift panels (see fig. 1). The upper lift panels will be mined along the roof-coalbed interface on advance, and the lower lift panels will be mined along the coalbed-floor interface on retreat, leaving a partition of coal about 5 feet thick in between the two lifts. Because the panels are to be mined along the roof and floor interfaces, the horizontal control for each lift would be easy to maintain. An ongoing advancing longwall panel has been incorporated in the multilift demonstration.

The seam's inclination in the demonstration area is 10 degrees. The headgate panel will be located at the dipping end, and the tailgate will be located at the top end. Both the headgates and tailgates will be driven on the strike of the seam.

The upper lift panel will be extracted first. After a waiting period of two to three years for the gob to consolidate, the lower lift development gateroads will be driven the full length of the panel in the destressed zone underneath the upper lift panel, and then the lower panel will be mined on retreat. In the advancing (upper) panel, there will be three roadways--two roadways located at the headgate (a separate roadway is required for the conveyor belt) and one located at tailgate. The gate roadways at the upper lift (see fig. 2) are formed by constructing 8-ft-wide packwalls as the longwall face advances and do not require any prior development work. The roof of these roadways will be supported according to the approved roof support plan. The lower lift development roadways will be driven in a typical dual-entry configuration. The entries to the lower lift development roadways will be ramped down from the upper panel entries (see fig. 3) after which the dual-entry drivage

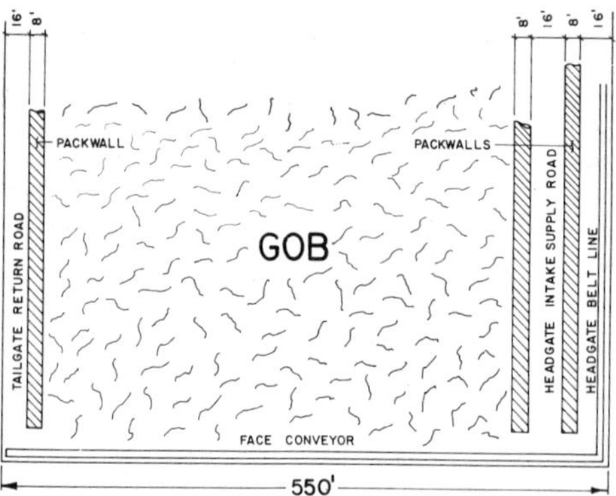

Fig. 2. Plan view of an advancing upper lift longwall panel.

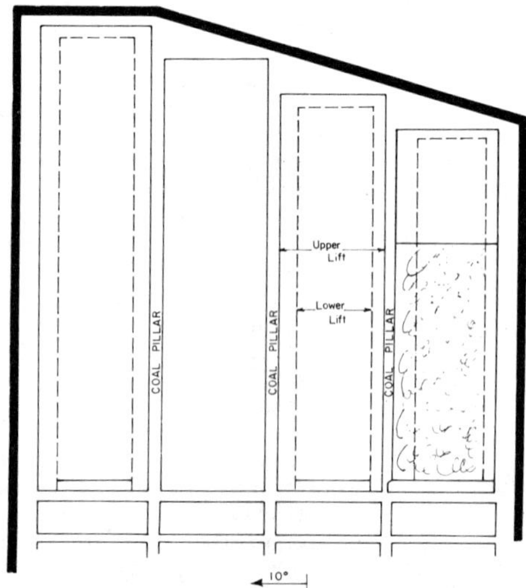

Fig. 1. Plan view of the proposed four upper lift and three lower lift panels.

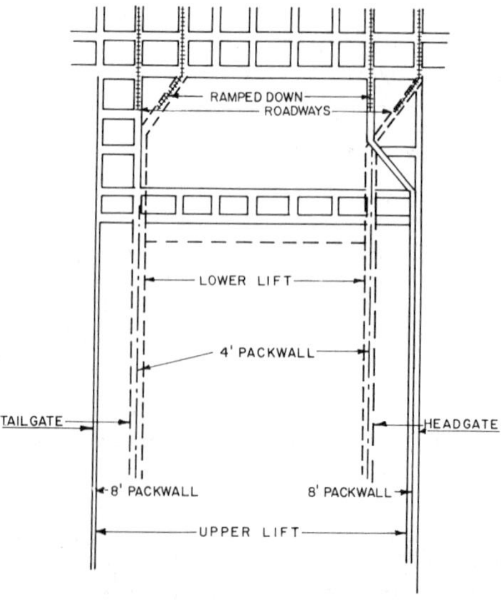

Fig. 3. Plan view of the rampded-down dual entry drivages, lower lift, and upper lift panels.

will be started. The dual entry will be created by mining a single place 26-ft wide and dividing it in the middle with a 4-ft-thick, packwall-filled continuous crib, thus creating two 11-ft-wide compartments (see fig. 4). The lower (retreat) longwalls will extend across the inside compartments of the headgate and tailgate roadways. The lower lift gate roadways will be positioned 65 ft from the rib side of the top panel packwall, to the gob side of the bottom panel roadways. Thus, 42 ft of pillar will be left between the upper lift and lower lift roadways so that the mining of the lower panel will not disturb the upper lift gate roadways. The lower lift panels are 100 ft shorter at each end than those of the upper lift and are totally within the upper lift panel. The upper lift panels will be separated by protective barrier pillars 100 ft wide.

The construction of packwalls and the use of an adequate methane drainage system are critical to the success of this demonstration. The methods of constructing the packwalls and the methane drainage system projected for use in the multilift demonstration will be similar to those currently being employed in the present advancing longwall panel.

The packwalls are constructed with an aggregate composed of minus 1/4-inch crushed washer-reject and Type 1 Portland cement. The crushed reject is transported underground in 10-ton cars to a slurry-pumping station bunker where, prior to its pumping, it is mixed with 2 percent gelatin bentonite and water and is slurried through 4-inch pipe to the packwall forms (see fig. 5). The binding agent (Type I Portland Cement) is pressure injected into the slurry pipe 3 ft from the end pipeline by means of an air pump. The mixture is poured into the packwall forms, that are constructed of 1-1/8-in-thick plywood cross-tied with 5/8-inch-diameter steel rods on approximately 3-ft centers. Each 10-ft-long form holds 20 cu ft of material and takes about 50 minutes to fill. Three to four hours are required to build each 10-ft-long packwall.

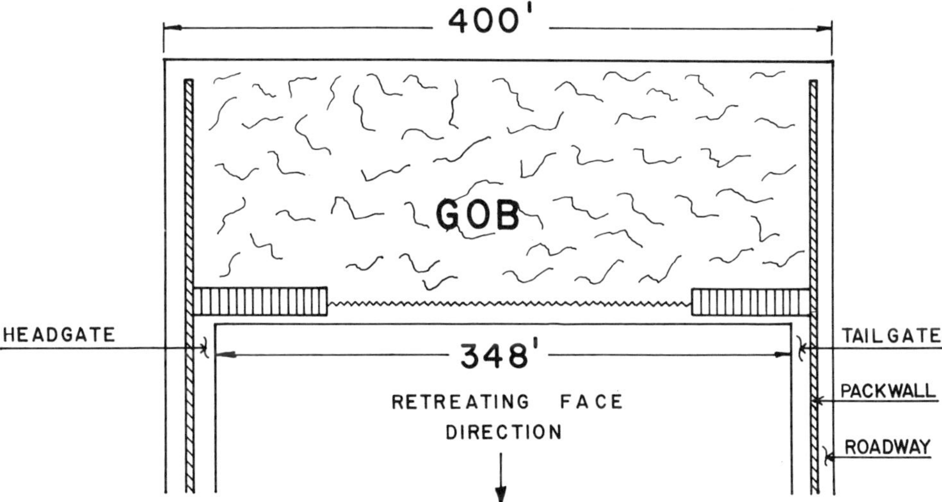

Fig. 4. Plan view of retreating lower lift panel.

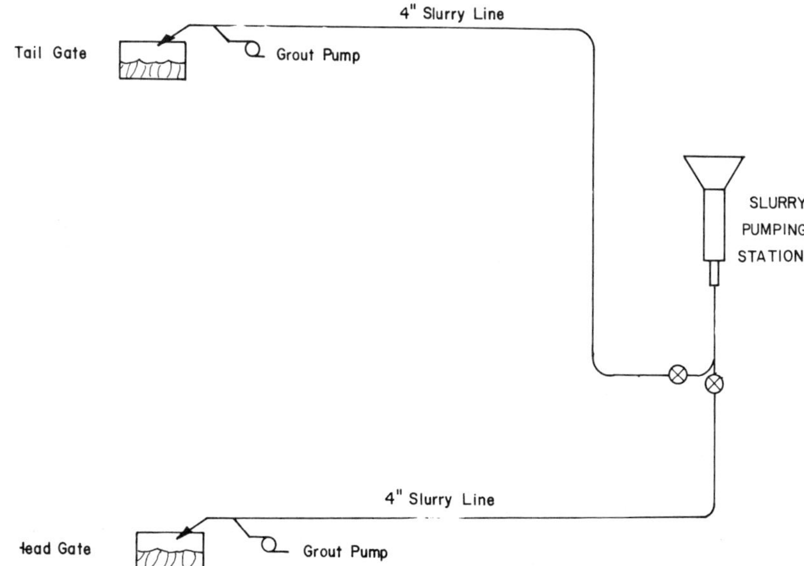

Fig. 5. Illustration of pump rack system to pump slurry into the packwall forms located at the longwall face.

The methane drainage system consists essentially of drilling a series of boreholes at intervals of not more than 75 ft along the advancing tailgate entry. The boreholes are drilled 160-ft deep and inclined at 60 degrees (see fig. 6). The first 60 ft of each borehole is sealed with a standpipe. The spacing of holes at 75 ft has been found effective and efficient for the drainage system. The last inby hole is always kept within 125 ft of the face. Presently, three 702 Nash vacuum pumps connected in parallel are located on the surface to provide the required vacuum to operate the system. The methane drainage pipeline from the pump station to the start of the longwall panel is 10 inches in diameter; the remaining line to the face is 8 inches in diameter (see fig. 7). The gas boreholes are connected to this 8-in-diameter pipeline. The system is monitored continuously by a 0 to 100 percent methanometer located at the surface station. The methane drainage system will be installed in the upper lift panels and will remain operational during the mining of the lower lift panels. It is expected that all the methane gas will be captured during the mining of upper lift panels.

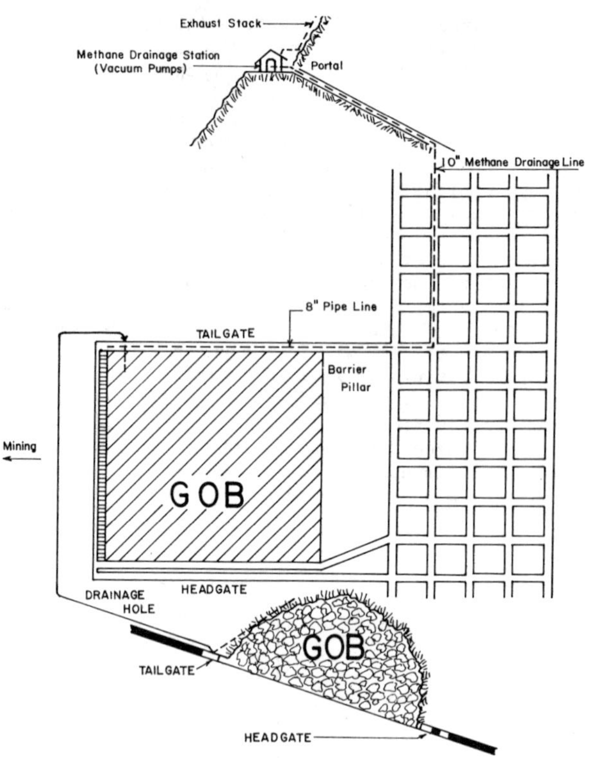

Fig. 7. Illustration of methane drainage system at advancing upper lift panel.

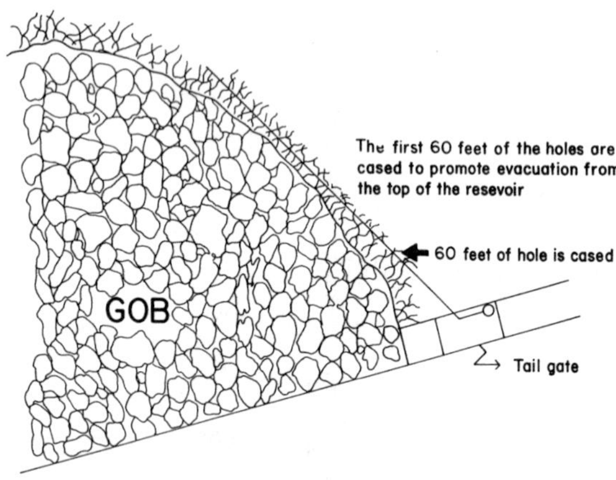

Fig. 6. End view of installed methane drainage pipe in the upper lift.

It is anticipated that this demonstration program will take about 10 years to complete. Two sets of longwall equipment will be needed. The Department of Energy portion of the cost-sharing agreement is the furnishing of one set of longwall equipment; the other set will be provided by Mid-Continent. The Government's equipment will be utilized in the extraction of one upper lift and all three of the lower lift panels. Mid-Continent's equipment will be used in the mining of three upper lift panels.

The mining technique and the design of the panels used in this project are quite similar to the Daw Mill and Newdigate mines in the United Kingdom, where the National Coal Board has mined seams of about the same thickness as the Coal Basin seam quite successfully.

The project is of special significance in that it is the first time a 28-ft-thick seam will be mined underground in the United States. It is anticipated that the two-lift mining method will increase the resource recovery of L.S. Wood No. 3 mine to at least 70 percent.

CONCLUSION

1. Although Mid-Continent's coal is a high quality metallurgical coal and commands a very good price in the metallurgical coal market, the construction cost of the packwalls and the methane drainage system, combined with the low coalbed recovery, adds significantly to the cost of mining coal by the advancing longwall methods.

2. The mining of several panels has proved that the advancing longwall mining system with packwalls can be worked successfully.

3. This demonstration is based on proven methods and, therefore, it is fair to anticipate that the two-lift mining method should be successful.

4. The Government-funded equipment has many unique features, including the ability to extend the cantilevers of the canopy under full load.

5. This demonstration, if successful, can provide an underground mining technology for working thick seams safely and with high coalbed recovery; this will contribute significantly to the development of more efficient systems in the United States.

REFERENCES

1. Bise, C. J. and R. V. Ramani. "An Evaluation of
   Underground Mining Technology for Western Thick
   Coal Seams." Pa. State Univ., SME Meeting,
   September 1975.

2. Anonymous. "Mining Plan." Phase 1 Task 2 Cooperative Agreement DE-FC-01-79ET10039, U.S. DOE and
   Mid-Continent Resources, Inc.

3. Reeves, J. A. "Encouraging Experiences and
   Future Commitment to Longwall Mining at Mid-
   Continent Resources." First Int. Symp. on
   Thick and Steep Seam Coal Mining, London, England,
   May 1980.

4. Ghose, A. K. and R. D. Singh. "Thick Seam Mining
   Methods--A Global Review." Proc. of Int. Symp.
   on Thick Seam Mining. Dhaubad, India, 1977.

Chapter 28

THE EFFECT OF MINING WIDER WEBS ON A LONGWALL FACE

Paul J. Guay - Senior Engineer
Jonathan Ludlow - Senior Engineer

Foster-Miller Associates, Inc.
350 Second Avenue
Waltham, Massachusetts   02154

ABSTRACT

Based on two studies that were funded by the U.S. Department of Energy, the authors provide an overview of the benefits expected from mining deeper webs on United States longwalls. The first study, completed in November 1979, identified the limits and potential gains from taking deeper webs with presently available equipment. The second on-going study discusses the conceptual design and benefits of a future system for taking very deep webs.

INTRODUCTION

The material presented in this paper is summarized from two research programs presently funded by the U.S. Department of Energy. The first study involved the potential for mining wider webs with presently available equipment and methods. This work was completed in November, 1979 and was published in two volumes (1). The study investigated the potential benefits of mining web depths up to 45 in. with conventional pan-mounted double ended ranging drum (DERD) shearers extracting 6.5 ft of coal in the Pittsburgh seam.

The second program involves the development of a very high production longwall system specifically designed to safely mine web depths beyond 45 in. This work involves the design, construction, and demonstration of such a system which is not expected to be completed until 1985.

Initial interest in the benefits of mining deeper webs was a result of the success experienced in England where the National Coal Board (NCB) began mining with 1 meter (39 in.) wide drums in 1974. The favorable results obtained in the Western Area of the NCB have been reported on at previous sessions of the American Mining Congress by Mr. Ray Hunter, the Director of the Western Area (2). Work in the United States originally began in 1976 under the auspices of the U.S. Bureau of Mines (USBM) as part of an overall program to promote the utilization of longwalling by maximizing its potential as a high production method of mining. This work has since been transferred to the DOE Division of Fossil Fuel Extraction.

At the present time, longwall mining represents less than 6 percent of the underground coal production in the United States. The reasons for the limited use of longwalling in this country appear to be the high cost of a longwall system, the time and cost of panel development and dust. Wide web studies were initiated with the expectation that the higher cost of equipment would be offset by substantial gains in productivity and that there would be a measurable reduction in dust per ton of coal produced on the face. However, it was also recognized that an increase in the rate of panel extraction would result in an additional strain on entry development and mine haulage. Solutions to

these problems were considered to be beyond the scope of effort structured for the wide web studies. Finally, based on the British experience, it is further expected that mining wider webs will also favorably impact both safety and working conditions on the face.

THE BRITISH EXPERIENCE

In the United Kingdom the depth of web mined is indirectly set by British regulations. These appear to have also established normal practices in the United States, as the British have historically been the principal influence on our use of the shearer. However, British practices, at least in some areas, appear to be changing in favor of mining wider webs.

Mining very deep webs is not new in Britain. In the early days of longwalling, 6 to 8 ft deep cuts were taken by machines like the Meco-Moore Miner shown in Figure 1. The roof was controlled by props and bars hand set behind the machine. To insure the safety of the miners, the British Inspectorate regulated the maximum spacing between props (prop density) and the maximum distance from the first line of props to the face (the prop-free front distance).

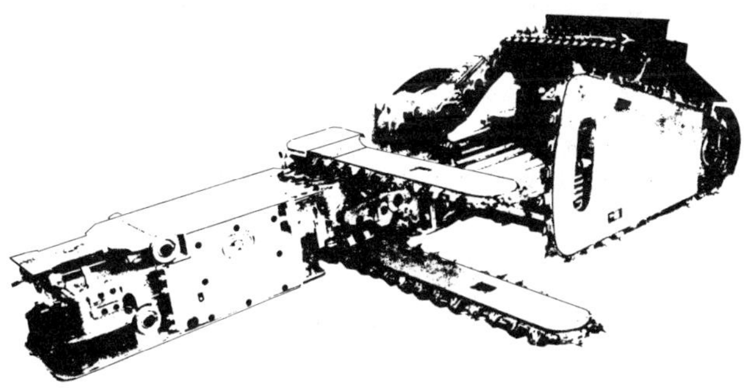

Fig. 1 - The Meco-Moore mechanized longwall miner

The introduction of the modern powered support obviated the need to regulate density. The prop-free front (PFF) regulation, however, could still be applied in principle and, therefore, remained in effect. As shown in Figure 2, the PFF distance must accommodate the width of the equipment, working clearances, and the width of the cutting drum (and, therefore, the depth of web).

In 1974 the standard NCB face was equipped with a 30-in. wide conveyor and a 200 hp DERD shearer set on a 6 ft, 6 in. PFF. However, this face could only accommodate a 21-in. wide drum, as shown in Table 1. As also shown, higher production faces set on a 6 ft, 9 in. PFF had been introduced in both 1970 and 1973.

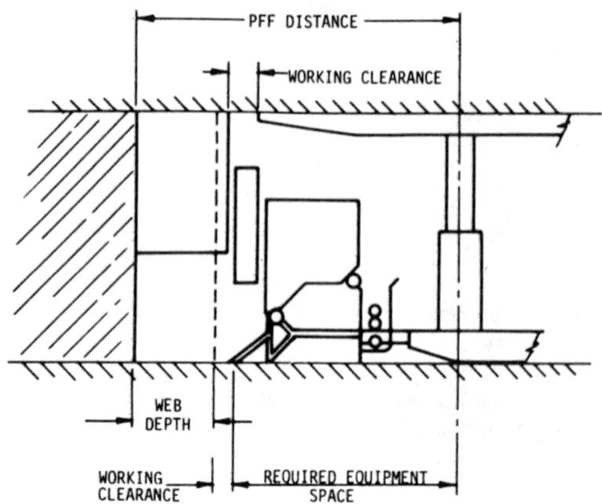

Fig. 2 – Effect of PFF regulation on web depth

To offset declining productivity, which had peaked in early 1970, Western Area of the NCB in early 1974 began studying techniques for mining 1 meter wide webs (39 in.) at the Holditch Colliery. Permission for the required 8 ft, 3 in. PFF was granted by the Inspectorate and mining commenced in November of 1974.

The initial success at Holditch was quickly followed by the placing of two 1-meter wide web faces at the Hem Heath Colliery in the same Area. The detailed results of these operations have been widely reported (3) (4) (5) (6) and can be summarized as follows:

a. There was an average increase in weekly production of 42 percent at Holditch and 28 percent on the first face at Hem Heath, as shown in Table 2.

b. There was no measurable increase in shearer power consumption per ton of coal mined at either site, confirming the results of preliminary surface trials.

c. A reported reduction in the rate of accidents, per 100,000 shifts, is shown in Table 3 (5).

d. Improved roof conditions credited to less stress cycling of the immediate roof strata.

e. Less dust on the face was reported but not conclusively confirmed by any known published results.

f. A shift in floor loading towards the face which caused some problems at Hem Heath where the floor was relatively soft.

g. Some problems with pan connectors due to the increased snaking distance.

The typical face arrangement used to mine the 1-meter wide webs at both Holditch and Hem Heath is shown in Figure 3. The supports were Gullick Dobson 240-ton, 6 leg chocks with extendable canopies (forepoling). The Anderson Mavor, 200-hp DERD shearers were equipped with 39-in. wide, 3-start spiral vane drums.

This face arrangement exemplifies the difference between British and American longwall equipment. Whereas the British extensively use 5 and 6 leg chocks of less than 200-ton capacity, Americans are presently using shields of 500 to 600-ton capacity.

TABLE 1. PROGRESSION OF LONGWALL FACE DEVELOPMENT IN THE UNITED KINGDOM WITH DERD SHEARERS

PFF	AFC	Shearer	Drum Width	Remarks
6 ft, 6 in.	30 in. twin 18 mm	200 hp DERD	21 in.	Standard face for DERD shearer starting in 1967
6 ft, 9 in.	30 in. twin 18 mm	200 hp DERD	24 in.	Face design initiated in 1970. Some 82 faces operating
6 ft, 9 in.	30 in. single 26 mm	200 hp DERD	24 in.	High production face introduced in 1973
7 ft, 3 in.	30 in. single 26 mm	200 hp DERD	27 in.	Latest design of high production face
8 ft, 3 in.	30 in. single 26 mm	200 hp DERD	39 in.	One meter web faces at Holditch and Hem Heath

TABLE 2. RELATIVE PERFORMANCE OF WIDE AND NARROW WEBS AT HOLDITCH AND HEM HEATH

Item	Holditch Colliery		First Face Hem Heath	
	Normal Web	Wide Web	Normal Web	Wide Web
Web depth (in.)	22.2	37.2	22.3	37.0
Working weeks	8	29	10	39
Production/week (tons)	4479	6358	5121	6585
Shears/week	22	19	18.5	17.3
Tons/shear	201	331	277	380
Tons/manshift	22.1	31.0	23.2	24.4
Percent increase in web	67		66	
Percent reduction in shears	13		6	
Percent increase in production	42		28	

TABLE 3. ACCIDENT STATISTICS, HOLDITCH COLLIERY, NORMAL AND WIDE WEB OPERATIONS

Dates	Pick Coverage (in.)	PFF	Accident rate per 100,000 shifts	Total Accidents	System
7/70 – 8/71	21-3/4	6 ft, 9 in.	272	77	Advance
10/71 – 9/72	24	6 ft, 9 in.	303	43	Retreat
10/72 – 9/74	21-3/4	6 ft, 9 in.	301	77	Advance
11/74 – 7/75	39 (1-m)	8 ft, 3 in.	70.7	6	Retreat

Average distance with narrow web = 20 to 23 in.
Average advance with wide web = 38 in.

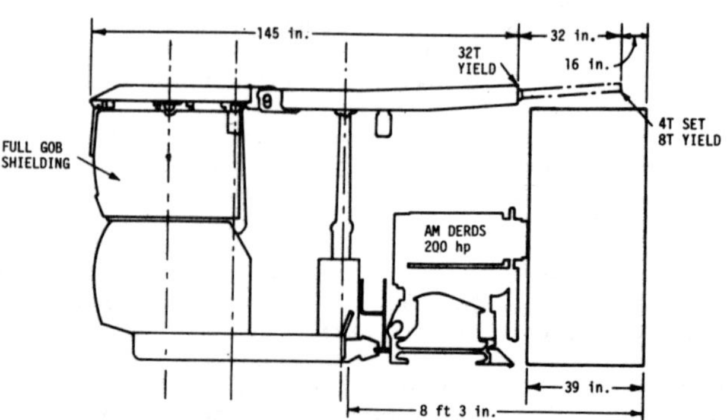

Fig. 3 – Equipment arrangement, first face Hem Heath Colliery

At the greater depths of mining the British long-wall of necessity whereas in the United States we longwall in the hopes of obtaining maximum production. This, in combination with our generally harder cutting conditions, results in our also using higher capacity shearers.

Although the reported benefits of mining 1-meter webs in the United Kingdom were under quite different conditions, there is no fundamental reason to believe that similar results could not be achieved in the United States. This belief was reinforced by the fact that there did not appear to be a logical reason for our mining with 30-in. wide drums, other than being influenced by British regulations which have historically fixed the design of equipment now being used in America.

THE IMPACT OF WIDER WEBS

The key to the performance of any longwall face is roof control. With wider webs, some aspects of roof control are expected to improve while other aspects will require special consideration. The factors which would improve conditions are believed to be:

a. Less stress cycling - defined as the number of times that the support canopy is released (advanced) and reset under a finite section of roof.

b. Rate of face advance which would be expected to increase with increasing web depth - this is likely to reduce the influence of upper strata convergence.

c. The effect of forward abutment pressure in the entries in advance of the face - wider webs are not expected to reduce abutment pressure but only reduce its detrimental influence, which is time related.

The factors which clearly require special attention are:

d. The shift in resultant roof load towards the face - this can result in instability of the support and problems with the floor as experienced at Hem Heath.

e. Tip loading at the face - this requires special attention because of our preference for one web back supports and the requirement to mine uni-directionally, using the half face or modified half face method.

f. Deterioration of the immediate roof in front of the support because of the greater span of unsupported roof.

The problems with support stability, floor loading, and tip loading at the face can be reduced by the proper design and application of supports.

The need to control the greater span of unsupported roof in front of the canopy is perhaps the factor that is the most difficult to define. The British solution was to employ forepoling with 3-ton tip capacity

(yield) at Holditch which had to be increased to 8 tons at Hem Heath. Improved designs of the canopy extensions have subsequently been developed by the NCB and support manufacturers.

The very limited American experience with wide web mining, with 36-in. wide drums, has depended on the integrity of the roof with no special provisions for forward support.

At one of the sites, there are presently plans to increase drum widths to 39 in. using the same approach. With this approach, undoubtedly the maximum web that could be mined at a majority of sites in the United States would be limited by the integrity of this span of roof. This will constrain the principal benefits that can be realized from mining deeper webs.

FMA's report concludes that evaluation of the potential impact of a wider web on both roof and floor conditions involves consideration of the inherent features of various types of supports, their application, and the need for special features.

Although the British reported an improvement in roof conditions, it is generally believed that the principal advantages to be gained in the United States will be increased productivity, a reduction in dust per ton of coal mined, and improved working conditions on the face.

This paper will also show that there should be a substantial reduction in the cost of mining on the face.

The increase in production is mostly related to the increased tonnage won per pass of the shearer. The results at both Holditch and Hem Heath, as shown in Table 2, indicate that the percent increase in tons per pass was substantially greater than the percent decrease in passes completed per week. This appears to represent a sharp decrease in the influence of the non-productive modes of the mining cycle (flitting and sumping).

There are several factors which are expected to favorably effect dust on the face, with the assumption that the shearer is the principal source. These factors are:

a. The increased coal production is from deep in the web. Dust from this tonnage produced by the leading drum can be better controlled before it enters the ventilating air stream.

b. For a given pick penetration and capacity of shearer (in tons per minute), both drum speeds and haulage speeds will be reduced with increasing web depth. Slower drum speeds are generally credited with a reduction in dust make.

c. Reduced haulage speed reduces the rate of cutting the free face - a likely source of troublesome dust which immediately enters the air stream.

Finally, working conditions and thus morale on the face are expected to improve because the man effort required per ton of coal won is reduced. This was reported by the British to be a real benefit which contributed to a reduction of both absenteeism and accidents on the face.

THE IMPACT ON EQUIPMENT

The mining of wider webs will principally effect the shearer and the roof supports. Requirements for modifications to the shearer can be more accurately defined than those required for the supports, because of our inability to quantify the effects of roof and floor loading. Powered supports have been developed on the basis of trial and error. The report, from which this paper is derived, has made an attempt to quantify the influence of wider webs on supports but only as to the relative effect of factors specifically related to web depths.

THE SHEARER

This discussion is limited to the mining of wider webs with conventional on pan DERD shearers. The effect of wider drums on in-web low seam shearers is different. The effects of wider drums on the requirements of a pan-mounted shearer fall into the following general categories:

a. Machine stability and structural modifications

b. Power requirements for cutting, loading, and haulage

c. Coal loading

d. Steering and control.

Two of the major shearer manufacturers (7) were surveyed as to the maximum widths of drum that could be mounted to existing shearers, the modifications that would be required to accommodate these drums, and the estimated cost of these modifications. It was their opinion that present shearers of 170 kW capacity could mount drums up to 42 in. wide, whereas the larger 300 kW shearers could be fitted with 48-in. wide drums. Drum widths beyond 48 in. would represent a new design of shearer. This paper, therefore, considers 45-in. webs, with 48-in. drums, to represent the potential for existing equipment. Web depths beyond 45 in. are considered to be in the realms of future equipment development.

Modifications to widen the machine underframe would be needed to improve stability. These wider underframes would be supported off the ramp plates and gob side furniture to reduce the effect of carrying the machine weight on the conveyor pans.

The wider web shearer will require larger ranging arms, and additional strength in the drum and ranging arm mountings. The additional strength and stiffness will be required to accommodate the increased overhung loads and the increase in torque delivered to the slower speed drums.

For a given capacity of machine (in tons per minute) drum speed is inversely proportional to web depth for a constant pick penetration, and a constant specific energy of cutting (in kilowatts per ton minute). The drum speed (n) in revolutions per minute can be defined as:

$$n = \frac{3387}{f\Delta}\left[\frac{kW.\ S_{pe}}{D_w\ H.}\right]$$

where:

f = the number of picks per line

Δ = pick penetration in inches

D_w = depth of web in inches

H = height of extraction in feet

S_{pe} = specific energy of mining in kilowatt per ton minute

kW = available shearer power.

For example, in good cutting conditions a 300 kW shearer can typically cut and load 16.7 tons/min (1000 tons/hr). In 6.5 ft of coal, this machine would mine a 27-in. web (30-in. drum) at a drum speed of 54 rpm, with 2 picks per line and a 3-in. pick penetration. The same drum would mine a 45-in. web (48-in. drum) at a reduced drum speed of 32 rpm. For the same power consumption, torque to this drum would increase 67 percent. Ranging arms, mountings, and gearing will be required to accommodate the loads associated with this increase in torque.

In the above example the haulage speed would be reduced from 26.8 ft/min to 16.1 ft/min with the deeper web. If it is assumed that the specific energy of cutting does not increase, then the haulage power requirement would not increase, but haulage thrust would. The increased thrust would come from the horizontal component of the coal reaction to the additional torque applied to the picks (8).

The shearer manufacturers felt that loading coal out of a 48-in. wide drum, with present technology in spiral vane drums design, would not be a problem. This appears to have been confirmed by surface trials conducted by the MRDE on an in-web shearer at Swadlincote. This Anderson Mavor machine was fitted with 48 in. wide by 48 in. diam drums. Surface trials, cutting a bench of simulated coal, showed efficient loading from the three-start spiral vane drums. The real test of this machine was scheduled to begin in November, 1980 when this shearer is put underground at the Marckham Colliery in North Derbyshire.

The potential does exist for an increase in the depth of step that would result from correcting the cut with a wider drum. However, offsetting this effect would be the slower speed of the shearer which may allow the operator to better follow the seam.

Finally, much of the discussion and relationships established above are based on there being no appreciable change in the cutting effort required as the web depths increase.

The known work, for which data has been published, is by Osterman (9) in Germany and Mutmansky (10) in the United States. The results of testing done by MRDE and the performance of the shearing machines at Hem Heath and Holditch, as reported in by Hunter (2), are also available for interpretation.

Osterman experimented with a small 80 kW single drum shearer with 1100 mm (43 in.) diameter drum working very hard coal (4300 lb/in.[2]). The web depth was varied from 15 to 30 in. using two drum speeds, with haulage speed varying, with web depth for constant output. The results of Osterman's findings are shown in Figure 4. At first, these results would indicate that there is a decided increase in the specific energy of cutting coal with increasing web depth, in the range 15 to 30 in. However, the curves also show an increase in specific energy with an increase in drum speed at a specific web depth. This represents an increase in energy with a decrease in pick penetration since haulage speed was nearly constant for a specific web depth. Osterman concluded that his findings showed that the specific energy of mining might increase with web but certainly increased with a decrease in pick penetration.

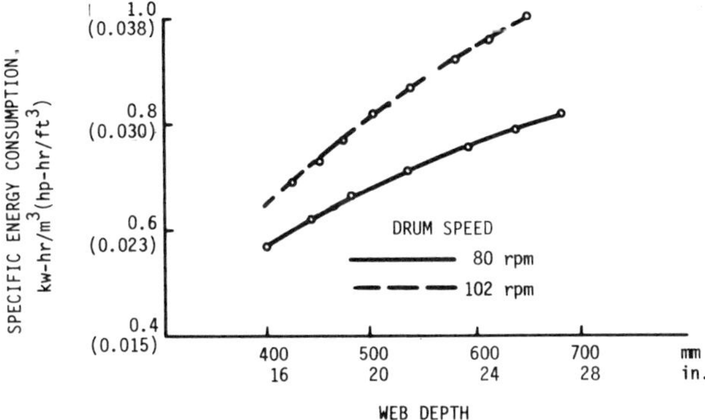

Fig. 4 - Shearer performance with increasing web depth - German test results by Ostermann

Mutmansky attempted to apply Osterman's results to typical coal cutting conditions in the United States by plotting the trend in Osterman's data through a known point. Mutmansky chose as his point of reference, for United States conditions, performance of a 300 kW shearer which typically can cut and load 1000 tons/hr at a web depth of 27 in. This curve is shown in Figure 5 for the span of web depths 15 to 30 in.

The surface tests conducted by MRDE involved a 200 kW shearer mounting drum widths of 32, 36, 42, and 48 in. This machine was used to cut simulated coal at varying web depths again with finite variations in haulage speed. However, the drum speed was essentially kept constant at 56 rpm, producing results similar to Osterman's. A few tests were run at an increased drum speed of 64 rpm (a 15 percent change) which resulted in a 25 percent increase in power consumption for a specific web depth and haulage speed. This also confirms the sensitivity of these results to pick penetration.

Finally, Hunter reported that in both surface tests and underground operations, the Hem Heath and Holditch shearers showed a constant cutting rate proportional to measured power consumption regardless of web depth.

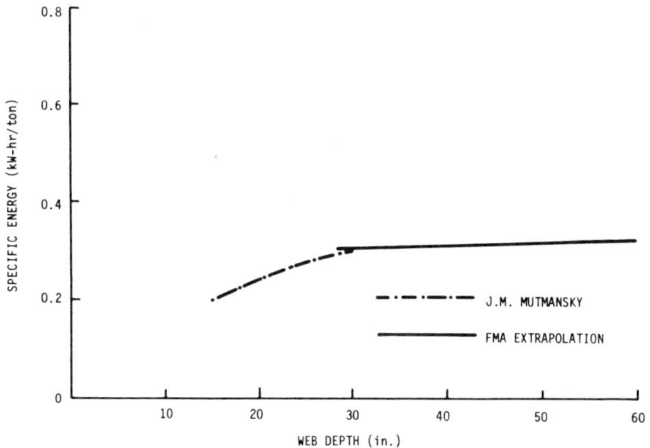

Fig. 5 - Specific energy versus web depth

The authors concluded that there was at least as much evidence to suggest that energy consumption was dependent on pick penetration as there was that it was dependent on web depth. The authors further conclude that state-of-the-art shearer drum design, with regard to both depth of cut and pick arrangement (density), suggests the possibility of three zones of hardness across the web depth as shown in Figure 6.

Zone A is easy to cut because of its proximity to the free face and the probability that it is partially fractured by the influence of the converging roof strata. The ease of cutting within this zone is probably most exemplified by the success of the plow. The plow is a passive cutting tool which "scrapes" off 3 to 8 in. of coal at speeds up to 300 ft/min.

Zone C is known to be hard cutting. State-of-the-art design of shearer drums situates a tight cluster of cutting picks in this area to overcome the hard cutting conditions. Zone B appears to be an intermediate hardness zone which is effectively cut by a density of picks that are typically one to two picks per line around the drum with a center-to-center spacing, along the drum axis, of 1-1/2 to 2 times pick width.

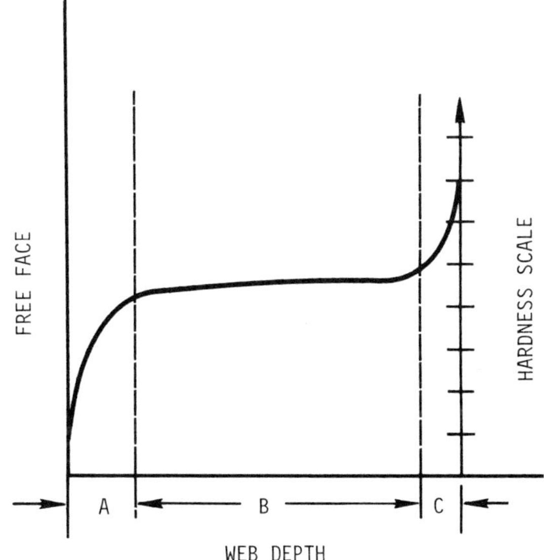

Fig. 6 - Suggested hardness zones across the web

If this theory of cutting zones is valid, then the specific energy of cutting with increasing web depth would be characterized by the following:

a. The narrower webs are strongly influenced by Zone A, thus explaining Osterman's steeply sloping curve for web depths 15 to 30 in.

b. Zone C may experience still harder cutting as the web depth increases. However, with increasing web depth, its influence is diminished since it appears to be a relatively narrow band.

c. The hardness of Zone B increases slightly with increasing web depth.

This model was used in extrapolating the curve of Figure 5 for web depths beyond 30 in. This curve was used in subsequent work projecting the capacity of and increasing productivity of mining deeper webs.

THE ROOF SUPPORTS

As previously mentioned, the total experience in the United States with mining wider webs is limited to two presently operating faces equipped with 36-in. wide drums. In both cases no special provisions were added to the existing supports, other than to provide for the increased stroke in the advancing rams. At one of these sites there are plans to extend the drum width to 39 in. having experienced no roof control problems with the 36-in. drums. This trial and error approach can be used to eventually establish the maximum web depth that can be mined, based on the limitations of existing supports and the integrity of the unsupported roof.

However, to obtain the maximum benefits from mining wider webs will require, in the future, a better approach to the roof control problem since it undoubtedly is the principal constraint on depth of web.

This study is the first known attempt in the United States to identify, qualify, and with the best available information quantify the effects of web depth on roof control. The obvious difficulty with making specific recommendations is the inability to adequately quantify geological conditions.

The criteria that were identified as being particularly relative to selecting a support for a wider web face are:

a. The need for increased capacity and/or load density

b. The stability of the various types of supports and the resulting floor loading

c. Control of the immediate roof strata in front of the supporting canopy

d. The influence of an increase in the PFF distance.

These criteria, of course, are not the only considerations to be used for the selection of a support but they are the factors that become more significant as the depth of web increases.

The capacity of the support was expected to increase, with an increase in web, because of the greater span of roof over the supports. However, it was also known that capacity is related to the nature

and the position of the competent strata above the seam, the height of extraction, and the caving characteristics of the roof. To assess the relative effect of a wider web on the required capacity of a support, a mathematical relationship developed by L.V. Wade (<u>11</u>) was employed. This relationship was used because it was the only known formulation that attempts to take into account those factors which are known to primarily influence the required capacity of a longwall support. Wade suggested that support density d_s (in tons per square feet) is:

$$d_s = 4\delta h \left[1 + \sum_{n=1}^{5} c_n \right]$$

where:

h = height of coal extracted in feet

δ = density of roof rock in tons per cubic feet

$\sum_{n=1}^{5} c_n$ = summation of multiplying coefficients

The coefficients c_1 through c_5 establish the influence of the caving characteristics of the roof, web depth, bridging to first fall, immediate roof weight and extended down time. Total support capacity thus becomes:

$$c_t \text{(tons)} = d_s \left(\frac{\text{tons}}{\text{ft}^2} \right) \times \text{canopy area (ft}^2)$$

Results typical of this analysis are shown in Figure 7 which relates the influence of web depth and different conditions of roof strata to required support density.

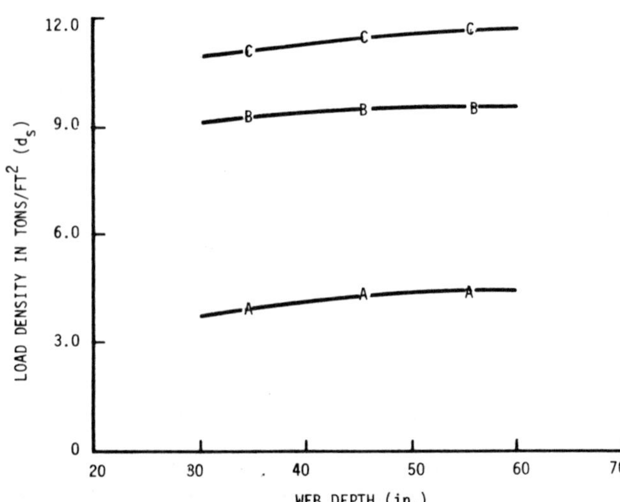

Fig. 7 - Comparative influence of roof strata and web depth on support density

These comparative results suggest that the size of supports is not appreciably affected by increasing web depth but is greatly influenced by the type of roof strata. It is, therefore, likely that larger supports will not be required for wider web longwalling. However, the selection as to type of support is very critical with regard to stability and floor loading.

The study investigated various support types with regard to comparative stability and floor loading at increasing web depths. The roof was modeled (12) to reflect the expected shift in roof load towards the face as the web depth increased. The results of this investigation can be summarized as follows:

a. Supports set one web back exhibit both instability and a rapid shift in floor loading towards the face. This effect was most prominent for the typical 2-leg shield but also could be a problem with the 4-leg shield at web depths beyond 39 in.

b. The 6-leg chock, similar to the British support of Figure 3, was particularly stable with fairly equalized floor loading.

c. A 4-leg shield, with provisions for adequate manway behind the front row of props, would be equally stable. The front row of props and base would be brought forward right up behind the conveyor. This would represent a modification to the 4-leg shield as presently used in the United States.

No satisfactory method was found to model the immediate roof bridging, the open span between the forward tip of the canopy and the face. It, therefore, could not be determined at what depth of web or under what conditions of roof that immediate support of this span would be required.

One web back supports also increase the PFF distance with increasing web depth. The resulting increase in the length of overhanging (cantilevered) canopy detrimentally affects tip loading at the face and the contact pattern of canopy and roof. Irresberger (13) points out that the first contact point can be as much as 50 cm (1.64 ft) from the front tip of the canopy on a narrow web plow face.

In summation, it is suggested that for mining web depths up to 48 in. the most universally applicable support would be the modified 4-leg shield fitted with an improved design of forepoling. The forepoling should feature tip loading in excess of 20 tons, maximum coverage of the exposed roof area, and provisions for maximizing contact with the roof. Forepoling requirements will be more demanding than designs applied to the meter webs in the United Kingdom because unidirectional cutting, with half face sumping, is generally practiced in the United States.

THE BENEFITS

The British experienced a substantial increase in production, by going to a meter web, because the percent increase in tonnage won per pass was greater than the decrease in passes completed per shift. There is no reason to believe that this would not also be the case on an American longwall.

However, to quantify the expected increase on United States longwalls required that projections be made based on our different mining conditions, methods, and equipment.

Therefore, the expected impact of wider webs on both shift and annual output were projected from a typical high production longwall in the eastern United States. This typical face was established as the average performance of five contemporary longwalls surveyed specifically for this study. These faces were mined by state-of-the-art equipment consisting of 300 kW DERD shearers, high capacity shield, and center strand conveyors with 300 hp drives. Four of these five faces were mining 6.5 to 7.0 ft of coal, and all were extracting a 27-in. deep web with 30-in. drums, using the half face method of sumping. The average performance of the five operations is summarized in Table 4.

A mathematical model was then developed to represent a shift of production as influenced by the tons of coal mined per cycle, lost time on and off the face, the capacity (in tons per hour) of the mining equipment, and the nonproductive modes of the operating cycle (sumping and flitting). This model is shown in Table 5 for the operation of a DERD shearer mining the half face sumping mode.

Applying this model to project the performance of the typical face, with increasing web depth, involved the following considerations:

TABLE 4.   PERFORMANCE PARAMETERS CONTEMPORARY HIGH CAPACITY LONGWALL

Panel	
Face width	500 ft
Panel length	3000 ft
Height extracted	6.5 ft
Web depth	27 in.
General	
Working days/year	220 days
Production shifts/day	2 shifts
Maintenance shifts/week	6 shifts
Total shift time	480 min
Travel time and miscellaneous	78 min
Available working time at face	401 min
Downtime Analysis Per Shift	
Total down or lost time	120.5 min
Armored face conveyor	37.1 min
Shearer	33.6 min
Stage loader	4.9 min
Roof supports	4.6 min
Outby haulage and miscellaneous	35.4 min
Geological and other	4.9 min
Panel Development	
Manshifts required	572 manshifts
Entry development rate	57.2 ft
Average tons/shift	268 tons
Shift Breakdown	
Tons/pass	329 tons
Passes/shift	4.0 passes
tons/shift	1316 tons
Cutting and loading rate (tons/hr)	517.1 tons
Shearer speed	
Cutting (ft/min)	13.1 ft/min
Cleaning (ft/min)	30.1 ft/min
Cycle time (min)	70.1 min
Shift Utilization	
Loading (min)	280.5 min
Cutting (min)	152.7 min
Cleaning (min)	66.7 min
Turnaround (min)	61.1 min
Scheduled maintenance (min)	0 min
Unscheduled maintenance and miscellaneous	120.5 min
Travel and miscellaneous	79 min
Annual Summary	
Working shifts/year	440 shifts
Maintenance shifts/year	300 shifts
Tons/panel	438,750 tons
Working shifts/panel	334 shifts
Panels/year	1.106 panels
Tons/year	485,050 tons

TABLE 5. SHIFT PRODUCTION FORMULA – HALF FACE SUMPING DOUBLE ENDED SHEARER

Tons per shift (TPS) = tons per cycle × cycles per shift

$$TPS = \frac{LWH}{266.7} \times \frac{T_t - t_t - t_\ell}{\frac{L}{S_{c\ell}} + \frac{L}{S_c} + 2t_e}$$

where:

T_t = total shift time in minutes (480 for an 8-hr shift)

t_t = lost time off the face in minutes (travel time, etc.)

t_ℓ = lost time on the face in minutes (down time related to machine system failure, geological conditions, etc.)

L = face length in ft

W = web depth in in.

H = height of extraction (seam height) in ft

$S_{c\ell}$ = shearer speed in ft/min while mining coal

S_c = shearer speed in ft/min while tramming (flighting)

t_e = turnaround time at the entries in min

The relationship between machine system capacity (C_c) in tons/hr and shearer speed in ft/min is:

$$S_{c\ell} = \frac{4.44 \, C_c}{WH}$$

The constants 4.44 and 266.7 are based on in situ specific weight of coal at 90 lb/ft^3.

NOTE: This formula applicable where cycle time is a function of shearer performance restricted by either shearer power, coal haulage system capacity or shearer control.

a. The capacity of the system (C_c), in tons per hour, as limited by either the capacity of the shearer at increasing web depths, or the capacity of the face conveyor.

b. Lost time on the face (t_ℓ) as affected by an increase in web depth as shown in Table 6. At a web depth of 27 in. this table represents the breakdown of lost time as averaged from the five-mine survey of Table 4. As shown, increases in web depth were expected to impact the performance of the face conveyor, shearer, roof supports and increase the lost time due to geological conditions. Performance of the stage loader, and outby haulage were considered to be not affected by web depth and therefore were treated as constants.

TABLE 6. LOST TIME* ON THE FACE (t_ℓ) WITH INCREASING WEB DEPTH

Item	Extracted Web Depth in Inches						
	27	30	36	39	42	45	48
Face conveyor	37.1	38.0	39.6	40.8	41.8	42.7	43.6
Shearer	33.6	34.4	36.1	36.9	37.8	38.6	39.4
Stage loader	4.9	4.9	4.9	4.9	4.9	4.9	4.9
Roof support	4.6	4.7	5.0	5.2	5.3	5.5	5.6
Outby haulage	35.4	35.4	35.4	35.4	35.4	35.4	35.4
Geological and other	4.9	5.1	5.4	5.6	5.7	5.9	6.0
Total	120.5	122.5	126.4	128.8	130.9	133.0	134.9

*All times are minutes lost per normal 8-hr shift.

c. Lost time off the face (t_t) shearer flitting speed (S_c), and turnaround time (sumping time t_e) were also considered to be not affected by web depth.

Applying this reasoning to the typical high production longwall resulted in the expected increases in shift production shown as Curve A in Figure 8. For this particular operation capacity of the system (C_c) was limited by the face conveyor to 700 tons/hr for all web depths. This is considered to be the continuous rated capacity of this type of conveyor as exemplified by the Eickhoff EKF-3. The capacity of the 300 kW shearer was never the constraint to rate of mining.

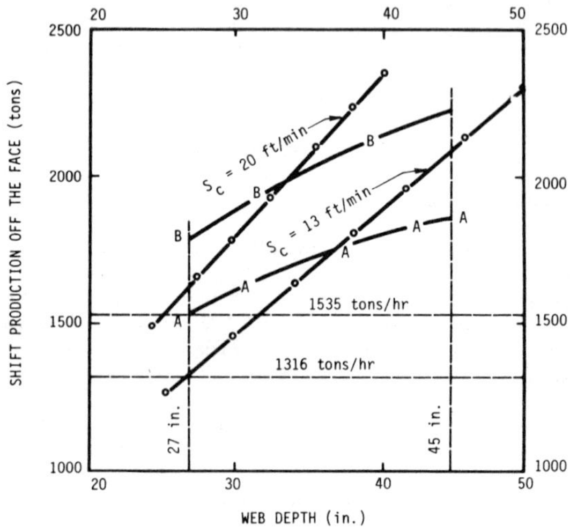

Fig. 8 - Shift production versus web depth in a 6.5-ft seam

The performance Curve B shows the results achieved when conveyor capacity is increased to 1000 tons/hr, which represents the latest development in conveyors. In this case, the constraint to production is the 300 kW shearer at web depths greater than 30 in. This is due to the projected increase in the specific energy of mining from Figure 5.

It should be noted that these projections are sensitive to the estimated increases in lost time on the face, the impact of changes in both sumping time and flitting time, and mining conditions on the face. For example, if flitting time and turnaround time were both increased 20 percent, there would be a decrease in 139 tons (7.5 percent) in shift production when the system of Curve A is cutting a 45-in. web.

Favorable and poor mining conditions are also represented in Figure 8, where poor conditions may limit the shearer cutting speed to 13 ft/min. Under these conditions, system A would be mining at less than machine capacity at all web depths less than 38 in. Even under good conditions, as defined, system B should not be mining web depths less than 34 in.

The curves of Figure 9 show the expected increase in annual production as a result of mining the wider webs. These curves reflect the constraints to annual production imposed by panel development and the lost time due to overhaul and movement of the longwall face.

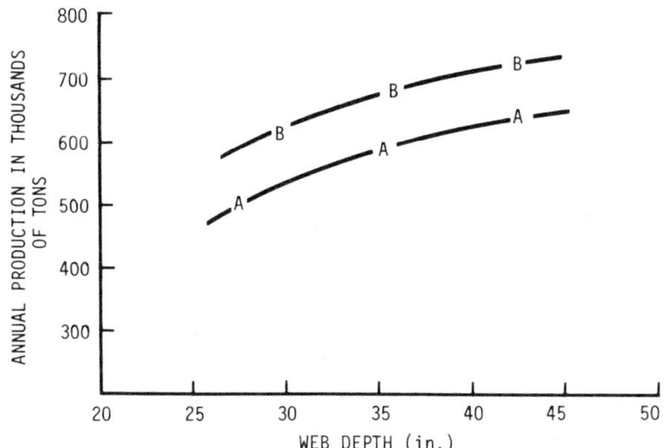

Fig. 9 - Annual production with variation in
web depth

Operations on the face and the number and skills
of the personnel are not expected to be affected by
mining deeper webs. Therefore, if the additional
cost of the equipment does not negate the increased
productivity, then the cost effectiveness should
improve with increasing web depth.

The estimated costs of the modifications and
special features, reflecting the depth of web, were
applied to the shearer, supports and coal haulage
system (when applicable) to arrive at the capital
cost per ton curve of Figure 10. System A is the
combination of 300 kW shearer and 700 tons/hr face
conveyor. This system was assigned the cost of 6-
leg chocks, with roof conditions defined as good.
System B employs the 1000 tons/hr conveyor and the
modified 4-leg shield, under roof conditions defined
as more difficult. Both support systems reflect the
estimated cost of sliding extensions, increased capacity, and extensions to the advancing rams. The
capital costs included the principal, interest, and
taxes depreciated over a 10-year period.

The cost savings demonstrated in Figure 10 would,
of course, be affected by the increased cost of
maintenance, power and water, panel development, mine
haulage, coal preparation, etc., associated with the

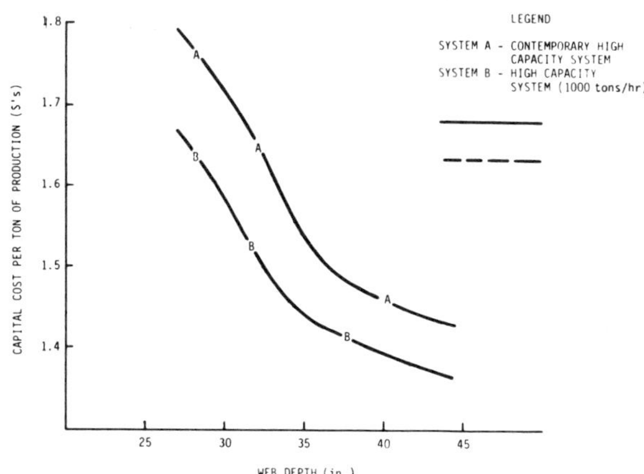

Fig. 10 - Capital cost per ton as a function of
web depth

increase in production. However, the impact on the
cost of equipment, earlier identified as one of the
principal constraints on the increased use of longwalling, is apparent.

THE FUTURE OF UNITED STATES WIDE WEB

Current United States Practice

As has been noted above, wide web longwall mining
has yet to be adopted in the United States. Two
companies, however, have accepted that this approach
offers potential advantages and have installed faces
with wider than standard webs of 36 in.

The organization with the widest experience of
wide web mining is the Fairmont Division of Consolidation Coal. At Loveridge mine a high output face
has been working with 36-in. drums mounted on an
Eickhoff EDW300L shearer. The only significant modification required was the adaptation of pushover
rams for the 250-ton Kolckner 2-leg IFS shields to
a 36-in. stroke.

This installation mining in the Pittsburgh seam
with good (for longwall) roof and floor conditions
has reportedly returned improved production and dust
levels. Shearer haulage speed, while cutting (tail
to head), is normally limited by outby haulage and
dust control considerations although speeds on the
order of 20 ft/min can be achieved with the wider
drums, which are of two-start construction and are
fitted with 4-in. gauge picks.

It is quite apparent that at Loveridge, contrary to European experience, no special consideration had to be given to controlling the larger area
of exposed roof. Indeed, it is understood that
Consol will soon be fitting slow speed ranging arms
that will further increase the PFF distance. In
addition, it is their intention to proceed in the
near future to a 42-in. drum without special modification to the roof supports.

The above represents a situation where a moderate
increase in drum width has been applied in a situation where exceptionally good roof conditions prevail.

A different situation exists at a Kentucky Carbon
mine where 36-in. wide drums have recently been
applied under a difficult roof consisting of sandstone
main roof over 2 to 4 ft of weak shale. Despite
these conditions, a short trial was achieved with the
wide drum before faults and rocky intrusions forced a
change to smaller diameter 30-in. wide drum.

On this face, Dowty 4-leg shields are employed
with an Anderson Mavor AB 16 shearer with mechanical
haulage. During the brief trial of the 36-in. drum,
which was hampered by entry roof problems, a modest
increase in production was reported. Kencar intends
to refit the wide drum when the face has passed the
faulated zone and is confident that a net increase
in productivity will result.

In both the cases above, the modest increases in
drum width, over that which is traditional, appear
to be offering potential for increased production.
In the first case, exceptional roof conditions would
appear to allow a modest further increase in web
without modification to the roof supports. In the
second case, roof conditions are more difficult and
36 in. appears to be the upper limit without special
equipment.

Wider Webs

It is likely, therefore, that under United States conditions a modest increase in web width may be achieved with unmodified equipment and that this will allow modest, but significant, benefits in productivity and dust levels. If, however, a wider web of the order of 42 to 48 in. is to be achieved suitably modified or specially designed roof supports should be used.

It would appear from European experience that the most important factor is the provision of means to control the exposed roof between the support tip and the new coal face. In the United States, immediate forward support (IFS) is achieved by the one-web back method. Application of one-web back methods to a 48-in. web would severely affect the stability of the support. It would, therefore, appear that the canopy extension will be the preferred method since it will allow the canopy length to be reduced (and thus increasing support density) while at the same time allowing the newly exposed roof to be covered rapidly.

Although other requirements of a wide web support, such as an improved traveling way and gob shielding, are required, improved forward roof support remains the most important factor in wide web mining. It is possible that the improvements in roof conditions that have been reported as a consequence of wide web mining may be due in part to the necessity of paying special attention to this problem (Hawthorne Effect).

Thick and Thin Seam Wide Web

The first Foster-Miller report, on the impact of variation of web depths on longwall face, was concerned primarily with the practicalities and economics of a system employing DERD shearer in a 6.5-ft extraction. If wide webs were to be employed in thicker seams, special attention would have to be payed to shearer stability since the moments causing the shearer to try to lift off the pan will be increased. In any case, thick seam mining is subject to special constraints such as face support, lump breakage and coal clearance that will not be reduced by the adoption of wider webs.

In thin seams, however, there is potential for significant improvement since under these conditions it is not roof support or outby coal clearance, but the difficulty of traveling the face and the time spent in face end operations that are the major constraint on production. As web width increases the number of passes required to achieve a given shift output is reduced as is the speed at which the operator must travel to keep up with the machine at a given production rate. Wider webs will reduce the impact of face end operations and improve working conditions on the face.

Very Wide Webs - The Future

All of the above is addressed to the application of existing equipment to facilitate the extraction of webs up to 48 in. If, however, a very much wider web is mined, it is likely that both the shearing machine and the roof support system must be designed with this in mind. As shown in Figure 8, shift production continually increases up to web depths of 48 in. This trend could be expected to be continued for web depths beyond 48 in. However, this would represent the application of yet to be designed equipment. The remainder of this paper is devoted to the description of one such system which is under development by Foster-Miller for the U.S. Department of Energy.

This system is referred to as the Kloswall Longwall Mining System and was developed from the theory that an optimized roof support coupled with a purpose designed shearer taking a very wide web (60 in.) would offer very significant advantages in terms of productivity and cost effectiveness while offering a chance to significantly reduce dust levels.

From the previous study, it was determined that the ideal support for very wide webs would be identified as a support having the following specific features:

a. Superior stability and floor loading

b. Instant forward support with near 100 percent roof coverage

c. Good tip loading (minimum PFF)

d. Minimum canopy length to provide both maximum support density and minimum stress cycling of the roof.

Such a support is illustrated in Figure 11 where the front hydraulic legs are moved forward of the conveyor. This support could be designed as either a 2-leg shield (as shown) or as a 4-leg shield.

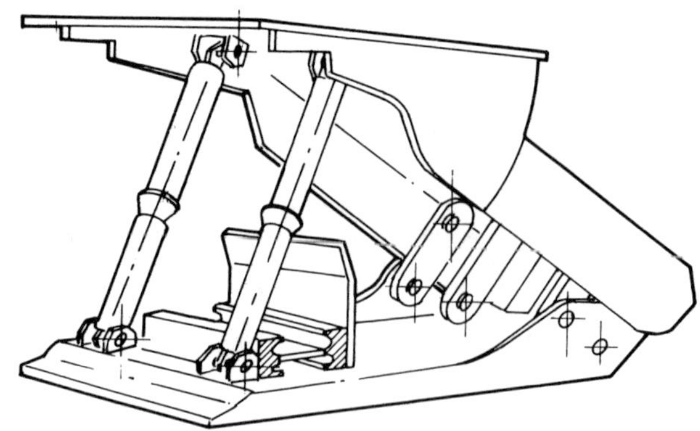

Fig. 11 - Proposed configuration, very wide web support

Such a support features a very short canopy (for maximized density) and a very broad base (to reduce floor loading). The support layout is almost the most stable possible since the roof loads are passed directly to the base by an A-frame structure. The main significance of this design is that by placing the AFC on the support base <u>behind</u> the hydraulic legs, it is possible for the support to be advanced immediately behind the shearer before the AFC is snaked. IFS is thus achieved without the need for a long canopy (minimum PFF) while providing maximum coverage of the roof.

The shearer conceived for this system is also a substantial departure from current practice, as shown in Figure 12. This face side view shows that the layout favored is that of a three-drum shearer with two fixed floor drums and a ranging upper drum. As a major departure the drum drives, both electric motors and speed reducers, are mounted within the drums which in the case of the floor drums are supported at both the face and gob sides. This latter feature is made possible by the floor drums being angled by some 10 deg from the normal to the face.

The major problem in the application of this concept is the loading of cut coal across the 60-in. wide web and through the support legs of the shield. For this reason the Kloswall shearer does not use spiral vane drums. The "paddle" style of drum was chosen to feed coal onto a high-speed loading conveyor which then propels the coal through the supports.

This arrangement should also allow for effective dust control since cutting is taking place well out of the airstream and coal is being passed through an almost totally enclosed chamber prior to discharge onto the AFC.

Work on this system began in October of 1976 with the development of conceptual designs. It was recognized that the principal problems in the development of such a system was the design of the coal loading subsystem. Full-scale design and testing of the coal loading hardware was begun in October of 1979. These critical tests are expected to be completed in the spring of 1981.

Successful completion of these tests will be followed by the construction and surface tests of the shearer and roof supports and an underground demonstration of the complete mining system.

CONCLUSIONS

This paper has sought to record the situation regarding the application of wider webs in the United States and in Europe. In doing this, the constraints on operating methods and equipment have been pointed out. It is the opinion of the authors that there exists a significant potential for modest increases in productivity if webs are increased to the maximum allowed by prevailing conditions without modification of shearer or roof supports. It is also concluded that greater benefits in terms of increased productivity and possibly dust control will result from the adoption of wider webs in the range of 40 to 48 in.

Both mining personnel and longwall equipment manufacturers, interviewed for the wide web report, agreed that in all probability wider webs would result in increased productivity. This opinion undoubtedly was a result of the reported success of the 1 meter faces in the United Kingdom. However, while wider web longwalling in the United States is virtually nonexistent, there are presently some 20 1-meter web faces being mined by the NCB.

If the American coal mining industry is going to be called on to substantially increase the nation's supply of coal in the next decade, then a substantial part of this increase will have to come from underground mines in the east, where longwalling has already proved to be widely applicable. In this scenario, the prospects for substantially increasing productivity and reducing dust by mining wider webs will be particularly attractive.

The alternatives to wider webs for increasing longwall shift production are not as readily applicable or desirable. Faster shearers, at present web depths, are likely to produce more dust. Decreasing lost time will require an increase in machine system reliability, which, although desirable, is difficult to achieve and maintain. Mining longer faces is certainly possible, but at the additional expense of more supports and a longer face conveyor.

Increasing the average web depth mined in the United States will require the combined effort of the mine operators and the equipment manufacturers. The existing potential offered by presently available equipment is considerable. The application of that potential at this time can lead to the eventual availability of very high production systems such as Kloswall.

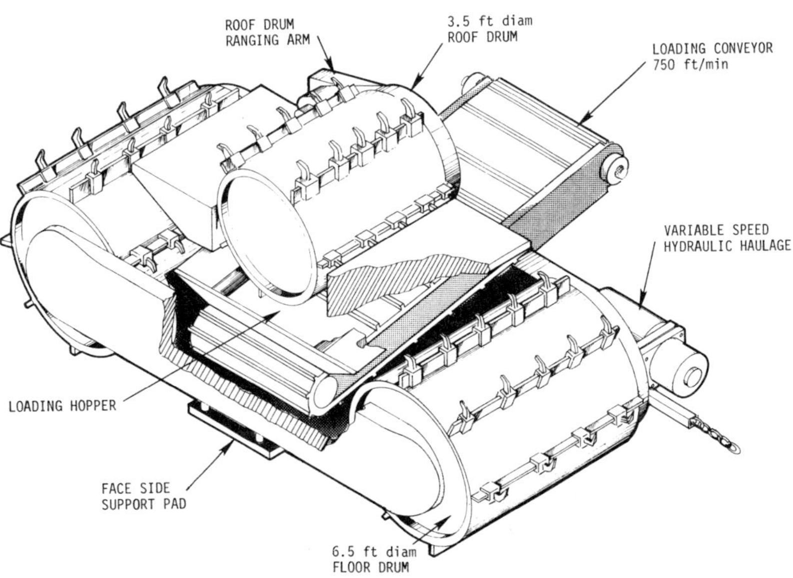

Fig. 12 - The Kloswall shearer

REFERENCES

1. Guay, P. and T. Hawkes, "Study of the Effects of Web Variation on the Performance of a Longwall Face," FE/9058-1 (Vol. I) and FE/9058-2 (Vol. II), available from the National Technical Information Service, U.S. Department of Commerce, Springfield, VA 22161.

2. Hunter, J.R., "Wide Web Working," The Institute of Mining Engineers (London) Annual Conference, May 1977.

3. Borne, W.J.W., "Advanced Technology Mining," The Institute of Mining Engineers (London) Annual Conference, May 1977.

4. "Mechanization Profile of the Mining Research and Development Establishment of the NCB," May 1977.

5. Hardman, D., "Wide Web Mining in the Ten-Foot Seams at Holditch Colliery." Presented to the Institute of Mining Engineers North Staffs Branch, November 1975.

6. Smales, T.E., "Dust Control and Planned Layout in the Ten-Foot Seam at Holditch." Presented at the Institute of Mining Engineers North Staffs Branch, September 1975.

7. Personal conversations with engineering personnel at Bochum W. Germany (Eickhoff) and Motherwell United Kingdom (Anderson Mavor).

8. "Mechanics of Cutting and Boring," Report 77-19, Cold Regions Research and Engineering Laboratory, Hanover, NH.

9. Ostermann, W., "Cutting Performance of Drum Shearer Loader as a Function of Web Depth and Drum Speed," Gluckauf, February 1966; PMSRC Translation, December 1976.

10. Mutmansky, J., "Analysis of the Effect of Changes in Cutting Width on Longwall Performance," Preliminary Report to U.S. Bureau of Mines, September 1976.

11. Wade, L.V., "Longwall Support Load Predictions from Geological Information," Society of Mining Engineers AIME Transactions, Vol. 262, p. 209, September 1977.

12. Wilson, A.H., "Support Load Requirements on Longwall Faces," General Meeting of Manchester Geological and Mining Society Branch of IMinE, Paper No. 4558, May 1974.

13. Irresberger, H., "Roof Control on Longwall Faces," Colliery Guardian International, p. 34, April 1978.

Chapter 29

DEMONSTRATION OF LONGWALL MINING IN THIN SEAMS

by

Ernest A. Curth
Mining Engineer
United States Department of Energy
Pittsburgh Mining Technology Center

Joseph A. Gill
General Manager
Leeco, Inc.
London, Ky.

ABSTRACT

The Government and Leeco concluded a cost-sharing agreement in 1976 to demonstrate longwall mining of a thin coalbed, 1 m or less, in a mine near Hyden, Ky.

A premining investigation laid the groundwork for mine design and equipment specifications. Four-legged shields, a face conveyor, a stage loader and an in-web shearer were selected to operate on a 183 m face. Longwall mining began in 1979 and initially was slowed by low mining height and equipment problems. A rock mechanics program provides early warning capability and criteria for future mine design. Monitoring of surface effects is included.

INTRODUCTION

In the eastern United States, 44.5 billion tons[1] or 29 percent of a coal reserve base of 152 billion tons minable by underground methods fall in the 0.71 m to 1.07 m (28 in to 42 in) range (1).[2] The underground reserve base includes bituminous and anthracite coalbeds more than 0.71 m (28 in) in thickness to a maximum depth of 305 m (1,000 ft). The tonnage in the low coal range is now of less economic importance than reserves in the above 1.07 m (42 in) category. Some of it is currently mined and all is considered a paramarginal source that will become of increasing interest in the future.

As a consequence of selective mining, thin coalbeds are often left out in favor of thicker overlying or underlying coalbeds. But mining thin coalbeds in sequence from top to bottom in the strata profile makes use of available resources and extends the life of mines that age economically as extraction moves away from openings. Another advantage of such planned mining is a better control of rock mass behavior and methane emission.

In thin coalbeds, large areas must be extracted to maintain a high rate of production. To produce 900 tons of material per shift, a 150 m (500 ft) face must advance 3.1 m (10 ft) per shift in 1.5 m (60 in) coal; 4.6 m (15 ft) in a 1.0 m (40 in) bed; 5.20 m (17 ft) in 0.9 m (36 in); and 6.1 m (20 ft) in 0.75 m (30 in). Entry development must keep pace with such face advances.

The human factor is of the highest importance (2). Personnel movement should be restricted to a minimum. Ergonomics, the human aspect of machine design, enters into consideration in assembling, taking apart, moving, maintaining and operating equipment.

Automation at progressive levels may be attempted to remove the workman from the face. Roof support units can be controlled individually by push button or in batches of up to 10 units by manual or shearer initiation, obviating valve manipulation. Remote control from the entry presents a high level of automation.

Under the Government's Advanced Mining Technology Program, Leeco, Inc. (a Kaneb Company) and the Government concluded a cost-sharing agreement in 1976, the objective of which was to demonstrate mining of thin coalbeds, 1 m (40 in) or less, by longwall methods capable of improving health and safety and increasing productivity and resource recovery. The scene of the demonstration is Leeco's No. 22 mine, located in the Kentucky Highlands near Hyden, KY. Mining is conducted in the Fireclay Rider coalbed. A cleaning plant, situated at the mouth of No. 22 mine, prepares an utility grade fuel through jigs for the coarse run-of-mine--sized above 13 mm (1/2 in)--and through hydroclones for the fines and middlings. Trucks carry the product to unit train loading facilities at Manchester and London, KY. Central shop, laboratory, and offices are located at London, KY.

The contract management plan called for a scope of work to be performed in the following phases:

Phase I: Premining Investigation
Phase II: Mine Design
Phase III: Equipment Procurement
Phase IV: Demonstration Mining

PREMINING INVESTIGATION

The Contractor selected an appropriate site, and J. T. Boyd Company conducted a geotechnical study of the site under subcontract. Little geologic information was available; therefore twenty core holes were drilled on a 300 m (1,000 ft) random grid to 3 m (10 ft) beneath the coalbed. Cores were obtained by diamond drill, using NQ size wire line tools (nominal 48 mm).

---

[1] One metric ton = 1 000 kg used in this text.

[2] Numbers in parenteses refer to items in the list of references at the end of this report.

Data obtained from the cores were used to trace coal thickness maps and isopachs of strata intervals and overburden (fig. 1).

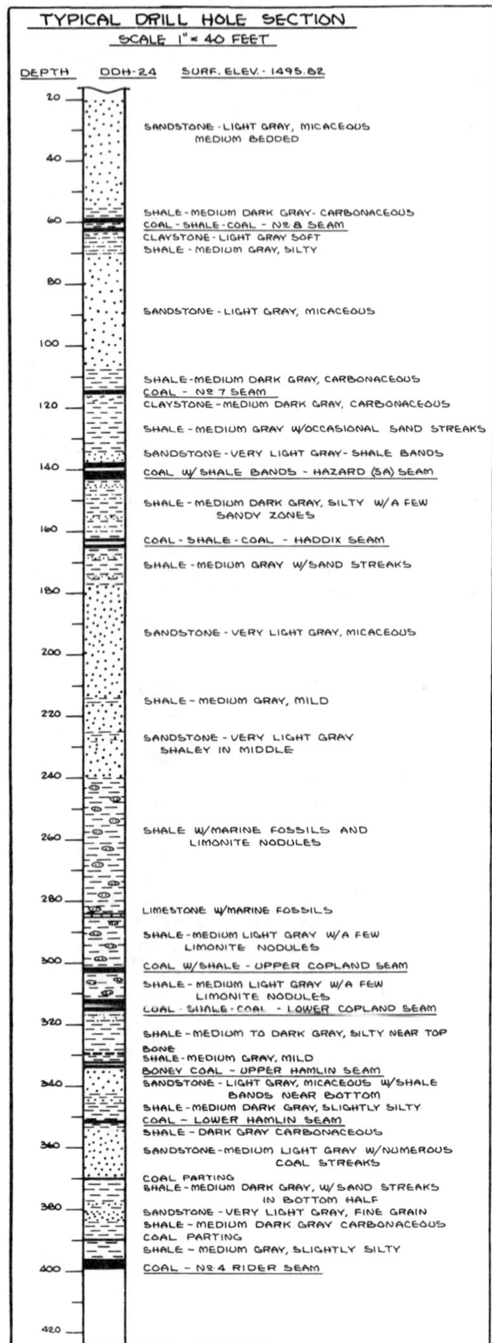

Fig. 1. Typical geologic column.

Selected samples from an interval of 17 m (50 ft) above the coalbed to 3 m (10 ft) below the coal were preserved by paraffin coating for physical testing. Determination of physical properties of the strata adjacent to the coalbed provided indicators of ground mass behavior. For instance, the mean load density a roof support system should offer can be estimated by considering strata separation and cantilever action in a stiff strata. The following rock properties were determined in the laboratory:

Unconfined compressive strength
Indirect tensile strength (Brazilian test)
Modulus of elasticity (Flexure)
Shore hardness

The premining effort indicated the following:

1. The selected site includes recoverable reserves of adequate quantity under a cover greater than 60 m (200 ft).

2. The Fireclay Rider coalbed (locally No. 4 Rider) ranges from 0.9 m (36 in) to 1.05 m (42 in) except in 5 boreholes where the coalbed is less than 0.9 m (36 in) in thickness. The seam dips less than 1% to the southwest.

3. Adverse roof indicators were determined, such as shale thinning out under sandstone channels and some shallow cover combined with rapid topographic change.

4. Water is expected to seep from the surface through the gob and will have to be pumped. A wet and muddy floor will cause inconvenience in a thin coalbed; however, the floor shale appears to be firm. The large Shamrock mine, 24 - 32 km southwest of the site, successfully practices pillar extraction in the same coalbed by retreating from the outcrop in increasingly thicker overburden. The generalized strata column of Shamrock indicates a cyclothem similar to Leeco; an immediate roof shale 0 - 3.0 m (10 ft); a sandstone main roof, 10 m (33 ft) in thickness; and a firm shale floor. Therefore, roof and floor conditions and cavability at the Leeco mine may be as favorable as at Shamrock even under less than 60 m (200 ft) of cover.

MINE DESIGN

Figure 2 shows a four-panel layout for longwall retreat as projected by J. T. Boyd Company. A fifth panel will be added. Six main entries are driven from the portal in a northeast direction. The outer entries, No. 1 and No. 6, are return air courses with fans driven by 112 KW motors and delivering air quantities of 23 to 70 m^3/sec (50,000 - 150,000 cfm). Equalizing overcasts are provided. The middle entries are intakes, and the No. 5 entry is the combination belt and track entry with a minimum vertical clearance of 1.25 m (50 in). The main belt is 1.20 m (48 in) wide, uses a rigid structure, and is driven by a 150 kW head drive. Rubber-rail vehicles, flat cars, and man riding battery locomotives provide men and supply transportation along the main line and in the panels. The roof in the entries is supported by 16 mm (5/8 in) roof bolts, 0.9 m (3 ft) and 1.2 m (4 ft) in length, on a 1.2 m (4 ft) grid. The portals are built of concrete with canopies to protect workmen from material falling from the highwall. A training building and a cleaning plant were added recently to the original surface facilities consisting of office, lamphouse, warehouse, substation, coal stacker, truck loading, and ponds for water storage.

Due to ample outcrop exposure, methane control in the Fireclay Rider coalbed is not expected to pose a problem. MSHA does not approve gob ventilation through the tailgate entries unless at least one entry is cribbed. Eight 250 mm (10 in) holes drilled from the surface into a bleeder extension and equipped with an exhaust fan will serve the same purpose with less cost.

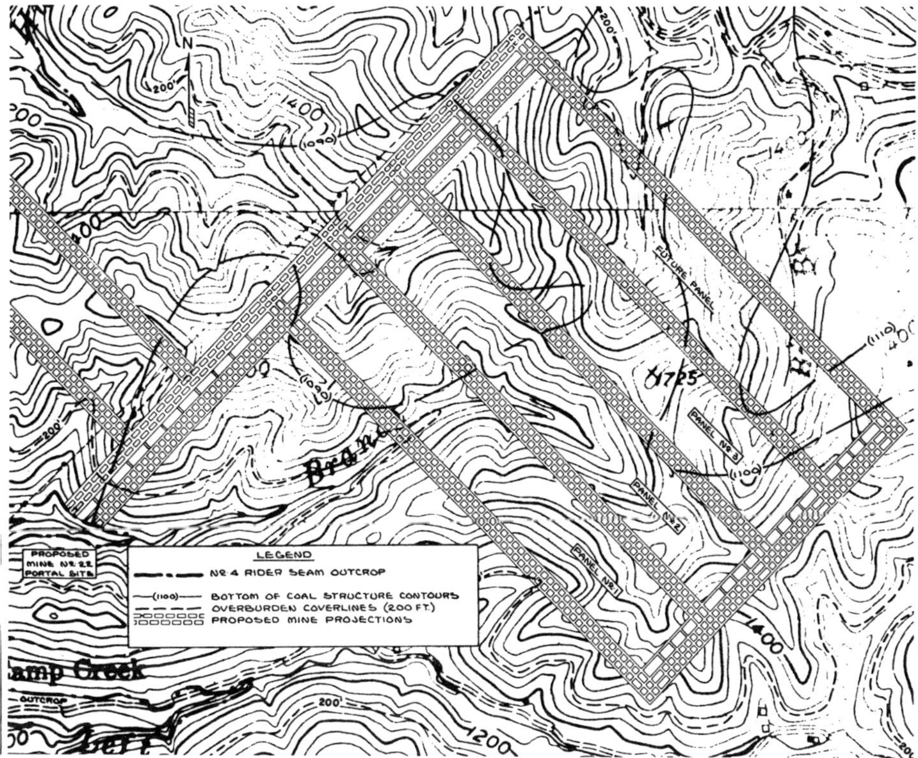

Fig. 2.   No. 22 mine projection.

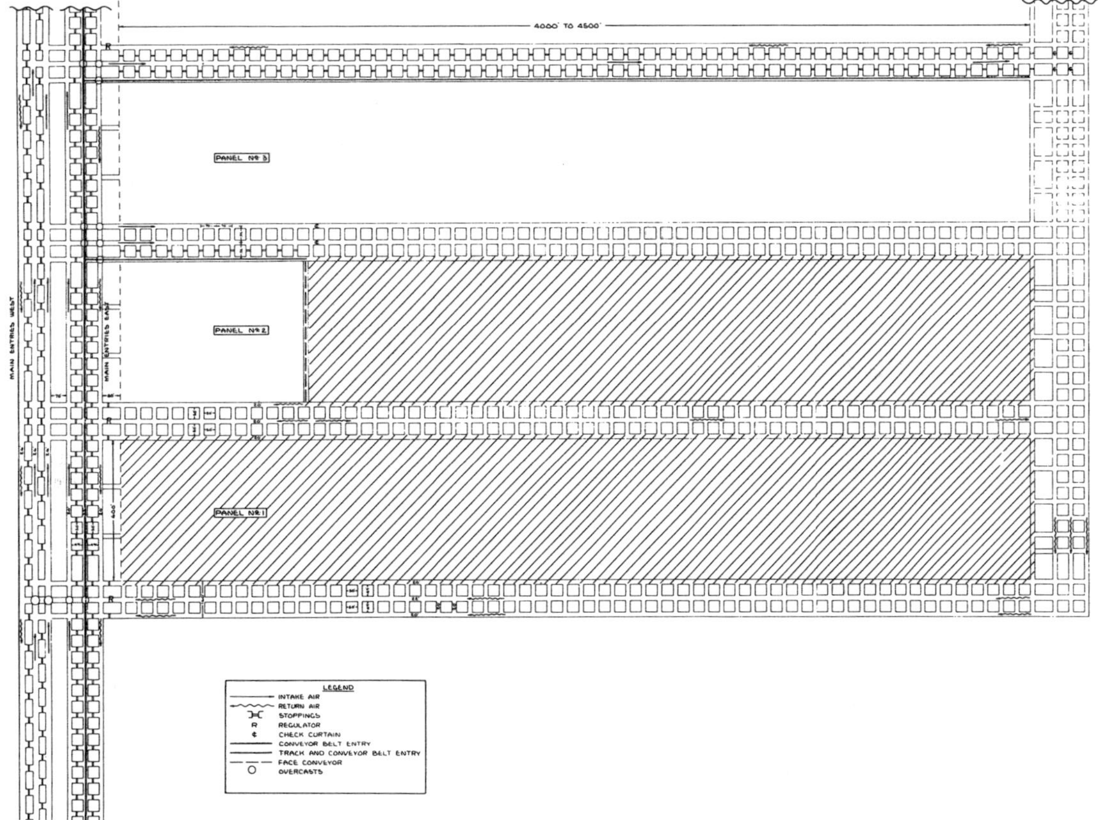

Fig. 3.   Longwall retreat mining plan.

Power is supplied to the substation at 37 kV and transformed to 4 160 V for mine distribution. Long-wall equipment is powered by 960 V, and continuous miner units and all the other equipment by 440 V.

Figure 3 shows the panel operation plan, including the gob area of the first panel, longwall face extraction in the second panel, and panel development in the third panel. During the development, one entry is the combination belt and track entry on a split with limited air velocity; the adjacent entry is the intake; and the third entry is the return aircourse. The panels are connected to a bleeder entry system. During panel extraction, both entries adjacent to the belt entry, which is now the maingate, are on intake air; and at the tail end, all three entries are on return. The panel belts are 0.91 m (36 in) in width and extensible by a length of 90 m (300 ft). They use rigid structure and 94 kW drives.

Development entries and breakthroughs are 6 m (20 ft) wide driven on 21 m (70 ft) centers, thus leaving 15 x 15 m (50 x 50 ft) chain pillar blocks. The following simple example shows how the Safety Factor for ground support is computed:

Assuming pillar dimensions of 15 x 15 m (50 x 50 ft) and 6 m (20 ft) wide entries and crosscuts, the extraction ratio is the following: (3)

$$441 \text{ m}^2 = \text{total area}$$
$$\underline{-225 \text{ m}^2} = \text{left in pillars}$$
$$216 \text{ m}^2 \text{ mined} = \text{extraction ratio of 49 percent}$$

Under 200 m (650 ft) of cover, which exerts a vertical load of approximately 4 550 kPA (650 psi), the weight of the original burden is distributed over 51 percent of the coal left in situ after first mining. Hence, the unit load on the coal rises to 4 550/0.51 = 8 900 kPa (1 275 psi). Assuming the compressive strength of the coal to be 14 000 kPa (2 000 psi), the Safety Factor is

$$S = 14\ 000/8\ 900 = 1.6$$

Selection of face length is guided largely by economic and technical considerations. Cost of equipment ownership increases with face length, but the cost of panel development per unit of mined coal decreases as face length increases (4) (fig. 4). Also, conveyor chain reliability sets a limitation to face length. A face length of 183 m (600 ft) was determined to be the most economic length compatible with conveyor reliability. A panel length of 1 340 m (4 400 ft) satisfied a minimum requirement of 45 m (150 ft) of cover and also allows for panel starts under more than 120 m (400 ft) of cover. Maximum cover is 200 m (650 ft).

EQUIPMENT PROCUREMENT

Along with Phase II - Mine Design, J. T. Boyd developed specifications for the longwall face equipment, including roof supports, a shearer, a face conveyor and a stage loader with all the electric and hydraulic accessories. After selection, the following face equipment was procured and delivered at the minesite in the fall of 1978 except for the shearer, which arrived in April 1979 after extensive testing at the National Coal Board facilities and modifications at the factory in the United Kingdom:

1. Roof Support

a. Recommendations resulting from the premining investigation led to the selection of roof shields. Factors for favoring shield over chocks are
   1. A sheltered working space requiring minimum cleanup work.
   2. Structural stability which allows advancing without delay even with brushing roof contact.

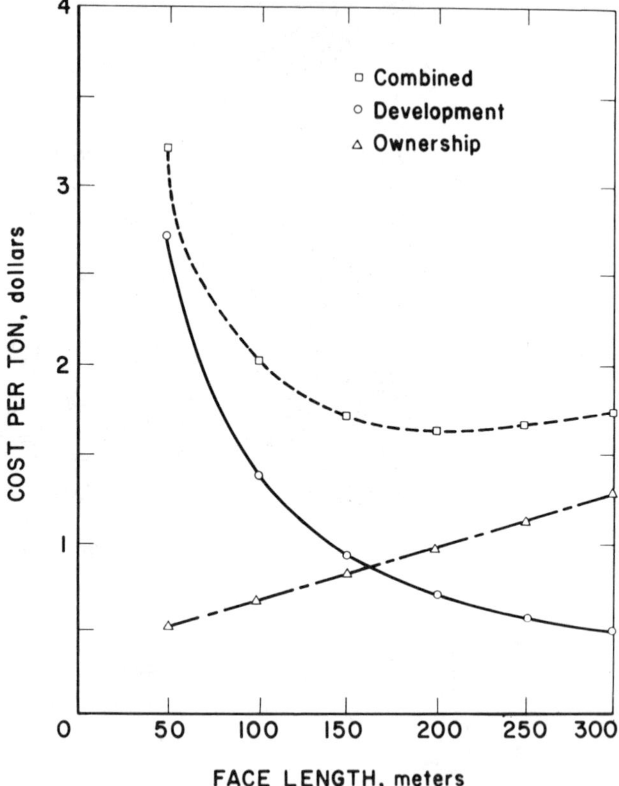

Fig. 4. Face length in terms of ownership and development costs.

A shield line of 124 four-legged shields, each leg yielding at 800 kN (81.65 tons) and the units spaced 1.521 m (60 in) to operate on a 187 m (612 ft) length, was procured. Four-legged shields were selected to provide the desired high tip load and support capacity, an evenly distributed floor pressure, and adequate ventilation area in the low mining height. The legs are double acting and double telescopic (fig. 5).

Closed height of unit: 650 mm (25.6 in)

Max. extended height: 1 450 mm (57.1 in)

Weight: 5 280 kg (5.28 tons)

Roof yield load at 914 mm working height: 2 470 kN (252 tons) at a yield pressure of 40 MPa (5 800 psi)

Roof setting load at the above working height: 1 927 kN (196 tons) at a system pressure of 31 MPa (4 500 psi)

Tip load at yield: 498 kN (51 tons)

Mean Load Density at 914 mm (36 in) height and a span of 305 mm (12 in) of exposed roof between canopy tip and face:

Fig. 5.   Roof shield.

At setting before shearer pass: 420 kPa (4.5 tons/ft^2)

At yield before shearer pass: 541 kPa (5.7 tons/ft^2)

Floor pressure at yield: 2 124 kPa (310 psi)

Advancing ram capable of 760 mm (30 in), 914 mm (36 in), and 1 000 mm (39 in) strokes and using differential

   pressure for conveyor push: 94.8 kN (9.6 tons)

   and ringside pressure for shield pull: 229   kN (23.4 tons)

One piece rigid canopy, split gob shield

Lemniscate linkage

Roof contact advance

Ventilation area at 914 mm (36 in) height: 1.58 m^2 (17 ft^2)

Crawl space for travelway at 914 mm height: 760 x 457 mm (30 x 18 in)

Dowval bidirectional adjacent control valve system, capable of progressive automation by addition of control equipment (fig. 6).

Base, self-cleaning so that debris and loose coal are passed into the gob.

Hydraulic pressure indicators and fittings for connection with gages and recorders.

Side shields and canopy dust suppression flaps with two rams.

Brackets for lighting.

The shields were tested for function, performance, and structural integrity at the manufacturer's facility in the United Kingdom. The National Coal Board Mine Research and Development Establishment certified that the tests conform to the Mines Inspectorate Detailed Standard Testing Procedure, which would enable this support to be used underground in the United Kingdom.

b. Two self-advancing entry roof supports consisting of a shield and a 6 230 kN (630 tons) chock.

c. Power Pack - with two five-piston high-pressure pumps, powered by 56 kW motors, each pump capable of supplying 106 L/min (28 gal/min) to the dual pressure system:

   31 MPa for leg setting

   20.7 MPa (3 000 psi) for all other functions.

A 760 mm (30 in) high mixing and storage tank holding 910 L (240 gal) of 5% oil-water-emulsion.

Fig. 6.   Bi-Di rotary valve.

d. One hundred handset props, yielding at 178 kN
    (18 tons) for additional roof support in the
    gateroads near the face ends; maximum spacing of
    1.8 m (6 ft) lengthwise along the entry for a
    distance of at least 9 m (30 ft) outby the face;
    height range 0.76 to 1.2 m (30 to 48 in).

2. Armored Face Conveyor

    On 187 m (612 ft) sprocket centers; 190 mm
    (7.5 in) high and 782 mm (31 in) wide, accept-
    ing a 19 mm DIN twin outboard chain, operating
    at a speed of 1.1 m/sec (212 ft/min) with a
    112 kW motor at each end. Head and tail
    terminal 825 mm (32.5 in) high.

    UMM chain tensioner.

3. Stage Loader

    A domestic product; swivel beam to accomodate
    face wander and to overlap 3 m (10 ft) over the
    mobile tailpiece of the extensible panel belt;
    tail end fixed to face conveyor delivery end;
    30 kW, 960 V motor.

4. Shearer

    Double drum ranging arm in-web shearer of 200 kW,
    fitted with chainless haulage at the face side
    of the conveyor (fig. 7); can be changed to
    chain haulage also at the face side.

Fig. 7.   In-web shearer.

2 sets of 2-start drums, 914 mm (36 in) wide, and
    1000 mm (39 in) in diameter; using 100 mm
    (4 in) picks instead of the regular 75 mm
    (3 in) picks will increase cutting diameters
    to 966 mm (38 in) and 1 060 mm (42 in),
    respectively (fig. 8);

Adapter to provide a 1 000 mm (39 in) web;

Five drum speeds variable by changing gear
    wheels: 74, 86, 98, 113, 120 rev/min;

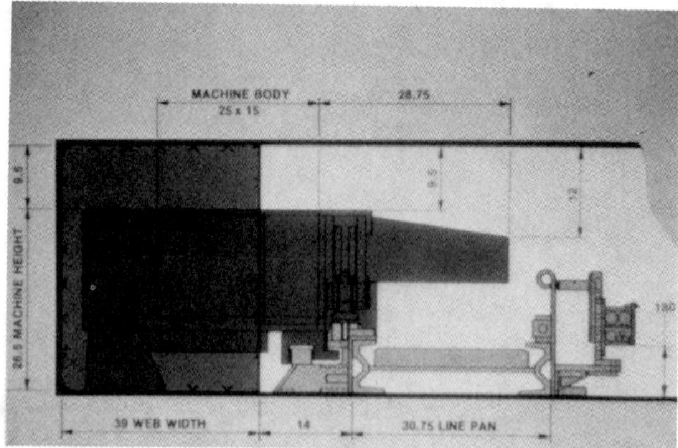

Fig. 8.   Shearer schematic cross section.

Horizon control by ranging arms and face side
    levelling jacks (fig. 9);

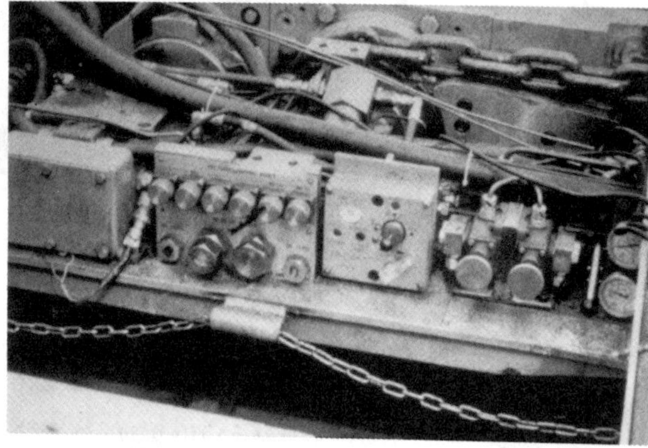

Fig. 9.   Shearer control canopy.

9-speed mechanical haulage: 1 - 12.25 m/min
    (3-40 ft/min) (fig. 10).

Ranging arms to cut by the drives of the face
    conveyor;

Weight: 16 300 kg (16.3 tons)

Length: 7.7 m (25 ft)

Width: 2 m (65.5 ft)

Fig. 10.   Haulage speed selection.

Dust suppression by pick face flushing (fig. 11) and hollow shaft ventilation by water driven venturi;

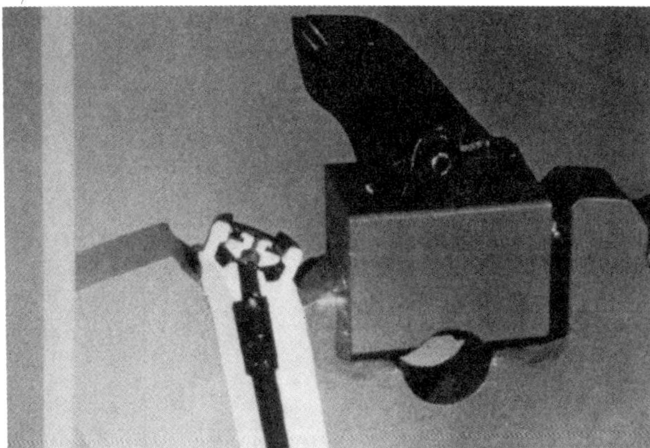

Fig. 11.   Pick face flushing.

Prestart warning by siren and by water issued from sprayheads of either of the drums in gear for a period of 7 seconds;

Perard spill plates, cable handlers, ramp plates, and drive racks;

Cable and water hose are laid into the cable trough along the panline by a shoe riding atop the roll top of the spill plate and kept from lifting out by spring loaded arms;

A fully assembled shearer with chainless haulage was tested on a simulated coal face at the Swadlincote site of the National Coal Board Mining Research and Development Establishment in the United Kingdom during a period of several months. The battery of tests included cutting, loading, haulage, energy consumption and ergonomics with both chainless and chain haulage, and several types of drums and lacing patterns. After modifications, particularly of the trapping system, the test indicated that chainless haulage is feasible and delivered adequate results on the simulated coal face.

5. Electric Package

Power center 4 160/960 V

Control package with cables; emergency cut-off keys placed every 7.5 m (24.5 ft).

Telecommunication system; page units placed every 15 m (49 ft)

Face lighting to conform with Mandatory Standards.

PANEL DEVELOPMENT

Work at the minesite started in December 1976 when the proposed portal was faced up. After the portals were built in concrete, the main entries were turned in June 1977. Figure 12 is a diagram that charts the progress of mine and longwall panel development from the portal through the production stage. The following types of continuous miners were used in the course of development work:

Code	Type	Face Haulage
CM 1	Drumhead	Ram or Shuttle Cars
CM 2	Low Seam Drumhead	Continuous with Double Bridge Conveyors
CM 3	Drumhead	Ram or Shuttle Cars
CM 4	Dual Auger	Continuous with Bridge Conveyors

A development crew of 10 handled following equipment:

    1 Continuous Miner
    1 or 2 Ram or Shuttle Cars or a Continuous
       Haulage
    1 Roof Bolter
    1 Scoop Tram
    1 Bulk Rock Duster

The mains started out with six entries, and one entry was added past the main gate development of Panel 1 (fig. 13). Panels are developed by three entries, and a minimum vertical clearance of 1.0 m (40 in) is maintained in the combination track and belt entry. Table 1 indicates the projected tons/m of development, and Table 2 the average monthly advance and production of each unit during development. Best results were achieved with units CM 1 and CM 3 combined with shuttle cars. CM 3 is a very rugged unit suited to cut rock to provide a minimum vertical clearance of 0.91 mm (36 in) where the coal is thinning out.

Driving of the main entries was stopped when they reached the projected location of Panel 5 in November 1979. Roof and floor conditions deteriorated due to shallow cover and water inflow under a ravine. Depth to coal in a nearby drill hole was 41.5 m (136 ft). The unit was moved to a location near the portal to turn panel entries to the left. Panel development was discontinued too when unit CM 3 driving Panel 2 Bleeders cut through with Panel 1 Bleeders in January 1980, because no further panel development was needed with one panel development completed ahead of the longwall retreat.

DEMONSTRATION

Prior to initiating the face operation, Leeco conducted an intensive longwall training program for face crews in a building erected on the surface for the purpose. The face conveyor with all electric switch gear was installed in the building, and in April 1979 the training installation was completed by adding the roof shields and the in-web shearer with modified cutting modules. Manufacturers' service personnel conducted hands-on training for two teams of future longwall crews during April and May 1979. Hazards peculiar to longwall mining were identified, such as flying particles of coal and rock and bursting hydraulic hoses and fittings.

The longwall is manned by a crew of eight and one or two foremen:

    1 shearer operator
    1 snaker
    2 jack setters
    1 tailgate man
    1 maingate man
    1 console man
    1 mechanic

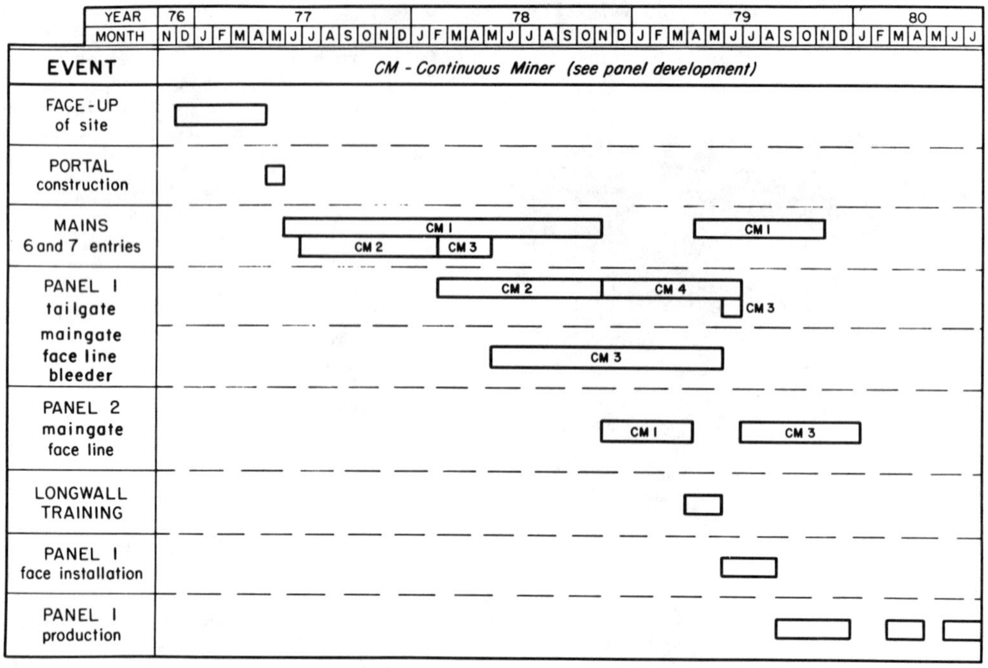

Fig. 12.　Chart of longwall development from mine opening through production.

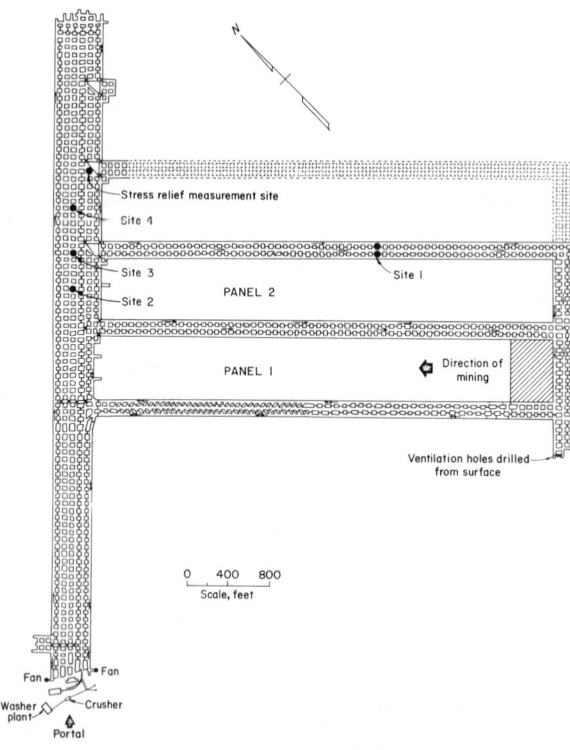

Fig. 13.　Mine map.

TABLE 1 - <u>Projected Tonnage per 1 m of Development at 0.94 m of Mining Height</u>

Openings	No entries	Centers entries	Centers crosscuts	Length advance	Width opening	Linear m advance	Tons per m entry	Tons per length advance	Tons per m advance
		m	m	m	m	m	tons	tons	tons
Mains	6	21	21	21	6	201[1]/	7.50	1 507	72
	7	21	21	21	6	237[2]/	7.50	1 778	85
Panel entries bleeders	3	21	21	21	6	91[3]/	7.50	683	33.5

[1]/ 6 entries   x 21 = 126 m   [2]/ 7 entries   x 21 = 147 m   [3]/ 3 entries   x 21 = 63 m
     5 crosscuts x 15 = <u> 75 m</u>   6 crosscuts x 15 = <u> 90 m</u>   2 crosscuts x 15 = <u>30 m</u>
                    201 m   237 m   93 m

TABLE 2 - <u>Average Monthly Unit Production During Development</u>

Openings	No. entries	Distance m	No. units	Type unit	No. months	Advance m/month	Tons/m advance	Tons/month and unit
Mains	6	1 019	2	CM 1	12	85	72	3 060
				CM 2, 3				
	7	323	1	CM 1	6	54	85	4 590
	7	561	1	CM 1	7	80	85	6 800
Panel 1 Tailgate	3	622	1	CM 2	9	70	33.5	2 327
	3	610	1	CM 4	7.5	81	33.5	2 732
	3	128	1	CM 3[1]/	1	128	33.5	4 304
Maingate Bleeders	3	1 574	1	CM 3	12	131	33.5	4 407
Panel 2 Maingate	3	610	1	CM 1	5	122	33.5	4 087
Bleeders		915	1	CM 3	7	131	33.5	4 047

[1]/ Turning off the bleeders and driving NW to meet CM 4 advancing SE.

During June, July and August 1979, the face equipment was moved from the surface to the Panel 1 staging area by rail (fig. 14). The face conveyor, in lengths of three pans at a time, was assembled in the staging area by means of Curry scoops. Roof shields also carried by scoops followed. The shearer was then installed on the panline, and the whole face installation was ready and passed the MSHA electrical inspection in August 1979. Figure 15 shows the face layout. Figure 16 indicates the bidirectional operation mode as follows:

1. Start — The shearer cuts out into the maingate; leading drum M cutting, rear drum T idling.

2. Snaking — The shearer reverses motion and snakes into the face; drum T, now in front, cutting, drum M idling.

3. Sumped in — The shearer completes the snake and past shield 20 reaches an attitude parallel to the face. The conveyor can be moved up to the face and straightened out.

4. Cutting out the wedge — The shearer, moving back into the maingate, cuts out the wedge of coal.

5. Cutting to tailgate — The shearer flits back past the snake position and continues a full cut, cutting out into the tailgate. As the shearer passes, the conveyor is moved up to the face. The shields follow.

The motion at the tailgate mirrors the above cycle. The shearer, moving toward the maingate, reaches the attitude parallel to the face past shield 100.

The coal in the face area measured approximately 0.7 m initially and then increased to 0.75 m (30 in) in height. The shearer drums armed with 75 mm (3 in) picks cut a mining height of 0.91 m (36 in). The drums are 0.91 m (36 in) in width. With this cutting web, the solid state load sensing device allowed a haulage speed of no more than 1.80 m/min (6 ft/min) without tripping. Therefore, the stroke of the

Fig. 14. Shearer on wheels.

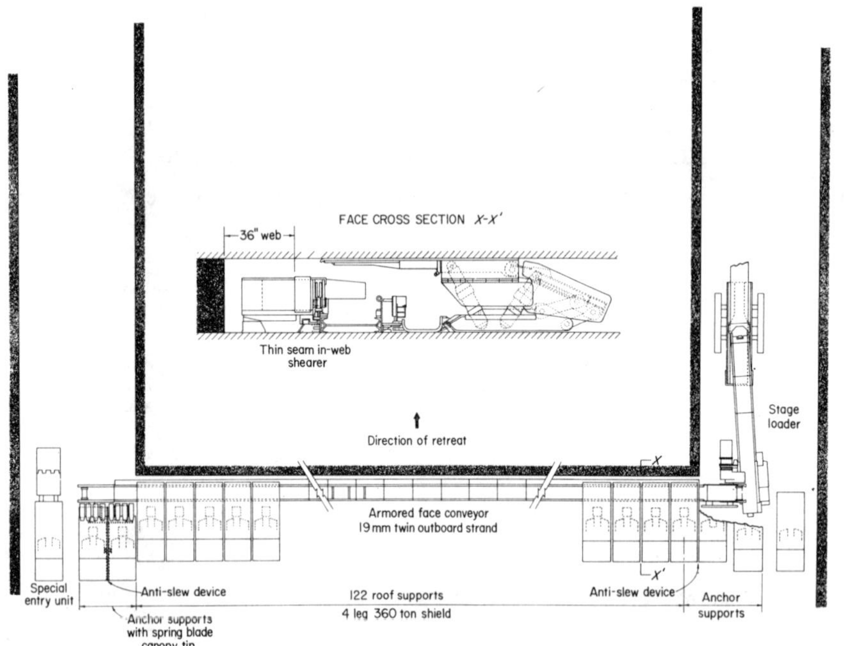

Fig. 15. Face layout.

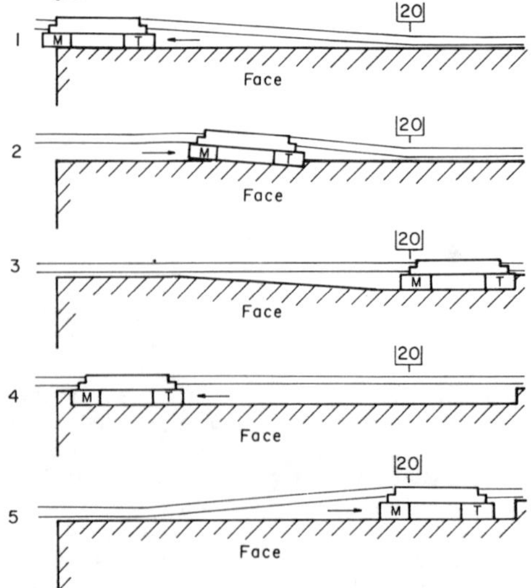

Fig. 16. Longwall operation mode.

shield advancing rams was reduced to 0.8 m (32 in) in order to narrow down the web and thus to increase the haulage speed to 2.40 m/min (8 ft/min). Even with this narrower web, the chainless haulage system was not adequate to the severe conditions at the Leeco mine, where the machine must cut its own floor to provide 0.91 m (36 in) of vertical clearance in a 0.75 m (30 in) coalbed.

In brief, the in-web shearer was the major source of face delays. With the chainless haulage, the main problems were breaking of the driving chain, frequent tripping and breakdown of the solid state load sensing device, and fouling of the captivation shoes. After the haulage module stripped some gear teeth in January 1980, the shearer was taken out of the mine and converted to chain haulage. The machine was out of service for most of 2 months.

Substituting a haulage chain for the chainless haulage brought on other problems, notably breaking of the haulage chain. Also, the debris have to be lifted now as high as 0.45 m (18 in) from the shearer track over the haulage chain into the panline versus a lift of only 0.25 (10 in) with the chainless system.

The shearer was again taken out of service during May 1980 to provide more effective tensioning of the haulage chain and to improve the efficiency of the loading process by fitting a horizontal tubular guidance system to the ramp plates of the panline, which the manufacturer claimed has proven successful in a United Kingdom mine. Production has improved since the new guidance system was introduced.

The face conveyor was next to the shearer in causing delays. At the very beginning, the twin outboard DIN 19 mm chain broke frequently and some of it had to be replaced. This failure was ascribed to improper tightening of the shackles; but chain breakage was still quite frequent during the course of the operation. Repairs are particularly time consuming if the break occurs in the bottom race. Also, the face conveyor was running for several months driven by the maingate motor only, pending replacement of a defective tail end motor. Such an operation is extremely damaging to the chain, too. The chain appeared to be wearing out quickly, and the manufacturer had tests performed which indicated that wear and failure of the chain can be attributed to excessive rubbing. A new chain was installed in August 1980.

Leaking rod seals occurred at several advancing rams of the shields as a consequence of inexperienced shearer operation in the beginning when steps were cut into the floor. The hollow ram rods buckled and damaged the seals. They were replaced with solid rods.

A minor rock fall occurred in front of four shields.

Failure to advance several shields in step with the shield line during two shearer passes resulted in these shields yielding until they became solid. They were, however, raised and advanced to the shield line by increasing the pressure in the hydraulic system above the yield pressure through the use of an hydraulic intensifier that is normally employed to test the performance of yield valve cartridges.

Leeco monitors the operation and prepares the following documentation;

Longwall Foreman's Daily Report
Longwall Operating Summary (monthly)
Longwall Maintenance Report (monthly)
Monthly Longwall Manpower Analysis
Longwall Operating Analysis

The "Longwall Operating Analysis" was developed by Leeco to tabulate the time of machine availability for cutting and the actual length of cut in feet achieved during each shift versus the potential length of cut calculated for assumed haulage speeds of 1.5, 2.4, and 3 m/min (5, 8, and 10 ft/min). Turnaround maneuvers at face ends make the shearer travel 244 m (800 ft) to cut a 183 m (600 ft) face. For instance, it takes a shearer 100 min of cutting at 3 m/min (8 ft/min) to complete a full pass on a 183 m (600 ft) face. Thus, the actual shearer speed of 2.4 m/min (8 ft/min) is reduced by a ratio of 0.75 to an apparent speed of 1.8 m/min (6 ft/min) to adjust for the turnaround time.

The Rock Mechanics monitoring program, proposed and conducted by the University of Kentucky, was designed to achieve the following objectives:

1. To provide early warning capability by measuring strata pressure changes and movement during mining so that corrective measures could be taken quickly should an unusual situation arise.

2. To evaluate the adequacy of the mine design and to derive design criteria for future longwall mining efforts.

The program includes the following tasks:

1. Measurement of pressure changes in selected chain pillars adjacent to the longwall panels as mining progresses.

2. Measurement of ground mass movement.

3. The instrumentation of two boreholes drilled from the surface through the coal to monitor the response to mining of the strata above the coal.

4. Remote monitoring of stresses in two of the underground chain pillars between longwall panels during the mining operation.

The original horizontal in situ field stresses in the rock before longwall mining at the level of the coal seam are measured by the borehole deformation method. Four sites will be instrumented (fig. 13). Changes in stress as mining progresses will be measured by stress meters of the vibrating-wire gage type. Movements of the rock mass underground will be determined by measuring convergence and horizontal movement between roof and floor in entries and by measuring differential roof sag by means of anchors set at different elevations in the roof.

The subsidence study, also proposed and conducted by the University of Kentucky, is designed to determine the characteristics of subsidence due to longwall mining of a thin coalbed under irregular surface topography. The work is accomplished as follows:

1. Establishment of a system of subsidence monuments and periodic surveys to determine subsidence characteristics over three mined longwall panels.

2. The installation of an Automatic Data Acquisition System (ADAS) line to determine the character
   istics of the subsidence traveling wave.

CONCLUSIONS

With a shearer haulage speed of 3.65 m (12 ft/min)
and adjusting for the turnaround times at the face
ends, the shearer could cut a 183 m (600 ft) face in
75 minutes. A shearer pass in 0.91 m (36 in) coal
with a web of 0.91 m (36 in) yields approximately
200 tons of coal. Discounting average delay times,
which may be extensive in thin coal, a cutting time
of 265 min per shift will allow for 3.5 passes per
shift or a potential production of 700 tons. At
present the operation is still far from reaching
this goal.

REFERENCES

1. Staff, BuMines. The Reserve Base of Bituminous
   Coal and Anthracite for Underground Mining in
   the Eastern United States, BuMines IC 8655,
   1974, 428 pp.

2. Sanders, G. Longwall in Thin Seams. Min. Cong.
   J., March 1970, pp. 57-64.

3. Curth, E. A. Relative Pressure Change in Coal
   Pillars During Extraction, A Progress Report,
   BuMines RI 6980, 1967, 20 pp.

4. Gales, T. Minimum Cost Strategies for Longwall
   Equipment Moves, BuMines Contract with Gates
    Engineering Co., J0166127, 1977. 125 pp.

Chapter 30

STUDY OF LONGWALL COAL MINING

APPLICATION IN THE PLAINS REGION, CANADA

Tony B. Szwilski

Department of Civil Engineering
University of Kentucky
Lexington, Kentucky

ABSTRACT

A short study has been made of the underground
mining potential in the Plains region, the Province
of Alberta, Canada. Particular reference is made to
the possible application of longwall mining: Presently
underground mining is limited to small scale room-and-
pillar methods. The problems anticipated with large
scale mining at shallow depths (30 m) are examined.
Production factors based upon longwall retreat mining
utilizing shield supports are also considered.

INTRODUCTION

The Plains region accounts for over 80% of the
reserves (inferred) of Alberta. The total measured
coal reserves of Alberta have been estimated at around
20 Gt of which the Plains accounts for about 50%
(measured). The three main coal areas of Alberta are
the Inner foothills, Outer foothills, and the Plains
region, Fig. 1.

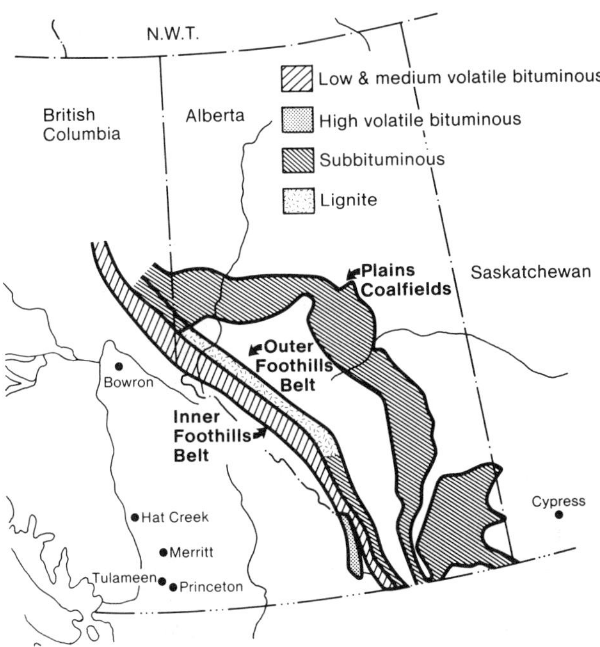

Fig. 1. Areas of coal deposits in Alberta Province.

The Plains region coal deposit is delineated on the
West by the Rocky Mountains coal belt and covers an
area of about 246 049 km^2. The coal strata occurs in
massive lateral zones associated with the Upper
Cretaceous Oldmen and Foremost formations. The

geological formation of the Plains region dip very
gently towards the southwest: consequently, horizontal mining can be considered. However, the
lateral continuity of the coal-bearing strata is
often interrupted by the sedimentary process and
effects of glaciation.

The depths of the coal seams vary from several
meters to over 650 m. However, it is not anticipated
that mining will exceed a depth of 250 m, giving a
mining range of about 25 to 250 m.

The coal is generally of low rank high volatile
bituminous and subbituminous of the Upper Cretaceous
age, the calorific value on a dry basis ranging from
15 119 to 28 032 kJ/kg and an ash content in the
Edmonton area of about 15%.

The workable coal seams on a large scale are of a
particular lensoid shape, the floor being gently
undulating. However, the resulting gradual changes of
the coal seam thickness should not affect coal extraction operations. The structure of the coal seam often
exhibits splitting or fingering with an increasing
thickness of clay partings which may affect the mine
ability of the seam by lessening the thickness of the
seam or decreasing the coal quality (dirty seams).
The coal seams in the Plains region are seldom associated with the stronger rock formation series of
sandstone or siltstone, the predominant lithology of
the coal strata being defined by thinly bedded clay
shales, silty shales, and sandstone, the clay cement
content varying.

The thickness of the individual coal seams may be
conveniently classified into three categories:

(1) Thin coal seams, less than 1.52 m, probably
not economical for underground mining (presently.)

(2) Medium thick coal seams 1.52 to 3.66 m, the
upper limit being described by the coal cutting
machine. This range generally offers efficient
longwall mining operation.

(3) Thick seams 3.66 to 5.1 m and over. This
category also includes the coal seams divided by
thick partings.

The majority of the workable seams are of the
medium-thick coal category. Fig. 2 shows a coal
section in the Plains region that represents five
coal seams (Anon.,1974)

Strip Mining: Presently, strip mining predominates
in the Alberta province. Most of the strip mines
operate at a stripping ratio of less than 10:1. Some
produce coal and supply a nearby power plant at a cost
of between $2 to $3/t. The economic limit is considered to be a stripping ratio of about 30:1. Fig. 3
gives coal strata sections at three strip mining sites
near Edmonton (Pearson, 1959). The overburden varies,
the limit of which is considered to be 120 m as
problems are encountered with the overburden dump,
especially in wet weather.

Market: The growth of the market for the subbituminous (steam coal) basically depends upon the internal

electrical power demand within the Alberta province itself. Transportation out of the province to the area of highest power demand in Ontario province is presently out of the question, due to the high costs involved. Also, as the moisture content of the Plains coal often approaches 30%, difficulties are introduced in long distance transport as the coal fragments when dry. There is no doubt that on an economic/productivity basis underground mining in Alberta is not able to compete with strip mining. However, the protection of valuable farming land and the environment have to be considered, especially near the principal cities where the electric power demand is the greatest.

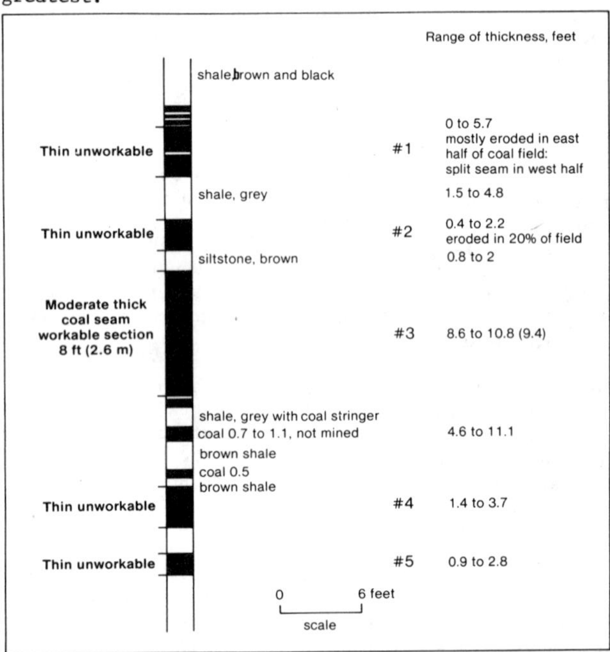

Fig. 2. Coal seam section in Plains coal field.

UNDERGROUND COAL MINING AT SHALLOW DEPTHS

The Plains coal may be described as variable, non-competent roof, hard subbituminous coal, well cleated, varying in thickness, and floor of medium competence. The coal seams have been subject to many small mining operations in the past, the majority of these small mines being limited to a depth of about 50 m working mainly room-and-pillar, although hand-filled longwall mining was often used.

The two main systems of mining considered are room-and-pillar and longwall. Room-and-pillar mining in Alberta has been found to be profitable and capital investment is lower than for longwall. However, good roof and floor strata conditions are essential for room-and-pillar mining to be a success. Also, leaving remnant pillars has been found to be hazardous to future farming or development of the land due to the often sudden occurrence of cavities or potholes due to collapse of the rooms.

Although capital investment for longwall is substantially higher than for room-and-pillar many advantages are offered. The surface subsidence is more effectively controlled and occurs more gradually, lessening the risk of fissures or the sudden appearance of cavities at the surface. Coal extrac-

tion using longwall mining is generally higher; 80+% extraction may be achieved compared with 40 to 60% for room-and-pillar mining. Also for longwall manpower requirements are less, environment conditions at the face are more effectively controlled, and the safety record is better.

The coal seams, the thickness of which varies up to 6 m, are well cleated and will facilitate coal cutting. However, due to the shallow depth the front abutment of the coal face will not be of a magnitude to significantly fracture the coal. The roof strata of the coal measures around the Edmonton area consists of incompetent muddy shales to more competent sandstone (Figs. 2 and 3) of uniaxial compressive strength 60 MN/m^2. In the Edmonton area the bearing capacities of the roof (with coal tops) and floor (saturated conditions), determined by the use of a hydraulic jack and 46 cm square plate, are about 4.2 MN/m^2 and 3.5 MN/m^2, respectively. Loading tests were carried out in saturated conditions to simulate harmful effects of an inundation of water from a flash-flood at the surface. A break-off line in the roof can effectively be achieved by a row of 101.6 mm diam round wood props 1 m apart producing a high degree of waste flushing. Where the overburden is at a minimum, about 20 m, the very nature of the weak strata has given rise to concern regarding possible harmful effects of high setting loads offered by the face-powered supports. The possibility of high setting loads causing direct shear of the immediate roof strata to the surface should be considered when selecting the face-powered supports and respective loading characteristics for such shallow depths of working. This subject possibly deserves further study.

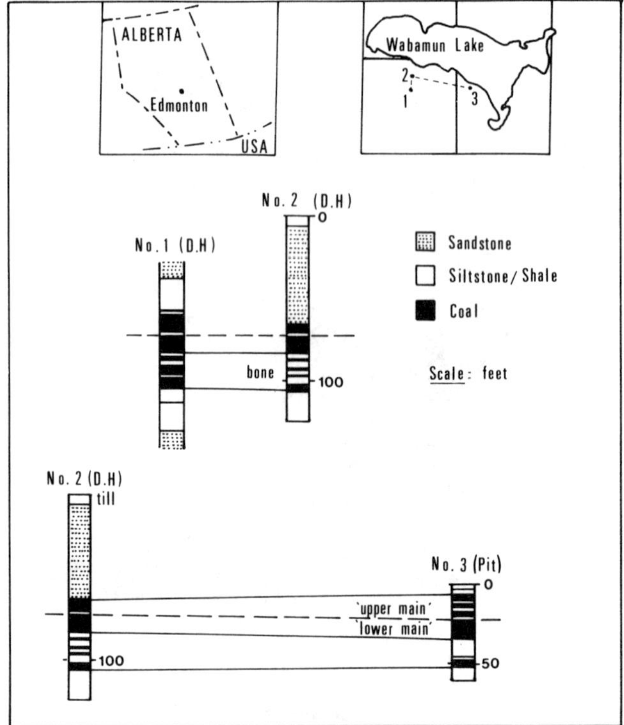

Fig. 3. Coal seam section at strip mining site, Edmonton, Alberta.

<u>Strata Control Aspects</u>: Strata control concerns the controlled movement of the strata rocks surrounding a mining excavation. Mining methods may be studied

for the most effective strata control, bearing in mind the economic constraints. Following underground mining excavation the strata pressures are redistributed around the opening (Whittaker, 1974), Fig. 4 (based on experience in the United Kingdom). Consequently, the need to support the excavation arises due to the localized fracturing and detachment of roof and wall rocks. The longwall method of mining is a most efficient method of mineral extraction in stratified deposits and is quite adaptable for high mechanization and high rates of production. Longwall mining is generally not limited by depth and can be employed in coal zones of weak strata. As the longwall faces advance (or retreat), the coal which once effectively acted as a support is extracted. The load is then transferred from the now gob area to the perimeter of the extracted area creating strata pressure abutments, Fig. 4. The gob (waste area) becomes destressed and this phenomena allows mining to progress. The unsupported roof in the waste area caves and in time reconsolidates and slowly accepts load from the overburden, the degree and rate of reconsolidation depending upon the depth of workings. Knowledge of the distribution of strata pressures produced by longwall mining is important in deeper mines when considering the design parameters of the longwall face and planning future workings.

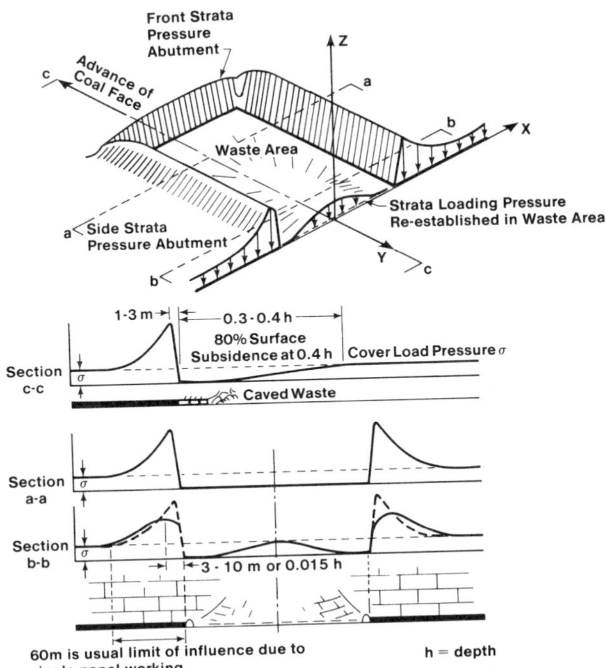

Fig. 4. Redistribution of strata pressure following coal panel extraction.

The main forms of interaction are generally associated with effects of high pressure zones created by nearby workings, remnant pillars, and rib edges, and strata displacements caused by undermining (King, et al., 1972); this is particularly highlighted in deep workings. However, the shallow coal seam conditions of the Plains coal field will give rise to lower strata pressures: for example, workings at a depth of 100 m will realize abutment pressures of about 8.4 to 10 MN/m^2. It is quite often standard practice to leave a rib pillar between successive

longwall faces giving suitable protection for the gate roadways. The width of the rib pillar should be calculated taking into account the depth, extraction height, and strength properties of the coal. Protection of the longwall workings against interaction may be afforded by choosing the direction of advance of the longwall face parallel to the old rib edges in other seams. Also, since strata loading and interaction effects are time-dependent, rapid retreating faces assist in avoiding roof control difficulties.

High Speed Retreat Faces: The length of longwall coal face generally varies from about 100 to 300 m. The main factors that influence the length of the longwall coal face are:

(1) Capital outlay to equip the coal face for a specific production rate.
(2) Strata conditions due to competence of coal face advance is a controlling factor.
(3) Coal face horizon control of the armored flexible conveyor (AFC) and powered supports.
(4) The length of coal face (width of panel) should be of such a magnitude as to allow effective caving. Good caving affords the dissipation of the vertical strata displacement and alleviates high strata abutment pressures.

Consequently, the longwall face should be long enough to provide good caving and maximize coal production by lessening the turnaround effects at the face, and short enough to satisfy the constraints of speed of advance, capital outlay, and horizon control.

Therefore, the Plains region coal seam conditions may be conducive to a longwall face length in the order of 10 to 150 m. The Star Key coal mine, Edmonton, is currently considering a longwall retreating system, electing a face length of about 100 m (using 66 powered supports). The prospects for effective caving are good and the competent hard clay/shale floor will be a large controlling factor in the stability of the roadways, especially at such shallow working depths, 20 to 25 m.

High Speed Development of In-Seam Roadways: Retreat mining demands efficient high speed development. However, doubts about the stability of the development roadway and the inconsistent drivage rates has dampened the popularity of retreat mining in the deeper mining conditions in Europe. High speed development in the thick Plains coal seams will necessitate the employment of a sound and effective rectangular support system. The Plains region coal is generally well cleated and should easily be machined, giving good development rates. However, the poor roof conditions will require the use of a fast and efficient bolting system. The choice of type and distribution of bolts is important. The system of support should take into account the basic costs of the supports and the stability required throughout the expected life of the development roadway.

POWERED FACE SUPPORTS

The object of the powered face support is chiefly to give direct support to the roof top, provide a mobile support, and separate the caved area from the coal face to provide a safe working place. The powered supports must provide a working area to allow the travel of mine personnel, cutting machine, and an efficient course for mine ventilation. They must also accommodate any peak roof loading conditions at the coal face. There are many variables to consider for efficient support operation, the two major factors being:

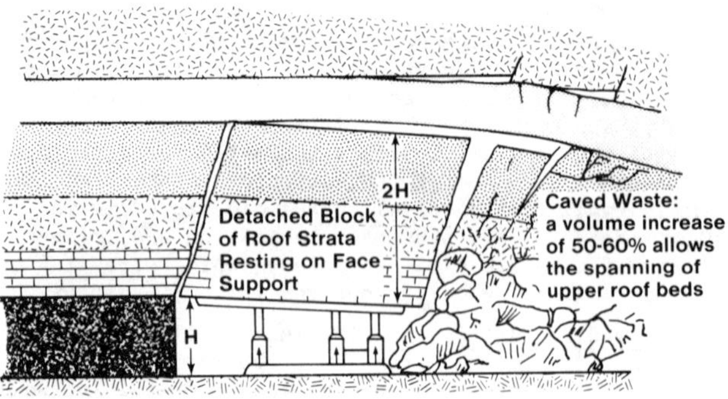

Fig. 5. Roof strata loading on powered face supports (after Ashwin).

(1) The expected strata loading of the supports, Fig. 5. The magnitude of strata loading is directly influenced by extracted seam height, strength characteristics of the roof strata, depth of workings, and time of residence of the support (rate of advance of the coal face).

(2) Caving and flushing characteristics of the roof strata. A range of power-set legs are available in the order of 30 to 200 t. They may be single or double telescopic, both single or double acting. The loading parameters are given in two categories: setting loads and yielding loads.

Setting Load: High setting loads may damage weak roofs by the direct load or the cyclic setting and releasing operations. However, low setting loads may allow the strata to generate higher loads on the powered supports.

Yield Load: The yield load gives a direct indication of the overall strength of the support unit. Where weak roof conditions exist caving has not been found to be a problem. However, strong sandstone roofs often require high shearing forces to promote caving. Also, powered supports must be designed to accept sudden (dynamic) loadings. Therefore, a safety factor $\underline{S}$ is applied in the design parameters of the support unit. Table 1 gives a summary of the recommended setting and yielding load densities, based on the relation (Ashwin, et al., 1978): minimum setting load density = 0.08H MN/m^2, Fig. 5, where the considered safety factor is $\underline{S}$ = 2.

Table 1—Summary of Recommended Setting and Yielding Load Densities

Seam Thickness, m	Min. Setting Load Density (SLD),* MN/m^2	Nominal SLD,† MN/m^2	Nominal Yield Load Density, MN/m^2
2	0.16	0.21	0.26
3	0.24	0.32	0.34
4	0.32	0.42	0.52

*Minimum setting load density (SLD) + 33% * = nominal (SLD)
†Nominal yield load density = nominal SLD x 1.25.
‡Losses in hydraulic system in setting load.

For the Edmonton area, it is anticipated that the yield load requirement for the shield support will be about 250 to 300 t. It has been estimated that given the worst strata loading conditions the shield support will induce a load of 1.05 to 1.4 MN/m^2 in the floor and 0.07 to 0.35 MN/m^2 in the roof. Double telescopic legs will be required to operate in a coal seam varying in thickness from 1.52 to 3.66 m.

SHIELD SUPPORTS

There are two principal types of powered face supports: chock and shield support. The shield support was originally developed in the USSR after World War II with the Sh-48 and later the OMKT models. The original objective of the shield support was to provide good roof support and isolate the coal face from the soft incompetent roof strata and high flushing waste. The author has made several visits to the principal coal fields of East and West Europe, observing numerous installations of shield support units. Traditionally, European mining practice has favored the shield support in incompetent friable roof strata where a high flushing factor is prevalent and the chock support for more harder competent sandstone type roofs. In the United States the shield has been found in many cases to be a flexible support that may accommodate a wide range of roof strata conditions. The shield support appears to be a favorable support for the coal strata conditions of the Plains region. Table 2 gives a summary of the loading characteristics of the shield supports.

Advantages:
(1) good safety record;
(2) maintenance costs are low compared with the chock support;
(3) there is considerable advantage for transport; this is important in the US with height restrictions on locomotive roads in coal mines of about 1.3 m; face installation is also easier;
(4) accomodates variation in seam thickness; better than that of the double telescopic chock support;
(5) short face to waste distance; In weak friable roofs this is an advantage;
(6) simple and rigid construction;
(7) complete separation of caved roof from the working place, protection of the face area against dust and heat flow from caved gob, and safety from roof falls at the face;
(8) the delay in setting the supports is low; and
(9) easier travelling along the face free from hydraulic hoses.

Disadvantages:
(1) high initial costs;
(2) the shields give a low cross-sectional area at the face, often making dust problems more difficult to control;
(3) as the shields are generally close together (skin to skin) it is often necessary to grade the floor of the seam, which may prove difficult with hard floors; and
(4) Gradients of up to 30° have been worked employing shield-powered supports; however, difficulties may be encountered with steeper seams.

Table 2—Loading Characteristics of Shield Supports*

	Chock Shield			2-Leg Caliper Shield			4-Leg Shield		
Height Range, m									
Closed	0.94	1.69	1.75	0.89	1.18	1.60	0.70	0.86	1.83
Open	1.93	2.79	3.40	1.80	2.68	3.61	1.60	2.16	3.66
Total yield load, t									
Minimum	280	300	300	189	226	310	183	250	290
Maximum	290	300	300	224	280	316	207	284	290

After Dowty.

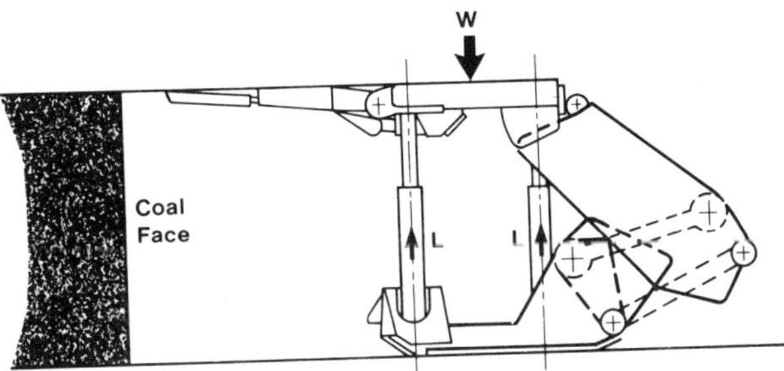

Fig. 6. Chock shield support.

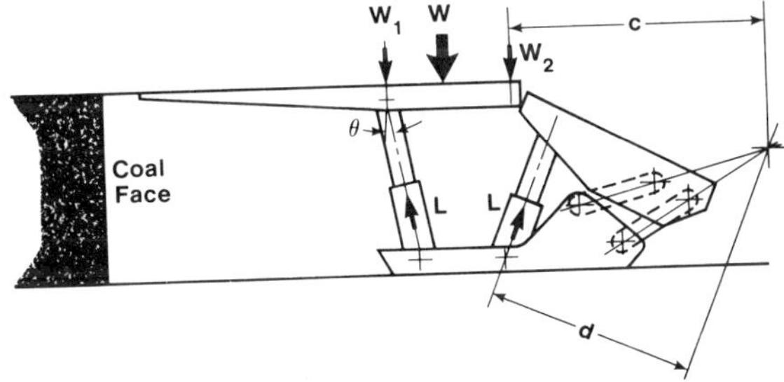

W - Total Roof Load

$$\text{Where } W = W_1 + W_2 = L \cdot \cos\theta + L \cdot \frac{d}{c} \text{ (approx.)}$$

Fig. 7. Four-leg shield support.

Two popular types of shield supports are the chock shield support (Dowty), Fig. 6, and 4-leg shield, Fig. 7. As the shields have a high resistance to lateral movement and good anti-flushing arrangements, they should be employed where the ground conditions make roof control difficult, thus minimizing loss of output. Consequently, employing shields may reduce the need to leave coal tops where leaving coal cannot be afforded. Typical yield load/height of shields are given for three types of shield support in Fig. 8.

COAL FACE CUTTING MACHINE

The most popular coal cutting machine currently employed in longwall mining is the shearer loader. In thick seams, the double ended ranging drum shearer (DERDS) would be the preferred machine. In the Edmonton area a DERDS should provide the flexibility necessary to operate in a coal seam varying in thickness from 1.52 to 3.66 m. Ideally, cutting machines should be so designed as to enable them to cut at low

drum speeds reducing maintenance to a minimum. Attention should be given to visual aspects to enable the operator to control and manipulate the machine with ease. Radio control has been used to assist the coal cutting operation by offering the operator a place of safety. There should be an emphasis on robustness and simplicity in design, with mechanical functions in place of hydraulics where possible. There should also be a facility for dust suppression.

Automatic sensing devices, such as nucleonic steering, have been incorporated on a fixed drum shearer to cut to the required level at the roof, although good results may be achieved by an experienced operator. The employment of shield supports may offer the extra room for the operator to maneuver so that he may concentrate on cutting the coal. However, this extra room will be a decided disadvantage if a haulage chain is used. The violent vibration or "whipping" of the haulage chain is always a potential hazard, and the breakdown of the chain under tension is also a source of danger. The employment of a horizon rack-a-track system may alleviate this hazard by eliminating the haulage chain.

At the present rate of development in machine design it is anticipated that the size of the coal face cutting machine may eventually exceed 746 kw, using 4160 V and weighing up to 50 t. The power of the machine will increase as the demand for good coal cutting at slower shearer drum speeds increases, and for cutting thicker webs 9 to 12 m. Drum speeds of 33 rpm have been used successfully; however, the optimum is considered to be 21 rpm. This optimum is a compromise between minimum dust make and speed of drum required to maintain a cutting advance of up to 0.11 m/s, although dust making in the Plains coal should not be a problem due to the high moisture content of the coal.

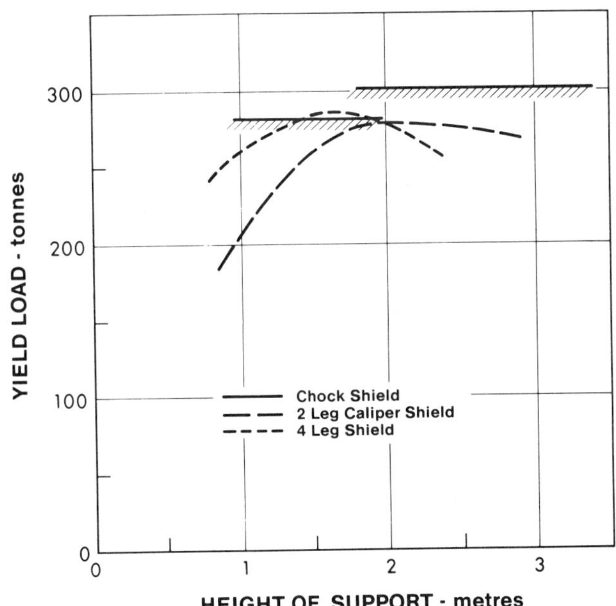

Fig. 8. Loading characteristics of shield supports.

ARMORED FLEXIBLE CONVEYOR (AFC)

The AFC should be designed to transport coal continously and must be strong enough to withstand the forces of the cutting machine and the push-over rams of the powered face supports. The large shearer can weigh as much as 20 to 25 t and can result in the warping of pan deck plates, especially with the 86 cm pans. However, the closed bottom pans and bottom plate welded across the width of the pan forms a strong box section which assists in preventing warping. Generally, the AFC is the weakest link in the production capacity of the longwall coal face. There are three factors to consider in selecting an AFC:

(1) The capacity of the cutting machine. For example, assuming a cutting machine rate of 0.05 m/s, a width of cut of 0.7 m and height of extraction of 3 m, this gives a production rate of about 780 t/hr.

(2) The conveyance capacity of the AFC. Table 3 gives the details of typical conveyance capacities of two standard sizes of conveyor.

Table 3—Typical Values of AFC Capacities

AFC Speed, m/s	Capacity in t/hr	
	AFC Size, 76 cm	AFC Size, 61 cm
0.86	455	300
0.95	500	330
1.08	570	375
1.20	640	420

(3) For a DERDS cutting in both directions along the coal face, the coal has to pass beneath the underframe of the shearer, Fig. 9. Also, the coal conveyance capacity depends upon the relative velocity of the DERDS and the AFC. That is, cutting toward the main gate (intake) and tailgate (return), the conveyance capacity of the AFC would be about $A(V_c - V_m)$ and $A(V_c + V_m)$, respectively (Fig. 9.). Assuming speeds of DERDS and AFC during cutting of 0.05 and 0.95 m/s, respectively, these relations would give (coal loading factors of) $A(V_c - V_m) = 0.9A$ cutting toward the main gate (MG) and $A(V_c + V_m) = 1.0A$ cutting toward the tailgate (TG).

In the United States, a record-breaking coal face produced over 12 000 t/d of coal. To achieve this figure an AFC would have to run for 24 hr continuously (nearly) at a capacity of 500 t/hr. Any stoppages would result in lost production as the AFC cannot run at over-capacity.

The current trend for AFC design is toward heavier duty pans and it is anticipated that further developments of the AFC will be open-section pans of 1.0 m wide chain section of 26 to 34 mm, and a conveyance speed of 1.37 m/s (V_c) resulting in a capacity of 1000 to 1500 t/hr. The pans will have to be strengthened with a center plate of 2.5 cm thickness. Dowty Meco is now developing a 0.86-m wide conveyor with a capacity of 1000 t/hr.

CONCLUSIONS

The Plains Coal region is a vast area of relatively untapped coal reserves. Presently, the majority of the coal produced in Alberta province is by strip mining. On the basis of coal productivity, underground mining cannot compete with strip mining. However, when taking into account several environmental factors, underground mining may become a viable option.

Longwall retreat mining utilizing shield supports is considered to be the preferred mining method. A new coal mining project, to be economically feasible, would probably exceed a coal production of 2 Gt/a. Two longwall retreating faces could achieve such a production target, each face being 100 m long and installed with 66 shield supports.

The longwall mining method is an efficient method

of coal extraction. The sequence of operation should be planned carefully so as to achieve near full extraction, avoiding interaction effects of nearby workings and detrimental effects due to surface subsidence.. High speed retreat faces have been used with success in many coal fields. Determing the principal design parameters of the longwall face requires a compromise between economic and strata control factors.

Coal production depends, to a large extent, on the conveyance capacity of the armored flexible conveyor (AFC) at the coal face and the outby conveyors. An efficient bunkerage system would serve to even out the various production peaks and afford continuous conveyance to the surface. With the current trend of cutting wider webs, more heavier duty coal face equipment will be necessary, including more powerful coal cutting machines, stronger AF conveyors, and heavy duty powered face supports to accommodate a prop-free-front distance (PFF) in the range of 2.3 to 2.6 m.

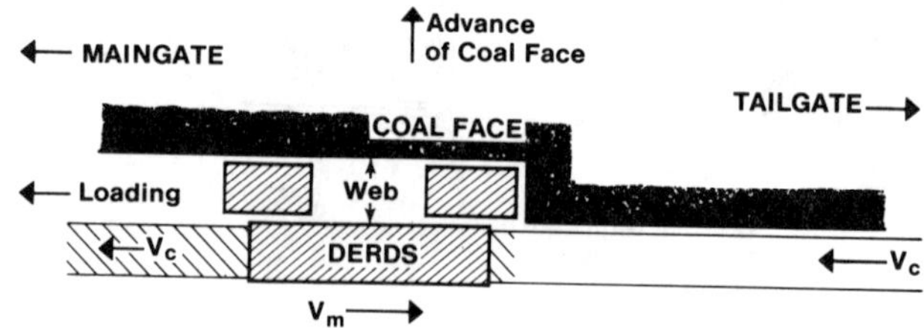

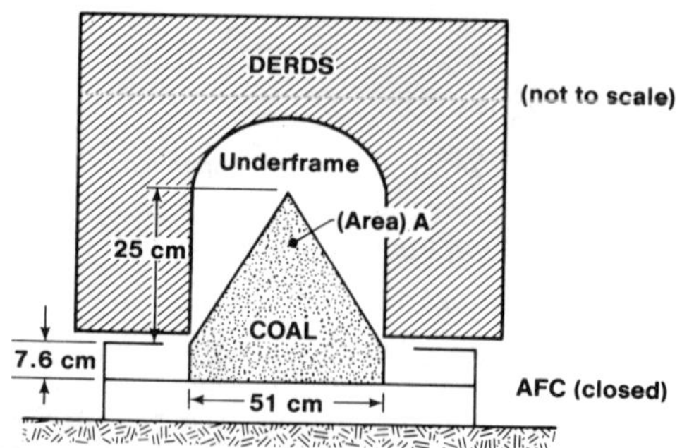

Fig. 9. Coal face production parameters, where DERDS denotes double-ended ranging drum shearer and AFC denotes armored flexible conveyor.

REFERENCES

Anon., 1974, "Northern Alberta Route (No. 14)," 10th
    Commonwealth Mining and Metallurgical Congress,
    Sept.

Ashwin, D.P., et al., 1978, "Some Fundamental Aspects
    of Face Powered Support Design," Mining Engineer,
    Vol. 129, No. 119, Aug. pp. 659-671

Graham, J.J., 1978, "A Review of Some Recent Powered
    Support Developments," The Mining Engineer, June.

King, et al., 1972, "The Effects of Interaction in
    Mine Layouts," 5th International Strata Control
    Conference, London.

Person, G.R. "Coal Reserves for Strip Mining, Wabamun
    Lake, District of Alberta," Preliminary Report 60-1,
    Research Council, Alberta.

Whittaker, B.N., 1974, "An Appraisal of Strata Control
    Practice," Mining Engineer, Oct.

Chapter 31

SHEARER AUTOMATION HORIZON CONTROL

E. R. Palowitch
U.S. Department of Energy, Pittsburgh, PA

and

P. H. Broussard, Jr.
NASA Marshall Space Flight Center, Huntsville, AL

INTRODUCTION

The United States aims to achieve some modicum of energy self-sufficiency through means such as conservation, increased domestic oil and gas production, substitution by the electric utilities of coal or coal-oil mixtures for oil and gas, and development of a viable synthetic liquid fuels industry. Coal's role in attaining this goal will necessitate doubling coal production by the end of the 1980's. And if the United States does in fact become the World's Saudi Arabia of coal, the total annual production to meet these domestic needs and export markets could double again by the year 2,000.

As only a small portion of our Nation's reserves are recoverable by state-of-the-art surface mining methods, most of the coal produced in the future will have to be mined underground (Averitt, 1975). But this projected increase in coal demand is occurring in the face of a decade of radically decreasing productivity. Specifically, average worker productivity in underground coal mines has decreased from 14.2 tonnes (metric tons) per workershift in 1969 to 7.2 tonnes per worker-shift in 1979; and 95 percent of coal produced underground during this period was produced by room-and-pillar methods (U.S. Dept. of Energy, 1980). Although there is no complete agreement on the impact of recent Federal legislation, there is strong correlation between the dramatic decrease in productivity in underground coal mining and the implementation of the recent Federal mine health and safety regulations (U.S. Congress, 1969).

A review of the wide range of coal mining research, development, and demonstration projects currently being funded by the federal government and by industry gives little encouragement for increasing worker productivity on room-and-pillar sections underground in the immediate future (Palowitch, 1978). Conversely, the opportunities for increasing production/productivity underground with longwall methods appear somewhat more encouraging. Although at least one longwall face in the U.S. did produce more than one million tonnes of coal in a single year, the national two-shift average production from one hundred or so longwall faces is about 1,000 tonnes per day (Huwood-Irwin Co., 1978 and Consolidation Coal Company, 1976). Clearly, there exists some real opportunities for improvement. One of these opportunities is to automate the cutter-loader.

THE CASE FOR AUTOMATION

There are at least five good reasons for automating the longwall shearer: (1) greater worker safety; (2) increased production/productivity; (3) increased coalbed recovery; (4) decreased dilution; and (5) decreased maintenance costs.

(1) Improved Safety - If the shearer can be traversed across the face while cutting, loading, and conveying coal without the need for "hands-on" operators, then many of the hazards to face workers and machine operators will be obviated. And although an insufficient data base exists to prove the reduced hazards inherent in the longwall method on an exposure-hour basis, this system, because of its higher productivity, is far safer than room-and-pillar methods on a per tonne basis.

(2) Increased production - As the volume (tonnage) of coal extracted is a product function of the height of coal cut, web depth, and haulage speed of the machine, coal output can be increased by increasing the height of coal cut, by increasing the web depth, and by increasing the speed of the machine. Nature fixes the thickness of the coalbed available for mining, and a recently completed study showed that the maximum practicable web depth for state-of-the-art shearers was about 100 cm (Foster Miller Assoc., 1979). Inasmuch as face conveyors for removing coal from the shearer at rates up to 1,000 tph are available, the limiting factor to high production on advanced technology faces is the ability to operate the machine at high speeds. The shearer speeds required to produce 1,000 tph in various heights of coal when cutting a 76-cm deep web are shown in Figure 1. Also shown in the figure are the maximum speeds at which miners were found able to travel through escapeways under emergency conditions (Stahl, W. R., 1962). By allowing about 0.3 m for the

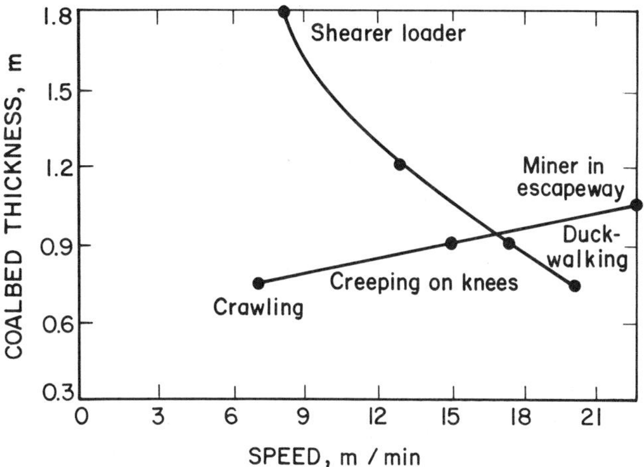

Figure 1. - Shearer speed required to produce 1000 tonnes per hour and miner escapeway travel speed as a function of coalbed thickness.

thickness of roof beams and conveyor push rams, it appears that the 1,000 tph capacity of the system cannot be realized by manual shearer operators in coalbeds less than about 1.2 m thick. The mining of thinner coalbeds and production rates higher than 1,000 tph can be achieved only through the use of an automated shearer.

(3) Increased Coalbed Recovery - A survey by A. D. Little Co., Inc., (1978) of the accuracy with which manual operators are able to cut to a given coalbed-roof interface and coalbed-floor interface showed that

much coal was inadvertently left unmined because of the operators' inability to control the proper cutting horizons. This unmined coal represented a significant loss in output per cut and a wasting of coal reserves.

(4) Decreased Dilution - The A. D. Little survey also showed that shearer operators frequently cut into the roof and into the floor unintentionally. This noncoal material mined with the coal dilutes its quality, puts an added burden on the preparation plant, and ultimately must be disposed of.

(5) Decreased Maintenance - If the roof and/or floor inadvertently cut is sandstone, limestone, or hard shale, the number of cutter bits consumed can be excessive, and the added wear on the machine can be significant.

THE CONTROL PROBLEM

Sophisticated guidance and control systems have been designed, installed, and operated over a spectrum of production, transportation, and experimental activities in environments far more hostile than the underground mining environment. But mining coal underground presents a somewhat unique problem. Coal being a sedimentary rock varies widely in chemical, physical, and mechanical properties and may contain copious quantities of methane, other gases, and water. Coalbeds often thicken or thin within short distances, undulate and, in places, are faulted, eroded, or wanting. Although the average thickness, stratigraphy, and chemical composition of a coalbed like the Pittsburgh coalbed may be remarkably consistent over hundreds of square miles, local variations in thickness, the occurrence of sulfur balls, and diverse stratigraphy frustrate any guidance and control concepts based on predictive modeling or assumptions of uniform planes.

Clearly, any system designed to mine a coalbed from a stratigraphic geologic section at depth must be able to control the extraction equipment by sensing the interfaces between the coalbed and its enclosing strata. The extractor must be capable of mining coalbeds that vary in physical properties, change in thickness, and undulate; the conveying system must follow closely the cutting operation and remove the coal from the face area cleanly and continuously; and the roof must be supported adequately and continuously as the face advances.

The double-ranging drum shearer meets all these extraction requirements. By adjusting the cutting horizons of the leading and trailing drums, a coalbed that varies in thickness and/or undulates can be accommodated easily. And because the cutting drums can move vertically independently of the shearer frame (and hence independently of the conveyor), the machine can follow reasonable coalbed perturbations.

The face conveyor, upon which the shearer travels, provides guidance in the horizontal plane and hauls coal from the face area continuously. As the face conveyor is pushed forward by the rams on the roof supports immediately following cutting and loading, it provides the added function of advancing the shearer incrementally (and essentially continuously).

At present, operation of a double-ranging drum shearer requires two operators: one to range the leading drum at the coal-roof interface and the other to range the trailing drum at the coal-floor interface. The machine operators must function as both error

detectors and controllers, i.e., they must observe the elevations of the cutting drums and operate controls to raise or lower the drums as required to cut at the desired interfaces. Because coal is opaque, the operators must rely on the noise, vibration, color of cuttings, marker bands in the coalbed, and the relationship of the present cut to the last cut. The efficiency of this closed-loop feedback system thus depends largely on the ability of human operators to observe errors and to take corrective action. Automated control of this cutting function would upgrade machine operators from "hands-on" operators to machine supervisors. While automating the cutting function may not be cost-effective in thick, flat-lying, easy-to-mine coalbeds, it becomes increasingly attractive as coalbeds become thinner and mining conditions deteriorate.

CONTROL SYSTEMS

Full automatic control of the longwall extraction process requires control in the three orthogonal planes. The vertical (pitch) control system adjusts the vertical position of the shearer drums in the direction of travel to insure that the cutting of the coal from the coalbed is taking place at a nominally specified distance from the coal/noncoal boundaries along the roof and the floor in accordance with specified error criteria. The use of human operators will be minimized, consistent with good practice, in the actual control of the machine; however, human oversight of face activities will always be required. Human operators will figure most prominently during the turnarounds at the end of each pass and whenever any unusual conditions might be encountered, e.g., mining through a fault.

Control of the roll about the shearer-loader's longitudinal axis is needed to prevent the formation of "saw toothed" steps in the floor and/or roof which could create serious problems for roof support and face conveyor advance. A control system based on a damped pendulum biased to the general inclination of the coalbed has been found adequate.

The heading of the shearer and the subsequent trajectory it executes is determined by the position of the individual roof supports and the extensions of the conveyor push rams. Errors caused by less than full ram extensions can accumulate and cause deviations in the conveyor azimuth and linearity which must be monitored and controlled. A technique based on measuring the variations in heading between successive conveyor sections with an angle transducer is being developed as a practical method for measuring deviations in yaw; automated roof control systems are available commercially and can be interfaced easily with this yaw control system. Only the development of a vertical control system will be discussed further.

Vertical Control System

The major technical impediment to developing a practicable vertical control system for longwall shearers is the lack of sensors to detect the location of the cutter bits relative to their desired location in the coalbed. Ideally such sensors would operate at the point of cutting, not require contact with the roof or floor, generate reliable real-time measurements continuously, have at least a 2- to 25-cm range with ±10 percent accuracy and 1-cm resolution, not require frequent or elaborate recalibration, operate on both roof and floor so as to allow bidirectional cutting, be mineworthy, and meet Mine Safety & Health Administration approval requirements.

In order to locate the boundary between a coal seam and its enclosing strata, some measurable contrast must exist. Thus, for discrimination, it must be possible to detect the difference in some material property. These differences may be of mechanical, magnetic, thermal, optical, or radioactive in nature.

The roof and floor strata enclosing most U.S. coalbeds are shales, sandstones, limestones, and clays, which usually vary in thickness and in physical, chemical, and petrographic properties. These changes often are subtle, and the transition from one rock type to another may be more a gradation than a discernable change, thus making discrimination by a sensor difficult. The coalbeds often contain such inclusions as streaks, bands, and partings of noncoal material and such concretions as sulfur balls and coal balls.

Because of the various geologic formations present, a wide range of coalbed boundaries, from discrete to diffuse, can be expected. Thus boundary detection often depends upon the physical property being measured. For example, mechanical properties of coal can approach those of shale, becoming harder and less friable closer to the interface; hence a mechanical sensor could not distinguish between them. Yet, the same interface might be detected by an optical sensor because of the difference in color.

A variety of coal-rock interface detectors have been studied in the laboratory, in the Bureau of Mines' Safety Research Coal Mine and at the Department of Energy's Mining Equipment Test Facility in Bruceton, PA (Pazuchanics, 1978 and Broussard, 1979). Of all the sensor concepts evaluated to date, only two appear to be practicable for underground use: (1) a natural radiation detector for measuring the amount of coal left unmined on the roof or floor and (2) a sensitized pick for identifying the type of material being cut.

Most underground coal mine workings receive low levels of natural radiation. These gamma rays are ejected during the natural disintegration of radioactive atoms of a potassium isotope in the strata enclosing coalbeds. As coal tends to disperse gamma rays, the reduction in natural radiation level as it passes through a layer of coal can be used as a measure of the thickness of that coal layer--the thicker the coal layer, the fewer the gamma rays that traverse the coal. A sodium iodide crystal with a photomultiplier tube that counts the number of gamma rays emitted by the naturally occurring radioactive material in the roof strata can be used as a natural radiation sensor.

To determine the usefulness of this concept, radiation levels were measured in samples of mine roof over the extent of the Appalachia Region from Birmingham, Alabama to Pittsburgh, Pennsylvania. Additional data were obtained from mines in New Mexico, Utah, Illinois and Colorado. The levels found were consistent over great distances within and between the mines sampled. Figure 2 shows the typical relationship between radiation count level and the thickness of a coal layer. Note that the 10- by 20-cm sodium iodide detection crystal provides a useful range of coal measurement from 0 to at least 25 cm.

The concept of instrumenting one or more of the cutting picks on a mining machine to discriminate between the material being cut is not new. But recent advances in instrumentation, signal processing, and the

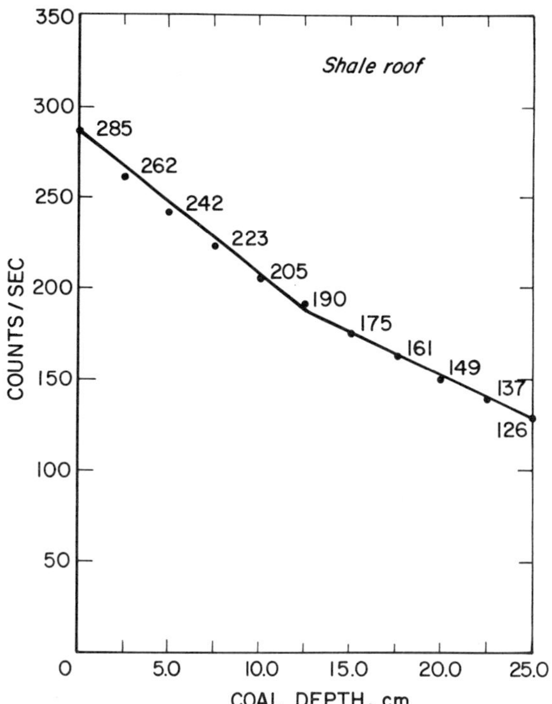

Figure 2. - Natural radiation as a function of the thickness of layer of coal.

development of mineworthy microprocessors have made it possible to improve the accuracy, reliability, and suitability of a sensitized pick for underground service.

By strain gauging one or more of the bits on the cutting drum, it is now possible to measure instantaneously and continuously the relative hardness of the material being cut at the interface. A minimum of two picks spaced 180 degrees apart normally is used. Based on a drum rotation speed of 45 rpm, a new suite of data is available every 0.33 s. The pick assumes that a force or rate of change of force threshold would be established for coal. The picks are interrogated generally only when they are in the arc containing the roof and floor interfaces. The decision to raise or to lower the cutting drum is based on outputs from the sensitized pick and a last cut follower.

It is not essential that the pick identify coal or rock correctly 100 percent of the time--correct readings as often as 70 percent of the time are sufficient for controlling the drums effectively. In fact, 100 percent accuracy is not achievable. For example, when the pick strikes a thin piece of rock, the output will signal erroneously that coal has been struck. Or, if the drum is raised rapidly while cutting coal, it is possible to generate a sufficiently large signal (due simply to kinematics) to indicate erroneously that rock has been struck. To accommodate these unusual situations, one decision rule might be such that no signal is sent to reposition the cutting drum until at least three successive like-signals have been received, e.g., coal-coal-coal or rock-rock-rock, but not coal-rock-rock, or coal-rock-coal. Another reason for requiring at least three like-signals is that excessive "hunting" by the shearer arms can result in overheating of the machine's hydraulic oil. Even though some response is lost, drum control is adequate.

The specific algorithm to be used in practice depends on the probability of successfully identifying rock or coal on each encounter; and this, to some extent, will depend on the mining strategy and on the geometry of the seam, i.e., how rapid the response must be to follow the roof contour acceptably. A simple calculation shows that if, for example, the probability of correct identification is 0.8, then requiring four successive like-signals would have a probability of 0.409. In this case the drum's response to a control command would be unacceptably slow.

There are two ways to transmit the signals generated by the sensitized pick to the microprocessor mounted on the shearer. One is to telemeter the signals via a transmitting antenna mounted on the drum to a receiving antenna mounted on the shearer; the other is to send the signals to the microprocessor via hardwire through a slip-ring arrangement in the hub of the rotating drum. Both methods have been proven practicable during tests on Department of Energy's mock longwall facility at Bruceton, PA, and the telemetered technique has proven successful on an operating longwall face.

GUIDANCE AND CONTROL REQUIREMENTS

The two sensors described above are adequate for designing a practical horizon control system. The particular sensor(s) used depends on whether coal is to be left on either the roof or floor; the natural radiation sensor can be used to control the thickness of coal to be left, if any, or the sensitized pick can be used to signal the absence of coal. Of the two, the natural radiation sensor is the simpler to install. The sensor is packaged inside a 20- by 20- by 10-cm rugged container which is mounted on the face side of the shearer by means of a simple bracket; it protrudes only a few inches above the top of the shearer and therefore does not interfere with its normal operation. An electrical cable connects the sensor either to the microprocessor for automatic control or to an operator's display for manual control.

The sensitized pick requires welding of the pick block assembly to the drum trailing immediately behind, and extending approximately 0.6 cm above, a standard pick. If telemetry is used, the transmitting antenna is welded to the drum so as not to interfere with the cutting or flow of coal.

Two basic configurations of the Vertical Control System (Mode I and Mode II) can be employed. Mode I is used when the cutting strategy is to leave a layer of coal unmined on the roof; this control scheme utilizes a natural radiation sensor and data stored during the previous pass. Mode II is used when the cutting strategy is to remove all of the coal to the roof; this control scheme requires only a sensitized pick. Under Mode I, the drum can be controlled by knowing either continuously, or at a suitable number of points, the thickness of coal left on the roof during the last pass and by knowing the elevation of the drum relative to the last cut surface at corresponding downface locations; under Mode II, the drum can be controlled adequately by a trailing coal depth sensor, provided the sensor does not trail behind the drum by more than about 0.6 m. These two modes are the only practical choices available when control of the drum is effected by using information in the vicinity of the drum. Mode I works because if the elevation of the drum is known relative to where is is supposed to be (coal thickness measurement on last pass versus desired depth) and if the elevation now relative to what it was then is known, then the drum can be commanded to assume a position that will decrease the error determined on the

last pass. Mode II works because if the depth of coal left behind the drum is known, and if the haulage speed is reasonable, ranging arm response is fast, and distance from the sensor to the cutting drum is small, then the drum cannot have moved sufficiently far to invalidate the error signal, i.e., the measured error and the drum height are still highly correlated.

While the context of these two modes has been discussed in terms of leaving coal in the roof, it should be noted that the cutting strategy to extract all coal from the roof using a sensitized pick constitutes a special case of Mode II wherein the distance between the drum and the trailing sensor is zero and where the presence of rock is determined rather than the thickness of coal left measured.

When coal thickness is being measured, one of the problems in using Mode II is finding a location for the sensor that follows closely the drum yet will not be damaged from falling debris and not interfere with normal operations. In this connection, coal depth can be measured closer behind the roof drum by tilting the sensor toward the roof drum. If the sensor is used only to augment manual control, then the operator's ability to estimate the relative height of the drum will allow locating the sensor further behind the drum.

The preceeding discussion identified the information needed to control the leading drum and the candidate sensors to be used. But nothing has been said about controlling the trailing drum. To date no feasible means has been developed for measuring the depth of coal left unmined on the floor. This lack of progress is caused by the fact that a large portion of the coalbed normally is left intact by the cutting of the leading drum. A sensor that could measure thicknesses of coal of possibly a meter and be able to ignore the accumulated debris lying on the floor would be required.

Investigations by A. D. Little (1978) have shown that the roof and floor profiles are closely correlated, i.e., they tend to undulate in unison and they generally have the same slopes. This condition allows an acceptable floor cut to be made in one of two different ways: (1) by measuring the distance from the trailing drum to the present roof cut with a present-cut follower and commanding the drum to maintain a preselected distance or (2) by storing the elevation (continuously or discretely) of the leading drum and commanding the trailing drum arm as it occupies corresponding positions to assume an elevation angle such that a cut of constant roof-to-floor height results. This is a purely geometrical approach.

Regardless of whether or not a portion of the coalbed is to be left unmined, an additional sensor, called a last-cut follower, is required. Its purpose is to measure the elevation of the horizon presently being cut relative to the horizon previously cut. It may be a mechanical, optical, electro-magnetic, or acoustical device. The mechanical follower is simpler but somewhat more vulnerable to damage than the other types of follower because the mechanical system must maintain physical contact between the drum hub and the last cut surface on the roof. A mechanical follower was found satisfactory during several tests on the mock longwall face at Bruceton. A small 35-GHZ radar, two types of small acoustic sensors (one using a standard piezo-electric device, the other using a Polaroid transducer) and an optical sensor which uses a laser light source also have been tested with acceptable results.

CUTTING TRIALS

The Joy double-ended ranging drum shearer on the mock longwall face at the Department of Energy's Mining Equipment Test Facility was retrofitted with a pair of sensitized picks, a present-cut follower, a last-cut follower and a down-face distance sensor. A block of "synthetic coal" 25 m wide by 15 m deep by 2 m high, consisting of a mixture of minus 25-cm coal, fly ash, cement, and water, was cast to simulate coal on a longwall face. This block was capped with 15 cm of concrete to simulate a mine roof. Figure 3 shows the shearer cutting the synthetic coal on this face.

FIELD TESTS

An Anderson-Mavor double-ended ranging drum shearer was retrofitted with a natural radiation sensor on one longwall face, and an Eickhof 300L double-ended ranging drum shearer was retrofitted with a pair of sensitized picks on another face to evaluate their performance underground.

The natural radiation sensor utilized a 10- by 20-cm rectangular sodium-iodide crystal with a lead colli-mating shield mounted in a rugged holder. As this sensor does not have to be in close proximity to the

Figure 3. — Shearer cutting synthetic coal on mock
longwall face.

While the synthetic coal and cement cap have a relatively smaller difference in hardness than true coal and roof rock, they are sufficiently different to provide a facility for testing shearer control components. This mock longwall face has been used as a test bed to evaluate individual sensors and a fully automated, computer-controlled, vertical control system.

The sensitized picks were mounted on the leading drum using a transmitter and antenna on the drum with telemetry and also a hard wire installation making use of slip rings. Mechanical, acoustical and radar last-cut and present-cut followers were mounted on the ranging arms. Potentiometers were installed on the ranging arm actuators to measure the elevations of the drums and a downface distance sensor was retrofitted into the gearing on the "Joytrack" to locate shearer position along the face.

Cutting of the artificial coal automatically has been accomplished in a completely satisfactory manner by feeding the outputs of these sensors to an on-board microprocessor. In fact, the physical application of the previously discussed conceptual algorithms presented no mechanical difficulty, and retrofitting the **automatic control system to the shearer proved straight-**forward.

In addition to proving the practicability of controlling shearers automatically, these tests provided baseline information needed to evaluate this system underground.

roof, it was installed on the shearer aft of the center and on the face side so that it protruded only a few inches above the shearer frame. A series of static roof measurements was made along the longwall face to ascertain that a uniform and measurable radiation level existed. It was found that sandstone replacement in the predominantly shale roof did cause the background radiation to vary. The count rate for sandstone, while sufficiently strong for a useable signal, is only about one-fourth that of shale. Thus, if the sensor is calibrated for shale, erroneous coal depth readings will be obtained when the sensor encounters sandstone, and the sensor will indicate a layer of coal thicker than actually is present. It is believed that this condition is somewhat unusual, i.e., roofs usually tend to be composed of a single rock. This sensor operated successfully during two operating shifts with no damage to the sensor and without interference to the normal operation of the shearer. The natural radiation sensor mounted on the shearer is shown in Fig. 4.

Although efforts to leave a few inches of top coal on the roof were not successful (the coal persisted in falling), the sensor recorded accurately the radiation levels from the roof under the harsh operating conditions found underground (see figure 2).

All of the laboratory experiments and field tests to date indicate that a rugged, MSHA-approved natural radiation sensor can be fabricated and retrofitted easily to existing longwall shearers to provide timely and accurate information on the thickness of coal left on the roof by the leading drum. Only in those instances where there were anomolous intrusions or other radical changes in the character of the rock types in the immediate roof did the radiation sensor fail to provide acceptable outputs.

Figure 4. - Natural Radiation Sensor mounted on Shearer.

Figure 5. - Sensitized Pick Mounted on Shearer Drum.

Figure 5 shows the sensitized pick mounted on the shearer. Because of the amount of roof rock being cut at the time the test was conducted, the sensitized pick was mounted virtually flush with and approximately in line behind the pick it was shadowing. While this arrangement afforded excellent protection to the sensitized pick, the responses of the pick to coal and shale were less dramatic than those which would have been obtained had the sensitized pick protruded about 0.5 cm above the pick it trailed.

Figure 6 shows output signals that are typically generated by the sensitized pick during successive revolutions of the cutting drum. The extremely high peak which occurs approximately 90° after the pick enters solid material was caused by the pick striking the "middleman" -- a 10- to 15-cm thick band of hard rock near the middle of the coalbed. This peak shows dramatically the increase in signal amplitude when rock is struck compared with the peaks of the signal where coal is cut. Statistical testing of the field data showed that

1. By looking at the magnitudes of the strains, the slope of the strains, and other measures, the response of the pick to coal differed significantly from the response of the pick to rock. Mean values for shale were as much as twice the mean values for coal.

2. By the selection of an appropriate threshold level, the sensitized pick output was correct 70 to 75 percent of the time.

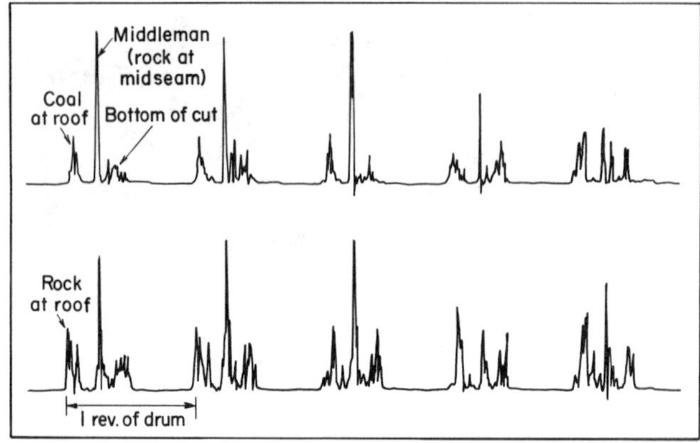

Figure 6. - Sensitized Pick Outputs.

Figure 7 shows a 35-GHz radar and Figure 8 shows an acoustic present-cut follower respectively, mounted near the cutting drums of the shearer.

Figure 7. - 35-GHz Radar Present-Cut Follower.

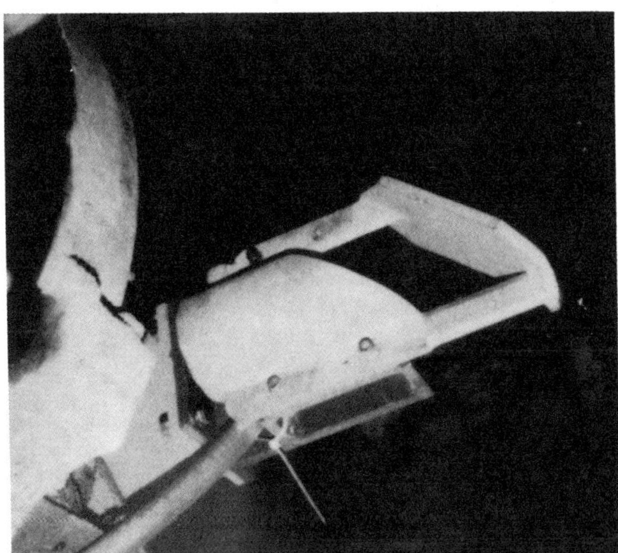

Figure 8. - Acoustic Present-Cut Follower.

CONCLUSIONS

Experience to date support the following conclusions:

1. Extensive mathematical modeling and physical simulation experiments all indicate conceptually that the cutting drums of longwall double-ended ranging drum shearers can be automated.

2. At least two types of sensors are needed--one for measuring coal thickness and one for recognizing the surface being cut. These sensors have been developed. A natural radiation sensor can measure the thickness of coal left unmined, and a sensitized pick can identify the material being cut.

3. Mechanical simulators have shown that the responsiveness and accuracy of these sensors and the machine dynamics of state-of-the-art shearers are such that a real time control system can be developed that will reduce the rms value of the cutting error characteristic of manual control by at least 50 percent.

4. Cutting trials at the Mining Equipment Test Facility have proven that rugged sensors and microprocessors that can function in the hostile underground environment can be developed.

5. Field trials have demonstrated that MSHA approved sensors can be retrofitted easily to commercial shearers and maintain the cutting drums at the desired elevations under automatic control.

6. There is no doubt that an automated vertical control system is practicable and that a fully automated longwall system can, in fact, be developed.

7. If a fully automated horizon control system on the shearer is not desired, it is possible to use only portions of the system, i.e., a

simple sensor - natural radiation or sensitized pick, depending on the application - and a display can be used in a man-in-the-loop system wherein the human operator remains as the controller. The display can be mounted either on the machine near the operator's controls or in the radio or umbilical cord hand-held remote control units.

ACKNOWLEDGMENTS

The authors are indebted to their colleagues in the Department of Energy and the National Aeronautics and Space Administration, to Messers. Ed Moore and Harry Elkins of Kaiser Steel Co., Mr. John Janes of Old Ben Coal Co., and to personnel from Foster Miller Associates, Inc., the General Electric Co., and the other contractors upon whose efforts this paper is based.

REFERENCES

Arthur D. Little, Inc., 1978, "Last Cut Survey," NTIS No. FE-8880, April.

Averitt, P., 1975, "Coal Resources of the United States," U.S. Geological Survey, Bulletin 1412.

Broussard, P. H. and E. R. Palowitch, 1979, "Control for Longwall Shearers," Annual Meeting AIME/SME, New Orleans, LA, Preprint No. 79-05, Feb., pp. 31-39.

Consolidation Coal Company, 1976, "Record Setting Longwall Crews Honored at Banquet," The Consol News, Feb.-March.

Foster Miller Assoc., 1979, "Effects of Wide Web Variation on the Performance of a Longwall Face," Vol. 1 and 2, NTIS No. FE-9058-1, -2, Nov.

Huwood-Irwin Co., 1978, "Off-the-Wall,", Vol. 1, No. 3, Aug.

Palowitch, E. R., 1978, "Underground Coal Mining Research," Mining Congress Journal, Vol. 64, No. 2, Feb., pp. 34-41.

Pazuchanics, M. J. and E. R. Palowitch, 1978, "Coal Interface Sensors for Automated Mining Machines," Proc. 4th West Virginia University Conference on Coal Mine Electrotechnology, Morgantown, WV, Aug., pp. 34-1 through 34-11.

Stahl, W. R., 1962, "Escapeways and Other Emergency Measures in Coal Mines," U.S. Bureau of Mines IC 8127, 13 pp.

U.S. Congress, 1969, "The Federal Coal Mine Health and Safety Act of 1969," Public Law 91-173.

U.S. Department of Energy, 1980, "Coal Data: A Reference," Report 0064(80), July, p. 34.

Chapter 32

PROGRESS WITH THE GUIDANCE OF ANDERTON SHEARER LOADERS IN THE UK

Peter G. Tregelles
Director of Mining Research and Development

and

Derek K. Barham
Head, Automation Engineering and Steering Branch, MRDE

INTRODUCTION

A successful step towards mechanisation of the collier's work was taken in 1954 when the first Anderton shearer loader was commissioned in Lancashire, and progress was reinforced in 1956 when the first powered supports were installed in Nottinghamshire. By 1970, virtually 100% of the coalfaces in the United Kingdom were mechanised and the industry's productivity had doubled.

In 1961 the National Coal Board recognised that the next move towards greater productivity and improved safety would be to automate the longwall system. What was not well enough understood, however, was that the basic prerequisites of automation - reliability and measurement - had not yet been achieved. Consequently, the schemes for automatic longwall faces and for an entirely automated colliery were abandoned at the close of the decade.

A review of mining research and development policy, commissioned by the Board in 1974, confirmed that the system of longwall mining had great potential for investment and that the short-term policy of automation must once again be pursued. Since then, aided by advances in computer technology, systems for automating the transport of coal from the face to the coal preparation plant have been successfully instituted. Applications of sub-systems dealing with the conveyor, coal preparation plant, and various items of fixed plant are now at the exploitation stage.

In the medium term, a number of key problems have been identified, which will arise once the productivity from the coalfaces has doubled, as it is now expected to do in the next five years. One of the most important of these problems is machine guidance, and considerable research and development work is now being carried out in this area.

This paper reviews the work on machine guidance prior to 1974 and explains the programme on which the National Coal Board's Mining Research and Development Establishment is currently working.

ORIGIN AND DEVELOPMENT OF MACHINE GUIDANCE SYSTEMS

Remotely Operated Longwall Face

Recognition of the need for machine guidance came in the early 1960s. It did not result from immediate production requirements, but from the long-term objective of the Chairman of the National Coal Board, Lord Robens, to develop a remotely operated longwall face (ROLF). As that project was studied, it became obvious that the manual horizon control of the Anderton shearer loader (ASL) would have to be replaced by an automatic system, not only to keep the machine in the seam but also to control the shape of the extraction so that other coalface equipment, such as roof supports and conveyors, could be remotely operated.

The three-dimensional aspects of machine guidance were not so clearly recognised then, and the problems of monitoring and controlling face alignment and advance were not given priority, although mechanical arrangements were included to preset the advance of the armoured flexible conveyor (AFC) by each individual ram. Similarly, the control of along-face machine speed and loading was not included in the initial programme.

Early ASL control development work in the late 1950s had concentrated on a horizon sensor, and a nucleonic backscatter sensor was developed and tested on a ROLF face in 1964 using a fixed drum shearer, then the most commonly used machine. The trial showed that improvements to the sensor were required, along with control engineering techniques to produce a stable system of steering.

These improvements were incorporated into a control system for steering a fixed drum shearer, using the information from a nucleonic horizon sensor and a machine tilt transducer, and the prototype was proved in several underground trials in 1969.

While this steering system was coming to fruition in the late 1960s, work began on an alternative horizon sensor. This was mounted in the drum and measured the force on a cutting pick as it cut the strata. In those seam conditions where there was a regular band or boundary of strata which produced an identifiable force level on the pick, it was possible to produce a stable steering system using the pick force information. Successful underground trials of the sensitized pick were conducted in 1969 and 1971, but further development was curtailed because staff resources were not available. There was also some doubt concerning the number of seams having a reliable and constant sensing band, but no instrument was available to carry out large scale surveys on the suitability of seams for pick force sensing.

The development of guidance systems for the ASL had originally been prompted by the introduction of the remotely operated longwall face, but during this period ROLF was not providing the revolution that had been expected. After six applications with varying degrees of success, the concept foundered, although the principle of remote operation of machine and power supports had been demonstrated. In the opinion of the authors, there were two main reasons for the decline.

Firstly, the industry had only recently come to grips with face mechanisation. By 1965, only half the faces in the country had power loaders and only a quarter had powered supports. Managers were not then prepared to make the major transition from mainly manual coalwinning to automated, mechanised methods of mining.

Secondly, the application of electronic control technology to general industry was still in its infancy, and to successfully apply this technology to the hostile and less controllable environment underground, more time and experience were required. Even by 1969, the end of the ROLF era, the principles were understood but solutions to the technological problems were only beginning to emerge.

As a result of the original experiments' limited success, and the same experiences on a larger and more public scale at Bevercotes - the world's first colliery planned for automatic operation - enthusiasm for automation was quenched. The effect of introducing automation at the wrong time was to set work back by a number of years, and it is interesting to note that at the end of the 1960s the number of staff directly working on R & D was only half the peak number working during the ROLF experiments.

However, the early trials of the vertical guidance system based on the nucleonic sensor had shown great benefits despite the failure of ROLF, and further development and exploitation of this as an objective in itself continued.

Vertical Guidance of the Fixed Drum ASL

Production versions of the fixed drum ASL steering equipment became available in 1972, and 80 systems have now been installed at an average rate of twelve per year, with approximately ten working at any one time.

Production Experience. Reports from the early installations indicated very favourable results. Production benefits include an increase in the percentage of seam extraction, since larger drums have been fitted, and a decrease in the amount of dirt in the unwashed product. The achievement of roof and floor profiles without steps greatly improves strata conditions, enabling a rapid rate of roof support advance with reduced maintenance costs. Controlled extraction, which excludes roof and floor stone, reduces respirable dust, and safety is improved since men do not have to work on the face side of the AFC in order to grade and manually steer the face.

On the debit side, the introduction of a relatively new, complex electronic and hydraulic system has created training and maintenance problems for the colliery. Although the maintenance is within the capabilities of the Board's craftsmen, it has generally been necessary during the introductory period for an engineer to devote a large part of his time to the system, and purpose training for craftsmen has been essential. The

system requires more maintenance than a standard machine, which counteracts the decrease in maintenance on other face equipment.

Estimates of the pay-off have been calculated from data collected at twelve collieries in 1972/3. The main sources of extra revenue have been increased bulk output, reduced ash, and lower maintenance and marketing costs. The total value of these benefits is estimated to be £365,000 per year at today's prices, for a 300,000 tonne face producing an annual income of £6.9 million. Capital outlay at each installation for machine modification and for sensing and control equipment is £35,000, plus an annual charge of approximately £10,000 for maintenance labour and spare parts.

A further investigation was carried out in 1976 by a manufacturer and the National Coal Board to identify the common factors in successful and unsuccessful applications. The findings showed that successful installations have in the main been associated with faces where a definite production or marketing problem has been clearly established and monitored, or where forward-looking management has wished to improve the performance of high-producing faces. Less success has usually occurred where machine guidance was applied as a last attempt to improve roof control on faces which already had poor geological conditions or production performance.

Limitations. The main limitations of the guided fixed drum ASL became evident during the early 1970s, the most important being the number of machines which can use the steering system. In 1969, 608 fixed drum ASLs were at work (90% of installed shearers). The new generation of ASLs, however, uses ranging drums on extended arms, and for control engineering reasons cannot be steered using this system. At present 66% of the machines at work are ranging drum ASLs (576 machines); there are now only 301 fixed drum ASLs on faces, and many of these are secondary machines.

A second restriction is the range of extracted seam thickness in which the fixed drum shearer can work, this being a minimum height of 1.07 m (42 in) to accommodate the horizon sensor, and generally a maximum of 1.83 m (72 in), which is the maximum drum diameter consistent with machine stability.

The characteristics of the nucleonic horizon sensor also prevent wide-scale application. Firstly, the roof coal thickness must be between 0.05 m (2 in) and 0.2 m (8 in), and there must be a distinct parting between roof coal and overlying stone. Secondly, any gap between the surface of the sensor and the coal roof causes errors in reading. For this reason, the sensor is mounted on an outboard hydraulically raised arm, which has been a continual source of delay due to maintenance and repairs. The sensor mounting can also impose a constraint on the mining system since it prevents roof sensing in both cutting directions. Although it has been installed on bidirectional faces, there is the risk of controlling the coal/stone interface on alternate cuts only.

Altogether, these restrictions mean that at the present time the original vertical guidance system can be used on only about 10% of faces.

There are other features in the machine and
guidance system design which can produce errors in
steering even in the most suitable conditions. The
mechanical steering mechanism cannot cope
immediately with large random vertical movements
caused by rapid convergence or loose coal under the
AFC, so the shearer occasionally cuts into the roof
until the face has advanced far enough for the
machine to restore the horizon. Also, there is no
means of automatically checking the monitored data
for faults or errors which might cause the machine
to be steered incorrectly.

Finally, a factor limiting large scale
exploitation of the system was touched on earlier.
The difficulty of introducing a novel electronic
control technique to both manufacturers and
operators was not entirely recognised at the
outset, and although methods of technology transfer
were introduced in 1972, they were insufficient to
break through the technological barrier.

CURRENT OBJECTIVES FOR GUIDANCE OF COALFACE MACHINES

In the Board's 1974 review of mining research and
development, past experience was examined in order
to formulate objectives and future work plans.
This overall look at R & D strategy determined that
the present system of coalmining would most likely
continue into the near future. The key short-term
problem, therefore, was to make the best possible
use of existing techniques by increasing the
reliability of components so that the system as a
whole could be automated.

The first step towards this has been the
computer control of coal transport which, when
complete, will eliminate one restriction on
coalface production. The very high outputs
currently being achieved from exceptional single
faces will then become the norm, provided that

obstacles to operating face equipment at this
faster rate can be overcome. If production is to
be increased even further, the guidance of
coalwinning machines will be vital. For this
reason, the following objectives for machine
guidance were adopted.

-   The system must be applicable to at least 80%
    of current coalfaces, considering geological
    conditions, machine type and mining system.
    This immediately identifies secondary
    objectives: applying the system to other
    machines, primarily the ranging drum ASL, and
    developing new horizon sensors to extend the
    range of roof coal thickness and extraction
    thickness.

-   The machine must be automatically guided or
    controlled in three dimensions: vertical;
    horizontal in line of face advance; and
    along-face.

-   The guidance system should be primarily a
    stand-alone system, but it must ultimately
    integrate with the automatic control of AFC and
    roof support advance and with the operational
    monitoring and control system of the mine. It
    should have internal diagnosis and data
    checking, and must also provide management with
    information at the surface.

-   When a successful guidance system has been
    developed, it must be rapidly exploited by
    introducing manufacturers of machine and
    control equipment to the work early in the
    development process, and by implementing
    technology transfer policies.

FIGURE 1. Analogue steering system with nucleonic probe on a ranging drum ASL

PROGRAMME FOR VERTICAL GUIDANCE OF THE RANGING
DRUM ASL

Problem of Control

The first objective was to increase the system's
range of application, so work was conducted on the
most commonly used machine, the ranging drum ASL.
To identify the control engineering problem, a
single-ended ranging drum shearer was tested in
1975/6, using an analogue control system similar to
that used for the fixed drum shearer (Figure 1).
The trial machine used a nucleonic sensor for
measuring roof coal, with additional instrumentation
to measure machine tilt and in particular to
measure the random position of the conveyor
opposite the drum, caused by the distance of the
ranging drum from the machine shoes.

The work demonstrated that the ranging drum ASL
steering was inherently unstable due to the random
and excessive vertical movement of the AFC when
advanced onto the newly cut floor. This effect was
caused by:

- offset of the drum from the shoes, which causes
  amplification and propagation of undulations;
  this becomes very important with the longer
  ranging arms now in use

- large steps generated by pitch steering of the
  drum, which can tilt the conveyor up excessively
  or, in some cases, allow loose coal to prevent
  the conveyor from moving down

- variation in advance and the relationship of the
  centre of gravity of the machine to a step,
  particularly where large steps have been
  generated.

In addition to the inherent instability caused
by the machine and conveyor construction, the
trials demonstrated that the vertical guidance
system was unable to stabilise the steering. It
was impossible with the analogue system and
nucleonic sensor to measure the drum position
within the seam, and the resulting delays between
cutting and sensing prevented the system from being
stable. When the conveyor moved randomly, the drum
position was altered relative to its last roof cut,
and unless corrections were made for this the
control of roof coal thickness was impossible.

The trials and associated computer and
quarter-scale modelling confirmed that it was
essential to measure or compute the drum's position
in the seam and its position in relation to the
roof of the previous shear in order to produce a
stable control system. A new research programme,
formulated in 1976, included the development of
sensors to monitor the horizon and also the
along-face position of the machine, since all
horizon measurements must be related to specific
locations along the face. The requirements of the
control strategy dictated the use of digital
computing techniques, and since the objective was a
stand-alone guidance system, the development of a
modular microprocessor system suitable for an
underground environment was also included in the
programme.

Development of Sensors

Coal Interface Sensors. Progress has been made on
two new sensors: one is a detector of naturally
occurring gamma radiation, developed from the
successful nucleonic backscatter sensor, and the
other is a redevelopment of the 'sensitized pick',
now called the pick force sensor. It is
anticipated that when these two sensors become
available, monitoring systems can be designed to
cover at least 80% of seam conditions.

The natural radiation sensor detects gamma
radiation from the stone above the seam and
measures the proportional change in radiation with
varying thicknesses of roof coal. Naturally
radiating isotopes are present almost universally
in shale overlying the seams in the UK. The
sensor has a range of 0 - 0.45 m (18 in) of coal
thickness between the extraction and overlying
shale and, most significantly from the operational
point of view, does not need to be closer than
0.25 m (10 in) to the roof. The principles and
response of the sensor are shown in Figure 2.

Various prototype sensors have been constructed
and tested underground for several months. The
performance has been very close to theoretical
expectations, and production versions are now
being manufactured by Salford Electrical
Instruments Ltd for use on current guidance system
development projects, as well as for replacing the
nucleonic sensors on the existing fixed drum ASL
guidance system and for monitoring coal heading
machines. A portable sensor for carrying out
surveys of prospective sites is also being
produced.

The main drawback to the sensor is that it can
only be mounted at some distance from the drum.
To be of use in a control system, the roof
thickness information must be stored until the
drum arrives at the relevant location.

The pick force sensor is being developed from
earlier work, and consists of a pick on the face
side of the drum, instrumented with a piezo
crystal or strain gauge, which produces electrical
signals proportional to the force the pick
experiences while cutting through individual bands
of varying hardness in the strata (Figure 3). The
sensor does not directly indicate the drum's
horizon but produces a pattern of forces related
to the pick's angular position in the drum. This
pattern is correlated with a master pattern
representing the correct drum position, and the
correlation indicates the error in position of the
drum within the seam (Figure 4).

This complex method is the only way currently
envisaged of coping with applications where
neither roof nor floor stone must be cut and
intermediate bands of strata within the seam are
irregular or inconsistent. There is still no
accurate way of independently surveying sites for
possible use of pick force sensors, and this
method increases the likelihood of successful
applications.

The pick force sensor has received simulated
underground trials on a shearer cutting an
artificially banded coalface at the surface test
site. It produced force patterns which, although

apparently dissimilar to the master pattern, responded to the correlation techniques to give an accurate drum position. A reliable method of transmitting signals from the rotating drum to the body of the machine has been proved and the equipment is ready for underground trials.

<u>Roof Height Sensors</u>. These sensors detect the drum position relative to the previously cut roof and have a range of 0.25 – 1.0 m (10 – 40 in). Two lines of development are being followed. An ultrasonic device, which relates the travel time of a beam of sound to the distance from the roof, has been proved in the laboratory and in simulated face conditions, and an underground prototype is being produced. A radar version, depending on the

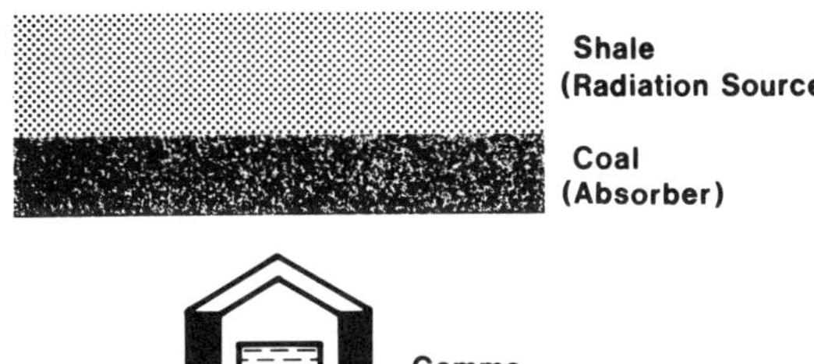

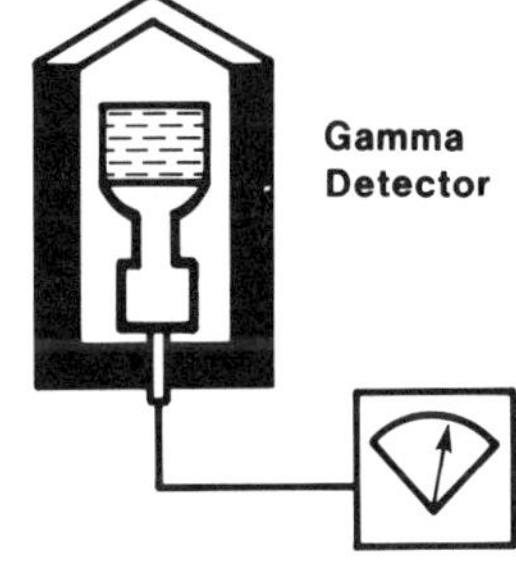

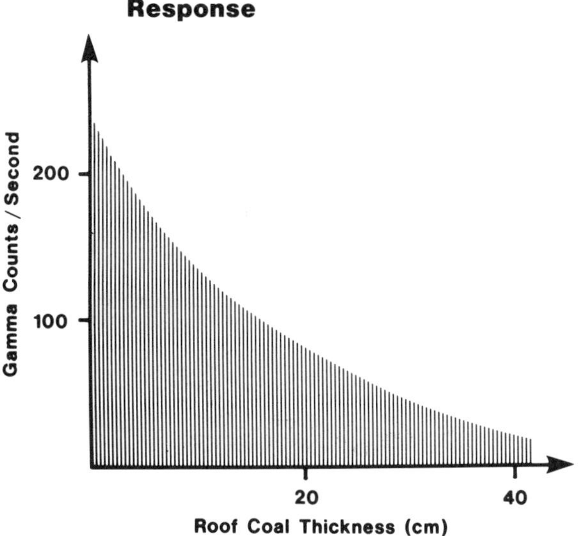

FIGURE 2. Principles and response of the natural radiation sensor

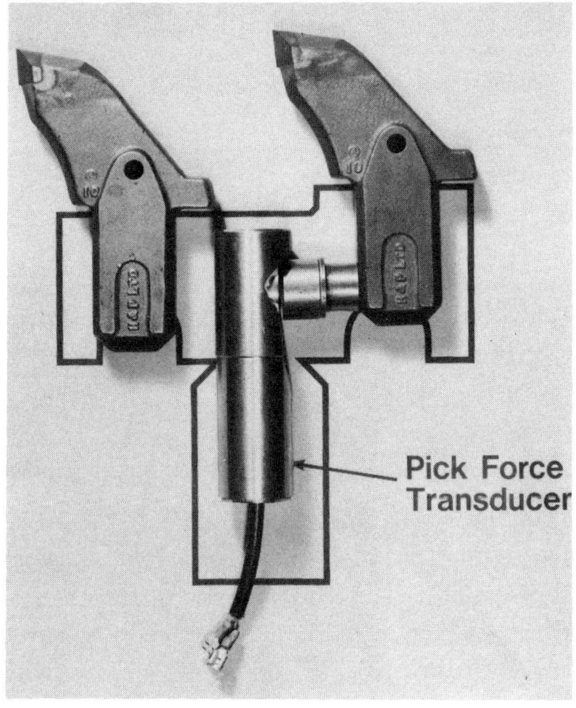

FIGURE 3. Pick force sensor

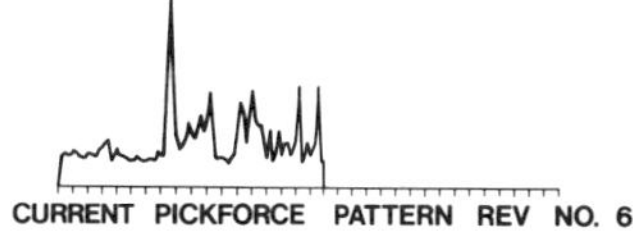

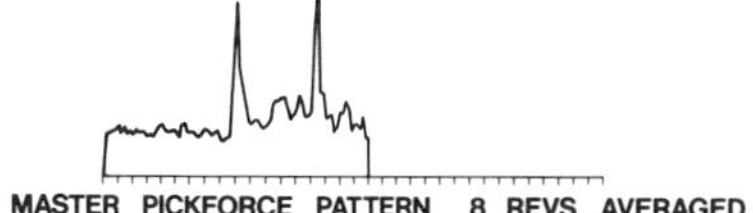

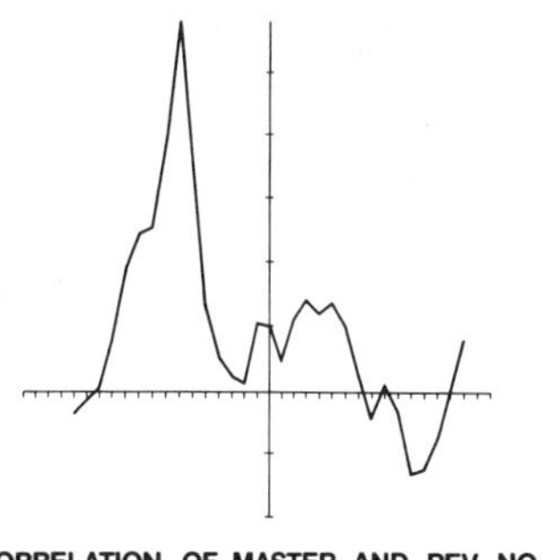

FIGURE 4. Pick force pattern and correlation

Whilst Cutting Roof

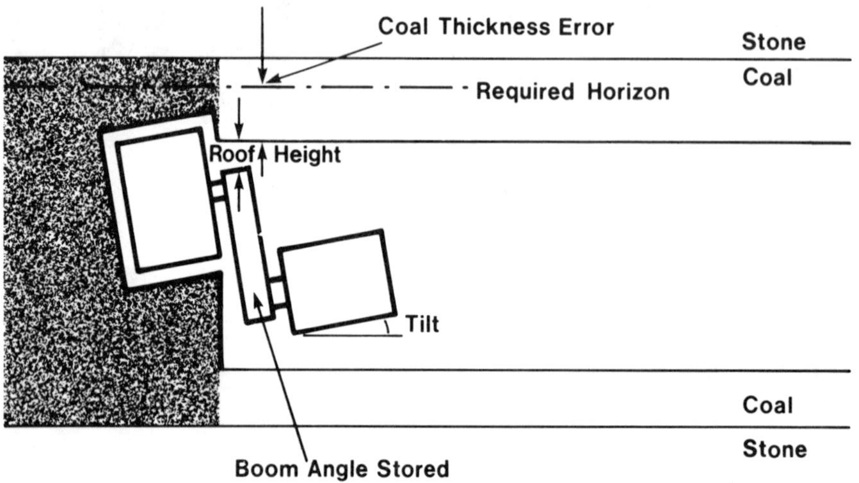

Whilst Cutting Floor

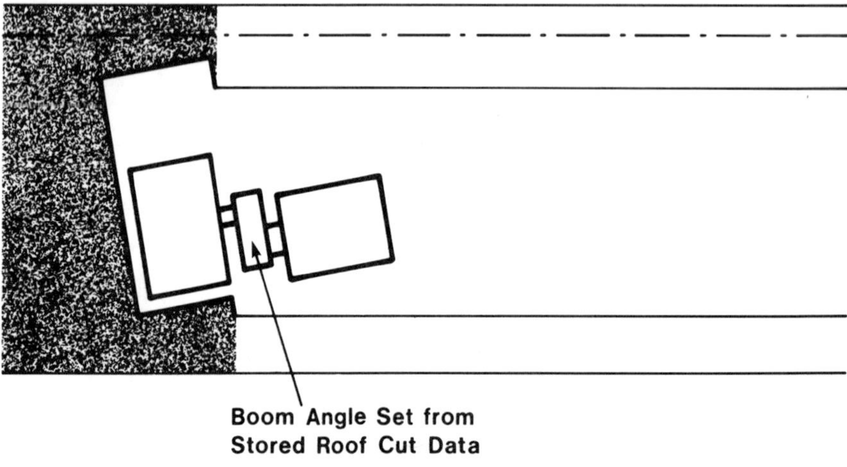

FIGURE 5. Guidance system for ranging drum ASL using natural radiation sensor

frequency change of a reflected microwave in relation to distance from the roof, has been demonstrated in the laboratory, and a surface trial unit is under construction.

The accuracy of both these sensors will be affected by dust and water on the coalface equipment, and techniques to combat this have been designed. It should be noted that these sensors will have another use on automatic faces as obstruction detectors, and can also be used for coal-flow measurement on conveyor systems.

Other Sensors. An along-face machine position sensor derives the machine's location from optical vane detectors measuring the rotation of a shaft in the haulage section of the ASL. It is a key development, since all horizon information must be related to distance travelled along the face.

Transducers to measure machine angle and boom position have also been developed, based on conventional techniques but modified to suit underground conditions.

Development of a Control System

Guidance of the ranging drum ASL has been the subject of considerable study at MRDE, which showed that even when practical coal interface sensors were available, complex control problems still existed. Several years' work and three underground trials were carried out before an acceptable solution was found.

The aim of the steering system is not only to ensure extraction from the correct part of the seam, but also to control recovery from displaced extractions caused by random AFC movements or periods of manual steering. The system must also allow for different levels of automatic control, depending on the availability of sensors and the existence of faults. These added requirements, which lead to more sophisticated electronic hardware, are the result of criticisms that the previous system lacked integrity and continuity of operation.

System Operation. The essentials of the guidance system are illustrated in Figure 5, which shows the application of the natural radiation sensor to a single-ended rotating drum ASL. During the first pass of the machine with the drum cutting the roof, the step size is measured in comparison to the previous shear and combined with the error of previously cut roof coal, which is established from memory. A correction is then applied to the drum position to achieve the required horizon. In this way the system compensates both for random disturbances in the current shear and for steering errors detected on the previous shear. The ranging arm angle is stored until the second pass with the drum cutting the floor. The drum-down position is then computed from the stored drum-up position at corresponding parts of the face, producing the required extraction.

The pick force sensor operates similarly, except that information about random conveyor movement and roof coal thickness is contained in the output of the pick force sensor, and the measurement of the roof height compared to the previous roof measurement is used to control the

rate of recovery from large errors or displacements from horizon.

The system can also be applied to the double-ended ASL, with both roof cut and floor cut being controlled on the same pass of the machine.

Extensive computer and quarter-scale modelling of the system has taken place, and optimum rates of correction and recovery have been established to ensure that guidance control remains stable and that the along-face and face advance profiles are suitable for the AFC and powered supports.

System Hardware. The requirements of the guidance system and sensor design dictated that the system hardware be computer-based. A machine-mounted modular microcomputer system has been developed, capable of receiving and processing sensor data, carrying out programs of control functions, indicating faults, transmitting and receiving information to and from other parts of the mine and surface, and linking with the machine operating equipment.

The system (Figure 6) is built around the executive microcomputer. This acts as the 'manager', collecting and routing information from the sensors or surface to the correct module for transmission, machine operation, or further processing. The pick force and steering analyser modules are identical microcomputers carrying out sensor correlation and control system functions.

The other modules are interfaces with operators, sensors or external equipment. The data access and display unit receives and displays sensor inputs and fault indication on demand from a robust operator keyboard. The machine interface module segregates the digital control signals from high-voltage solenoid driving circuits, and the data transmission module utilises the pilot core or ancillary core in the machine power cable to send and receive data via the gate. The data received in the gate can be similarly accessed and transmitted to the surface, where trial and managerial information can be processed and displayed.

The total system hardware is built into the machine and, excluding power supplies, fills a volume of approximately 0.04 m^3 (1.4 ft^3). All the modules have been built and tested in prototype form, and versions suitable for underground use are now under construction by Hawker Siddeley Dynamics Engineering, Whiteley Electrical Radio Co Ltd, and Marconi Space and Defence Systems Ltd.

System Software. The writing of software for the system has proved to be a major task, and has highlighted the complexity introduced by trying to cover all the design objectives for a flexible and reliable guidance system. The essential task of the software is to instruct the microcomputers to carry out the following actions:

- to communicate with each other and with the data access and display units and to implement the machine control

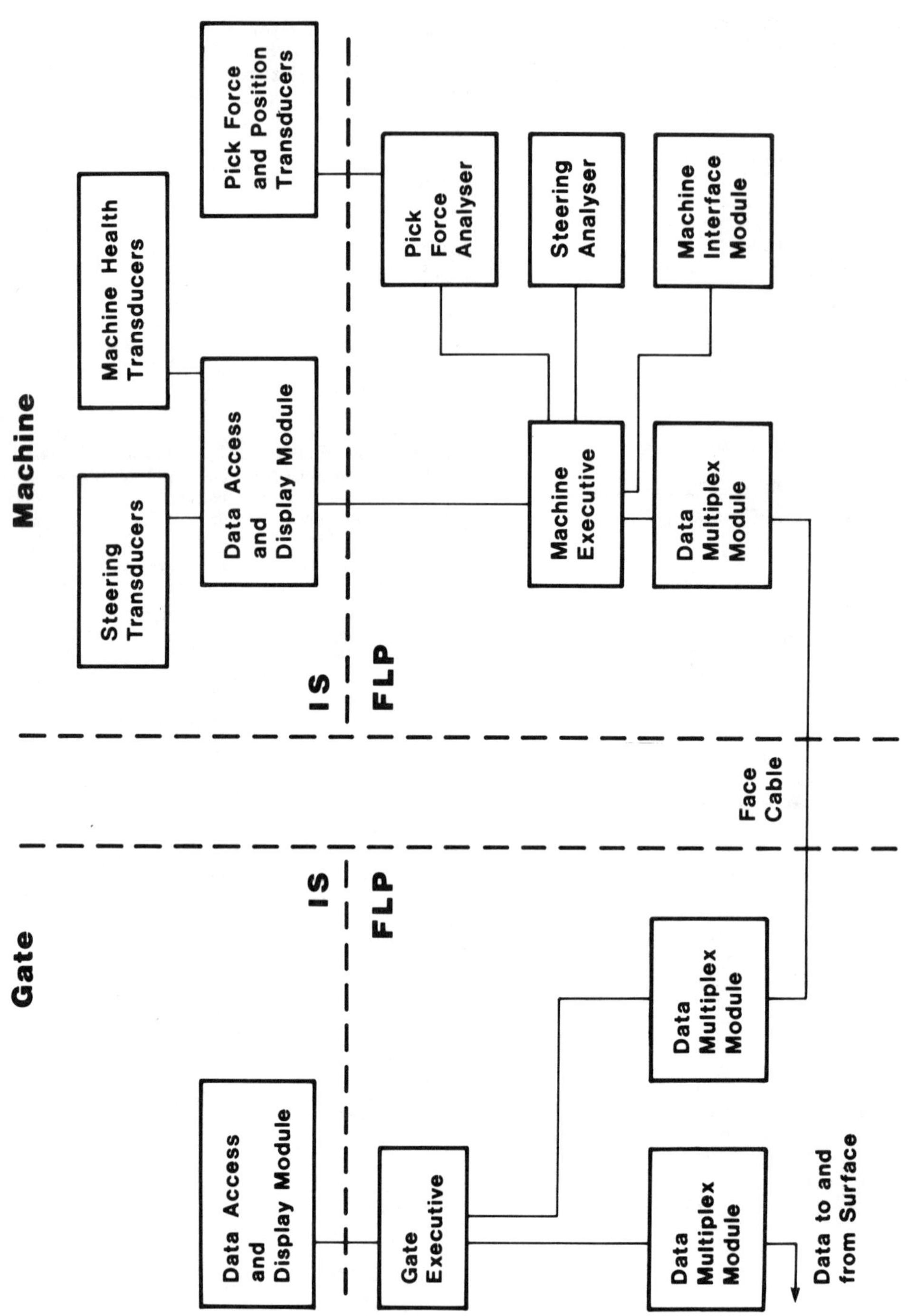

FIGURE 5. Digital electronics for steering system

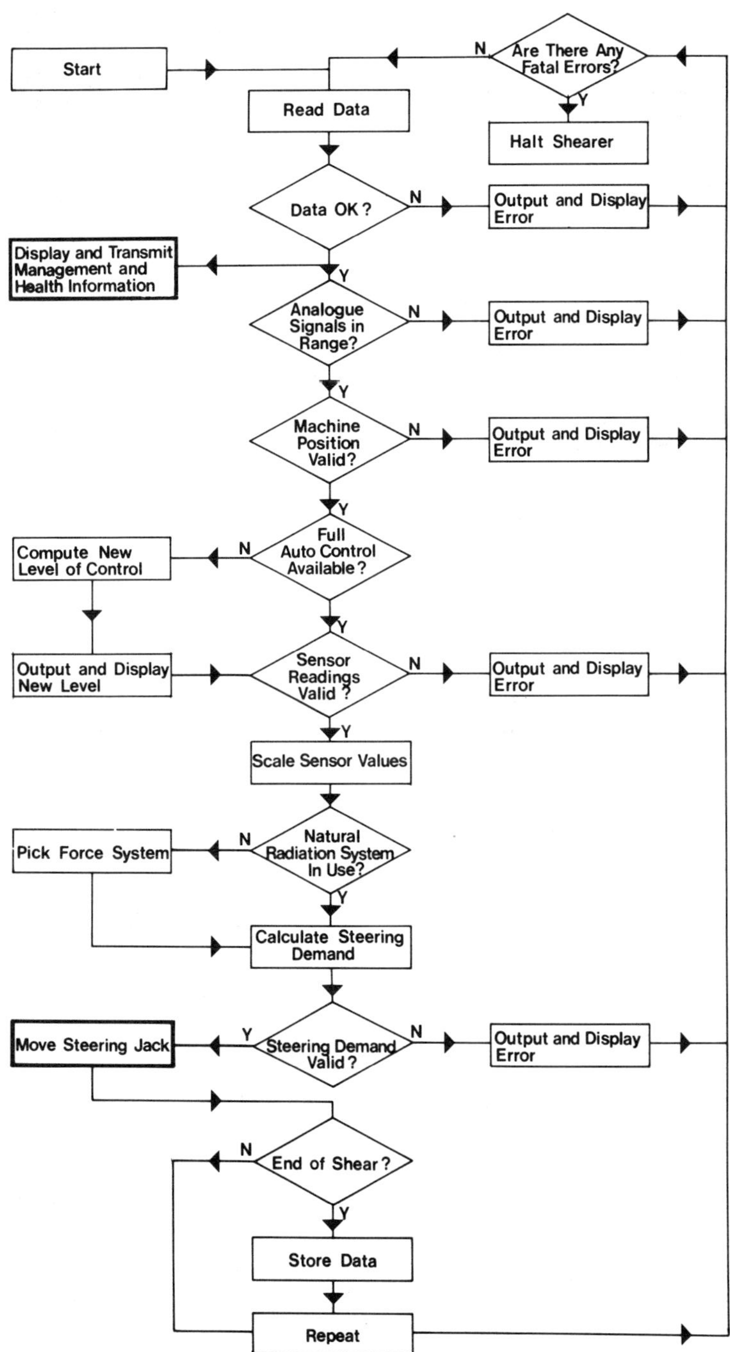

FIGURE 7. Simplified flow chart for steering system software

- to carry out the calculations on all steering information, store any relevant data, and send out control signals

- to recognise errors and faults and display these at the machine, gate and surface.

A very simplified flow chart for the steering system software is shown in Figure 7, illustrating actions in all the above categories. Some impression of the size and complexity of the software for machine vertical guidance control can be obtained from the fact that it was found necessary to provide program storage of 46,000 x 12 bit words on the machine, and a further 6000 x 12 bit memory in the gate.

Underground Trials of the Digital Machine Guidance System

A site for underground trials has been finalised and a machine is being suitably modified and assembled by British Jeffrey Diamond, a division of Dresser Europe. A diagrammatic view of the machine showing the location of sensors and microcomputer equipment is shown in Figure 8. Work on the face is planned to start in April 1979 using a step-by-step introduction of the system, beginning with further sensor trials in May and working up to a trial of the microcomputer system in August.

HORIZONTAL GUIDANCE OF THE ASL

Concept

Prior to 1975, control of horizontal guidance had only been attempted in a piecemeal fashion, with attempts to monitor and control the rams for extending the AFC. Some research into the use of inertial guidance was also carried out in the late 1960s.

The identification of new objectives and policy for machine guidance led to a revitalised interest in the problem of horizontal guidance. The nature of the problem was analysed into three aspects, illustrated in Figure 9. Firstly, misalignment of the face between the gates causes poor strata control and damage to machines and other face equipment. Secondly, lack of control of the face angle to the gates causes delays and extra costs because the AFC and powered supports creep towards the gates. Finally, the correction of alignment and angle and the lack of control of AFC pushover lead to loss of face advance. Evidence indicates that at least 5% of output per shear is lost because the available width of the drum is not fully utilised.

It was recognised that the first stage of the solution was to find reliable and accurate methods of monitoring all aspects of horizontal steering. Two alternative solutions are to have a mobile survey instrument which traverses the face (most suitably on the coalwinning machine), or to have a number of static instruments at intervals along the face, sufficiently close together to give the resolution required for control purposes. Both solutions are currently within the work programme.

Coalface Surveyor

Following a feasibility study in 1975, in which machine-mounted equipment based on inertial

(gyroscopic) platforms and lasar or infrared optical techniques were investigated, it was decided to opt for the latter. The development of the Coalface Surveyor has been carried out under contract by Marconi Space and Defence Systems Ltd. The Surveyor carries out continuous triangulation computations to plot the path of the machine relative to a series of reflectors suspended from the supports or AFC along the face. The optics instrument emits a beam of low-intensity infrared light which is reflected back and focused, such that the angle at the machine subtended by any two reflectors is continually measured. These angles, together with the distance travelled by the machine, enable the system to compute the alignment of the face. Figures 10 and 11 show the principle of the system, which is similar to the human ability enabling a car driver at night to detect bends in the road ahead by observing the apparent change in spacing of roadside reflectors.

The equipment consists of an optics unit with light-emitting and detecting features, and a small microcomputer on the machine which outputs the reflector positions and machine position. The information is then transmitted to the gate, where a minicomputer calculates the profile of the face, displays it locally, transmits it to the surface, and interfaces with an AFC advance control system.

Laboratory equipment has been built, and tests on an ASL at the surface trial site proved that the profile can be as accurate as 1 m in 200 m (3 ft in 650 ft) and can cope with a radius of curvature of 25 m (82 ft) over short lengths of face. Prototype underground equipment is now being constructed and trials are expected to begin in September 1979, using a shearer built by Anderson Strathclyde Ltd.

The major drawback to the Coalface Surveyor is that no practical way has been found to set a reference to the equipment on each shear. Hence, it computes the alignment of the face but not the face angle or face advance. It is therefore necessary to incorporate at least two direct measurements of face advance in order to orientate the Coalface Surveyor profile.

Direct Methods of Measuring Coalface Advance

Investigation of existing German and French instruments, which measure face advance by paying out wire or punched tape into the waste, indicated difficulties concerning life of the wires, storage of information during power down or electronic failure, and data format which would, in the UK case, have to interface with the Coalface Surveyor and AFC advance control equipment. It was therefore decided to develop a similar instrument suitable for UK needs, and trials indicated that terylene cord would give the best life. The instrument is mounted on the powered supports and the advance of face is measured by rotation of a pulley geared to a shaft encoder, with data transmission providing digital information at the gate. The prototype unit is under construction and underground trials will begin in May 1979. The objective is to have up to 64 advance instruments on the face, with a gate-end profile display and interface to the AFC advance control equipment.

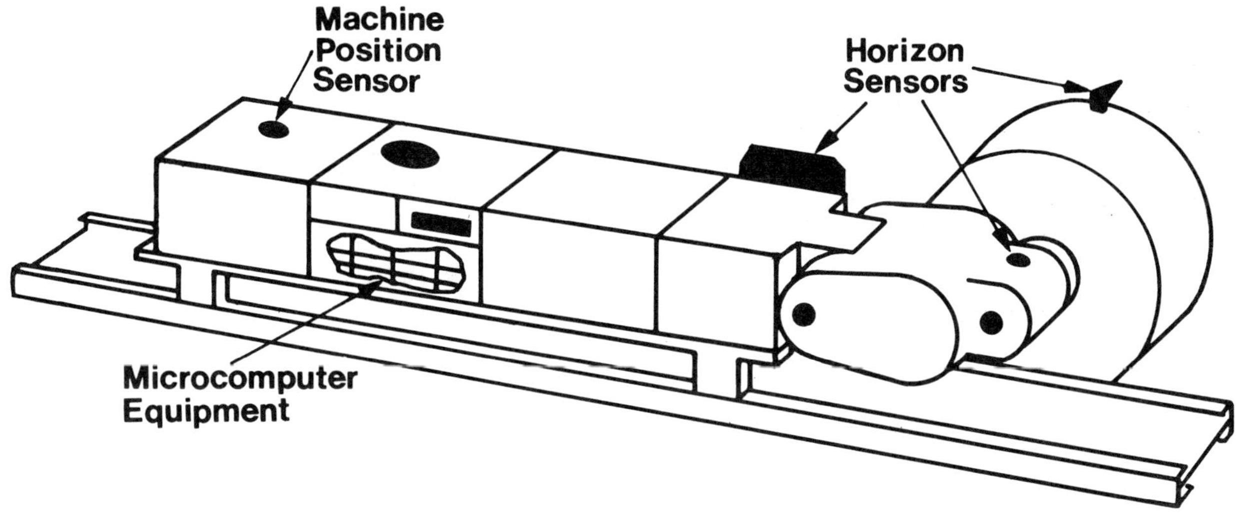

FIGURE 8. Location of sensor and microcomputer on underground trials machine

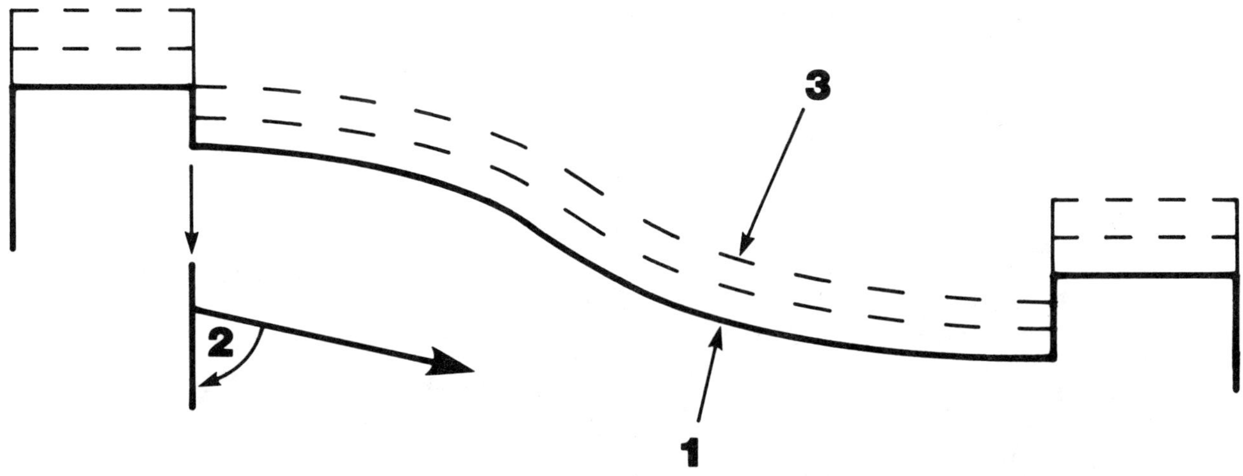

FIGURE 9. Aspects of the horizontal guidance problem. The objective is to maximise advance and machine running time by monitoring and control of 1) face alignment, 2) face angle, 3) face advance

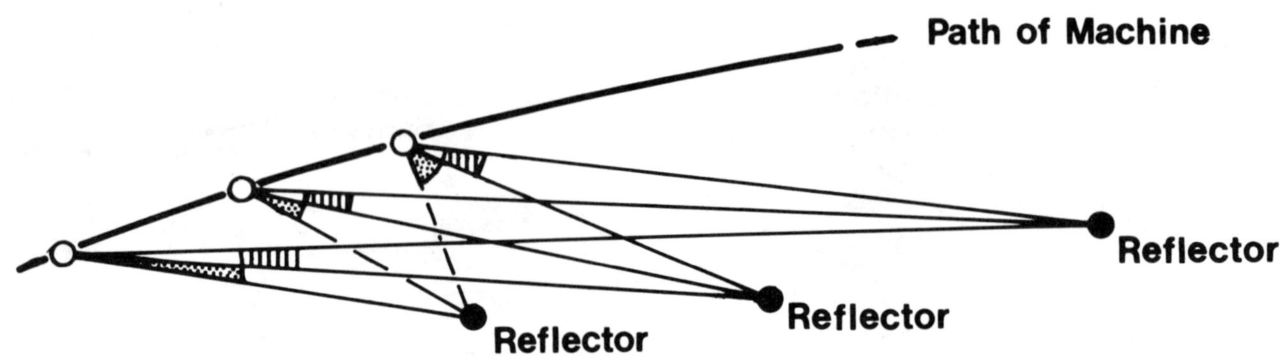

FIGURE 10. Principles of the Coalface Surveyor

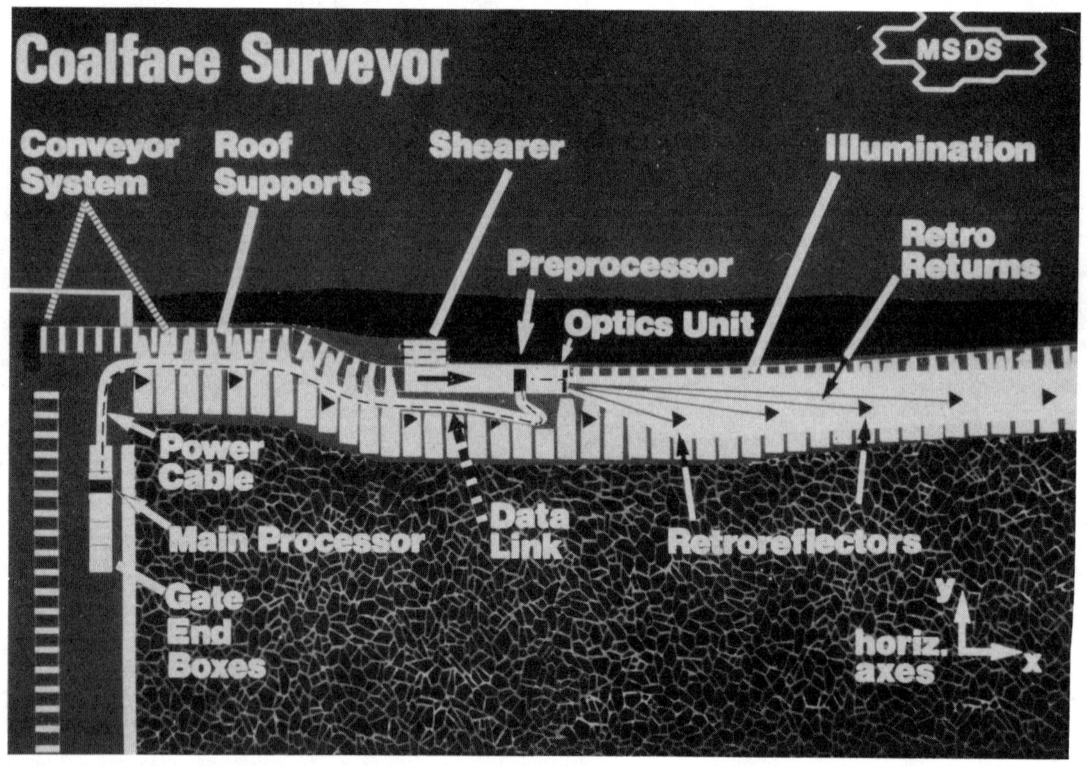

FIGURE 11. Layout of the Coalface Surveyor

An alternative method of direct measurement, studied by Dowty Ltd, is now entering the feasibility stage. It measures the advance of a pair of powered supports relative to each other by measuring the change of angle of a connecting strut with either end fixed on adjoining supports. This method will be applicable in the region of mechanised packholes, where cords may be fouled by packing machinery.

It should be noted that the face surveyor is a technologically complex single instrument, while direct measurements of advance are provided by a large number of relatively simple instruments, with the consequent installation and maintenance problems. Only experience will show which of the two approaches is superior, or whether a combination of the two will provide reliable monitoring of face alignment, angle and advance.

Control of Face Advance

Although it is intended that monitoring should be proved before control is attempted, some early steps to integrate the two have started. This is because of the long lead time on developments and because the automatic operation of roof supports developed for the ROLF have recently been revived and successfully demonstrated in underground trials.

A development contract has commenced with Gullick Dobson Ltd, a major power support manufacturer, which will integrate, in stages, horizontal guidance monitoring with control of face advance. The first step will provide on-face digital readouts at ramming points of the distance to be advanced, and the operator will use this information to assist his manual operation. The development will finally close the control loop by automatically advancing the AFC by the distance which will optimise face advance consistent with maintenance of face alignment. It has been recognised that this may not be simple, and a control engineering approach is being planned to simulate a complete face profile and its influence on the advance distance required at every point. Trials for the first stage of this control system for horizontal guidance are expected in 1980.

Finally, it has been recognised that if the maximum web depth of the ASL is to be fully achieved, any misalignment must be corrected by holding back the parts of the face which are in advance. Some consideration is being given to installing a horizontal steering mechanism on the ASL, so that for a short length of face the drum could be advanced into the parts of the face which are behind. A prototype ASL now under construction by Mining Supplies Ltd will not only have ranging drum vertical steering, but will also have the ability to advance the drum by up to 0.16 m (6 in) from its normal cutting position.

RESOURCES APPLIED TO MACHINE GUIDANCE R & D

The increased emphasis on the importance of automating coalfaces has made it necessary to greatly increase the number of scientists and engineers working directly and indirectly on the problem of guiding underground mineral-extracting machines. In 1974 there were approximately 21 R & D staff working on projects related to guidance, with an external expenditure of £45,000. There are now 35 people working on it with an external budget of £900,000 per year, most of which is used to obtain the expertise of specialised staff from private companies, mainly in the aerospace industry.

The vital feature of this team of engineers and scientists is its inclusion of many types of specialist to cope with specification, analysis and simulation of control systems, development contract management, underground trials, and exploitation in a production environment. This last is particularly important, since many of the manufacturers who have recently become involved, especially those from the aerospace field, have little or no knowledge of the coalmining industry.

CONCLUSION

The programme of work on machine guidance described in this paper is exceptionally demanding of mining R & D resources. Whilst the short-term work on reliability, measurement and automation of coal transport absorbs 10% of the Mining Research and Development Establishment's £20 million budget, this single medium-term project has an annual expenditure of about £1.5 million.

There is now no doubt that the project will succeed. The problem which will then follow will be helping the industry to absorb the very high technology employed. However, this is a difficulty the industry already faces in the short term, and for which good plans have been made.

Technology transfer is being achieved in a number of ways. Firstly, the industry, through highly authoritative major development committees, endorses analysis of the problem and provides support to the problem solvers. Secondly, manufacturers are involved in the programme from the early stages of the development phase so that they can identify with the work, supply technical needs, and guide the prototypes towards feasible production models.

Thirdly, the demonstration phase, mainly financed by MRDE, ensures that new techniques are explained to the industry and that they can accommodate operator idiosyncracies. Finally, a new centre at MRDE provides facilities for training staff to install, operate, maintain and manage new technology. It also produces teaching packages which the mining and manufacturing industries are able to utilise.

The authors recognise the far-sightedness of the National Coal Board which has enabled the machine guidance programme to be developed and to be supported through the short-term problems with which the industry is faced, and wish to thank Mr L.J. Mills for permission to publish this paper.

Chapter 33

FUTURE R&D LONGWALL PROGRAMS PROPOSED BY BITUMINOUS COAL RESEARCH, INC.

R. L. Mundell, Supervising Engineer
F. du Breuil, Assistant Manager, Mining Research
Bituminous Coal Research, Inc.
Monroeville, Pennsylvania

ABSTRACT

Longwall methods to mine coal in the United States provide the potential advantages of greatly increased productivity, improved safety, and a higher percent recovery. However, use of longwall systems during the past 10 years has increased much more slowly than was anticipated.

This paper discusses some of the problems inhibiting the increased use of longwall systems and a proposed research program to address one of the most critical problem areas.

INTRODUCTION

Multiple-entry retreat longwall mining systems in the United States have disadvantages inherent to their operation. Probably the most significant disadvantage is the lack of flexibility of the system. The necessity of mining through gas wells and geologic unconformities such as rock faults, clay veins, seam variations, etc., may present very difficult operational problems. Capability of selective mining of higher grade coal is reduced. Also, increased methane liberation and dust control on some operations may create additional problems. However, several distinct advantages outweigh the disadvantages: improved safety, increased productivity, and increased coal recovery. In addition, some seams, because of adverse mining conditions, can be mined economically only through the use of longwall mining systems.

Recent mine accident records indicate that on longwall mining operations accident frequency rate is increased (primarily resulting from minor injuries involving material handling, equipment failure and repair, and fall of rock or coal between roof supports). However, a significant decrease in disabling and fatal injuries has been recorded. Longwall systems appear to have a very high potential for dramatically reducing roof fall accidents—the continuing number one cause of underground mine injuries and fatalities.

Currently, longwall mining accounts for approximately 5 to 7 percent of the total U.S. underground coal production. Recent production records indicate that longwall systems can increase productivity (tons/man-shift) by a factor of four. Improved coal recovery is inherent with the longwall system, particularly with three or less development entries on each side of the panel.

Mechanized longwall mining began in the United States a little over 20 years ago (about 1960). By 1970, approximately 30 systems were in operation. Currently, approximately 100 systems are operating. Conservative projections throughout industry and government indicate that the number of U.S. longwall operations should increase by more than 50 percent and account for approximately 10 to 15 percent of underground coal production in the next five years.

Although the number of systems placed in operation each year has increased steadily during the last several years, the number of longwall systems operating currently is considerably less than was anticipated in 1970.

Because of the formidable advantages of longwall mining, Bituminous Coal Research, Inc. (BCR), formed a Longwall Mining Working Committee in 1979, comprised of industry members representing nearly every operating longwall system in the U.S. One objective of this committee was to identify and prioritize the problems inhibiting the increased use of longwall mining in the U.S. The committee developed recommendations for the most critical problem areas to be addressed by government and other research organizations in order to facilitate the increased use of longwall mining.

PROBLEM AREAS

One critical problem area identified by the committee was directly related to federal regulations regarding mine ventilation requirements, specifically 30 CFR 75.326 Air Courses and Belt Haulage Entries, and 30 CFR 75.1707 Escapeways, Intake Air, Separation from Belt and Trolley Haulage Entries. The committee felt that the Mine Safety and Health Administration should consider re-evaluating these regulations.

The regulations were developed and formulated primarily for room and pillar mining and should be re-evaluated in terms of considering the use of belt-entry air on the longwall face. However, this modification may require more stringent standards regarding the fire-resistant characteristics of all materials utilized in the belt entry.

To comply with these regulations, without obtaining MSHA-approved exceptions, also necessitates the development of three or more panel entries. Re-evaluation, with possible modification of the regulations, to facilitate companies' long-range planning incorporating single- and/or double-entry longwall panel systems, is recommended. Mining companies are most hesitant to design and plan a mining operation utilizing a $7- to $11-million longwall installation dependent on being granted a variance from the federal regulations. Fewer longwall development entries provide the following advantages:

Development time to bring the longwall face into full production is reduced.

The total coal recovery is increased through reduction in the number of chain pillars.

The percentage of total mine manhours spent on the longwall face is increased, taking advantage of the improved safety and productivity (four entries on each side of the panel result in approximately 40 percent of the mined coal from each panel coming from the development entries).

Roof control in the head and tailgate entries is possibly improved.

This brings us to another problem area identified by the committee--roof control. Problems include:

Difficulty in controlling the roof in the head-gate and tailgate entries

Premature roof collapse at the coal face

Uncontrolled fall of roof behind the supports due to varying roof strata.

Another problem area recommended for additional research effort was coal transport. Specifically, conveying systems of increased capacity and systems requiring less maintenance are needed to more efficiently transport the tonnage produced by double-drum shearers. In addition, equipment needs to be developed to effectively handle the large lumps of coal produced in the higher western coal seams.

A research program should also be initiated to study, develop, and design equipment and procedures to significantly reduce the time required to move longwall equipment from one panel to the next in low-coal seams.

Without question, though, the most critical problem area recognized by the committee as requiring an immediate, concentrated, and cooperative government-industry effort was respirable-dust control on double-drum longwall shearer operations. Considerable reduction in respirable-dust exposures has been achieved during the past 10 years on most longwall operations. For plows and single-drum shearers, engineering controls have been effective. For double-drum shearers, on the other hand, engineering controls have not been sufficient, and a significant reduction in respirable-dust exposures has been achieved only by implementing a "modified cutting sequence" which involves cutting primarily in one direction. By employing this cutting sequence, most of the face personnel work on the intake-air side of the shearer. However, most longwall operators feel unidirectional cutting results in a loss in productivity because the double-drum shearers are designed to cut bidirectionally. Although this rationale sounds logical, cutting sequence information and productivity data obtained during a 1978 MSHA survey indicated that an increase in productivity occurred on many longwall shearer sections when bidirectional cutting was replaced with unidirectional cutting. This question regarding double-drum shearer productivity vs. face-cutting sequence and the continuing problem of respirable-dust control, because of their direct relationship, formed the basis for the committee recommending that a comprehensive research program be developed to address this problem area. This was followed by a series of meetings with MSHA, USBM, and DOE representatives to discuss the recommendation of the committee. The meetings resulted in the BOM and DOE developing a research-funding proposal titled "Mine Demonstrations of Longwall Dust Control Techniques." This research program is the result of excellent cooperation between the coal-mining industry and government mining regulatory and research agencies. Since this approach is recognized as an expedient procedure to apply technological advancements throughout the coal-mining industry, continued cooperative efforts are anticipated in the future.

PROPOSED LONGWALL RESEARCH PROGRAM AT BCR

Introduction and Background

Since passage of the Federal Coal Mine Health and Safety Act of 1969, extensive domestic and foreign research programs have been directed at the problem of longwall respirable-dust control. As would be expected, without exception, these programs pursued a particular approach to solve a specific aspect of the problem. Based on the results of this research effort, combined with original approaches to the problem, the U.S. coal industry has, to date, tested nearly all respirable dust-control equipment and techniques applicable to double-drum longwall shearers. This has included ventilation variations primarily for improved dilution and airflow control, auxiliary air-intake entries, different types of water sprays mounted on cutting drums, cowls, the shearer, and roof supports along the face, hang-on dust collectors, and, to a limited extent, some variations in machine-cutting parameters such as increased depth of cut, improved vane angle, and decreased drum rpm. Unfortunately, most of these tests did not produce the documentation necessary to accurately evaluate the effectiveness of the dust-control equipment and/or technique tested. This could be attributed primarily to the fact that most coal-producing companies do not have the research personnel, equipment, and/or time to conduct the comprehensive test program required to accurately quantify the effectiveness of individual dust controls. This results in possibly discontinuing significantly effective controls and vice versa. In addition, dust-control equipment and/or a technique may be considerably more effective in one mine than another. To date, comparative effectiveness in different mines also has not been documented.

Similarly, during the past 10 years a number of face-cutting sequences have evolved on double-drum shearer operations. These include various bi-directional and unidirectional sequences, half-face cutting, etc. As indicated earlier, unidirectional cutting is mandated by MSHA for the majority of the double-drum longwall operations to reduce respirable-dust exposures. Much controversy exists between the industry and MSHA regarding the effect on productivity of the face-cutting sequence. Therefore, BCR's most immediate proposed longwall research program encompasses a system's concept for respirable-dust control, combined with a comprehensive evaluation to determine the effect of face-cutting sequences on productivity. The entire program will cover 45 months, and involve the cooperation of as many as 19 participating coal companies.

Respirable Dust Control Evaluations

Proposed Program

The first part of the program will be the evaluation of several compatible dust-control techniques concurrently or in succession on each longwall section. The program will include, after an initial upgrading of the water system where necessary, first, a base-test period to determine the existing conditions and engineering parameters on the longwall section, then, a number of similar test periods to evaluate, successively, dust-control measures agreed upon by each cooperating coal company. The initial upgrading of the water system will be to ensure that an adequate quantity of water, at sufficient operating pressure, will be delivered to the shearer to provide effective spray operation. This may consist of installation of higher capacity pipes and/or hoses, pressure pumps, improved filtration systems, and, where wetting agents are used, proportioning injection pumps. Recommendations will follow a three-day, in-mine survey to observe the face operation and obtain continuous recordings of water pressure and volume in the section water line, as well as to monitor operating pressure at the spray nozzles. Respirable-dust concentrations and exposures will also be recorded using both gravimetric and instantaneous dust-measuring instruments.

Upon completion of this upgrading of the water system by mine personnel, the evaluations of specific dust-control measures will be accomplished by conducting comparative tests; i.e., alternating test sequences with and without using the control measure under evaluation. In order to avoid, as much as possible, uncontrollable variables (such as geological changes in the coal seam), test sequences will be alternated several times and at short time intervals. Practical considerations, however, limit how often a particular control method or piece of equipment will be taken in and out of operation. For example, water distribution between the two cutting drums can be changed at every pass by a proper valve arrangement; this would permit a comparative test sequence to consist of one pass of the shearer. A wetting agent, however, will have to be used for half a shift and another agent, or none, during the second half. It cannot be changed and evaluated at each pass because of the time required to flush the wetting agent from the water lines and from the surface of the coal. A comparative test sequence will probably be half-a-shift duration. At the other extreme, changing cutting drums on a shearer is time-consuming and best accomplished during a weekend, so that frequent alternation is not feasible.

Continuous, automatic, full-shift recordings will be obtained for the following: (1) intake-air velocity, (2) return-air velocity, (3) spray-water flow to face, (4) spray-water pressure, (5) power to shearer, (6) respirable-dust level approximately 20 feet downwind of the shearer, and (7) respirable-dust level in the return airway. In addition, respirable and total airborne dust concentrations will be measured in the intake and return airways, respectively.

A time study of shearer and face operations will be conducted to permit correlation with the recorded variables and identification of face events related to dust levels. The time study will also define the cutting sequence employed at the face and provide coal-cutting rates. All operational delays will be noted, particularly those attributable to the dust-control equipment.

The continuous recording of respirable-dust levels will be obtained using SIMSLIN instantaneous respirable dust sampling instruments, developed by the National Coal Board. This unit was selected because of its compactness and moderate weight for an instrument capable of continuous recording. The memory unit data will be transferred to a strip chart; the average respirable-dust concentration can be obtained for the shift or any portion of the shift. To correlate the time study to the respirable-dust levels, the face events will be noted on the strip chart so that actual occurrences can be related to respirable-dust levels. The major elements of the cutting sequence will be recorded, with down-time and production-related events to provide the information necessary to explain the various peaks and valleys in respirable-dust concentrations occurring during the different specific phases of the face-cutting sequence. Hand-held tape recorders will be used in conjunction with conducting the time study to ensure that all operational-event information is recorded that is necessary to fully analyze the data. A continuous recording of the power to the shearer will supplement the evaluation to provide an insight into the actual cutting rate, or the actual work expended to remove the coal on a continuous basis during the different phases of the cutting sequence.

Face Cutting Sequence vs. Productivity

As indicated in the Introduction, the cutting sequence at many longwall faces is mandated for the purpose of reducing respirable dust exposures; and much controversy exists as to the effect this has on productivity. The effect of cutting sequence on productivity depends on a number of factors, including:

Type and operational features of equipment

Seam height

Face length

Mine conditions (bottom and roof characteristics)

Many mine operators utilizing double-drum shearers report productivity losses ranging from 5 to 50 percent when employing unidirectional rather than bidirectional cutting. However, proponents of unidirectional cutting cite the following advantages attributable to increased production on their double-drum shearer faces:

Easier to maintain face alignment

Improved face conditions (better clean-up)

Less time lost turning cowls at headgate and tailgate

Improved crew morale because of reduced respirable-dust exposure

The objective of this second part of the overall program is, therefore, to determine the effect cutting sequences have on productivity as well as on respirable-dust control.

To accomplish this objective, full-shift time studies will be conducted with particular emphasis on detailed monitoring of the specific phases of each cutting sequence. Normal time-study practices will be followed, with all delays categorized and identified. The power-recorder chart will complement the recording of the shearer operating time; it will also be used to relate the cutting effort of the machine to its maximum capacity.

Data will be obtained during the course of the program on several cutting sequences at the same face, which will be of great value from the standpoint of evaluating both productivity and dust control. For example, two methods of unidirectional cutting, head-to-tail and tail-to-head, will be compared. It appears that this should not change productivity, as the cutting sequences are identical except reversed relative to the head and tailgate. However, the direction of the conveyor relative to the direction of

shearer travel is, of course, different. The lead
drum generally cuts over 50 percent of the face
height--possibly 65 to 75 percent. If the shearer
travels in the same direction as the conveyor, the
coal cut by the lead drum (and that which falls off
the face ahead of the machine) is transported by the
conveyor without passing under the machine. Consequently, one common bottleneck, that of coal lumps
jamming under the machine, is eliminated; and productivity may be increased. The data obtained during the
study on this face will quantify any increase in
productivity.

SUMMARY

In summary, the dust-control evaluation procedures
of this research program will provide well-
substantiated documentation not currently available on
the effectiveness of most dust-control equipment and
techniques applicable to double-drum longwall shearer
operations. In addition, sufficient information and
data will be provided to accurately determine the
effect of face-cutting sequences on productivity for a
variety of longwall equipment operating in a range of
seam heights, face lengths, and mine conditions. If
this research program is followed by an effective
transfer of the results to the industry, a substantial
reduction in respirable-dust exposures on these
operations may be achieved. The results, if applied,
will also have the potential to significantly increase
the productivity of double-drum longwall shearer
operations.

VI. Census of US Longwall Installations

Census of Longwall Installations Operating in United States Coal Mines During 1979

(Compiled by Mary Ann Gross, Department of Energy, Bruceton, Pa. Supplemental data from Coal Age indicated by (*). Reprinted by permission from *Coal Age* magazine.)

Company/Mine/Location	Seam	Cutting Height (in.)	Panel Width (ft)	Panel Length (ft)	Over-burden (ft)	Support Entries	Depth of Cut (in.)	Cutter/Loader	Haulage System*	Face Conveyor	Face Supports (Capacity: Legs/Tons)	Capacity* (tons/shift)	Year of Initial Face Installation*
American Electric Power Company — Price River Coal Company (formerly Braztah Corporation)													
Price River No. 3 Helper, Utah	Subseam No. 3	75	504	3200	1400-1600	HG-2 TG-3	30	Anderson Mavor S DD 300	Chain Acco	Halbach & Braun 30 SS 2x150	Chocks 4/500 Dowty	750	1976
American Electric Power Company — Southern Ohio Coal Company													
Martinka No. 1 Fairmont, W.V.	Lower Kittanning	52	500	3000	200-500	4	30	Eickhoff-340L S DD 544	Chain Eickhoff	Halbach & Braun 30 SS 2x150	Chocks 6/510 Gullick Dobson	1000	1977
Meigs No. 2 Athens, Ohio	Clarion No. 4-A	57	504	5310	300	3	30	Anderson Mavor S DD 500	Chainless Roll Rack	Westfalia 26 TIB 3x175	Shields (4-leg) 610 Dowty	800-1000	1978
Armco Inc.													
Walhonde No. 7 Montcoal, W.V.	No. 2 Gas	66	460	5700	500-1000	3	30	Eickhoff-170L S DD 272	Chain Halbach & Braun	Halbach & Braun 26 SS 2x150	Shields (2-leg) 370 Kloeckner Becorit	—	1975
Barnes & Tucker Coal Company													
Lancashire No. 20 Barnesboro, Pa.	Lower Kittanning	60	550	5000	550-600	3	30	Anderson Mavor S DD 300	Chain	Dowty Meco 26 TIB 3x200	Chocks 4/700 Dowty	1700	1977
Lancashire No. 24-B&D Barnesboro, Pa.	Lower Freeport D	44	480	4600	350-400	3	30	Eickhoff-130L S SDF 208	Chain Dowty Meco	Dowty Meco 18 TOB 2x150	Chocks 5/275 Gullick Dobson	615	1967
	Lower Freeport D	42	480	4600	350-400	3	30	Eickhoff-130L S SDF 208	Chain Dowty Meco	Dowty Meco 18 TOB 2x150	Chocks 5/275 Gullick Dobson	615	1968
Lancashire No. 25 Barnesboro, Pa.	Lower Freeport D	40	530	4000	400-500	3	30	Eickhoff-130L S SDF 208	Chain Dowty Meco	Dowty Meco 18 TOB 2x150	Chocks 6/275 Gullick Dobson		
Bethlehem Mines Corporation													
Cambria Slope, No. 33 Ebensburg, Pa.	Lower Kittanning B	54	600	5000	500-800	3	27	Anderson Mavor S SDF 300	Wire Rope Bethlehem	Huwood Irwin 18 TS 2x150	Chocks 4/280 Huwood-Irwin	1175	1968
	Lower Kittanning B	52	600	5000	500-800	3	27	Anderson Mavor S SDF 300	Wire Rope Bethlehem	Dowty Meco 18 TS 2x150	Chocks 4/170 Huwood-Irwin	1175	1968
	Upper Kittanning	50	600	3500	500	3	27	Anderson Mavor S SDF 300	Wire Rope Bethlehem	Huwood-Irwin 18 TS 2x150	Chocks 4/280 Huwood-Mastabar	1284	1975
Somerset No. 60 Eighty Four, Pa.	Pittsburgh	66	585	3000	525-600	3	30	Eickhoff-170L S DD 272	Chainless Eicotrack	Halbach & Braun 30 SS 2x200	Shield (2-leg) 352 Kloeckner-Becorit	1100	1978
No. 131 Van, W.V.	Powellton	72	560	1800	200-1200	3	30	Anderson Mavor S SDF 270	Chain Crosby	Huwood-Irwin 22 TOB 2x150 1x100	Chocks 4/400 Huwood-Irwin	600	1978
Bethlehem Mines Corporation — Beth Elkhorn Division													
Pike No. 26 Jenkins, Ky.	Elkhorn No. 2	48	455	2600	400-1100	3	28	Anderson Mavor S SDF 270 S SDR 270	Chain Huwood-Irwin	Huwood-Irwin 18 TS 2x150	Chocks 4/280 Huwood-Irwin	500	1973
CF & I Steel Corporation													
Allen Weston, Colo.	Allen	66	432	3500	450-1500	4	30	Eickhoff-170L S DD 272	Chain	Halbach & Braun 30 SS 2x125	Chocks 4/500 Gullick Dobson	—	
	Allen	72	450	3300	1500-1700	3	30	Anderson Mavor S SDF 300	Chain	Halbach & Braun 30 SS 2x150	Chocks 4/500 Hemscheidt	—	

* Data from Coal Age 1980 Survey. 1. Production capacity measured in clean coal. 2. Mine idled August 1979. 3. Longwall now dismantled and removed.
LEGEND—CUTTING MACHINES: S-shearer, DD-double drum, SDR-single drum ranging, SDF-single drum fixed, P-plow. FACE CONVEYORS: SS-single strand, TOB-twin strand outboard, TIB-twin strand inboard, TS-triple strand.
EXPLANATION CUTTING MACHINES: "S SDR 300" means "shearer, single drum ranging, 300 hp." FACE CONVERYORS: "30 SS 2x125" means "30 mm chain size, single strand, two 125-hp motors." (All horsepower given in American terms.) FACE SUPPORTS: three digit number for shields is average yield pressure.

Company/Mine/Location	Seam	Cutting Height (in.)	Panel Width (ft)	Panel Length (ft)	Over-burden (ft)	Support Entries	Depth of Cut (in.)	Cutter/Loader	Haulage System[*]	Face Conveyor	Face Supports (Capacity: Legs/Tons)	Capacity[*] (tons/shift)	Year of Initial Face Installation[*]
Cannelton Industries Lady Dunn No. 105 Cannelton, W.V.	Powellton	48	342	2200	800	3	27	Sagem S SDF 230	Chain Acco	Westfalia 18 TS 2x125	Chocks 5/200 Dobson	1000	1965
Carbon Fuel Company Morton No. 34A Portal Winifrede, W.V.	No. 2 Eagle	58	500	3000	800	3	30	Anderson Mavor S DD 300	Chain Acco	Westfalia 18 TS 2x125	Chocks 4/460 Westfalia	500	1980
	No. 2 Eagle	54	450	3000	800	3	30	Anderson Mavor S DD 300	Chain Acco	Dowty Meco 22 TOB 1x200 1x150	Chocks 4/500 Dowty	500	1978
Clinchfield Coal (Division of The Pittston Company) Camp Branch No. 1 Dante, Va.	Lower Banner	51	200	2500	800	3	27	Sagem S DD 400	Chain	Westfalia 26 TIB 3x125	Chocks 4/700 Westfalia	—	—
Moss No. 4 Duty, Va.	Tiller	60	195	2700	400	3	27	Sagem S DD 400	Chainless Peratrack	Westfalia 26 TIB 3x175	Shields (4-leg) 577 Westfalia	—	—
Consolidation Coal Company — Blacksville Operations Blacksville No. 1 Wana, W.V.	Pittsburgh	84	545	3000	700-1100	4	30	Eickhoff-170L S DD 272	Chainless Eicotrack	Halbach & Braun 26 TIB 2x200	Shields (4-leg) 500 Dowty	—	1979
Blacksville No. 2 Wana, W.V.	Pittsburgh	78	450	2800	750-1100	4	24	Eickhoff-170L S DD 272	Chain	Westfalia 18 TS 2x125	Chocks 4/460 Westfalia	—	1972
Consolidation Coal Company - Fairmont Operations Loveridge No. 22 Fairview, W.V.	Pittsburgh	96	500	2800	1100	4	30	Eickhoff-300L S DD 480	Chain	Halbach & Braun 26 SS 2x150	Shields (2-leg) 352 Kloeckner Ferromatic	—	1977
Loveridge No. 22 Miracle Run Portal Fairview, W.V.	Pittsburgh	96	500	3000	1500	4	36	Eickhoff-300L S DD 480	Chainless Eicotrack	Halbach & Braun 30 TIB 2x300	Shields (2-leg) 352 Kloeckner Becorit	—	1978
O'Donnell No. 20 Four States, W.V.	Pittsburgh	88	567	3500	500-800	4	30	Eickhoff-300L S DD 480	Chain	Halbach & Braun 26 TIB 2x200	Shields (2-leg) 352 Kloeckner Ferromatic	—	1976
Robinson Run No. 95 Shinnston, W.V.	Pittsburgh	82	537	3000	600-800	4	30	Eickhoff-300L S DD 480	Chain	Halbach & Braun 26 TIB 2x200	Chocks 6/510 Gullick Dobson	—	1972
	Pittsburgh	80	587	4000	400	4	30	Eickhoff-300L S DD 480	Chain	Halbach & Braun 26 TIB 2x200	Shields (2-leg) 352 Kloeckner Ferromatic	—	1975
Consolidation Coal Company - Morgantown Operations Humphrey No. 7 Mt. Morris Portal Osage, W.V.	Pittsburgh	86	383	3800	400-900	4	30	Eickhoff-300L S DD 480	Chainless Eicotrack	Halbach & Braun 30 SS 2x200	Shields (2-leg) 352 Kloeckner Becorit	—	1979
Consolidation Coal Company - Moundsville Operations Ireland Moundsville Portal Moundsville, W.V.	Pittsburgh	66	584	5400	400-800	4	30	Eickhoff-150-2L S DD 480	Chainless Eicotrack	Halbach & Braun 30 SS 2x200	Shields (2-leg) 352 Kloeckner Becorit	—	1977
McElroy Moundsville, W.V.	Pittsburgh	70	584	7000	500-600	4	30	Eickhoff-150-2L S DD 480	Chainless Eicotrack	Halbach & Braun 30 SS 2x200	Shields (2-leg) 400 Joy	—	1979

[*] Data from Coal Age 1980 Survey. 1. Production capacity measured in clean coal. 2. Mine idled August 1979. 3. Longwall now dismantled and removed.
LEGEND—CUTTING MACHINES: S-shearer, DD-double drum, SDR-single drum ranging, SDF-single drum fixed, P-plow. FACE CONVEYORS: SS-single strand, TOB-twin strand outboard, TIB-twin strand inboard, TS-triple strand.
EXPLANATION CUTTING MACHINES: "S SDR 300" means "shearer, single drum ranging, 300 hp." FACE CONVERYORS: "30 SS 2x125" means "30 mm chain size, single strand, two 125-hp motors." (All horsepower given in American terms.) FACE SUPPORTS: three digit number for shields is average yield pressure.

Company/Mine/Location	Seam	Cutting Height (in.)	Panel Width (ft)	Panel Length (ft)	Over-burden (ft)	Support Entries	of Cut (in.)	Cutter/Loader	Haulage System*	Face Conveyor	Face Supports (Capacity: Legs/Tons)	Capacity* (tons/shift)	Year of Initial Face Installation*
Shoemaker Moundsville, W.V.	Pittsburgh	67	584	4000	650	4	30	Eickhoff-170L S DD 272	Chainless Eicotrack	Halbach & Braun 30 SS 2x200	Shields (2-leg) 352 Kloeckner Ferromatic	—	1975
	Pittsburgh	67	584	7000	650	4	30	Eickhoff-170L S DD 272	Chainless Eicotrack	Halbach & Braun 30 SS 2x200	Shields (2-leg) 352 Kloeckner Becorit	—	1979
Consolidation Company - Washington Operation Westland No. 1 Washington, Pa.	Pittsburgh	66	600	3700	320-360	3	30	Eickhoff-170L S DD 272	Chain	Halbach & Braun 30 SS 2x200	Shields (2-leg) 352 Kloeckner Becorit	—	1977
Eastern Associated Coal Corporation Federal No. 1 Grant Town, W.V.	Pittsburgh	78	434	3000	500	3	30	Sagem S DD 460	Chain	Westfalia 22 TIB 2x125	Chocks 6/510 Gullick Dobson	750	—
Federal No. 2 Fairview, W.V.	Pittsburgh	96	536	5400	750	HG-4 TG-3	27	Sagem S DD 460	Chain	Westfalia 22 TIB 2x175	Chocks 4/460 Westfalia	1350	—
	Pittsburgh	96	436	5400	750	HG-4 TG-3	27	Sagem S DD 460	Chain	Westfalia 26 TIB 3x175	Shields (2-leg) 352 Westfalia	1350	—
Harris No.1 Bald Knob, W.V.	Eagle	60	480	4500	200-1200	4	4	Westfalia P Anbau 2x125	—	Westfalia 18 TS 2x125	Frames 4/560 Westfalia	750	—
	Eagle	60	480	4500	200-1200	4	30	Sagem S DD 400	Chainless Peratrack	Westfalia 26 TIB 3x175	Shields (4-leg) 572 Westfalia	750	—
Harris No. 2 Bald Knob, W.V.	Campbell Creek	72	500	4000	500-2000	4	30	Sagem S DD 400	Chain	Westfalia 26 TIB 3x175	Shields (2-leg) 426 Westfalia	900	—
Keystone No. 1 Keystone, W.V.	Pocahontas No. 3	54	400	2000	600-700	4	30	Sagem S SDR 230	Chain	Westfalia 18 TS 2x125	Frames 4/560 Westfalia	450	—
Keystone No.2 Herndon, W.V.	Pocahontas No. 3	60	345	2400	500-1200	4	30	Sagem S SDR 230	Chain	Westfalia 18 TS 2x125	Chocks 4/460 Westfalia	1000	—
Keystone No. 4 Stotesbury, W.V.	Pocahontas No.3	40	380	5000	600-700	5	4	Westfalia P Gleithobel 2x125	—	Westfalia 26 TIB 2x125	Shields(4-leg) 440 Westfalia	600	—
Keystone No.5 Sophia, W.V.	Pocahontas No. 3	40	400	4000	600	4	4	Westfalia P Hook 2x125	—	Westfalia 13 TS 2x125	Chocks 4/700 Dowty	650	—
Kopperston No. 1 Kopperston, W.V.	Eagle	60	600	6000	1000-1200	3	30	Joy 1LS S DD 235	Chainless Joyrack	Joy 26 TIB 2x150	Chock Shields 500 Joy (4-leg)	120	—
Wharton No. 2 Barrett, W.V.	Hernshaw	44	250	4500	500-800	3	27	Sagem S SDR 230	Chain Acco	Westfalia 18 TS 2x125	Chocks 4/700 Westfalia	600	—
Florence Mining Company Florence No. 1 Seward, Pa.	Lower Kittanning	78	450	4700	300-400	4	30	Anderson Mavor S DD 300	Chainless Rackatrack	Mining Supplies 30 SS 3x125	Shields (2-leg) 575 Thyssen	—	—
Gateway Coal Company Gateway Clarksville, Pa.	Pittsburgh	72	485	3400	500-700	3	30	Anderson Mavor S DD 300	Chain	Halbach & Braun 26 SS 2x125	Shields (2-leg) 352 Kloeckner Ferromatic	—	—
Greenwich Collieries North Mine No. 1 Barnesboro, Pa.	Lower Freeport D	48	460	4800	350-480	HG-4 TG-5	30	Eickhoff-170L S SDR 272	Chain Acco	Halbach & Braun 26 SS 2x125	Chocks 6/510 Gullick Dobson	—	—

* Data from Coal Age 1980 Survey. 1. Production capacity measured in clean coal. 2. Mine idled August 1979. 3. Longwall now dismantled and removed.
LEGEND—CUTTING MACHINES: S-shearer, DD-double drum, SDR-single drum ranging, SDF-single drum fixed, P-plow. FACE CONVEYORS: SS-single strand, TOB-twin strand outboard, TIB-twin strand inboard, TS-triple strand.
EXPLANATION CUTTING MACHINES: "S SDR 300" means "shearer, single drum ranging, 300 hp." FACE CONVEYORS: "30 SS 2x125" means "30 mm chain size, single strand, two 125-hp motors." (All horsepower given in American terms.) FACE SUPPORTS: three digit number for shields is average yield pressure.

Company/Mine/Location	Seam	Cutting Height (in.)	Panel Width (ft)	Panel Length (ft)	Over-burden (ft)	Support Entries	of Cut (in.)	Cutter/Loader	Haulage System*	Face Conveyor	Face Supports (Capacity: Legs/Tons)	Capacity* (tons/shift)	Year of Initial Face Installation*
South Mine No.2 Barnesboro, Pa.	Lower Freeport D	48	460	4800	350-480	3	30	Eickhoff-170L S SDR 272	Chain Acco	Halbach & Braun 26 SS 2x125	Chocks 6/510 Gullick Dobson	—	—
Island Creek Coal Company — Virginia Pocahontas Division													
Virginia Pocahontas No. 1	Pocahontas No. 3	48	580	5500	1200-2000	4	4	Westfalia P Hook 2x125	—	Westfalia 26 TIB 2x175	Frames 4/560 Westfalia	450	1969
Keen Mountain, Va.	Pocahontas No. 3	48	580	5500	1200-2000	4	4	Westfalia P Gleithobel 2x175	—	Westfalia 26 TIB 2x175	Shield (4-leg) 640 Westfalia	700	1968
	Pocahontas No. 3	48	580	5500	1200-2000	4	4	Westfalia P Hook 2x125	—	Westfalia 22 TIB 2x125	Frames 4/560 Westfalia		
Virginia Pocahontas No. 2	Pocahontas No. 3	48	510	5000	1200-2000	4	4	Westfalia P Gleithobel 2x175	—	Westfalia 22 TIB 2x175	Chocks 4/700 Westfalia	—	1977 [2]
Keen Mountain Va.	Pocahontas No. 3	48	510	5000	1200-2000	4	4	Westfalia P Gleithobel 2x175	—	Westfalia 26 TIB 2x175	Shield (4-leg) 640 Westfalia	—	—
Virginia Pocahontas No. 3	Pocahontas No. 3	60	580	5500	1200-2000	4	4	Westfalia P Hook 2x125	—	Westfalia 22 TIB 2x175	Frames 4/560 Westfalia	700	1974
Keen Mountain, Va.	Pocahontas No. 3	60	580	5800	1200-2000			Halbach & Braun Kompact 2x200	—	Westfalia 30 SS 2x200	Chocks 4/560 Hemscheidt	650	1972
	Pocahontas No. 3	60	580	5500	1200-2000	4	4	Westfalia P Hook 2x125	—	Westfalia 22 TIB 2x175	Frames 4/560 Westfalia	—	—
Island Creek Company — Northern Division													
North Branch Bayard, WV.	Upper Freeport	108	450	3000	400-900	3	30	Eickhoff-300L S DD 480	Chainless Eicotrack	Halbach & Braun 30 TIB 2x300	Shields (2-leg) 440 Kloeckner Becorit	1650	1978
Island Creed Coal Company — Beatrice Pocahontas Company													
Beatrice Keen Mountain, Va.	Pocahontas No. 3	60	580	5500	1800-2000	4	4	Westfalia P Hook 2x125	—	Westfalia 26 TIB 2x175	Frames 4/560 Westfalia	640	1967
	Pocahontas No. 3	60	580	5500	1800-2000	4	4	Westfalia P Gleithobel 2x125	—	Westfalia 26 TIB 2x175	Shield (4-leg) 640 Westfalia	600	1979
	Pocahontas No. 3	60	580	5500	1800-2000	4	4	Westfalia P Hook 2x125	—	Westfalia 22 TIB 2x175	Frames 4/560 Westfalia	—	—
Itmann Coal Company													
Itmann No. 3 Itmann, WV.	Pocahontas No. 3	48	480	5500	800	4	27	Anderson Mavor S SDF 300, S SDR 200	Chain	Dowty Meco 18 TS 2x150	Chocks 4/700 Dowty	—	1971
	Pocahontas No. 3	44	490	5000	800-1000	4	30	Anderson Mavor S DD-Inweb 300	Chain	Dowty Meco 26 SS 2x150	Chocks 4/700 Dowty	—	—
Jewell Ridge Coal Corporation													
Jewell No 12A Jewell Valley, Va.	Lower Jewell	36	380	3300	600	4	4	Westfalia P Gleithobel 2x125	—	Westfalia 18 TS 2x125	Chocks 4/460 Westfalia	—	—
Jones & Laughlin Steel Corporation													
Vesta No. 5 California, Pa.	Pittsburgh	78	450	3000	550-600	3	30	Anderson mavor S DD 300	Chain	Halbach & Braun 30 SS 2x200	Shields (4-leg)400 Dowty	—	—

* Data from Coal Age 1980 Survey. 1. Production capacity measured in clean coal. 2. Mine idled August 1979. 3. Longwall now dismantled and removed.
LEGEND—CUTTING MACHINES: S-shearer, DD-double drum, SDR-single drum ranging, SDF-single drum fixed, P-plow. FACE CONVEYORS: SS-single strand, TOB-twin strand outboard, TIB-twin strand inboard, TS-triple strand.
EXPLANATION CUTTING MACHINES: "S SDR 300" means "shearer, single drum ranging, 300 hp." FACE CONVEYORS: "30 SS 2x125" means "30 mm chain size, single strand, two 125-hp motors." (All horsepower given in American terms.) FACE SUPPORTS: three digit number for shields is average yield pressure.

Company/Mine/Location	Seam	Cutting Height (In.)	Panel Width (ft)	Panel Length (ft)	Over-burden (ft)	Support Entries	of Cut (In.)	Cutter/Loader	Haulage System*	Face Conveyor	Face Supports (Capacity: Legs/Tons)	Capacity* (tons/shift)	Year of Initial Face Installation*
Kaiser Steel Corporation													
Sunnyside No. 1 Sunnyside, Utah	Lower Sunnyside	108	300	3000	2000	HG-2 TG-1	30	Eickhoff-170L S DD 272	Chain	Halbach & Braun 30 SS 2x125	Chocks 4/450 Dowty	—	—
	Lower Sunnyside	72	500	3000	2500	2	30	Eickhoff-170L S DD 272	Chain	Halbach & Braun 30 SS 2x125	Chocks 4/500 Dowty	—	—
York Canyon Raton, N. M.	York Canyon	84	580	670	550-700	2	30	Eickhoff-300L S DD 480	Chain	Halbach & Braun 30 SS 2x300	Shields (2-leg) 353 Hemscheidt	—	—
	York Canyon	116	500	2600	600-800	HG-3 TG-2	30	Eickhoff-300L S DD 480	Chain	Halbach & Braun 30 SS 2x225	Shields (2-leg) 353 Hemscheidt	—	—
	York Canyon	84	576	6200	550-700	2	30	Anderson Mavor S DD 300	Chainless Roll Rack	Dowty Meco 26 TIB 2x300	Shields (2-leg) 375 Dowty	—	—
Kentucky Carbon Corporation													
Kencar No. 1 Phelps, Ky.	Elkhorn No. 2	54	360	4600	300-500	3	27	Anderson Mavor S DD 300	Chain Acco	Dowty Meco 22 TOB 2x150	Chocks 4/500 Dowty	500	1976
	Elkhorn No. 2	60	480	4600	300-500	3	32	Anderson Mavor S DD 300	Chain	Dowty Meco 22 TOB 2x150	Shields (4-leg) 650 Dowty	500	1979
Leeco, Incorporated													
Leeco No. 22 London, Ky.	Hazard No. 4 Rider	34	600	4000	150-600	3	30	British Jeffrey Diamond S DD-Inweb 300	Chainless Peratrack	Dowty Meco 19 TOB 2x150	Shields (4-leg) 360 Dowty	—	1979
Mid-Continent Coal and Coke Company													
Dutch Creek No. 1 Carbondale, Colo.	Coal Basin B	Top 96	550	4000	2400	HG-2 TG-1	30	Anderson Mavor S DD 300	Chain	Dowty Meco 22 TOB 2x150	Chocks 4/500 Dowty	—	—
L.S. Wood No. 3 Carbondale, Colo.	Coal Basin B	Top 96	600	2000	2000	HG-2 TG-1	28	Anderson Mavor S DD 400	Chain	Halbach & Braun 26 SS 2x200	Chocks 6/510 Gullick Dobson	—	—
Old Ben Coal Company													
Old Ben No. 21 Sesser, Ill.	Herrin No. 6	104	480	2000	660	3	30	Eickhoff-300L S DD 480	Chainless Eicotrack	Halbach & Braun 30 SS 2x200	Shields (2-leg) 527 Thyssen	2000	1978
Old Ben No. 24 Benton, Ill.	Herrin No. 6	104	480	1750	550-600	3	24	Joy 2LS S DD 565	Chain Campbell	Halbach & Braun 30 SS 2x150	Shields (2-leg) 642 Thyssen	2000	1976
Old Ben No. 25 Marion, Ill.	Herrin No. 6	104	500	4000	600	3	30	Eickhoff-300L S DD 480	Chainless Eicotrack	Halbach & Braun 30 SS 2x250	Shields (2-leg) 527 Thyssen	2000	1979
Old Ben No. 26 Sesser, Ill.	Herrin No. 6	104	500	2000	600	3	30	Eickhoff-300L S DD 480	Chainless Eicotrack	Halbach & Braun 30 SS 2x200	Shields (2-leg) 500 Hemscheidt	2000	1979
Quarto Mining Company													
Powhatan No. 4 Powhatan Pt. Ohio	Pittsburgh No. 8	87	490	3000	700	3	30	Eickhoff-300L S DD 480	Chainless Eicotrack	Halbach & Braun 30 SS 2x200	Shields (2-leg) 350 Hemscheidt	—	—
	Pittsburgh No. 8	78	500	5000	450-1100-	3	30	Anderson Mavor S DD 500	Chainless Roll Rack	Mininc Supplies 26 TIB 2x200	Shields (2-leg) 375 Thyssen	—	—
Powhatan No. 7 Powhattan Pt., Ohio	Pittsburgh No. 8	56	486	3200	350-650	3	30	Anderson Mavor S DD 300	Chain	Dowty Meco 26 TIB 2x200	Shields (4-leg) 500 Dowty	—	—

* Data from Coal Age 1980 Survey. 1. Production capacity measured in clean coal. 2. Mine idled August 1979. 3. Longwall now dismantled and removed.
LEGEND—CUTTING MACHINES: S-shearer, DD-double drum, SDR-single drum ranging, SDF-single drum fixed, P-plow. FACE CONVEYORS: SS-single strand, TOB-twin strand outboard, TIB-twin strand inboard, TS-triple strand.
EXPLANATION CUTTING MACHINES: "S SDR 300" means "shearer, single drum ranging, 300 hp." FACE CONVEYORS: "30 SS 2x125" means "30 mm chain size, single strand, two 125-hp motors." (All horsepower given in American terms.) FACE SUPPORTS: three digit number for shields is average yield pressure.

Company/Mine/Location	Seam	Cutting Height (in.)	Panel Width (ft)	Panel Length (ft)	Over-burden (ft)	Support Entries	of Cut (in.)	Cutter/Loader	Haulage System*	Face Conveyor	Face Supports (Capacity: Legs/Tons)	Capacity* (tons/shift)	Year of Initial Face Installation*
Rochester and Pittsburgh Coal Company													
Emilie Schelocta, Pa.	Upper Freeport	42	355	2000	200-300	4	30	Eickhoff-150L S SDR 240	Chain	Halbach & Braun 26 SS 2x125	Chocks 4/440 Kloeckner Ferromatic	—	[3]
Jane Schelocta, Pa.	Lower Freeport D	56	365	900	275-340	4	30	Eickhoff-150l S SDR 240	Chanin	Beitish Jeffrey 18 TS 2x125	Chocks 4/308 Kloeckner Ferromatic	—	[3]
Lucerne No. 6 Homer City, Pa.	Upper Freeport	52	350	2000	350	4	4	Westfalia P Gleithobel 2x25	—	Westfalia 28 TIB 2x125	Frames 4/500 Westfalia	—	[3]
United States Steel Corporation — Gary District													
Gary No. 9 Filbert, WV	Pocahontas No. 3	50	400	3200	200-600	3	4	Westfalia P Hook 2x125	—	Westfalia 18 TS 2x125	Frames 4/500 Westfalia	600'	1970
	Pocahontas No. 3	54	500	3200	200-600	3	4	Westfalia P Hook 2x125	—	Westfalia 18 TS 2x125	Frames 4/700 Westfalia	600'	1970
Gary No. 50 Pineville, WV	Pocahontas No. 3	58	360	3200	350-800	3	4	Westfalia P Gleithobel 2x125	—	Westfalia 26 TIB 2x175	Shields (2-leg) 353 Hemscheidt	700'	1977
Utah Power and Light Company — Emery Mining Corporation													
Deer Creek Hunington, Utah	Blind Canyon	144	480	2600	1600	3	30	Eickhoff-300L S DD 480	Chainless Eicotrack	Halbach & Braun 30 TIB 2x150	Shields (2-leg)420 Hemscheidt	3000	1979
Jim Walter Resources, Inc.													
Blue Creek No. 3 Brookwood, Ala.	Blue Creek	90	434	2900	1300-1500	3	30	Anderson Mavor S DD 300	Chainless Rackatrack	Mining Supplies 26 TIB 1x300 1x200	Shields (2-leg) 352 Kloeckner Becorit	1000	1979

Longwall Faces Installed in United States Coal Mines in 1980
(Compiled by Coal Age)

Company/Mine Location	Seam	Seam Height (in.)	Panel Width (ft)	Panel Length (ft)	Over-burden (ft)	Cutter/Loader	Haulage System	Face Conveyor	Face Supports (capacity: legs/tons)	Capacity (tons/shift)
American Electric Power Company—Southern Ohio Coal Company										
Meigs No. 1 Athens, Ohio	Clarion No. 4-A	54	500	3000	300	Sagem S DD 400	Chainless Peratrack	Westfalia 26 TIB	Shields (4-leg) 610 Dowty	800-1000
Meigs No. 2 Athens, Ohio	Clarion No. 4-A	54-60	500	4000	300	Sagem S DD 400	Chainless Peratrack	Westfalia 26 TIB	Shields (4-leg) 600 Westfalia	1000
Barnes & Tucker Coal Company										
Lancashire 25 Barnesboro, Pa.	Lower Freeport D	42	550	4000	350	Eickhoff-130L S SDF 200	Chain Dowty Meco	Dowty Meco 26 TIB	Chocks 6/475 Dowty	900

* Data from Coal Age 1980 Survey. 1. Production capacity measured in clean coal. 2. Mine idled August 1979. 3. Longwall now dismantled and removed.
LEGEND—CUTTING MACHINES: S-shearer, DD-double drum, SDR-single drum ranging, SDF-single drum fixed, P-plow. FACE CONVEYORS: SS-single strand, TOB-twin strand outboard, TIB-twin strand inboard, TS-triple strand.
EXPLANATION CUTTING MACHINES: "S SDR 300" means "shearer, single drum ranging, 300 hp." FACE CONVEYORS: "30 SS 2x125" means "30 mm chain size, single strand, two 125-hp motors." (All horsepower given in American terms.) FACE SUPPORTS: three digit number for shields is average yield pressure.

Company/Mine Location	Seam	Seam Height (in.)	Panel Width (ft)	Panel Length (ft)	Over-burden (ft)	Cutter/Loader	Haulage System	Face Conveyor	Face Supports (capacity: legs/tons)	Capacity (tons/shift)
Carbon Fuel Company Morton No. 34A Portal Winifrede, W.V.	No. 2 Eagle	48	500	4000	600-1100	Anderson Mavor S DD 300	Chain Acco	Dowty Meco 26 TIB	Shields (4-leg) 610 Dowty	500
Consolidation Coal Company Blacksville No.2 Wana, W.V.	Pittsburgh	84	575	3000	500-800	Eickhoff-170L S DD 228	Chainless Eicotrack	Halbach & Braum 26 TIB	Shields (4-leg) 500 Dowth	
Island Creek Coal Company Virginia Pocahontas No. 5 Keen Mountain,Va	Pocahontas No. 3	72	600	5000	1400-2200	Eickhoff-150 2L S DD 400	Chainless Eicotrack	Westfalia 26 TIB	Shields (4-leg) 570 Westfalia	900
Old Ben Coal Company Old Ben No. 25 Marion, Ill.	Herrin No. 6	90	515	4600	600	Eickhoff-300L S DD	Chainless Eicotrack	Halbach & Braun 26 TIB	Shields (2-leg) 500 Thyssen	900
Old Ben No. 27 W. Frankfort, Ill.	Herrin No. 6	84	420	2600	600	Eickhoff-300L S DD	Chainless Eicotrack	Halbach & Braun 26 TIB	Shields (2-leg) 500 Thyssen	2000
Quarto Mining Company Powhatan No. 7 Powhatan Point, Ohio	Pittsburgh No. 8	56	500	3200	350-650	Anderson Mavor S DD 300	Chain	Dowty Meco 26 TIB 2x200	Shields (2-leg) 350 Hemscheidt	N.A.
Utah Power and Light Company—Emery Mining Corporation Deer Creek Huntington, Utah	Blind Canyon	102	550	3750	1800	Eickhoff-300L S DD 465	Chainless Eicotrack	Halbach & Braun 30 TIB	Shields (2-leg) 350 Hemscheidt	N.A.
Wilberg Huntington, Utah	Hiawatha	117	595	4090	2000	Eickhoff-300L S DD 465	Chainless Eicotrack	Halbach & Braun 30 TIB	Shield (2-leg) 460 Hemscheidt	2500
Jim Walter Resources, Inc. Blue Creek No. 3 Adger, Ala.	Blue Creek	78	482	4250	1500	Eickhoff-150 2L S DD 500	Chainless Eicotrack	Mining Supplies 26 TIB	Shields (4-leg) 610 Dowty	1000
	Blue Creek	78	482	3900	1500	Joy ILS S DD 245	Chainless Joyrack	Mining Supplies 26 TIB	Shields (2-leg) 352 Kloeckner Becorit	1000
Western Slope Carbon Inc. Hawk's Nest E & W Somerset, Colo.	E	72-120	400	5000	2200	Anderson Mavor S DD 500	N.A.	Halbach & Braun 30 TIB	Shields (4-leg) 440 Dowty	1500
United States Steel Corporation Cumberland Kirby, Pa.	Pittsburgh	78	500	4000	900	Eickhoff-300L S DD 403	Chainless Eicotrack	Halbach & Braun 30 SS	Shields (4-leg) 360 Dowty	1300
Gary No. 50 Pineville, W.V.	Pocahontas No. 3	52-66	500	3800	600	Eickhoff S DD 230	Chainless Eicotrack	Halbach & Braun 26TIB	Shields (2-leg) 320 Kloeckner Becorit	700
Maple Creek No. 2 Eighty Four, Pa.	Pittsburgh	66	600	3100	350	Joy S DD 295	Chainless Joyrack	Joy 26TIB	Shields (2-leg) 450 Joy	1200

* Data from Coal Age 1980 Survey. 1. Production capacity measured in clean coal. 2. Mine idled August 1979. 3. Longwall now dismantled and removed.
LEGEND—CUTTING MACHINES: S-shearer, DD-double drum, SDR-single drum ranging, SDF-single drum fixed, P-plow. FACE CONVEYORS: SS-single strand, TOB-twin strand outboard, TIB-twin strand inboard, TS-triple strand.
EXPLANATION CUTTING MACHINES: "S SDR 300" means "shearer, single drum ranging, 300 hp." FACE CONVERYORS: "30 SS 2x125" means "30 mm chain size, single strand, two 125-hp motors." (All horsepower given in American terms.) FACE SUPPORTS: three digit number for shields is average yield pressure.

INDEX

D

E